National Academy Press

The National Academy Press was created by the National Academy of Sciences to publish the reports issued by the Academy and by the National Academy of Engineering, the Institute of Medicine, and the National Research Council, all operating under the charter granted to the National Academy of Sciences by the Congress of the United States.

The Effects on Populations of Exposure to Low Levels of Ionizing Radiation: 1980

COMMITTEE ON THE BIOLOGICAL
EFFECTS OF IONIZING RADIATIONS
Division of Medical Sciences
Assembly of Life Sciences
National Research Council

NATIONAL ACADEMY PRESS

Washington, D.C. 1980

The work presented in this report was supported by the Office of Radiation Programs, Environmental Protection Agency, under contract 68-01-4301.

PUBLISHER'S NOTE: This edition differs from the typescript edition that was released in July 1980 in that a number of changes made by the Committee (necessitated chiefly by correction of a programming error) have been incorporated.

Library of Congress Catalog Card Number 80-81659

International Standard Book Number 0-309-03095-1

Available from:
NATIONAL ACADEMY PRESS
2101 Constitution Avenue, N.W.
Washington, D.C. 20418

Printed in the United States of America

July 22, 1980

Mr. Douglas Costle
Administrator
Environmental Protection Agency
401 M Street, S.W.
Washington, D.C. 20460

Dear Mr. Costle:

I am pleased to transmit the report "The Effects on Populations of Exposure to Low Levels of Ionizing Radiation" prepared under contract 68-01-4301 with EPA's Office of Radiation Programs.

The report, familiarly known as BEIR III (after its authoring Committee on the Biological Effects of Ionizing Radiations), has had a troubled history. In May 1979, a version of the report was publicly released. But when it was learned that a significant number of committee members believed that the somatic effects section of the report did not adequately reflect the full range of committee opinion generated by the admittedly incomplete data base, further distribution was discontinued.

It is not unusual for scientists to disagree on the interpretation of data. Generally, the sparser and less reliable the data base, the more opportunity for disagreement. In this case, there are sufficient data concerning the effects of exposure to high doses of ionizing radiation, but little reliable information concerning the consequences of exposure to lower doses, especially those low doses to which a human population might be exposed. Upon the issue of how one may extrapolate from the high doses to the low, scientific argument turned on the question of how one may validly extrapolate from the measured effects of high doses to the most probable effects of low doses.

The BEIR III report exhibits the range of opinion concerning how this extrapolation may be performed. Many committee members believe that the data best support a linear quadratic model for estimating risk; others, however, believe that the linear or pure quadratic models provided better estimates. The report presents all of these views, in balanced fashion. The committee as a whole, despite individual preferences, has agreed that the report treats each of the possible interpretations in a fair manner. Two members have not found it possible to endorse the report. The dissenting statement by Dr. Radford espouses the linear model; that by Dr. Rossi favors the pure quadratic model. Both models are included. The polarity of these two views best illustrates the degree to which scientists disagree on this subject in absence of sufficient evidence to compel conclusion.

We believe that the report will be helpful to the EPA and other agencies as they reassess radiation protection standards. It provides the scientific bases upon which standards may be decided after nonscientific social values have been taken into account. If social values dictate a conservative approach, the report's linear model risk estimates may serve as a guide. If one wishes to accept scientists' best judgment while recognizing that the data simply will not permit definitive conclusions, one may select risk estimates using the linear quadratic model as a guide. Other considerations may lead to use of the pure quadratic risk estimates.

We regret that the transmittal of this report has been delayed so long. The Academy believes that the delay was necessary to permit time for restating the report so as to display all of the valid opinions rather than distribute a report that might create the false impression of a clear consensus where none exists.

Sincerely yours,

PHILIP HANDLER
  President, National Academy of Sciences

COMMITTEE ON THE BIOLOGICAL EFFECTS OF IONIZING RADIATIONS

PAUL B. SELBY, Biology Division, Oak Ridge National Laboratory, Oak Ridge, Tennessee

MARGARET H. SLOAN, Division of Cancer Control and Rehabilitation, National Cancer Institute, Silver Spring, Maryland

BENJAMIN K. TRIMBLE (deceased), Department of Medical Genetics, University of British Columbia, Vancouver, British Columbia, Canada

†EDWARD W. WEBSTER, Division of Radiological Sciences, Massachusetts General Hospital, Boston, Massachusetts

HENRY N. WELLMAN, Nuclear Medicine Division, Indiana University School of Medicine, Indianapolis, Indiana

Staff

ALBERT W. HILBERG, *Principal Staff Officer,* Division of Medical Sciences, Assembly of Life Sciences, National Research Council

DAVID A. MCCONNAUGHEY, *Senior Staff Officer,* Division of Medical Sciences, Assembly of Life Sciences, National Research Council

NORMAN GROSSBLATT, *Editor,* Assembly of Life Sciences, National Research Council

*Arthur Upton was chairman from inception of the study until his departure to become director of the National Cancer Institute in 1977.

† Served on a consultative group to restate the somatic-effects section first distributed in May 1979.

SUBCOMMITTEE ON GENETIC EFFECTS

DEAN R. PARKER, *Chairman*, Austin, Texas

SEYMOUR ABRAHAMSON, Department of Zoology, University of Wisconsin, Madison, Wisconsin

MICHAEL A. BENDER, Medical Department, Brookhaven National Laboratory, Upton, Long Island, New York

CARTER DENNISTON, Laboratory of Genetics, University of Wisconsin, Madison, Wisconsin

LIANE B. RUSSELL, Biology Division, Oak Ridge National Laboratory, Oak Ridge, Tennessee

WILLIAM L. RUSSELL, Biology Division, Oak Ridge National Laboratory, Oak Ridge, Tennessee

PAUL B. SELBY, Biology Division, Oak Ridge National Laboratory, Oak Ridge, Tennessee

SUBCOMMITTEE ON SOMATIC EFFECTS

EDWARD P. RADFORD, *Chairman*, Department of Epidemiology, University of Pittsburgh Graduate School of Public Health, Pittsburgh, Pennsylvania

GILBERT W. BEEBE, Clinical Epidemiology Branch, National Cancer Institute, Bethesda, Maryland

A. BERTRAND BRILL, Medical Department, Division of Nuclear Medicine, Brookhaven National Laboratory, Upton, Long Island, New York

REYNOLD F. BROWN, Office of Environmental Health, University of California, San Francisco, California

STEPHEN F. CLEARY, Department of Biophysics, Virginia Commonwealth University, Richmond, Virginia

CYRIL L. COMAR (deceased), Electric Power Research Institute, Palo Alto, California

JACOB I. FABRIKANT, Donner Laboratory, University of California, Berkeley, California

MARYLOU INGRAM, Institute for Cell Analysis, University of Miami School of Medicine, Miami, Florida

CHARLES E. LAND, Environmental Epidemiology Branch, National Cancer Institute, Bethesda, Maryland

CHARLES W. MAYS, Radiobiology Laboratory, University of Utah Medical Center, Salt Lake City, Utah

DADE W. MOELLER, Harvard University School of Public Health, Boston, Massachusetts

HARALD H. ROSSI, Radiological Research Laboratory, Columbia University College of Physicians and Surgeons, New York, New York

LIANE B. RUSSELL, Biology Division, Oak Ridge National Laboratory, Oak Ridge, Tennessee

MARGARET H. SLOAN, Division of Cancer Control and Rehabilitation, National Cancer Institute, Silver Spring, Maryland

BENJAMIN K. TRIMBLE (deceased), Department of Medical Genetics, University of British Columbia, Vancouver, British Columbia, Canada

EDWARD W. WEBSTER, Division of Radiological Sciences, Massachusetts General Hospital, Boston, Massachusetts

HENRY N. WELLMAN, Nuclear Medicine Division, Indiana University School of Medicine, Indianapolis, Indiana

ROBERT L. BRENT, *Consultant*, Department of Pediatrics, Jefferson Medical College of Thomas Jefferson University, Philadelphia, Pennsylvania

JOHN T. LYMAN, *Consultant*, Division of Biology and Medicine, Lawrence Berkeley Laboratory, University of California, Berkeley, California

BERNARD E. OPPENHEIM, *Consultant*, Nuclear Medicine Division, Indiana University School of Medicine, Indianapolis, Indiana

ROY E. SHORE, *Consultant*, Department of Environmental Medicine, New York University Medical Center, New York, New York

Preface

In the fall of 1976, the Office of Radiation Programs, Environmental Protection Agency, asked the National Academy of Sciences for current information relevant to an evaluation of effects of human exposure to low levels of ionizing radiation. This report, prepared by the Committee on the Biological Effects of Ionizing Radiations (BEIR Committee) and its subcommittees, in the Division of Medical Sciences of the National Research Council's Assembly of Life Sciences, is in response to that request. It deals with the scientific basis of effects of low-dose radiation and encompasses a review and evaluation of scientific knowledge developed since the first BEIR report (published in 1972) concerning radiation exposure of human populations.

The BEIR Committee endeavored to ensure that no sources of relevant knowledge or expertise were overlooked in its study. To this end, it established liaison with appropriate national and international organizations and solicited the opinions and counsel of individual scientists. We hope that the information contained herein will serve not only as a summary of present knowledge on the effects of ionizing radiation on human populations, but also as a scientific basis for the development of suitable radiation protection standards. It should be noted that the members of the Committee and its subcommittees acted as individuals, not as representatives of their organizations.

We extend our gratitude to the consultants who contributed to the development of this report, many of whom gave unstintingly of their time and thought.

We want to make special note of the contributions of Dr. Arthur C. Upton, who served as chairman of the Committee from November 1976 to July 1977 and resigned when he became the director of the National Cancer Institute, and Dr. Benjamin K. Trimble, of British Columbia, who served on the Subcommittee on Genetic Effects until his untimely death in November 1977. We also note the contribution of Dr. Cyril Comar, not only to this report, but also to the study of radiobiologic effects. Dr. Comar died in June 1979.

The BEIR Committee especially wishes to thank the scientists who have aided it in its work, particularly Drs. Robert L. Brent, John T. Lyman, Bernard E. Oppenheim, and Roy Shore, who not only contributed their time, but also gave considerable assistance in the preparation of some sections of the report.

A special note of appreciation is extended to Division of Medical Sciences staff members Dr. Albert W. Hilberg, whose knowledge and counsel were invaluable to the Committee, and Dr. David A. McConnaughey.

Mr. Norman Grossblatt, of the Assembly of Life Sciences, edited this report.

The preparation of this report required information from several scientific disciplines, and most sections were prepared by members who had particular expertise. Chapter IV was prepared by the Subcommittee on Genetic Effects, and Chapters V and VI, by the Subcommittee on Somatic Effects. The other chapters were prepared by various members of the Committee with direction and advice from the entire Committee.

There were unresolvable differences among the members of the Subcommittee on Somatic Effects concerning the methods of interpretation of human data to arrive at an estimate of health risks of low-dose, low-LET, whole-body radiation exposure. A draft final version of this report was distributed in limited number in May 1979. The somatic-effects sections of that version have here been restated. The restatement was drafted by a subgroup of the Committee, with discussions led by Dr. Jacob I. Fabrikant. The entire Committee has reviewed the report that follows.

Dissenting statements prepared by individual members of a National Research Council committee are not subject to the normal review processes of the National Academy of Sciences; nor are they subject to committee or staff editing or review. They appear exactly as the dissenting committee members prepare them. The NAS-NRC neither endorses nor takes responsibility for the content of the statements.

Contents

xiii

Summary and Conclusions

This report is intended to bring up to date the report of the Committee on the Biological Effects of Ionizing Radiations issued in 1972. In carrying out this intent, we have concentrated primarily on the long-term somatic and genetic risks to people exposed to ionizing radiation at low doses—the condition of principal concern with respect to risks to large population groups.

The major sources of the ionizing radiation to which the general population is exposed continue to be natural background (with a whole-body dose of about 100 mrems/yr) and medical applications of radiation (which contribute similar doses to various tissues of the body). For a given person, the dose from natural background varies with altitude and geographic location, as well as with living habits. Workers in nuclear and other industrial facilities in which radioactive material or x-ray equipment is used are occupationally exposed to levels of radiation that may exceed background severalfold, and the number of such workers is increasing.

The Committee cautions that the risk estimates presented here should in no way be interpreted as precise numerical expectations. They are based on incomplete data and involve a large degree of uncertainty, especially in the low-dose region. These estimates may well change as new information becomes available. Whatever the magnitude of these risks to society and to the individuals exposed, they must be kept in perspective if society is to derive benefits from the use of ionizing radiation. The Committee has no responsibility to recommend regulatory limits, nor does it address cost-benefit issues involving the use of ionizing radiation. These issues are beyond the scope of the task or expertise of this Committee.

RISK OF SOMATIC EFFECTS FROM RADIATION

1. Cancers arising in a variety of organs and tissues are the principal late somatic effects of radiation exposure. Organs and tissues differ greatly in their susceptibility to cancer induction by radiation. Induction of leukemia by radiation stands out because of the natural rarity of the disease, the relative ease of its induction by radiation, and its short latent period (2–4 yr). When the total risk of radiation-induced cancer is considered, however, it is clear that the risk of induced solid tumors (such as breast, thyroid, and lung cancers) exceeds that of leukemia.

2. The Committee recognizes that there is great uncertainty in regard to the shape of the dose-response curve for cancer induction by radiation, especially at low doses. Estimates of risk at low doses depend more on what is assumed about the mathematical form of the dose-response function than on the data themselves. Wherever possible, in estimating the cancer risk from low doses of low-LET* radiation, the Committee has used a linear-quadratic dose-response model that is felt to be consistent with epidemiologic and radiobiologic data, in preference to more extreme dose-response models, such as the linear and the pure quadratic.† The Committee recognizes that some experimental and human data, as well as theoretical considerations, suggest that, for exposure to low-LET radiation at low doses, the linear model probably leads to overestimates of the risk of most cancers, but can be used to define upper limits of risk. Similarly, the Committee believes that the quadratic model may be used to define the lower limits of risk from such radiation. For exposure to high-LET radiation, linear risk estimates for low doses are less likely to overestimate risk and may, in fact, underestimate risk.

3. There is now considerable evidence from human studies that age, both at exposure to radiation and at observation for risk, can be a major determinant of radiation-induced cancer risk. For this reason, the Committee has expressed risk in age-specific terms wherever possible.

4. The Committee's most difficult task has been to estimate the carcinogenic risk of low-dose, low-LET, whole-body radiation. It recognized that the scientific basis for making such estimates is inadequate,

---

*X rays and gamma rays are types of low-LET radiation. Neutrons and alpha particles are types of high-LET radiation.
† In this regard, this report differs substantially from the 1972 BEIR report.

but it also recognized that policy decisions and the exercise of regulatory authority require a position on the probable cancer risk from low-dose, low-LET radiation. Accordingly, the Committee decided that emphasis should be placed on the assumptions, procedures, and uncertainties involved in the estimation process, and not on specific numerical estimates. For the lifetime risk of cancer mortality induced by low-LET radiation from a single whole-body absorbed dose of 10 rads, based on the linear-quadratic model, the estimates of increase in risk range from 0.5 to 1.4% of the naturally occurring cancer mortality, depending on the projection model.* For continuous lifetime exposure to 1 rad/yr, the estimates range from 3 to 8%. Other dose-response models produce other risk estimates; the linear estimates are higher and the quadratic lower than the linear-quadratic. For example, for a single exposure to 10 rads of low-LET radiation, the linear and quadratic differ from each other by an order of magnitude. For incidence, the corresponding estimates of excess risk, expressed as percentages of lifetime cancer incidence, are broadly similar.

5. The Committee does not know whether dose rates of gamma or x rays of about 100 mrads/yr are detrimental to man. Any somatic effects at these dose rates would be masked by environmental or other factors that produce the same types of health effects as does ionizing radiation. It is unlikely that carcinogenic and teratogenic effects of doses of low-LET radiation administered at this dose rate will be demonstrable in the foreseeable future. For higher dose rates—e.g., a few rads per year over a long period—a discernible carcinogenic effect could become manifest.

6. Reductions in dose rate may decrease the observed radiation effect per unit dose, particularly for large doses of low-LET radiation, but not for doses in the linear portion of the linear-quadratic dose-response model and not for high-LET radiation. There appear to be mechanisms, however, pertaining especially to exposure to high-LET radiation, that increase the observed effect per unit dose when the dose rate is reduced. The Committee recognizes that dose rate may affect the risk of cancer induction, but believes that the information available on man is insufficient to adjust for it.

7. A notable development since the 1972 BEIR report is the increasing recognition that there are human genotypes that confer both increased susceptibility or resistance to DNA damage and increased cancer risk

* In interpreting the percentage increases in cancer risk above the naturally occurring rate, the following is an example: If the naturally occurring lifetime cancer risk is 160,000 cases per million persons, the rate is 16%. An increase due to radiation equal to 0.5% of the natural rate will result in an increase of 160,000 × 0.005, or 800 cases—that is, 160,800 total cases will occur. This represents a rate of 16.08% after radiation.

after exposure to carcinogenic agents, including ionizing radiation. The role of constitutional susceptibility to cancer induction is not well enough understood, however, for it to be used as a factor to modify risk estimates. Inasmuch as the risk estimates developed for this report are averages for large populations that presumably include many genotypes, it is unlikely that these risk estimates would be notably altered if data representing very small subsets of abnormally radiosensitive persons could be recognized and excluded from the calculations. If population subsets can be identified as being at substantially greater risk of radiation carcinogenesis, their risk will require separate estimation.

8. The developmental effects of radiation on the embryo and fetus are strongly related to the stage at which exposure occurs. Most information on such effects is derived from laboratory animal studies, but the human data are sufficient to indicate qualitative correspondence for developmentally equivalent stages. In laboratory animals, some developmental abnormalities have been observed at doses below 10 rads. Atomic-bomb data for Hiroshima show that the frequency of small head size was significantly increased by acute air doses in the range of 10–19 rads kerma (average fetal dose, gamma rays at 5.3 rads plus neutrons at 0.41 rad) received during the sensitive period. At Nagasaki, where almost the entire kerma was due to gamma rays, there was no significant increase in the frequency of small head size at air doses below 150 rads kerma. Because a given gross malformation or functional impairment probably results from damage to more than a single target, the existence of a threshold radiation dose below which that effect is not observed may be predicted. There is evidence of such thresholds, but they vary widely, depending on the abnormality. Observed dose-rate effects may also be the result of the multitarget causation of these abnormalities. Furthermore, exposure protraction can reduce dose effectiveness by decreasing to below the threshold the portion of the dose received during a particular sensitive period. Where a developmental effect is measured in terms of damage to individual cells, as in oocyte-killing, a threshold for this effect may be absent.

9. For somatic effects other than cancer and developmental changes (e.g., cataracts, aging, and infertility) considered in this report, the available data do not suggest an increased risk with low-dose, low-LET exposure of human populations.

RISK OF GENETIC EFFECTS FROM RADIATION

1. Because radiation-induced transmitted genetic effects have not been demonstrated in man and because of the likelihood that adequate

information will not soon be forthcoming, estimation of genetic risks must be based on laboratory animal data. This entails the uncertainty of extrapolation from the laboratory mouse to man. However, there is information on the nature of the basic lesions, which are believed to be similar in all organisms; and several physical and biologic variables of radiation mutagenesis have been experimentally explored. For these reasons, some of the uncertainties encountered in the evaluation of somatic effects are absent in the estimation of genetic risk. Human data have been used for estimation of genetic effects resulting from gross chromosomal aberrations.

2. In evaluating genetic risks, the Committee has considered new data on the incidence of genetic disease in human populations. In addition, recent theories of curvilinear dose-response functions and information on dose-rate effects for radiation of different qualities have been reviewed. For low doses and dose rates, a linear extrapolation from fractionated-dose and low-dose-rate laboratory mouse data continues to constitute the basis for estimating genetic risk to the general population. The Committee's genetic-risk estimates are expressed as effects per generation per rem, with appropriate corrections for special situations, such as exposures of small groups to high-LET radiation.

3. Although the Committee used a new method of estimating genetic effects expressed in the first generation, the present estimates of genetic effects are not notably different from those of the 1972 BEIR report. In the first generation, it is estimated that 1 rem of parental exposure throughout the general population will result in an increase of 5–75 additional serious genetic disorders per million liveborn offspring. Such an exposure of 1 rem received in each generation is estimated to result, at genetic equilibrium, in an increase of 60–1,100 serious genetic disorders per million liveborn offspring.

4. The ranges of the risk estimates given in the preceding paragraph emphasize the limitations of current understanding of genetic effects of radiation on human populations. Within this range of uncertainty, however, the risk is nevertheless small in relation to current estimates of the incidence of serious human disorders of genetic origin—roughly 10% of liveborn offspring.

5. Genetic-risk estimates have been restricted to persons with induced disorders judged to cause a serious handicap at some time during life. Even in that category, some disorders are obviously more important than others. In contrast with induced somatic effects, which occur only in the persons exposed, induced genetic disorders occur in descendants of exposed persons and can often be transmitted to many future generations. The major somatic-risk estimates considered in this report are concerned with induced cancers. Although many of these are fatal,

some, such as most thyroid cancers, are curable, but entail the risk and costs of medical care and disability. Somatic effects also include developmental abnormalities of varying severity caused by fetal or embryonic exposure. It is important to recognize that comparisons of genetic and somatic effects must take into account ethical or socioeconomic judgments that are beyond the scope of the Committee's responsibility. As an example of the problem, it is extremely difficult to compare the societal impact of a cancer with that of a serious genetic disorder.

I
Introduction

The potential effects of ionizing radiation on human populations have been a concern of the scientific community for several decades. The oldest of the scientific bodies that now have responsibility in this field are the International Commission on Radiological Protection (ICRP), formed in 1928, and the National Council on Radiation Protection and Measurements (NCRP), a U.S. organization formed in 1929 as the Advisory Committee on X-Ray and Radium Protection. Both continue to study radiation-protection problems that are of special relevance to the work of the present Committee on the Biological Effects of Ionizing Radiations.

The establishment of the U.S. Atomic Energy Commission and its program, in the 1940s, was accompanied by recognition of the need for more precise information on the biologic hazards of radiation, and large-scale animal experiments were initiated. In the early 1950s, the testing of nuclear weapons provoked public concern about the potential effects of ionizing radiation on human populations. In response to this concern, the president of the National Academy of Sciences (NAS) in 1955 appointed a group of scientists to conduct a continuing appraisal of the effects of atomic radiation on living organisms. That study, entitled "Biological Effects of Atomic Radiation," was supported by funds from the Rockefeller Foundation and led to a series of reports by six committees, which were issued from 1956 to 1963 and are generally referred to as the "BEAR reports."

Also in 1955, the General Assembly of the United Nations established the UN Scientific Committee on the Effects of Atomic Radiation

(UNSCEAR), which, among other tasks associated with monitoring and assembling reports of radiation exposure throughout the world, was "to make yearly progress reports and to develop a summary of reports received on radiation levels and radiation effects on man and his environment."[2] In accordance with that objective, the periodic reports issued by UNSCEAR (the most recent was released in 1977) have served as reviews of worldwide scientific information and opinion concerning human exposure to atomic radiation.

In 1959, the Federal Radiation Council (FRC) was formed to provide a federal policy on human radiation exposure. A major function of the FRC was to "advise the President of the United States with respect to radiation matters, directly or indirectly affecting health, including guidance for all federal agencies in the formulation of radiation standards and in the establishment and execution of programs of cooperation with States." To that end, the FRC published eight reports.

At the request of the FRC, the National Academy of Sciences–National Research Council (NAS-NRC) in 1964 established the Advisory Committee to the Federal Radiation Council in the NRC Division of Medical Sciences. The Advisory Committee, now called the Committee on the Biological Effects of Ionizing Radiations (BEIR), continues to review and evaluate available scientific evidence bearing on a variety of problems of radiation exposure and protection and continues to issue reports of its deliberations.

The BEAR reports provided a basis for public understanding of the expected effects of the testing of nuclear devices that had occurred so far and introduced the important concept of regulation of average population doses on the basis of genetic risk to future generations. They also emphasized the diagnostic and therapeutic use of x rays in medicine and dentistry as the greatest source of man-made radiation exposure of the population. However, in the later 1960s and the 1970s, concern arose that developing peacetime applications of nuclear energy, particularly the growth of a nuclear-power industry for production of electricity, could cause serious exposure of human populations to radiation. In February 1970, the FRC asked the NAS-NRC Advisory Committee to consider a complete review and reevaluation of the existing scientific knowledge concerning radiation exposure of human populations. This request from the FRC came about because of a natural concern on the part of the Advisory Committee that there had been no detailed overall review since the BEAR reports; new factors that might need to be considered, such as optional methods of producing electricity and the presence of environmental contamination different from types previously encountered; and a growing number of allegations made in the public media and before Congressional committees that current radiation-protection guides were inadequate to protect the health of the general population.

The NAS-NRC and the Advisory Committee accepted the task proposed by the FRC. On October 2, 1970, the Environmental Protection Agency (EPA) was established by the President's Reorganization Plan No. 3 of 1970. On December 2, 1970, the activities and functions of the FRC were transferred to the EPA Office of Radiation Programs. In concert with this change, the NAS-NRC Advisory Committee requested a change in its title, and the president of the NAS renamed it the Advisory Committee on the Biological Effects of Ionizing Radiations; the Committee's functions, activities, and staffing were not changed. The BEIR Committee produced its report in November 1972: *The Effects on Populations of Exposure to Low Levels of Ionizing Radiation* (BEIR I).[1]

The NAS-NRC and the BEIR Committee were asked by the EPA in early 1973 to review methods for health benefit-cost analysis that might be applicable to ionizing-radiation exposure. The Committee completed its report in 1976, and it was published in 1977: *Considerations of Health Benefit-Cost Analysis for Activities Involving Ionizing Radiation Exposure and Alternatives* (BEIR II).

In the fall of 1976, the EPA Office of Radiation Programs asked the NAS-NRC and the BEIR Committee to update the 1972 BEIR report on the basis of newly developed scientific information. The task before the Committee was specified in detail in the contract agreement between the NAS and the EPA signed on September 30, 1976:

The Contractor shall review the current state of knowledge on somatic and genetic effects of ionizing radiation. Under this review phase the Contractor shall consider the following:

(a) The extent to which animal data, particularly from inbred strains, is pertinent to estimating somatic radiation effects in human populations.

(b) Recent theories of curvilinear dose response functions for both high and low LET radiations for somatic and genetic effects.

(c) The effects of dose rate and protraction on the incidence of radiation effects from high and low LET radiations for somatic and genetic effects.

(d) The appropriateness of using relative risk estimates *vis à vis* absolute risk estimates for specific radiation related cancers based on a consideration of age related changes in patterns of radiocarcinogenesis. Particular emphasis on late results from *in utero* and childhood exposures would be extremely useful.

(e) The probable extent of synergistic interactions between ionizing radiation and other environmental and occupational promoters of carcinogenesis.

The Contractor shall make such recommendations to EPA on the potential risks from ionizing radiation as may be justified on the basis of current published scientific information. In particular, the Contractor shall provide recommendations on:

(a) The various ranges of dose and dose rates for which different numerical risk estimates are appropriate for both low LET and high LET radiations.

(b) The difference in human risk (somatic and genetic) that reasonably may be expected following acute and chronic exposures.

(c) Based on a consideration of these factors, numerical estimates of the somatic and genetic risks to humans from low dose rate ionizing radiations.

To carry out the required review and analysis, two subcommittees were formed to deal with the somatic effects and the genetic effects of low-level ionizing radiation.

The present BEIR Committee not only used the 1972 BEIR report as a guide in its review, but also quoted extensively from it when there was no apparent need for a change in wording. The 1972 BEIR report is no longer readily available, and the Committee felt that the extensive use of sections of it in the present report might allow the reader to gain a more complete view of the subject matter discussed.

REFERENCES

1. National Research Council, Advisory Committee on the Biological Effects of Ionizing Radiations. The Effects on Populations of Exposure to Low Levels of Ionizing Radiation. Washington, D.C.: National Academy of Sciences, 1972.
2. United Nations Scientific Committee on the Effects of Atomic Radiation. Report A/7613. General Assembly Official Records. 24th Sess. Suppl. No. 13. New York: United Nations, 1969.

II
Scientific Principles in
Analysis of Radiation Effects

The purpose of this chapter is to provide background information on the scientific principles involved in the measurement and evaluation of the biologic effects of ionizing radiation. That the literature on the biologic effects of radiation is extensive indicates the concern that has been manifest among governmental and other groups about the potentially harmful effects of a great expansion of nuclear technology and other applications of radiation. Indeed, it is fair to say that we have more scientific evidence on the hazards of ionizing radiation than on most, if not all, other environmental agents that affect the general public. Especially important is the evidence that has been obtained from studies of human populations that have been exposed to radiation for various reasons; however, the large body of experimental evidence on cell systems and experimental animals is also important for our understanding of radiation effects on living systems.

The following discussion summarizes briefly some aspects of ionizing radiation and its biologic effects, with special reference to concepts that we believe to be important to our present understanding of these effects, especially at low radiation doses. This discussion is clearly not an exhaustive review of the voluminous literature, but rather highlights general considerations that are pertinent to the detailed information in later chapters that form the basis of risk estimates ultimately derived in this report.

The units of radiation used in this report are those in common use. The main ones are the rad, the unit of absorbed dose (1 rad = 100 ergs/g = 0.01 joule/kg), and the rem, the unit of equivalent dose for different types of radiation (1 rem = 1 rad $\times$ a correction factor to equalize biologic effects). However, the reader should be aware that new units have been proposed and may well come into general use—in particular, the gray (1 Gy = 100 rads = 1 J/kg) and the sievert (1 Sv = 100 rems).

Radiation doses in this report are expressed in units used by the original authors. For comparative purposes, the conversion or modifying factors are specified in each case. Other units used in this report are defined at the place of first use.

Radiation effects have been classified traditionally as "somatic" if manifested in the exposed subject and "hereditary" or "genetic" if manifested in the descendants of the exposed subject. However, the term "genetic" is also applicable to effects that involve changes produced in the informational macromolecules of cells. Thus, some somatic effects of radiation may be mediated by genetic mechanisms that affect a wide range of body cells, whereas genetic effects involve only germ cells in the gonads.

The term "stochastic" is used to describe effects whose probability of *occurrence* in an exposed population (rather than their severity in an affected individual) is a direct function of dose. Stochastic effects are commonly regarded as having no threshold—that is, any dose, however small, has some effect, provided that the population exposed is large enough. Hereditary effects and some somatic effects, such as cancer induction, are considered to be stochastic. The term "nonstochastic" is used to describe effects whose *severity* is a function of dose. For these effects, there may be a threshold—that is, there may be a dose below which there is no effect. Examples of nonstochastic somatic effects are cataracts, nonmalignant skin damage, hematologic deficiencies, and impairment of fertility.

PHYSICAL ASPECTS OF THE BIOLOGIC EFFECTS OF IONIZING RADIATION

INTERACTION OF IONIZING RADIATION WITH CELLS

All ionizing radiation affects cells by the action of charged subatomic particles, which dislodge electrons from atoms in the irradiated material, thus producing ions. By this mechanism, energy is transferred from the radiation to the material, and the amount of energy absorbed per unit mass of the material is the *absorbed dose*, *D*.[11]

Radiation exposure occurs from many sources, described in Chapter III. Energetic charged particles may arise, for example, from radioactive substances that are inside or outside the irradiated material, or they may have been produced by a variety of processes involving high-energy radiation, such as x rays or neutron beams. Radiation is *directly ionizing* if it carries an electric charge that directly interacts with atoms in the tissue or medium by electrostatic attraction or repulsion. *Indirectly ionizing* radiation is not electrically charged, but results in production of charged particles by which its energy is absorbed. This kind of radiation produces high-velocity fragments of the atoms of the irradiated material; and these fragments become the source of energetic charged particles, which then act to ionize other atoms. It takes about 34 electron volts (eV) of energy to produce one ionization. Most human exposures to radiation are at energies of 0.05–5 million electron volts (MeV)—energies at which many ionizations occur as the radiation passes through cells.

A fundamental characteristic of charged particles produced directly or indirectly is their *linear energy transfer* (LET), which is the energy loss per unit of distance traveled, usually expressed in kiloelectron volts (keV) per micrometer (μm). The LET, which depends on the velocity and the charge of the particle, can vary from about 0.2 to more than 1,000 keV/μm.

Some particles expend virtually all their energy at linear energy transfers of less than a few kiloelectron volts per micrometer. In human exposures, the most significant of these particles are μ-mesons (muons), which are the principal components of primary cosmic radiation, and electrons, especially those emitted by beta radiation. Such high-energy electrons, as well as the indirectly ionizing radiation that produces them (that is, x rays and gamma rays), are referred to as low-LET radiation. This radiation is responsible for most of the absorbed doses received by the general population and by radiation workers, but high-LET radiation also contributes. The most important directly ionizing high-LET radiation is alpha radiation emitted by internally deposited radionuclides. Neutron radiation is the principal kind of indirectly ionizing high-LET radiation; neutrons interact mainly by producing recoil protons. Low-energy electrons are produced by both direct and indirect ionizing radiation and are intermediate in LET.

Ionizing radiation interacts with matter along more or less straight charged-particle tracks, but the deposition of energy is not uniform, especially if small volumes and low absorbed doses are considered. In the latter case, the energy is delivered to this volume in only a small number of discrete interactions (i.e., only a few particle traversals). The nuclei of the cells in the human body, which are the loci believed to be primarily affected by ionizing radiation at low doses, have an average diameter of roughly 5 μm. At radiation levels that are of interest in human exposure, the energy absorbed in these structures can vary greatly and, thus, differ substantially from the mean. It is therefore necessary to consider the microdosimetric quantity *specific energy, z,* which, like the absorbed dose, D, is defined[11] as energy divided by mass, but denotes values of this quotient in a localized region (in this case, the cell nucleus). The importance of this quantity becomes apparent if one determines the values of z in cell nuclei that have received about 1 yr of background radiation. This produces an absorbed dose of about 100 mrads of (mostly) low-LET radiation. In about two-thirds of the nuclei, $z = 0$, that is, no ionizations have occurred; in the remainder, z varies over several orders of magnitude, with an average value of about 300 mrads. If the same dose, D, were delivered by fission neutrons, z would differ from zero in only about 0.2% of the nuclei; however, in these affected nuclei, it would average 50 rads, i.e., 500 times the average dose. It is evident that the heterogeneity of energy deposition depends greatly on radiation type.

RELATIVE BIOLOGIC EFFECTIVENESS

Because $\bar{z}$, the average value of z, is always equal to D, microdosimetric considerations would be of little interest if the biologic effect* of radiation were simply proportional to z. In this case, the biologic effectiveness of radiation would be independent of LET, which is contrary to experience. The relative biologic effectiveness (RBE) of high-LET radiation relative to low-LET radiation is defined as D_L/D_H, where D_L and D_H are, respectively, the absorbed doses of low- and high-LET radiation required for equal biologic effect. The RBE is generally larger than 1, and values in excess of 50 have been reported for some types of cell effects at low absorbed doses. That is, high-LET radiation requires lower doses to produce equivalent effects. In general, increasing energy concentration in the cell results in a more than proportionally increased probability of effect. Probable exceptions to this are some effects on the genetic material that produce point mutations or cell transformations. However, for some genetic, as well as somatic, effects, the cell may respond to radiation energy in a nonlinear manner. Experimental evidence indicates that the response in these cases can be characterized as quadratic[13,18,20,24] and is consistent

---

*Although it may sometimes be difficult to provide a precise scale of *severity* of effect, it is possible to define the fraction of the exposed population that exhibits a specified degree of damage. The term "effect" is used here in this meaning.

with dependence on the square of the specific energy, z. The quadratic dependence on specific energy might be due to a mechanism whereby biologic effects result from misjunction of pairs of broken DNA molecules. However, this interpretation must still be regarded as hypothetical, and we use here a conservative terminology that states that the basic action is one in which pairs of *sublesions* combined to form *lesions*.

If it is estimated that the average range of interaction of sublesions is roughly 1 μm and it is assumed that the yield of sublesions is proportional to the mean value of specific energy, i.e., to the absorbed dose, then E, the frequency of effects (numbers or probabilities of lesions that depend on the combination of two sublesions), is proportional to the square of the specific energy. Thus,

$$E = K\overline{z^2}. \tag{II-1}$$

It can be shown[13] that $\overline{z^2}$, the mean value of z^2, is given by

$$\overline{z^2} = \zeta D + D^2, \tag{II-2}$$

where ζ is a microdosimetric quantity.* Thus,

$$E = K(\zeta D + D^2). \tag{II-3}$$

In this model, if the critical specific energy is deposited in sites of 1-μm diameter, the applicable values of ζ would range from 12.5 to 25 rads for low-LET radiation. Larger values would apply for smaller sites. The value of ζ for high-LET radiation on the basis of microdosimetry would typically be 100 times larger than the value for low-LET radiation. Therefore, the linear term would be much more important for high-LET radiation.

When $D = \zeta$, the linear and quadratic terms in Equation II-3 are equal. When D is less than 0.1 (i.e., the absorbed dose is low), the quadratic term becomes negligible, and the energy is deposited by single particles. Consequently, the fraction of the cells receiving energy is proportional to absorbed dose, and this energy is independent of dose and dose rate.

Equation II-1 implies that the RBE should vary from approximately 1 at high absorbed doses to the ratio of the ζ values of high- and low-LET radiation at low absorbed doses.[13] If this ratio were substantially larger than 1, there should be a considerable range of absorbed doses at which the RBE would be inversely proportional to the square root of the absorbed dose of

---

*ζ is the ratio of the second and first moments of the frequency distribution of specific energies produced by single events.

high-LET radiation, down to the doses where both the high- and the low-LET responses would be linear with dose. This behavior of the function relating RBE to the absorbed dose of high-LET radiation, including RBE values up to 100, has often been observed experimentally for fission neutrons.[23] It should be noted that, even if biologic effects depended on some power other than 2 for specific energy, the RBE would vary inversely with high-LET dose, provided that this power were larger than 1.

EFFECTS OF RADIATION ON AUTONOMOUS* CELLS

The above considerations and conclusions briefly summarize the theory of dual radiation action on autonomous cells. This simple form is, however, subject to qualifications and modifications.[9,12] According to the simplified theory, at low absorbed doses any radiation effect on autonomous cells must be proportional to absorbed dose and independent of absorbed dose rate. This conclusion applies even if there is a variation in radiation sensitivity among the cells and even if repair processes are operative, and whether or not there is a quadratic response. On the average, an event in the nucleus carries a probability of producing a given effect, and the fraction of cells affected is the product of this probability and the fraction of nuclei that could be affected. The latter fraction is proportional to the absorbed dose at low doses. However, when the absorbed dose is large enough for there to be an appreciable probability of multiple events, proportionality between absorbed dose and effect can no longer be expected, even for autonomous cells. According to Equation II-3, for a dose of $n(\zeta)$ rads, the effect will be $[n(n + 1)]/2$ times greater than the effect at ζ rads.

The relation given by Equation II-3 is shown in a logarithmic presentation in Figure II-1, which indicates the magnitude of the error that can occur in linear extrapolation. The unit of absorbed dose is ζ, i.e., the absorbed dose where the linear and quadratic components are equal, and the effect is plotted in units relative to the linear contribution at $D = \zeta$. It can be seen from Figure II-1 that there are about 2 decades of absorbed dose between the point where the slope of the curve is 1.1 and the point where it is 1.9. Precise radiobiologic experiments covering a hundredfold range of absorbed dose are rare, and it is thus not surprising that the entire transition from a linear to a quadratic dependence has rarely been observed, although this has been approached with low-LET radiation.[30]

---

*The term "autonomous" is applied to cells whose response to radiation is unaffected by the irradiation of other cells or by any other entities (e.g., individual cells in cell culture).

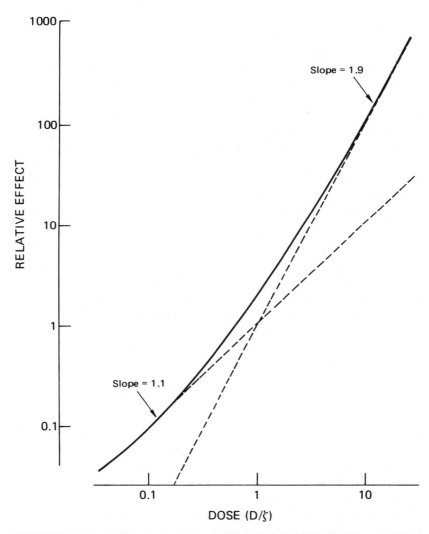

FIGURE II-1 Dose-effect relationship according to Equation II-3, plotted on logarithmic scales. The two dashed lines represent the linear and quadratic contributions to the effect, and their sum is the solid line. The dose must be varied by a factor of 100 for the full effect of the quadratic factor to become expressed. At low doses, the quadratic term is unimportant.

In general, data for yields, E, of cell effects can be satisfactorily fitted empirically to an expression of the form, similar to Equation II-3,

$$E = aD + bD^2 + C, \qquad (II-4)$$

where C is the zero-dose incidence, and a and b are empirically determined coefficients. There is disagreement, however, over the meaning of the coefficients a and b, at least in the form in which they are determined by simple fitting of Equation II-4 to the experimental data points. The classical radiobiologic view is that these coefficients accurately measure the admixture of one- and two-track events. The theory described above would ascribe these values to the physical nature of radiation absorption, with the measured damage resulting from the interaction of two sublesions, which may come about as an effect of either a single track or two separate tracks. In this view, a and b would vary according to the LET of the radiation.*

If the spontaneous rate is taken into account, Equation II-3 reduces to Equation II-4 if $a = K\zeta$ and $b = K$. The virtue of either the empirical form of this equation (Equation II-4) or the theoretical form (Equation II-3), as seen by their advocates, is that good data will yield accurate values of these coefficients, which will lead to precise estimation of the effects that would be produced at very low doses and low dose rates. In either Equation II-3 or Equation II-4, the time over which the dose is given is not included as a variable. Radiobiologic theory does include this in a correction factor for two-track events:[13]

$$G = 2(\tau/T)^2(T/\tau - 1 + e^{-T/\tau}), \qquad (II-5)$$

where G is a correction factor for yield of two-track events, τ is the average elapsed time between breakage and restitution (i.e., between lesion induction and lesion repair), and T is the duration of treatment. From the equation from which Equation II-5 is derived, the relation of yield for two-track events is $E \propto D^2G$—similar to Equation II-1, but with a coefficient, G, that depends on dose rate.

The maximum approached by G is unity when T approaches 0. In the range where T and τ are approximately equal, the value of G approaches 0.736—for about a one-fourth reduction in yield below simple, two-track expectations. Although this correction factor is usually invoked only in relation to the use of the dose-rate effect to estimate the mean longevity of

---

*Also, in this view, the coefficients a and b in Equation II-4 are related to the coefficient ζ in Equation II-3 as follows: $a/b = \zeta$.

lesions, it is obvious that it can also result in different errors for each dose point in dose-response curves, where total dose is varied by varying time, rather than by varying the dose rate. It is important to note that *this correction is not dose-dependent.*

An alternative interpretation is that the end points in question—for example, mutations—may depend on the operation of more than one mechanism. That is, there may be more than one biologic mechanism involved in addition to the presumed "dual-action" mechanisms of physical absorption. There may be more than one class of events involved in point mutations, as discussed in BEIR I. Furthermore, the end point, mutation, may result from the operation of both repair and damage mechanisms and may involve a variety of lesions. From this standpoint, it might be argued that the best estimate of damage at very low doses would be a linear extrapolation between the yield at the lowest dose for which there are reliable data and the yield at zero dose. Such an estimate would not differ appreciably from that based on the quadratic relationship, provided that the value of bD^2 at the lowest measured dose is not appreciably different from zero.

There is yet another viewpoint, perhaps more pertinent to the kinetics of induction of two-break rearrangements of chromosomes than to gene mutation, but not strictly limited to such rearrangements: the *observed* rates of damage may not reflect the rates of *induced* damage in any simple way, because of the nature of the process by which the end points are detected—for example, in the detection of chromosomal abnormalities. In consequence, it can be argued that the values of a and b obtained from Equation II-4 lack real biologic meaning, that is, that they neither describe the real mechanisms of damage nor serve as useful indicators of the low-level effects that are to be expected. Statistical and sampling complexities are not properly taken into account by a direct fitting of data to a simple quadratic expression. As a result, the derived values of a and b obtained may differ markedly from their true values. Furthermore, because the estimations of a and b based on observed data are not independent, an overestimation of one is accompanied by a compensatory underestimation of the other; this leads to an even greater error when it is their quotient, a/b (an estimation of ζ), that is considered.

A further complication at large absorbed doses is that radiation may produce a variety of effects. Because it has been assumed that each of these results from particular groupings of sublesions, it may be expected that, as the number of these increases, competition between effects may alter the dose dependence for one particular effect. An example of considerable practical importance concerns the interplay between malignant cell transformation and cell-killing within the same cell. Evidently, transformed cells cannot initiate tumors if they also have suffered reproductive death, which becomes increasingly probable at higher absorbed doses. Thus, dose-response data may show a decrease in effect at high doses—the so-called "cell-killing" effect.

Recent experiments[16] on radiation-induced transformation of cells in cell culture have yielded dose-effect curves whose slopes decrease between the linear and quadratic regions shown in Figure II-1. This example illustrates the fact that the dose-effect curves for autonomous cells can have complex shapes and that extrapolation from high doses can lead to an *underestimate* of the effect of low doses. The effect can be explained in terms of competition for sublesions in which the alternative effect is not cell-killing, but one of a variety of possible nonlethal cell alterations. A related finding is that, if the total dose is given in several successive fractions, rather than all at once, the transformation rate is *unchanged* in the linear region at the lowest doses, *reduced* in the quadratic region at the highest doses, but *increased* in the intermediate region where the slope of the curve is less than 1. This is to be expected, if there is no interaction between the dose fractions. Finally, in such systems, the RBE could be less than would be deduced from the ratio of ζ values. If single high-LET particles produce increments of ζ that are comparable with the range of absorbed doses for which there is a relatively constant transformation rate, the RBE might be considerably less than expected on the basis of the considerations presented above.

RELATION BETWEEN RADIATION EFFECTS ON CELL SYSTEMS AND MUTAGENESIS OR CARCINOGENESIS IN MAN

Some radiation effects are apparently due to damage to individual autonomous cells. In human radiation exposure, the most important example might be the mature gametes in the gonads. Other effects, such as cataractogenesis, are due to injury of several cells. Here, one would not expect proportionality between dose and effect, whether or not the cells involved in the response were autonomous.

For the most important somatic radiation hazard, carcinogenesis, it is often assumed because the number of cells at risk is very large that transformation of an individual cell does not necessarily result in cancer. Among the various inhibitory mechanisms that have been considered is a requirement that several contiguous cells be transformed, or the action of immunologic or other host defenses be impaired. In the former case, a multicellular interaction would be involved; in the latter, the response of individual cells may not be autonomous—for example, if the effectiveness of the defense mechanisms is limited by the number of cells transformed.

In both situations, the dose-effect curve could have various forms at low absorbed doses. For example, a downward curvature of the dose-response relationship has been observed for radiation-induced mammary neoplasms in one strain of rat[27] at absorbed doses of neutrons that are clearly much less than ζ, which indicates that this malignancy is not due to an autonomous-cell response. In this system, however, the RBE increases inversely with neutron dose in the same manner as observed for autonomous single cells. However, for both high-LET and low-LET radiation in dose ranges where the single-cell response is linear, a multicellular mechanism for cancer induction would theoretically produce a dose-effect relation with upward curvature (slope increasing with dose). Many dose-response curves for experimental carcinogenesis induced by low-LET radiation in mammals show such upward curvature (e.g., Ullrich *et al.*[32]). Although it is not clear what mechanisms are involved in this response, it cannot be assumed with any certainty that there is a dose-proportional, dose-rate-independent induction of cancer even at low absorbed doses of any radiation.

In view of the complexities of cancer production, especially in human populations potentially exposed to a multitude of environmental factors that may interact with radiation-induced effects, a theory based on studies of autonomous cells may represent a great oversimplification with regard to dose-response data. From biophysical considerations, at low absorbed doses any effects on individual autonomous cells are proportional to absorbed dose and independent of absorbed dose rate. The RBE of high-LET radiation is likely to be greater than 1 and to increase with decreasing effect until limiting values of RBE are obtained at low doses that are large for many types of effects on cells and organisms. Linear extrapolations from high absorbed doses are likely to result in overestimates of the risks of low absorbed doses, especially when high dose rates and low-LET radiation are involved.

APPLICATION OF DOSE-RESPONSE FUNCTIONS TO OBSERVED DATA

On the basis of the above theoretical considerations, some mathematical procedures have been applied to data obtained not only on individual cells, but also on whole animals and man. Because we do not yet have an adequate theory of cancer induction, the most important somatic effect of radiation, it is not possible to derive a theoretical basis for dose-response data for these effects from first principles. Nevertheless, because a genetic transformation in the cell nucleus is considered to be involved in cancer induction, as well as in genetic effects, theoretical approaches have been used primarily to develop some understanding of the effects of low doses of radiation.

The functional forms fitted to dose-response data from the studies considered in this report, when these data are detailed enough to permit it, are special cases of the general form (modified from Equation II-4):

$$E = F(D) = (\alpha_0 + \alpha_1 D + \alpha_2 D^2)\exp(-\beta_1 D - \beta_2 D^2), \quad \text{(II-6)}$$

where D is the radiation dose in rads, $F(D)$ is the incidence of effects (e.g., cancer) at dose D, the parameters α_0, α_1, α_2, β_1, and β_2 have positive values, and α_0 is the control or spontaneous rate of the effect under study. This functional form, which has been discussed by Upton,[33] can be viewed as a basically linear function (α_0 and α_1 are the only parameters relevant to risk at very low doses), with modifications that allow the fitted curve to express upward (positive) curvature at low doses (α_2) and downward (negative) curvature at high doses (β_1 and β_2) to take account of cell-killing effects. Depending on which of these coefficients vanish, the general form reduces to several simpler models—namely, the linear, the pure quadratic, and the linear-quadratic (quadratic with a linear term) models (see Figure II-2).

The curve-fitting procedure is an iterative weighted least-square procedure; technical details have been published.[14] On any given iteration, the weight corresponding to the observed rate (simple or age-standardized) at dose D is assumed to be the number of person-years (PY) at risk of the effect at that dose (usually the number of PY corresponding to a dose interval with average dose D), divided by the current value of the fitted function at dose D. That is, the rate times the PY is assumed to correspond to a Poisson variate with mean equal to PY times $F(D)$ at each dose—the number expected from the fitted curve.

The mathematical functions discussed above assume that there is no threshold dose below which there is no excess risk. On statistical grounds, however, the existence or nonexistence of a threshold dose is practically impossible to determine, unless there is a marked increase in risk for doses only slightly greater than the presumed threshold. That is because the sample size required to estimate or test an (absolute) excess is approximately inversely proportional to the square of that excess. For example, if the excess is truly proportional to dose, and if 1,000 exposed and 1,000 control subjects were required to test adequately the excess at 100 rads, then about 100,000 in each group would be required at 10 rads, and about 10,000,000 in each group would be required at 1 rad. On these grounds, it may be possible to assert that there is no threshold for an effect

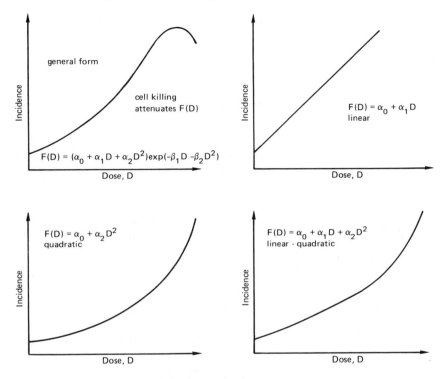

FIGURE II-2   Alternative dose-response curves.

above a given dose, but it can never be stated that there is none at any dose. In other words, empirical determination of the presence or absence of effects at very low doses is extremely difficult, except for biologic effects that may show very great sensitivity to radiation.

BIOLOGIC FACTORS IN RADIATION EFFECTS

Ionizing radiation interacts with cells in a manner that can be described on the basis of the physical or chemical reactions produced, but the step from these reactions to an eventual biologic effect is not fully understood. When we are concerned with long-term effects in complex organisms, the problem of relating deposition of radiation energy to the effects that appear much later is even more difficult. Furthermore, not only do individual cells vary in their responses to radiation, but tissues contain many different types of cells and many biologic interactions occur within and among tissues, so we may expect the effects of cell damage to be very complex indeed. This section considers some of the biologic factors that may influence responses to radiation.

CELL DIVISION

An important effect of radiation, which accounts for the symptoms and causes of death from exposure to large doses of whole-body irradiation, is suppression of cell division.[34] Nearly all lymphoid, bone-marrow, and intestinal epithelial cells responsible for rapid replacement of short-lived mature cells cease to be able to divide, and in these and many other tissues a substantial fraction of cells that would otherwise be capable of division die without further reproduction. If the organism is to survive these effects, the remaining stem cells must repopulate the tissues to overcome cell loss. An example of this process is the disappearance of granulocytes, as a result of suppression of cell division of precursor cells in the bone marrow, in the blood of persons irradiated at relatively high doses. Recovery may require days or weeks.

Cell-killing and suppression of cell division are nonstochastic effects of radiation—the ultimate biologic effects depend markedly on the fraction of cells affected. At low radiation doses, only a small fraction of the dividing cells may be damaged, and in tissues this damage may lead to no detectable change in function. In tissues with rapid cell turnover, interference with normal function will occur only when the affected cells constitute a large fraction of those available for replenishment of cell stores. We anticipate that host factors play an important role in determining the fraction of cells required to produce serious physiologic or biochemical abnormalities in association with this disturbance in cell replacement, especially in the intestinal tract and in the population of white blood cells. Such host factors include general nutritional status (e.g., availability of nutrients important in cell growth), the presence or absence of preexisting infection, or exposure to chemicals or drugs that have effects on cell division similar to those of radiation.

Nevertheless, because these effects are observed at high doses of radiation, they are of limited interest in this report. An exception is the irradiation of the developing fetus. In this case, especially during organogenesis early in pregnancy, cell division is occurring extremely rapidly, and normal development may depend on the integrity of relatively few cells from which the tissues will eventually develop. Only a small fraction of such cells need to be affected by radiation to interfere with proper organ development, so radiation at relatively low doses may lead to detectable teratogenic effects. Whether such effects occur depends critically on the number of stem cells available, as well as on the stage of fetal development.[8]

CELL MUTATION OR TRANSFORMATION

The genetic effects of chief concern in this report arise from radiation-induced dominant or recessive mutations in the DNA or chromosomal abnormalities of the germ cells. Similar types of changes in all other body cells are generally accepted as constituting an important step in the development of cancer, the major somatic effect of radiation applicable to low doses. To produce a carcinogenic effect, lesions produced in the DNA from radiation energy deposited in the cell nucleus must survive in cells that are not otherwise so damaged that they no longer have the capability of dividing. These changes may be localized to specific regions of DNA and may be induced by single-track events from radiation exposure; thus, they are considered to be stochastic, with a probability of occurrence proportional to radiation dose.

It is known that cells can repair some types of lesions in DNA,[26] and this repair may modify radiation damage. The repair processes are themselves under control of other portions of the cellular DNA. In some organisms, as a result of genetically transmitted autosomal recessive mutations, the repair mechanisms are deficient in the homozygote.[6,19,22] For at least one mutation-induced disease, ataxia telangiectasia, there is evidence that the defect in DNA repair makes the subject sensitive to ionizing radiation.[31] Disturbances in DNA repair might be expected to affect the risk of radiation-induced genetic effects, as well as the risk of cancer production.

Because these abnormalities are so rare in the human population, and because the affected persons should be kept from exposure by individualized protection measures, any special risk of their exposure to low levels of radiation is of little relevance to the risks of the general population. Similar considerations apply to persons with chromosome-21 trisomy (Down's syndrome) or with various other chromosomal abnormalities, whose cells are reported to be abnormally sensitive to radiation induction of chromosomal aberrations.[7,25] Our knowledge of biologic factors that might modify sensitivity to genetic effects is still limited to these rare conditions. Those who are heterozygous for the ataxia telangiectasia gene may also have a deficiency in DNA repair.[5] If an increased radiation sensitivity is demonstrated in the heterozygotes for the known DNA-repair-deficient conditions, the population at special risk of genetic or carcinogenic effects of radiation could be significant.

HOST FACTORS IN RADIATION CARCINOGENESIS

Present evidence indicates that cancer induced by chemical or physical agents, such as ionizing radiation, involves a multistage process, with evolution of molecular and cellular changes leading to changes in the tissue as a whole. The earliest stage of this process is the so-called initiation phase, in which events leading to lesions in the DNA occur in a single cell or in a small group of cells. These cells have the capability of transforming into a neoplastic process—that is, normal growth constraints are altered in these cells. There are control mechanisms in tissues that act to prevent development of transformed cells into a malignant tumor. These regulatory processes involve the normal cells adjacent to the transformed cells, as well as hormonal, immunologic, and other influences in the tissue or the body. Inherited traits can influence all stages of cancer by modifying tissue responses to initiation, as with the DNA repair mechanism, or by variations in the regulatory mechanisms.

The process that affects the regulatory control exerted on the transformed cell or cells in a way that permits them to begin uncontrolled growth leading to a cancer is referred to as "promotion." Some physiologic disturbance of the tissue frees the potentially rapidly dividing cell or cells from constraints on cell division. Such disturbances may include repeated damage to normal tissue, stimuli to cell proliferation (such as hormonal effects), or disturbances in recognition of immunologically transformed cells by immune processes.

This is a brief statement of the two-stage theory of carcinogenesis.[2] The first stage is initiation, associated presumably with eventual alteration in the cell genome, which causes loss of normal control of cell division in transformed cells. The second stage is promotion, a process by which a transformed cell is able to grow into a detectable cell mass identifiable as a cancer. These two stages may be separated by many years, a factor accounting at least in part for the long latent periods often observed in man between exposure to a carcinogen and development of a cancer.

Both the initiating and the promoting steps can be modified by biologic factors, including those characteristic of the host, acting in concert with a carcinogen, such as radiation. The probability of an initiating event may be affected, for example, by whether the cell nucleus already contains viral nucleoproteins incorporated into the DNA. In this sense, viral infection may play a permissive role in the induction process—a necessary but not sufficient condition for carcinogenesis.

It is clear, however, that host factors are especially important in the promoting stage, where relatively nonspecific alterations of normal tissue function may be important. Hormonal influences, which clearly exert great effects on cell proliferation in normal tissues, are one factor of considerable significance, at least in some cancers. The importance of hormones is determined by the tissue type; for example, sex hormones regulate growth in the sex organs, and pituitary hormones influence cell proliferation in the gonads, as well as endocrine glands, such as the thyroid. The immunologically active lymphoid cells, which may suppress or destroy transformed cells if they are recognized as immunologically "foreign" to the host, may also be important. The immunologic surveillance theory[4] of defense against cancer is now recognized as not applicable to all cancer types, but persons whose immune mechanisms are suppressed by drugs have increased risk of some neoplasms, notably reticulum cell sarcoma.[10]

Another factor in cancer promotion is the alteration of normal tissue integrity by a wide range of conditions, including irritant chemicals that reach epithelial structures, vitamin A deficiency, viral infection of the respiratory tract, and trauma. The precise role of any of these factors is not well understood in human carcinogenesis, but at least under experimental conditions their importance has been demonstrated for some neoplasms.

Finally, changes associated with the aging process have been postulated as predisposing to cancer through deterioration of tissue repair and loss of vitality of the normal cell complement.

This brief summary of mechanisms of carcinogenesis has been presented, because it is apparent that circumstances leading from cellular radiation effects to cancer involve many factors that may be highly variable in an exposed population. For this reason, we may expect sensitivity to cancer induction by radiation to be variable from individual to individual, as well as from time to time in the same individual. Thus, data on radiation dose versus cancer response obtained in cell systems or even in experimental animals must be applied to human populations with considerable caution.

EPIDEMIOLOGIC STUDIES AS THE BASIS OF RISK ESTIMATES FOR EFFECTS OF IONIZING RADIATION

In assessing somatic effects of ionizing radiation, the BEIR I report placed primary emphasis on studies of exposed human populations. In contrast, estimates of risks of hereditary effects on human populations have depended principally on evidence from animal experiments. However preferable it may be to have firm evidence of hereditary changes based on exposures of human populations to ionizing radiation, detection of increases in human mutations due to the action of any environmental agent is still difficult. For somatic abnormalities induced *in utero* by radiation, the position is somewhat intermediate—that is, some human data have been obtained, but we also depend on animal data.

The emphasis on human studies for determining the somatic effects of ionizing radiation remains valid, although theoretical and experimental studies continue to be important in extending our basic knowledge. For most types of health effects occurring in those exposed to radiation, we now have considerable human experience, as the balance of this report shows. Moreover, in terms of establishing human risk estimates, it is a well-recognized principle in the field of environmental toxicology that results obtained in animal experiments are not necessarily translatable directly to human populations. For example, the fact that the human population is genetically heterogeneous, with widely varying individual physiologic and biochemical characteristics, makes it likely that there are subpopulations at special risk from radiation exposure. It is difficult to simulate this kind of heterogeneity in animal populations, other than by inferences drawn from species variation in responses or from differences in susceptibility between strains of a given species.

We lack adequate information on the effects of low radiation doses in human populations, and in this regard we still depend on concepts that have been developed on the basis of experimental studies. In this report, these studies are discussed in some detail.

Although epidemiologic studies constitute our principal source of information on somatic effects of ionizing radiation in human populations, one must recognize that there are problems in their use. The first problem arises from the fact that generally the group has been exposed to radiation because of some particular characteristic and thus may not be representative of the population at large. The reasons why those exposed to radiation are not typical of the general population may not affect radiation sensitivity, but an appropriate comparison group is nonetheless required.

The epidemiologic technique to deal with the scientific problem of a potentially biased sample is to obtain a control group matched as nearly as possible to the exposed persons. In radiation epidemiology, considerable effort has been made to deal with the question of the suitability of a control group. For example, in the Japanese atomic-bomb survivors, the zero-dose groups (those in the cities at the time of the bombing, but so far away from the bomb detonation that they were not exposed) are useful controls, although in the Nagasaki sample they are comparatively few. An alternative method has been to consider the regression of effects (such as cancer rates) on radiation dose. Systematic differences in rates of cancer not related to radiation exposure, for example, might be expected to be uniform throughout all dose categories; thus, any trend associated with radiation dose would indicate a radiation-induced effect. The care with which control samples may be selected is exemplified by the most recent followup study by Shore and colleagues of women in northern New York who were given x-ray treatment for postpartum mastitis.[28] To eliminate possible sources of bias, three control groups were used: sisters of the patients given x-ray treatment; patients who had postpartum mastitis, but did not receive x-ray treatment; and sisters of those patients. All three control groups had a greater risk of breast cancer than would be expected from the New York State Cancer Registry, but there were no significant differences among the three control samples. Careful attention to the selection of control samples greatly increases the reliability of the breast-cancer risk estimates from this study. Similarly, in the study of late effects of radiation treatment for ankylosing spondylitis in Britain,[29] the suitability of comparing those patients' cancer risks with general cancer-mortality statistics for England and Wales, as in earlier reports, was questioned. Recently, however, a followup study of mortality in a smaller group of patients with the same disease and drawn from the same clinics, but not given x-ray therapy, has shown that their cancer-mortality experience is very similar to that anticipated from mortality statistics for Britain and Wales.[21] In other words, the fact that the patients had ankylosing spondylitis did not make their cancer risk unusual. In contrast, mortality from other causes in this group deviates markedly from the expected rates in the British population.

A second problem in studies of radiation effects on human populations arises because most of them are retrospective—that is, exposure to radiation has occurred in the distant past, so the exact dose of radiation delivered to individuals or to a group is often not known. This problem is common to all retrospective studies of effects of environmental agents on human populations. In the case of radiation exposures, it has often been possible to estimate the radiation dose after the fact. For example, for the Japanese atomic-bomb survivors, great efforts have been made to determine the radiation dose-distance relationships of the Hiroshima and Nagasaki bombs, to locate the site of exposure of each person in the city at the time of the bombing, and to determine the degree of shielding by buildings or terrain that may have reduced the radiation exposure.[1] In the case of groups irradiated by medical x-ray machines, it has sometimes been possible to operate the same machines with the original technical characteristics to determine the doses. In some instances, it is not possible to obtain a reliable estimate of dose—for example, for practicing radiologists whose mortality experience has been studied.[15] Despite these problems with radiation dosimetry in retrospective studies, determination of excess cancer is generally of value, even in groups lacking dose estimates. Such studies may give the first indication that the rate of a particular cancer has increased or that there is consistency among several studies of the types of cancer observed. Finally, information can sometimes be obtained about the latent period. Studies that produce inconsistent results suggest that radiation exposure is not a principal causative factor or that other factors have a role in carcinogenesis. A degree of consistency of results in a large number of studies constitutes major support for defining somatic risks.

A third problem in the use of epidemiologic data arises from the very long latent periods that may separate exposure to radiation and the development of effects in man. This is a problem especially if the latent period is influenced by demographic variables. For example, for some solid tumors, the latent period for cancer development may be longer for persons exposed to radiation when they are younger. A minimal latent period as long as 30 yr or more after exposure means that the true health risk of radiation exposure can be assessed only with extremely long followup of the populations under study. In general, followup of irradiated groups has not proceeded this long, so the extent to which risks of radiogenic cancer have been identified is not clear. This is one of the principal reasons why risks based on current followup studies may be underestimated, especially for persons irradiated at earlier ages. Therefore, to use the epidemiologic evidence in human studies available for any particular followup interval, it is necessary to make some assumptions about the way in which further cases are likely to appear in later years. When the BEIR I report was written, there was still little information on which to base estimates of long-term risk; most of the studies of solid tumors appearing in man were of relatively short duration.

Accordingly, two models for projecting the effect of radiation exposure at a particular level were used by the original BEIR Committee. The first of these was the so-called absolute-risk model. According to this model, if a population was irradiated at a particular dose, either all at once or over some period, expression of the excess cancer risk in that population would begin at some time after exposure (the latent period) and continue at a rate in excess of the expected rate for an additional period, the "plateau" or expression period, which may exceed the period of followup. In this model, the absolute risk is defined as the number of excess cancer cases per unit of population per unit of time and per unit of radiation dose, and, although it may depend on age at exposure, it does not otherwise depend on age at observation for risk.

In the second model adopted in BEIR I, the so-called relative-risk model, the excess cancer risk for the interval after the latent period was expressed as a multiple of the natural age-specific cancer risk for that population. The chief difference between the two models is that the relative-risk model took account of the differing susceptibility to cancer related to age at observation for risk. For the entire period of actual observation, the risk estimates derived from the absolute-risk and relative-risk models are arithmetically consistent, and the choice of one or the other is a matter of convenience. For the period beyond that from which the estimates were derived, both models make assumptions that may or may not be appropriate. This problem is especially significant for persons exposed either *in utero* or in childhood, at a time when at least some kinds of cancer appear to be more likely to be induced by radiation than in adults. The assumption of a risk that persists over the life span of a person becomes an important determinant of the total risk, especially if the number of excess cases is proportional to the number of spontaneous cases, which may, for example, increase markedly with increasing age. With the additional evidence now available, we are better able to evaluate the applicability of these two models to the information at hand. It should be noted that, if epidemiologic followup through the entire lifetime is complete, both models will give the same result for lifetime risk.

Support for interpretation of risks as an absolute number of cases of cancer arising from radiation exposure came initially from the analysis of leukemia risks in the Japanese atomic-bomb survivors. It was found by the late 1960s that the number of excess cases of leukemia had risen to a peak about 8 yr or so after the radiation exposure in 1945 and was declining toward the expected leukemia rate in a nonirradiated population. By the early 1970s, the excess risk of leukemia had nearly disappeared in this population. Later analysis of the leukemia excess in the Japanese population has shown that the number of cases per unit of population is a function of the age of the people irradiated. The time course of the development of excess risk appears independent of age at irradiation for chronic granulocytic leukemia, whereas that for acute forms of leukemia, considered as a group, appears to be different for different age cohorts, although most of the excess appeared within 20 yr or so after exposure.

The time of development of radiogenic leukemia cited above for the Japanese has also been observed among the British patients with ankylosing spondylitis given x-ray therapy;[29] it appears that the effect of radiation in producing leukemia can be considered to be ended by about 30 yr after the beginning of the expression of excess cancer. The earliest excess of myeloid leukemia occurred 2 yr after exposure; thus, the expression period for leukemia is 2 to about 30 yr after irradiation. The excess of bone cancer from radium-224 exposure has an expression time of 4 to about 30 yr.

For virtually all other types of cancer arising from radiation exposure, it is apparent with longer followup times that the excess cancer risk remains well beyond 30 yr. Indeed, some types of cancer may not even appear in excess 20 yr or more after exposure. Therefore, the question in determining final risk estimates is: For how long a period after exposure does an excess risk continue to accumulate? It is clear that the total number of excess cases that will be considered to arise from radiation is influenced by this period of expression, called in BEIR I the "plateau" and in this report the "expression time" of the radiation insult. Although for development of leukemia, and bone cancer arising from radium-224 exposure, we may be able to give reasonable estimates of the expression time for cancer production, for virtually all the other radiogenic cancers this is not yet possible.

The relative-risk concept assumes that the risk of radiation-induced cancer varies by age at observation and is proportional to the risk of spontaneous development of cancer in the population. An immediate problem, of course, is the question of what constitutes the natural cancer risk in a population. For example, in the case of bronchial cancer, do we accept the spontaneous risk as the current risk of lung cancer in a population containing a substantial proportion of cigarette-smokers, or is it more proper to use the nonsmoking population as the basis for calculating the risk estimates? Related to this question is the extent to which radiation will either add to or multiply the effects of other cancer-causing agents in the environment.

A second question is whether the relative hazard of radiation applies also to groups that may on other grounds be susceptible to cancer. BEIR I pointed out that some hereditary diseases characterized by chromosomal fragility were associated with increased risk of leukemia and other cancers. These conditions are relatively rare, but the list of recognized genetic abnormalities associated with increased cancer risk is growing,[17] and many of these may involve interactions of a susceptible karyotype with environmental exposures to carcinogens. There is indirect evidence that some cancer-prone groups are at increased risk of cancer from radiation exposure; that is, their radiation sensitivity is greater than that of others.

From the Tri-State Survey of childhood leukemia, evidence has been presented[3] that children irradiated *in utero* have a greater likelihood of developing leukemia if they have had allergies or childhood diseases, especially viral diseases, diagnosed before the development of leukemia. The presence of these other childhood factors increased the leukemia risk independently of radiation exposure, particularly in the group aged 1–4 yr. The added effect of radiation is, within the limits of statistical accuracy, consistent with an excess risk proportional to the risk in unirradiated persons.

The most important factor influencing the risk of spontaneous cancer is age. If the relative-risk model applies, then the age of exposed groups, both at the time of exposure and as they move through life, becomes very important. There is now considerable evidence in nearly all the adult human populations studied that persons irradiated at higher ages have in general a greater excess risk of cancer than those irradiated at lower ages, or at least they develop cancer sooner. Furthermore, if they are irradiated at a particular age, the excess risk after the latent period tends to rise *pari passu* with the risk in the population at large. In other words, the relative-risk model with respect to cancer susceptibility, at least as a function of age, evidently applies to some kinds of cancer that have been observed to result from radiation exposure. It should be emphasized, however, that this last conclusion depends on how long the populations have been studied; whether the risk remains proportional to the risk of spontaneous cancer in the older cohorts is still uncertain. And especially uncertain is whether the increased risk of cancer observed to be associated with irradiation in childhood or *in utero* will continue into adult life, as either an absolute or a relative risk.

Some important practical conclusions arise from considerations of the above kind. The first is that, whether a risk is ultimately expressed as a total number of cancers that will arise from a specified radiation exposure or as a percent increased risk over what would be expected without radiation exposure, it is evident that the numbers developed will depend on how long one assumes that the risk will remain increased. Because of limitations thus far on the duration of followup in epidemiologic studies, we can evaluate the total risk to an irradiated population for its entire life span only by making assumptions as to the future course of somatic effects that are likely to appear. It is therefore highly important that these assumptions be clearly stated.

A notable development since the 1972 BEIR report is the increasing recognition that there are human genotypes that confer both increased cancer risk and abnormal cellular sensitivity on carcinogenic agents, including ionizing radiation. In any case, before a susceptible population can form the basis of a separate risk estimate, it must be shown to be a significant fraction of the total population and the sensitivity of this population to radiation must be substantially greater than that of the population at large. There is no evidence that these two conditions are applicable to cancer risks determined from epidemiologic studies.

The role of constitutional susceptibility to cancer induction is not well-enough documented and understood to be used as a factor for modifying risk estimates for radiation carcinogenesis. In any event, the risk estimates developed for this report are unlikely to be significantly affected by such susceptibility, because both the observed incidence and the risk estimates are averages for large populations presumably having similar distributions of sensitivities. To the extent that substantial population subsets can be identified in the future as being at particularly greater risk of radiation carcinogenesis, their risk will require separate consideration.

In this report, we have calculated the sex-specific risk of cancer by site in each observed group, preferably for a limited exposure-age range (e.g., by decade of age), if the epidemiologic data permitted. In deriving the risk estimates that are applied to an entire population, the observations are extended into older groups with the appropriate assumptions stated (that is, the duration of cancer expression, whether the temporal expression of risk is relative to the normal age-specific rate, etc.). Finally, wherever possible, the total effect of radiation on a population is calculated from the age-specific excess risk of cancer per unit of dose.

REFERENCES

1. Auxier, J. A. *ICHIBAN*. Radiation Dosimetry for the Survivors of the Bombings of Hiroshima and Nagasaki. Technical Information Center, ERDA, Oak Ridge, Tenn., March 1977.
2. Berenblum, I. Sequential aspects of chemical carcinogenesis. Skin, p. 323. In Cancer. A Comprehensive Treatise. Vol. 1. Etiology. Chemical and Physical Carcinogenesis. New York: Plenum Press, 1975.
3. Bross, I. D. J., and N. Natarajan. Leukemia from low-level radiation. Identification of susceptible children. N. Engl. J. Med. 287:107–110, 1972.
4. Burnet, F. M. The concept of immunological surveillance. Prog. Exp. Tumor Res. 13:1–27, 1970.
5. Chen, P. C., M. F. Lavin, C. Kidson, and D. Moss. Identification of ataxia telangiectasia heterozygotes, a cancer-prone population. Nature 274:484–486, 1978.
6. Cleaver, J. E., and D. Bootsma. Xeroderma pigmentosum. Biochemical and genetic characteristics. Ann. Rev. Genet. 9:19–38, 1975.
7. Countryman, P. I., J. A. Heddle, and E. Crawford. The repair of x-ray-induced chromosomal damage in trisomy 21 and normal diploid lymphocytes. Cancer Res. 37:52–58, 1977.
8. Dekaban, A. S. Abnormalities in children exposed to x-irradiation during various stages of gestation. Tentative timetable of radiation injury to human fetus. J. Nucl. Med. 9:471–477, 1968.
9. Goodhead, D. T. Inactivation and mutation of cultured mammalian cells by aluminum characteristic ultrasoft x-rays. III. Implications for theory of dual radiation action. Int. J. Radiat. Biol. 32:43–70, 1977.
10. Hoover, R. Effects of drugs. Immunosuppression, pp. 369–379. In Origins of Human Cancer. Book A. New York: Cold Spring Harbor Laboratory, 1977.
11. International Committee on Radiation Units. Report 19. Radiation Quantities and Units, July 1971.
12. Kellerer, A. M., and H. H. Rossi. A generalized formulation of dual radiation action. Radiat. Res. 75:471–488, 1978.

13. Kellerer, A. M., and H. H. Rossi. The theory of dual radiation action. Curr. Top. Radiat. Res. Q. 8:85–158, 1972.
14. Knott, G. D. MLAB—A mathematical modelling tool. Computer Prog. Biomed. 10: 271–280, 1979.
15. Matanoski, G. M., R. Seltser, P. E. Sartwell, E. L. Diamond, and E. A. Elliott. The current mortality rates of radiologists and other physician specialists. Specific causes of death. Am. J. Epidemiol. 101:199–210, 1975.
16. Miller, R. C., and E. J. Hall. X-ray dose fractionation and formations in culture mouse embryo cells. Nature 272:58–60, 1978.
17. Mulvihill, J. J. Congenital and genetic diseases, pp. 1–37. In J. F. Fraumeni, Jr., Ed. Persons at High Risk of Cancer. New York: Academic Press, 1975.
18. Neary, G. J., J. R. K. Savage, and H. J. Evans. Chromatid aberration in *Tradescantia* pollen tubes induced by monochromatic x rays of quantum energy 3 and 1.5 keV. Int. J. Radiat. Biol. 8:1, 1964.
19. Paterson, M. C., B. P. Smith, P. H. M. Lohman, A. K. Anderson, and L. Fishman. Defective excision repair of γ-ray damaged DNA in human (ataxia telangiectasia) fibroblasts. Nature 260:444–447, 1976.
20. Powers, E. L., J. T. Lyman, and C. A. Tobias. Some effects of accelerated charged particles on bacterial spores. Int. J. Radiat. Biol. 14:313, 1968.
21. Radford, E. P., R. Doll, and P. G. Smith. Mortality among patients with ankylosing spondylitis not given x-ray therapy. N. Engl. J. Med. 297:572–576, 1977.
22. Remsen, J. R., and P. A. Cerrutti. Deficiency of γ-ray excision repair in skin fibroblasts in patients with Fanconi's anemia. Proc. Natl. Acad. Sci. 73:2419–2423, 1976.
23. Rossi, H. H. The effects of small doses of ionizing radiation. Fundamental biophysical characteristics. Radiat. Res. 71:1–8, 1977.
24. Sacher, G. A. Dose, dose rate, radiation quality, and host factors for radiation-induced life shortening, pp. 493–517. In K. C. Smith, Ed. Aging, Carcinogenesis, and Radiation Biology. New York: Plenum Publishing Corp., 1975.
25. Sasaki, M. S., and A. Tonomura. Chromosomal radiosensitivity in Down's syndrome. Jap. J. Hum. Genet. 14:81–92, 1969.
26. Setlow, R. B. Repair deficient human disorders and cancer. Nature 271:713–717, 1978.
27. Shellabarger, C. J., R. D. Brown, A. R. Rao, J. P. Shanley, V. P. Bond, A. M. Kellerer, H. H. Rossi, L. J. Goodman, and R. E. Mills. Rat mammary carcinogenesis following neutron or x-radiation, pp. 391–401. In Biological Effects of Neutron Irradiation. Symposium on the Effects of Neutron Irradiation upon Cell Function, Munich, 1973. IAEA-SM-179/26, 1974.
28. Shore, E., L. H. Hempelmann, E. Kowaluk, P. S. Mansur, B. S. Pasternack, R. E. Albert, and G. E. Haughie. Breast neoplasms in women treated with x-rays for acute postpartum mastitis. J. Natl. Cancer Inst. 59:813–822, 1977.
29. Smith, P. G., and R. Doll. Age and Time Dependent Changes in the Rates of Radiation Induced Cancers in Patients with Ankylosing Spondylitis Following a Single Course of X-ray Treatment. Paper No. IAEA-SM-224/711, International Atomic Energy Agency Symposium, March 1978.
30. Sparrow, A. H., A. G. Underbrink, and H. H. Rossi. Mutations induced in *Tradescantia* by small doses of x-rays and neutrons. Analysis of dose-response curves. Science 176:916, 1972.
31. Taylor, A. M. R., D. G. Harnden, C. F. Arlett, S. A. Harcourt, A. R. Lehmann, S. Stevens, and B. A. Bridges. Ataxia telangiectasia. A human mutation with abnormal radiation sensitivity. Nature 258:427–429, 1975.

32. Ullrich, R. L., *et al.* The influence of dose and dose rate on the incidence of neoplastic diseases in RFM mice after neutron irradiation. Radiat. Res. 68:115-131, 1976.
33. Upton, A. C. Radiobiological effects of low doses. Implications for radiological protection. Radiat. Res. 71:51-74, 1977.
34. Wald, N. Radiation injury, pp. 66-72. In P. B. Beeson and W. McDermott, Eds. Textbook of Medicine. 14th ed. Philadelphia: W. B. Saunders, 1975.

III

Sources and Rates of Radiation Exposure in the United States

NATURAL BACKGROUND RADIATION

Although mankind has produced many sources of radiation, natural background remains the greatest contributor to the radiation exposure of the U.S. population today. Background radiation has three components: terrestrial radiation (external), resulting from the presence of naturally occurring radionuclides in the soil and earth; cosmic radiation (external), arising from outer space; and naturally occurring radionuclides (internal), deposited in the human body.

TERRESTRIAL RADIATION

The rate at which a person receives radiation from natural background is a function of the person's geographic location and living habits. For example, the dose-equivalent (DE) rate from terrestrial sources varies with the type of soil in a given area and its content of naturally occurring radionuclides. The penetrating gamma radiation from these radionuclides produces whole-body exposure.

In general, the conterminous United States can be divided into three broad areas, from the standpoint of terrestrial whole-body DE rates (see Figure III-1): the Atlantic and gulf coastal plain, where terrestrial DE rates range from 15 to 35 mrems/yr; the northeastern, central, and far western portions, with DE rates ranging from 35 to 75 mrems/yr; and the Colorado plateau area, in which terrestrial DE rates range from 75 to 140 mrems/yr.[12]

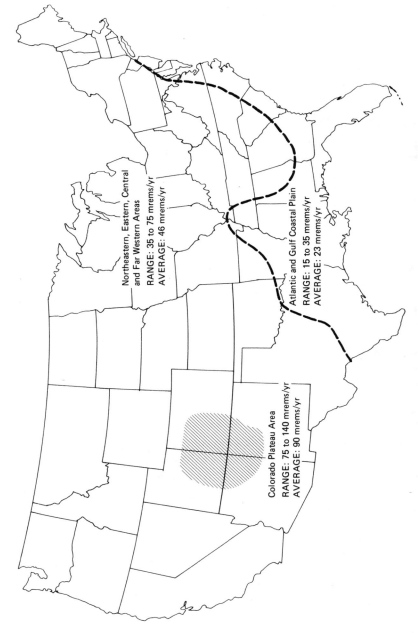

Northeastern, Eastern, Central
and Far Western Areas
RANGE: 35 to 75 mrems/yr
AVERAGE: 46 mrems/yr

Atlantic and Gulf Coastal Plain
RANGE: 15 to 35 mrems/yr
AVERAGE: 23 mrems/yr

Colorado Plateau Area
RANGE: 75 to 140 mrems/yr
AVERAGE: 90 mrems/yr

FIGURE III-1   Terrestrial dose-equivalent rates in the conterminous United States. Modified from Oakley.[12]

Combining the data shown in Figure III-1 (and more definitive data where available) with data on the geographic distribution of the U.S. population (based on the 1970 census), D. T. Oakley (personal communication) has developed the histogram shown in Figure III-2, which depicts the range of population whole-body DE rates from terrestrial sources in the United States today. As may be noted, the average DE rate to the U.S. population from terrestrial sources (disregarding structural shielding) is estimated to be 40 mrems/yr. (As will be seen later, when the DE received by various internal body organs from terrestrial sources is estimated, this value is generally reduced by 20% to account for structural shielding provided by buildings and then reduced by a second 20% to account for shielding provided by outer tissues of the body.)

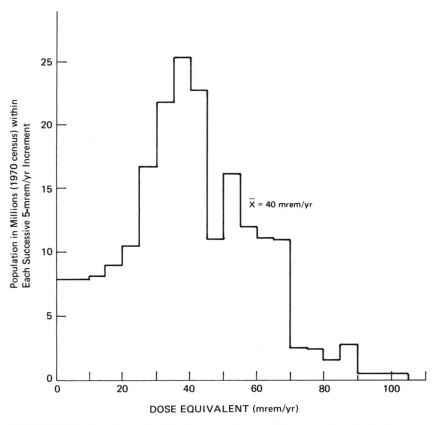

FIGURE III-2   Population distribution versus dose-equivalent rate of radiation from terrestrial sources. From D. T. Oakley (personal communication).

COSMIC RADIATION

Cosmic radiation includes both the energetic particles of extraterrestrial origin that strike the atmosphere of the earth (primary particles) and the particles generated by these interactions (secondary particles). By virtue of these interactions, the atmosphere serves as a shield against cosmic radiation and, the thinner this shield, the greater the DE rate. Thus, the cosmic-radiation DE rate increases with altitude. For example, the dose rate at 1,800 m is about double that at sea level. Because of variations in the earth's magnetic field, with which cosmic radiation also interacts, the DE rate also varies with latitude. Finally, the cosmic-radiation dose rate varies owing to solar modulation. For the United States, variations in the cosmic-radiation dose rate due to the latter two influences amount to less than 10%.[8] Because the components of cosmic radiation that reach the population are highly penetrating and are an external source, they result in whole-body irradiation.

Figure III-3 shows a plot of long-term average values of the cosmic-radiation DE rate in the United States against altitude. These data have been combined with information on the distribution of the U.S. population with altitude (Table III-1), to yield an estimated average DE rate to the U.S. population from cosmic radiation of about 31 mrems/yr (disregarding shielding).[8] (As will be seen later, when the DE received by the population from cosmic radiation is estimated, these values are generally reduced by about 10% to account for the fact that people spend a large fraction of their time indoors, protected by the structural shielding of buildings.)

NATURALLY OCCURRING RADIONUCLIDES DEPOSITED IN THE BODY

The deposition of naturally occurring radionuclides in the human body results primarily from the inhalation and ingestion of these materials in air, food, and water. Such nuclides include radioisotopes of lead, polonium, bismuth, radium, radon, potassium, carbon, hydrogen, uranium, and thorium, as well as a dozen or more extraterrestrially produced radionuclides. The heavier radionuclides are of particular interest, in that they are widespread in the biosphere and they, or many of the shorter-lived members of their decay series, are alpha-emitters.

Through measurements of the concentrations of these radionuclides in various body organs, it is possible to estimate the resulting DE rates to the U.S. population. Values of DE for selected body organs or components from specific beta- and gamma-emitting and specific alpha-emitting

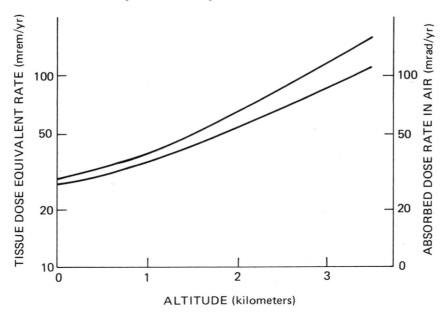

FIGURE III-3   Long-term average dose rates from cosmic radiation. The charged-particle absorbed dose rate in air or tissue is shown in the lower curve, and the total DE rate (charged particles plus neutrons) is shown in the upper curve for a depth of 5 cm in a 30-cm-thick slab of tissue. A quality factor of 2–10 was assumed for the range of energies within the neutron component. Reprinted with permission from National Council on Radiation Protection and Measurements.[8]

naturally occurring radionuclides are shown in Tables III-2 and III-3, respectively. In calculating these DE rates, a quality factor of 1 was assumed for beta radiation and a quality factor of 10 for alpha radiation.[8]

SUMMARY OF DE FROM NATURAL BACKGROUND

Table III-4 summarizes the average DE rates to the U.S. population from various sources of natural background radiation. As previously pointed out, the quoted values include a 10% reduction in the DE rate from cosmic radiation and a 20% reduction in the DE rate from external terrestrial radiation to account for the shielding effects of buildings and an additional 20% reduction in the DE rates from external terrestrial sources to account for shielding effects in the human body.[8,12]

TABLE III-1   Distribution of U.S. Population with Altitude and Accompanying DE Rates from Cosmic Radiation

Elevation, 10³ ft (km)	Population[a]	Cumulative Population, %	Approximate Cosmic Radiation DE Rate,[b] mrems/yr
0–0.5 (0–0.2)	86,600,000	48.3	26–27
0.5–1 (0.2–0.3)	63,000,000	83.4	27–28
1–2 (0.3–0.6)	19,700,000	94.5	28–31
2–4 (0.6–1.2)	5,300,000	97.4	31–39
4–6 (1.2–1.8)	3,900,000	99.6	39–52
6–8 (1.8–2.4)	618,000	100.0	52–74
8–10 (2.4–3.0)	71,000	100.0	74–107
>10 (>3.0)	14,000	100.0	107

[a] Data based on 1960 census, from Oakley.[12]
[b] Data from Figure III-3; DE rates adjusted to allow for 10% reduction owing to structural shielding from buildings.

TABLE III-2*   Annual Internal Beta and Gamma DE (mrem/yr) in Tissue from Internally Deposited Naturally Occurring Radionuclides[a]

Radionuclide	Soft Tissues (Gonads)	Cortical Bone Osteocytes	Cortical Bone Haversian Canals	Trabecular Bone Surfaces[b]	Trabecular Bone Marrow
³H	~0.001	~0.001	~0.001	~0.001	~0.001
¹⁴C	0.7	0.8	0.8	0.8	0.7
⁴⁰K	19	6	6	15	15
⁸⁷Rb	0.3	0.4	0.4	0.6	0.6
TOTAL	20.0	7.2	7.2	16.4	16.3

* Reprinted with permission from National Council on Radiation Protection and Measurements.[8]
[a] UNSCEAR (1972) [United Nations Scientific Committee on the Effects of Atomic Radiation, *Ionizing Radiation: Levels and Effects.* Vol. 1 (United Nations, New York)] gives the data as absorbed dose in tissue in mrad/yr.
[b] Cells close to surfaces of bone trabeculae.

TABLE III-3 Annual Alpha DE Rates (mrem/yr) from Naturally Occurring Radionuclides*

| Radionuclide | Concentration | | Dose Equivalent Rates | | | | |
| | | | Cortical Bone | | | Trabecular Bone | |
	In Air, pCi/m^3	In Bone,^{a} pCi/kg	Gonads	Osteocytes	Haversian Canals^b	Surfaces^c	Marrow
238–234U^d	6.9	0.8	12.4	7.7	4.8	0.9	1.2
226 Ra^d	—	7.8	0.2	16.4	10.2	6.6	1.0
228 Ra^d	—	3.8	0.3	19.0	11.0	8.0	0.4
222 Rn^e	150	—	0.4	0.2	0.2	0.4	0.4
220 Rn^e	1	—	0.01	0.1	0.1	0.2	0.2
210 Po^d	—	60	6	60	36	24	4.8
TOTAL			8	110	65	44	8.5

* Reprinted with permission from National Council on Radiation Protection and Measurements.[8]

^{a} The alpha-emitting nuclides are assumed to be uniformly distributed in mineral bone, although this may not be the case (ICRP, 1968) [International Commission on Radiological Protection, *A Review of the Radiosensitivity of the Tissues in Bone*, ICRP Publication 11 (Pergamon Press, Oxford)].

^{b} Cells lining the Haversian canals.

^{c} Cells close to surfaces of bone trabeculae (dose) averaged over the first 10 μm.

^{d} Calculated by the method of Spiers (1968) [*Radioisotopes in the Human Body: Physical and Biological Aspects* (Academic Press, New York)].

^{e} Derived from UNSCEAR (1972) [(see a in Table III-2)].

TABLE III-4   Summary of Average DE Rates from Various Sources of Natural Background Radiation in the United States[a]

| Radiation Source | Average DE, mrems/yr | | | | |
| | | | Bone | | G.I. Tract |
	Gonads	Lung	Surfaces	Marrow	
Cosmic radiation[b]	28	28	28	28	28
Cosmogenic radionuclides	0.7	0.7	0.8	0.7	0.7
External terrestrial[c]	26	26	26	26	26
Inhaled radionuclides[d]	—	100–450[e]	—	—	—
Radionuclides in body[f]	27	24	60	24	24[g]
TOTALS (ROUNDED)	80	180–530	115	80	80

[a] Derived from National Council on Radiation Protection and Measurements.[8] Quality factor for cosmic and terrestrial low-LET radiation assumed to be 1; quality factors for internal emitters were 1 for beta radiation and 10 for alpha radiation.

[b] Assuming 10% reduction to account for structural shielding.

[c] Assuming 20% reduction for shielding by housing and 20% reduction for shielding by body.

[d] Dose rates to organs other than lung included in "Radionuclides in body."

[e] Local DE rate to segmental bronchi.

[f] Excluding cosmogenic contribution, which is shown separately.

[g] Excluding contribution from radionuclides in intestinal contents.

RADIATION IN THE HEALING ARTS

X RADIATION

Patient Doses

Extensive studies on the development of indexes for evaluating the potential public-health effects of the use of x rays in the healing arts have been conducted by personnel of the Bureau of Radiological Health (BRH) of the Food and Drug Administration, U.S. Department of Health, Education, and Welfare.[13,15,16] These studies show that such use is the largest source of exposure of the U.S. population to man-made radiation. For example, it is estimated that over 300,000 x-ray units are being used in the United States for medical diagnosis and therapy—about 170,000 by dentists and about 130,000 by physicians, chiropractors, and podiatrists. The latest figures show that 39% of the medical units are in hospitals, 30% in physicians' offices (including those of osteopaths), 9% in chiropractors' offices, 7% in clinics, 4% in podiatrists' offices, and 4% in other facilities (such as for education and research); and 267 units are still being used in mobile x-ray survey vans.[7] An additional 7% are used in veterinary offices.

On the basis of a nationwide survey conducted in 1970, the BRH estimated that 65% (129 million) of the people in the United States were exposed to x rays for medical or dental purposes that year. The distribution of the examinations and treatments was as follows:

Radiographic procedures	75 million
Dental diagnosis	59 million
Fluoroscopy	9 million
X-ray therapy	0.4 million

Because of the extent of these exposures, the BRH has for some years attempted to develop an indicator for estimating the population dose from medical x rays. In one of its initial efforts, it conducted, in 1964, a nationwide survey of x-ray use and used the resulting data to calculate a factor called the "genetically significant dose" (GSD).[15] The GSD would have been an adequate and valid index of population dose, and thus an indirect measure of the biologic hazard from medical x rays, if genetic effects were the only, or the primary, biologic end point of concern. With increasing emphasis in recent years on the somatic effects of radiation, however, the shortcomings of the GSD as an overall biologic indicator have become more and more apparent. For example, some examinations that may contribute very little to the GSD may contribute substantially to the bone-marrow dose. The BRH has therefore recently been developing dose models for organs other than the gonads.[13]

The value for the GSD as quoted in BEIR I[10] was 55 mrems/yr. The original dose model has since undergone extensive review, and several errors have been discovered that caused the gonadal doses to be incorrectly estimated for some examinations. On the basis of a revised dose model, the BRH has estimated that the average GSD rate to the U.S. population related to the use of x rays in the healing arts in 1964, the year of the first survey, was actually 17 mrems/yr. Calculations based on a later survey in 1970 resulted in an estimated GSD rate of 20 mrems/yr.[15] The difference between the estimates for 1964 and 1970, however, was not judged to be statistically significant.

As mentioned above, more recent efforts have been directed to the calculation of absorbed-dose rates related to other organs of the body. The BRH has estimated that the average absorbed-dose rate for the bone marrow of the adult U.S. population from medical x rays was 83 mrads/yr in 1964 and 103 mrads/yr in 1970.[13] Estimates are that medical radiographic procedures contributed approximately 77% of this dose rate, and

TABLE III-5   Mean Active Bone Marrow Dose to the Adult Population from Medical and Dental X-Ray Procedures, 1970[a]

Examination	Mean Dose to Total Red Marrow per Examination, mrads[b]	Annual per Capita Examination Rate[b]		Annual per Capita Dose, mrads $\pm$ SE
Head and neck				
Skull	78	0.020		1.6 $\pm$ 0.1
Cervical spine	52	0.022		1.2 $\pm$ 0.2
Other	—	—		0.6 $\pm$ 0.2
Thorax				
Chest, photofluoro.	44	0.073		3.2 $\pm$ 0.3
Chest, radiographic	10	0.306		3.2 $\pm$ 0.1
Thoracic spine	247	0.010		2.5 $\pm$ 0.4
Ribs	143	0.009		1.3 $\pm$ 0.2
Other	—	—		1.9 $\pm$ 0.4
Upper abdomen				
Upper GI series	535			24.3 $\pm$ 4.7
Radiographic	294	0.046	13.5 $\pm$ 4.3	
Fluoroscopic	241	0.045	10.8 $\pm$ 1.9	
Scan	167			
Spot film	74			
Lumbar spine	347	0.023		8.1 $\pm$ 0.8
Gall bladder	168			3.7 $\pm$ 0.4
Radiographic	129	0.027	3.5 $\pm$ 0.3	
Fluoroscopic	39	0.006	0.2 $\pm$ 0.3	
Scan	29			
Spot film	10			
Small bowel series	422	0.002		1.0 $\pm$ 0.3
Other	—	—		2.1 $\pm$ 1.0
Lower abdomen				
Barium enema	875			21.2 $\pm$ 1.8
Radiographic	497	0.024	11.9 $\pm$ 1.0	
Fluoroscopic	378	0.024	9.3 $\pm$ 1.5	
Scan	268			
Spot film	110			
IVP	420	0.024		10.1 $\pm$ 0.6
Lumbosacral spine	450	0.013		5.7 $\pm$ 0.7
Abdomen (kidneys, ureters, and bladder)	147	0.020		2.9 $\pm$ 0.4
Other	—	—		0.4 $\pm$ 0.2
Pelvis				
Pelvimetry	595	0.002		1.4 $\pm$ 0.5
Pelvis	93	0.012		1.1 $\pm$ 0.2
Hip	72	0.009		0.7 $\pm$ 0.1
Other	—	—		1.2 $\pm$ 0.7
Extremities				
Femur	21	0.002		0.04 $\pm$ 0.02
Dental	9.4	0.312		2.9 $\pm$ 0.2
			TOTAL:	103 $\pm$ 5

[a] Data from Shleien *et al.* [13]

[b] Values have been independently rounded.

TABLE III-6   Per Capita Mean Active Bone-Marrow
Dose for Specific Age Groups from Medical X-Ray
Procedures in 1970[a]

Age, yr	Per Capita Mean Active Bone-Marrow Dose, mrads
15–24	52
25–34	81
35–44	107
45–54	120
55–64	143
65+	151

[a] Data from Shleien *et al.* [13] Each figure in second column represents
the product of the number of examinations of a specific type in an age
group and the mean active bone-marrow dose for the examination
(see Table III-5) divided by the number of persons in the specified age
group.

fluoroscopic and dental examinations about 20% and 3%, respectively.
Tables III-5 and III-6 summarize the sources and extents of the absorbed-
dose rates for specific portions of the body. [13]

Occupational Doses

Estimates of occupational doses associated with the use of x rays in
medicine and dentistry are limited, in that little more than film-badge
data are available and various agencies and organizations define occupational exposures differently. However, the Environmental Protection
Agency estimated that in 1968 about 195,000 persons were occupationally
exposed in the operation of medical x-ray equipment and that about
171,000 persons were similarly engaged in the operation of dental x-ray
equipment. The mean annual doses to these two groups were estimated to
be 320 and 125 mrems, respectively. [4] More recent data based on film-
badge measurements of dental personnel during 1975 are summarized in
Table III-7. The data show an average DE of about 50 mrems for that
year.

RADIOPHARMACEUTICALS

Over 10,000 U.S. physicians are licensed to administer radiopharmaceuticals to patients for diagnostic and therapeutic purposes. It has been
estimated that some 10–12 million doses are administered each year.

TABLE III-7   Distribution of Film-Badge Dose Data
for Dental Personnel, 1975[a]

Film-Badge Dose, mrems	Fraction of Personnel, %	Mean Dose, mrems
Nondetectable	84	—
100	12	41
100–250	1.7	175
250–500	0.4	300
500–750	0.4	600
750–1,000	0	—
1,000–2,000	0	—
2,000–3,000	0	—
3,000–4,000	0	—
4,000–5,000	0.4	4,300
5,000–6,000	0.4	5,200

[a] Data provided by Scientific Committee 45, NCRP, Washington, D.C.

Patient Doses

Data collected by the BRH show the following information on the use of radiopharmaceuticals in the United States (B. Shleien, personal communication):

• About 90% of the reported procedures involved five organ systems. Specifically, 24.1%, 20.3%, 18.1%, 16.5%, 10.9%, 3.2%, and 2.5% of the procedures involved brain, liver, bone, lung, thyroid, kidney, and heart, respectively.
• Radiopharmaceuticals labeled with technetium-99m were by far (81%) the most commonly used. Iodine-131, xenon-133, gallium-67, and iodine-123 were used in 7%, 4%, 3%, and 1% of the procedures, respectively.
• Approximately 14% of the patients were under the age of 30, and 69.6% were over the age of 44. Specifically, 2.7%, 11.5%, 16.2%, 36.3%, and 33.3% of the patients were 0–14, 15–29, 30–44, 45–64, and over 64 yr old, respectively.

A summary of the radiopharmaceuticals used and the range and average of the activity administered is given in Table III-8. These data are from a pilot study conducted in 1975 by the BRH.[5] Estimates of the patient doses per radiopharmaceutical administration are summarized in Table III-9. These data are based on an expansion and updating of the information provided by the sample covered in the pilot study. Although the pilot study was limited in scope, it indicated an average annual growth rate in the application of nuclear-medicine procedures of over 17%; it further indicated that there had been increases in the average whole-body and gonad radiation doses per radiopharmaceutical administration in 1975, compared with national data for 1966. Because the sample was so small, however, those conducting the pilot study cautioned that the data "cannot be said to be representative of nuclear medicine practice for all United States hospitals."[5] The EPA Office of Radiation Programs had estimated that whole-body patient doses from the diagnostic use of radiopharmaceuticals represented about 20% of the patient doses resulting from medical diagnostic radiology.[4]

It might be pointed out that the increasing use of radiopharmaceuticals is to be encouraged, particularly if the shorter-lived radionuclides and modern, sensitive counting equipment can be used. Because radiopharmaceutical procedures are often conducted on an outpatient basis, however, it must be recognized that the people to whom radioactive materials have been administered, particularly therapeutically, can be a source of exposure to family members and others.[3] The overall importance of this source to the general population is not known.

Occupational Doses

In 1968, there were some 80,000 medical radionuclide and radium workers.[4] Today, this total is undoubtedly much greater. The EPA has estimated that medical radionuclide workers receive a mean annual dose of about 260 mrems and radium workers about 540 mrems.[4]

Data on film-badge records for hospital radionuclide and x-ray personnel show that the mean annual dose for 1975 was 350 mrems (see Table III-10). This indicates close agreement with the EPA estimates for radionuclide and radium workers.

PRODUCTION AND USE OF NUCLEAR ENERGY

ATMOSPHERIC WEAPONS TESTS

During the 1950s and 1960s, when extensive testing of nuclear devices was conducted in the atmosphere, large quantities of man-made radioactive materials were produced and distributed to the environment throughout the world in the form of fallout. Although much of this debris has since decayed, the small amounts that remain will be a source of exposure of the

TABLE III-8 Procedure, Percent Radiopharmaceutical Used, Range of Activity, and Average Activity Administered[a]

Procedure	Radiopharmaceutical	Fraction of Procedures, %	Activity Administered, mCi		
			Low	High	Average
Bone imaging (total body)	Tc-99m EHDP	56.3	2.1	27.0	17.3
	[99mTc]technetium polyphosphate	3.1	10.0	15.0	12.7
	[99mTc]technetium pyrophosphate	40.5	2.1	30.0	17.2
Brain imaging	Tc-99m DTPA	7.4	3.0	20.0	19.2
	[99mTc]pertechnetate	92.4	6.0	30.0	17.6
	[99mTc]technetium pyrophosphate	0.2	10.0	30.0	16.1
Liver imaging	Tc-99m sulfur colloid	100.0	0.25	21.5	4.8
Lung perfusion	Tc-99m MAA	86.9	1.0	31.4	4.9
	Tc-99m HAM	12.2	1.0	15.0	4.2
	Other	0.9	—	—	—
Lung ventilation	Xe-133 gas	66.0	3.1	40.6	18.1
	Xe-133 in saline	34.0	4.0	15.0	7.1
Myocardial imaging	[81Rb]rubidium chloride	19.5	1.90	24.7	14.6
	[99mTc]technetium pyrophosphate	80.5	5.0	15.0	14.7

Renal imaging	[197Hg]chlormerodrin	4.4	0.2	0.2	0.2
	[131I]iodohippurate	0.7	0.05	0.5	0.35
	Tc-99m DTPA	49.3	1.1	19.7	6.0
	[99mTc]technetium glucoheptonate	44.2	15.0	15.0	15.0
	Tc-99m DMSA	1.4	2.0	5.0	3.9
Renogram	[197Hg]chlormerodrin	5.2	0.025	0.025	0.025
	[203Hg]chlormerodrin	7.7	0.025	0.1	0.05
	[125I]iodohippurate	0.5	0.2	0.2	0.2
	[131I]iodohippurate	86.1	0.02	0.33	0.168
	Tc-99m DTPA	0.5	15.0	15.0	15.0
Thyroid imaging	[123I]sodium iodide	25.0	0.04	1.05	0.33
	[131I]sodium iodide	45.4	0.015	1.0	0.059
	[99mTc]pertechnetate	29.6	0.5	4.0	1.361
Thyroid uptake	[123I]sodium iodide	26.8	0.029	1.05	0.33
	[131I]sodium iodide	72.5	0.005	1.00	0.064
	[99mTc]pertechnetate	0.7	0.5	3.0	1.6
Total body, soft tissue	[67Ga]gallium citrate	98.0	0.15	7.1	3.0
(tumor localization)	Ga-67 iron DTPA complex	1.3	3.0	3.0	3.0
	[131I]sodium iodide	0.7	1.07	5.0	3.0

a Data from McIntyre et al.[5]

TABLE III-9   Estimated Radiation Dose per Diagnostic
Radiopharmaceutical Administration, 1975[a]

Radiopharmaceutical	No. Administrations Covered in Pilot Study	Average Radiation Dose per Administration, mrads		
		Whole Body	Gonad	Bone Marrow
[131 I]sodium iodide	814	28	7	12
Other 131 I	317	210	204	106
[123 I]sodium iodide	326	12	9	10
99m Tc	11,014	177	245	258
133 Xe	608	5	5	5
Other	507	1,020	1,020	2,130
TOTAL	13,586	189[b]	242[b]	292[b]

[a] Based on McIntyre et al. [5]
[b] Weighted average.

TABLE III-10   Distribution of Film-Badge Dose Data
for Hospital Radiation Personnel, 1975[a]

Film-Badge Dose, mrems	Fraction of Personnel, %	Mean Dose, mrems
Nondetectable	43.6	—
100	25.2	41
100–250	12.6	159
250–500	9.0	354
500–750	3.45	618
750–1,000	2.0	867
1,000–2,000	2.53	1,391
2,000–3,000	0.8	2,416
3,000–4,000	0.25	3,391
4,000–5,000	0.19	4,435
5,000–6,000	0.08	5,457
6,000–7,000	0.04	6,500
7,000–8,000	0.03	7,443
8,000–9,000	0	—
9,000–10,000	0	—
10,000–11,000	0	—
11,000–12,000	0	—
12,000–	0.13	128,425

[a] Data provided by Scientific Committee 45, NCRP, Washington, D.C.

U.S. population for some time to come. In addition, periodic atmospheric tests of nuclear devices by nations that were not signatories to the limited Test Ban Treaty of 1963, such as the People's Republic of China, continue to add fresh fission-product debris to the worldwide inventory. The U.S. population dose from fallout from such tests has been estimated by the EPA.[14]

Table III-11 summarizes the estimated 50-yr dose commitment for several organs of the body in people in the north temperate zone due to atmospheric nuclear tests conducted before 1971. Table III-12 summarizes projections of the annual whole-body DE for the U.S. population from global fallout through the year 2000. As may be noted, the projected annual average whole-body DE rate for the U.S. population from these sources is 4–5 mrems/yr.

NUCLEAR POWER REACTORS

As of April 30, 1979, 70 nuclear power reactors had been licensed for operation in the United States. By the year 2000, as many as 250 units

TABLE III-11   50-Yr Dose Commitment from Nuclear Tests Conducted Before 1971, North Temperate Zone[a]

	Dose Commitment, mrads		
Source of Exposure	Gonads	Bone-Lining Cells	Bone Marrow
External exposure			
Short-lived radionuclides	65	65	65
Cesium-137	59	59	59
Krypton-85	2×10^{-4}	2×10^{-4}	2×10^{-4}
Internal exposure			
Hydrogen-3	4	4	4
Carbon-14	12	15	12
Iron-55	1	1	0.6
Strontium-90	—	85	62
Cesium-137	26	26	26
Plutonium-239[b]	—	0.2	—
TOTALS[c]	170	260	230

[a] Data from U.S. Office of Radiation Programs.[22]
[b] Dose commitment to bone-lining cells has been taken to be equal to integrated dose over 50 yr to bone.
[c] Totals rounded to two significant figures.

TABLE III-12   Projections of Annual Whole-Body DE
to U.S. Population from Global Weapons Testing
Fallout[a]

Year	Per Capita DE, mrems
1963	13
1965	6.9
1969	4.0
1980	4.4
1990	4.6
2000	4.9

[a] Data from U.S. Office of Radiation Programs.[22]

could be in operation. In addition, there are 73 nonpower reactors being used for tests, research, and university applications; about 80 nonpower nuclear reactors being operated by the U.S. Department of Energy; and 174 reactors in operation or under construction by the military services, most of them being operated under the auspices of the U.S. Navy as propulsion units for submarines and surface ships.[17]

Population Doses

Supporting these reactor operations are a variety of activities ranging from the mining and milling of uranium through the fabrication of reactor fuels to the storage of spent fuel or high-level radioactive wastes (depending on whether the spent fuel is chemically processed). Several hundred uranium mines are in operation in the United States, and they employ about 5,000 men. There are also 20 uranium mills and 21 fuel-fabrication plants in operation;[17] another 21 mills and one fuel-processing plant have ceased to operate. Although there have been problems with radionuclide releases from uranium mills, in the main the discharge of radionuclides into the environment from commercial nuclear power plants has been well controlled. Current regulations of the U.S. Nuclear Regulatory Commission, for example, limit whole-body DE rates for the general population from routine releases from commercial nuclear power plants to about 8 mrems/yr; the DE rate limit for individual body organs, such as the thyroid, is 15 mrems/yr.[21] Regulations promulgated by the EPA limit whole-body DE rates for the general population from planned releases from all sources originating in the nuclear-power industry to 25 mrems/yr; the DE rate limit for the thyroid is 75 mrems/yr.[18]

Of the specific radionuclides produced in fission, two that are of significance in terms of potential population dose, particularly in case of a major reactor accident, are strontium-90 and cesium-137. Three radionuclides of significance from the standpoint of routine operation of nuclear power plants are tritium (hydrogen-3), carbon-14, and krypton-85, all of which are somewhat difficult to remove from waste streams and confine. Projections of future annual whole-body DE rates for the U.S. population from these three nuclides are summarized in Table III-13. Although the release of iodine-131 is also of interest, current techniques appear to be adequate to restrict releases of this nuclide in normal nuclear power-plant operations to very low amounts. Overall estimates show that the DE rate for the average person in the United States from environmental releases of all radionuclides from nuclear operations is currently less than 1 mrem/yr.[11]

Occupational Doses

Information on occupational doses to personnel associated with commercial nuclear power plants and supporting activities, such as processing and fabrication, is tabulated and published on an annual basis by the Nuclear Regulatory Commission,[20] including exposures in industrial radiography performed with Commission-controlled radioactive materials. Summaries of these data are presented in Tables III-14 and III-15. Table III-14 shows the distribution of annual whole-body exposures by licensee category; Table III-15 shows the total man-rem accumulation by licensee category. Data on exposures of transient workers for the years 1960–1976, which have been subject to considerable discussion, are summarized in Table III-16.

Similar data on occupational radiation DE received by personnel as-

TABLE III-13   Projected Annual DE to the U.S. Population from Specific Nuclides[a]

Radionuclide	Body Organ	Per Capita DE, mrems				
		1960	1970	1980	1990	2000
Hydrogen-3	Whole body	0.02	0.04	0.03	0.02	0.03
Carbon-14	Whole body	0.3	0.6	0.6	0.6	0.6
	Bone	0.5	1.0	1.0	1.0	1.0
Krypton-85	Whole body	0.0001	0.0004	0.003	0.01	0.04
	Skin	0.005	0.02	0.1	0.6	1.6
	Lung	0.0002	0.0006	0.005	0.02	0.06

[a] Data from U.S. EPA.[4,22]

TABLE III-14   Distribution of Annual Whole-Body Exposures, by
Licensee Category, 1976[a]

Covered Categories of NRC Licensees	No. (%) Persons Monitored Within Each DE Range (in rems)						
	Total	Less Than Measurable	<0.10	0.10– 0.25	0.25– 0.50	0.50– 0.75	0.75– 1.00
Power reactors	66,800	30,085	13,859	5,277	4,192	2,537	2,036
	100%	45%	21%	8%	6%	4%	3%
Industrial radiography	11,245	5,023	2,184	1,208	887	544	353
	100%	45%	19%	11%	8%	5%	3%
Fuel processing and fabrication	11,227	5,942	2,815	959	580	307	221
	100%	53%	25%	9%	5%	3%	2%
Manufacturing and distribution	3,501	1,525	906	413	170	94	53
	100%	44%	26%	12%	5%	3%	2%
TOTALS	92,773	42,575	19,764	7,857	5,829	3,482	2,663
	100%	46%	21%	8%	6%	4%	3%

[a] Data from U.S. Nuclear Regulatory Commission.[20]

signed to tenders, bases, and nuclear-powered ships in the U.S. Navy are presented in Table III-17. These exposures are those which result from work related to the operation and maintenance of naval nuclear-propulsion plants. Data on the occupational radiation DE received by shipyard personnel from work related to naval nuclear-propulsion units are presented in Table III-18. As may be noted, the collective occupational DE from such operations, including both groups of personnel, reached a maximum of about 22,000 (3,529 + 18,763) person-rems in 1966, but has been considerably reduced; today the collective DE is well under 10,000 (2,812 + 5,207) person-rems/yr.

Another source of occupational exposure in the nuclear industry is the research and development work conducted in the national laboratories of, and by contractors to, the Department of Energy (DOE). A summary of whole-body dose received by contractor employees of DOE and its predecessor agencies from 1964 through 1975 is shown in Table III-19.

RESEARCH ACTIVITIES

High-voltage x-ray machines and particle accelerators are familiar features of research laboratories in universities and similar institutions. Today, almost 1,000 cyclotrons, synchrotrons, van de Graaff generators, and betatrons are in operation.[7] Although estimating is difficult, it can be

TABLE III-14
Continued

1-2	2-3	3-4	4-5	5-6	6-7	7-8	8-9	9-10	10-11	11-12	>12
4,882	2,355	789	487	188	70	26	11	1	0	0	0
7%	4%	1%	1%								
660	210	100	41	15	10	3	2	0	2	0	3
6%	2%	1%									
237	77	47	25	17	0	0	0	0	0	0	0
2%	1%										
148	77	52	31	16	10	5	2	0	0	0	0
4%	2%	1%	1%								
5,927	2,719	987	584	236	90	34	15	5	3	0	3
6%	3%	1%	1%								

conservatively calculated that some 10,000 people are occupationally exposed in the operation of these machines.

Other x-ray equipment used in research includes about 10,000 diffraction units and 3,000 electron microscopes. Studies have shown that a substantial number of radiation injuries have resulted from accidents involving diffraction equipment.[1] The number of people occupationally exposed in the operation of electron microscopes has been estimated at 4,400, with annual whole-body DE rates of 50-200 mrems.[9]

An emerging source of machine-produced radiation is the neutron generator; the total number in use is estimated at some 500. Data on the number of people involved in the operation of these units and the range of exposures are not available.

CONSUMER AND INDUSTRIAL PRODUCTS

A variety of consumer and industrial products yield ionizing radiation or contain radioactive materials and can therefore cause radiation exposure of the general population—e.g., television sets, luminous-dial watches, airport luggage x-ray inspection systems, dental prostheses, smoke detectors, high-voltage vacuum switches, electron microscopes, static eliminators, cardiac pacemakers, tobacco products, fossil fuels, and building materials. A summary of DE rates of the more important of these is pre-

TABLE III-15 Person-Rems Accumulated, by Category of Covered Licensees, 1973–1976[a]

Covered Categories of NRC Licensees	Calendar Year	No. Licensees Reporting	No. Persons Monitored	No. Persons with Measurable Exposure	Total No. Person-Rems	Average Exposure per Person (Based on Total Monitored), rems	Average Exposure per Person (Based on Measurable Exposures), rems
Commercial power reactors	1976	62	66,800	36,715	26,555	0.40	0.72
	1975	54	54,763	28,034	21,270	0.39	0.76
	1974	53	62,044	21,904	14,083	0.23	0.64
	1973	41	44,795	16,558	14,337	0.32	0.87
Industrial radiography	1976	321	11,245	6,222	3,629	0.32	0.58
	1975	291	9,178	4,693	2,796	0.30	0.60
	1974	319	8,792	4,943	2,938	0.33	0.59
	1973	341	8,206	5,328	3,354	0.41	0.63
Fuel processing and fabrication	1976	21	11,227	5,285	1,830	0.16	0.35
	1975	23	11,405	5,495	3,125	0.27	0.57
	1974	25	10,921	4,617	2,739	0.25	0.59
	1973	27	10,610	5,056	2,400	0.23	0.47
Processing and distribution of byproduct material	1976	24	3,501	1,976	1,226	0.35	0.62
	1975	19	3,367	1,859	1,188	0.35	0.64
	1974	24	3,340	1,827	1,050	0.31	0.57
	1973	34	4,251	1,925	1,177	0.28	0.61
TOTALS	1976	428	92,773	50,198	33,240	0.36	0.66
	1975	387	78,713	40,081	28,379	0.36	0.71
	1974	421	85,097	33,291	20,810	0.24	0.63
	1973	443	67,862	28,867	21,268	0.31	0.74

[a] Data from U.S. Nuclear Regulatory Commission.[20]

TABLE III-16　Dose Equivalent Received by Transient Workers, 1969–1976[a]

	1969	1970	1971	1972	1973	1974	1975	1976
No. workers terminating employment with two or more employers in one quarter	8	29	11	69	157	354	714	1,055
Collective DE, person-rems	5.4	14.6	2.8	61.3	135.5	175.9	507.1	745.3
Average individual DE, rems	0.68	0.50	0.25	0.89	0.86	0.50	0.71	0.71

[a] Data from U.S. Nuclear Regulatory Commission.[20]

TABLE III-17   Occupational Radiation Exposures Received by Personnel Assigned to Tenders, Bases, and Nuclear-Powered Ships from Operation and Maintenance of Naval Nuclear Propulsion Plants, 1955–1977[a]

Year	No. Persons Who Received Exposures in Specified DE Ranges (in rems)						No. Persons Monitored	Collective DE, person-rems
	0–1	1–2	2–3	3–4	4–5	5[b]		
1955	90	11	0	0	0	0	101	25
1956	108	10	4	0	0	0	122	50
1957	293	7	1	0	0	0	301	60
1958	562	11	3	0	0	0	576	100
1959	1,057	41	8	3	0	0	1,109	200
1960	2,607	88	8	4	3	1	2,711	375
1961	4,812	105	31	4	4	0	4,957	680
1962	6,788	182	75	31	17	1	7,094	1,312
1963	9,188	197	39	14	3	1	9,442	1,420
1964	10,317	331	93	35	15	3	10,795	1,964
1965	11,883	592	224	96	30	24	12,849	3,421
1966	18,118	541	156	95	44	28	18,982	3,529
1967	21,028	339	139	48	11	0	21,565	3,084
1968	24,200	373	103	20	2	0	24,698	2,463
1969	26,969	577	127	39	6	0	27,718	2,918
1970	26,206	610	134	30	0	0	26,980	3,089
1971	26,090	568	122	31	2	0	26,813	3,261
1972	33,312	602	180	13	1	0	34,108	3,271
1973	30,852	600	102	15	1	0	31,570	3,160
1974	18,375	307	65	2	0	0	18,749	2,142
1975	17,638	330	28	1	0	0	17,997	2,217
1976	17,795	369	56	9	0	0	18,229	2,642
1977	20,236	346	95	36	3	0	20,716	2,812

[a] Data from U.S. Department of the Navy. [b] Data obtained from summaries, rather than directly from original medical records. However, it is expected that the large effort to compile comparable data from original medical records would show differences no greater than 5%. Collective DE was determined by adding actual exposures for each person during the year.

[b] The occupational dose-rate limit in the naval nuclear propulsion program was reduced to 5 rems/yr late in 1966.

TABLE III-18   Occupational Radiation Exposures Received by Shipyard Personnel from Work Associated with Naval Nuclear Propulsion Plants, 1962–1977[a]

| Year | No. Persons Who Received Exposures in Specified DE Ranges (in rems) | | | | | | No. Persons Monitored | Collective DE, person-rems |
	0–1	1–2	2–3	3–4	4–5	>5[b]		
1962	11,409	657	548	486	164	123	13,387	5,600
1963	19,568	445	164	73	35	28	20,313	2,711
1964	19,367	751	413	199	143	30	20,903	5,132
1965	21,434	1,895	1,108	726	623	600	26,386	14,735
1966	22,787	1,787	1,252	794	1,038	486	28,144	18,763
1967	26,941	1,737	1,131	826	733	1	31,369	13,876
1968	30,948	1,277	755	499	289	0	33,768	8,665
1969	25,846	1,689	1,031	636	373	0	29,575	11,033
1970	21,319	1,968	1,326	723	492	0	25,828	11,974
1971	20,214	1,801	1,029	641	240	0	23,925	10,647
1972	17,390	1,668	845	139	5	0	20,048	6,998
1973	13,095	1,379	605	203	6	0	15,288	6,110
1974	12,447	1,452	746	310	50	0	15,005	7,209
1975	12,833	1,115	598	81	42	0	14,669	5,303
1976	13,057	1,270	633	30	0	0	14,990	5,309
1977	13,900	1,277	586	25	0	0	15,788	5,207

[a] Data from U.S. Department of the Navy. [b] Data obtained from summaries, rather than directly from original medical records. However, it is expected that the large effort to compile comparable data from original medical records would show differences no greater than 5%. Exposures from radiation sources licensed by Nuclear Regulatory Commission or a state have been excluded as far as practicable. Collective DE was determined by adding actual exposures for each person during the year.
[b] The occupational dose-rate limit in the naval nuclear propulsion program was reduced to 5 rems/yr late in 1966.

sented in Table III-20. The estimated average whole-body DE rate for the U.S. population from these sources is 4–5 mrems/yr. About three-fourths of this arises through external exposures—exposures to naturally occurring radionuclides in building materials.

In recent years, increasing attention has been given to radiation exposures of the U.S. population from natural sources whose dose rates have been increased because of technologic developments. One example, cited above, is the population DE due to naturally occurring radionuclides in building materials. Another source of exposure that has recently been recognized is airborne radon and radon daughter products that evolve

TABLE III-19 Whole-Body Radiation Exposure History for DOE and DOE Contractor Employees[a]

No. Employees in Each DE Range (in rems)

Year	0-1[b]	1-2	2-3	3-4	4-5	5-6	6-7	7-8	8-9	9-10	10-11	11-12	>12	No. Employees
1964	122,711	3,583	1,823	575	176	43	20	10	7	6	10	1	—	128,965
1965	128,360	4,158	1,704	515	294	40	32	26	25	22	6	2	—	135,214
1966	130,562	3,706	1,630	597	313	88	47	24	6	2	—	—	1	137,939
1967	102,510	3,472	1,572	555	168	35	29	23	17	4	1	—	—	108,386
1968	103,206	2,799	1,408	425	144	3	1	—	—	—	—	—	—	107,986
1969	98,625	2,554	1,313	335	86	4	—	—	—	1	1	—	—	102,918
1970	92,185	2,698	1,329	279	158	5	4	2	—	—	—	—	—	96,661
1971	90,640	2,380	888	275	118	8	3	—	—	1	1	—	2	94,315
1972	86,077	2,130	929	219	95	8	2	—	—	—	—	—	—	89,460
1973	89,071	1,944	727	172	60	2	1	—	—	—	—	—	—	91,977
1974	75,706	1,689	692	149	40	4	—	—	1	—	—	—	—	78,232
1975	85,451	1,846	753	232	142	—	—	—	1	—	—	—	—	88,425

[a] Data from U.S. EPA.[22]

[b] In 1975, approximately 65% of these employees received a DE less than measurable.

TABLE III-20   DE Rates from Selected Consumer Products[a]

Product	Body Portion Considered	Average Annual DE Rate, mrems/yr	
		For Persons Using Product	For Average Person in U.S. Population
Luminous compounds			
Wristwatches	Gonads	1–3	0.2
Clocks	Whole body	9	0.5
Television sets	Gonads	0.3 (females)	0.1 (females)
		1 (males)	0.5 (males)
Construction materials	Whole body	7	3.5
Combustion of fossil fuels			
Coal	Lungs	0.25–4	0.05–10
Oil	Lungs	0.002–0.04	0.004
Natural gas			
Cooking ranges	Bronchial epithelium	6–9	5
Unvented heaters	Bronchial epithelium	22	2
Tobacco products	Bronchial epithelium	8,000[b]	2,000[b]

[a] Data from NCRP.[9]
[b] Hypothetical maximum at highly localized points.[9]

from groundwater supplies used in the home. Approximately half the radon present in household water supplies becomes airborne.[2] In fact, concentrations in bathroom air after the spraying of radon-rich groundwater through a shower head can approach occupational limits. The significance of such sources is being investigated.

MISCELLANEOUS SOURCES

Several sources of radiation exposure of the general population do not fit into the categories just outlined. One is the added exposure from cosmic radiation that results from commercial airline travel, and another is the exposure that results from the transportation of radioactive materials.

COSMIC RADIATION DOSE TO AIRLINE PASSENGERS

Data for 1973 show that the U.S. public made about 281 million domestic flights.[23] In all, about 25% of the adult population, or 35 million people, flew at least once during that year. On the average, however, each person who flew made about 10 flights during the year. The average flight was at an altitude of 9.47 km and lasted 1.4 h. The average dose rate was 0.2 mrem/h, resulting in an average passenger DE of 2.8 mrems. For the population as a whole, this resulted in a cumulative dose of about 100,000 person-rems. These data, as well as those on cosmic-radiation DE to cabin attendants and aircraft crew members, are summarized in Table III-21.

DOSE DUE TO TRANSPORTATION OF RADIOACTIVE MATERIALS

The Nuclear Regulatory Commission has recently completed a detailed study of the population doses associated with the transportation of radio-active materials in the United States.[19] The potential magnitude of this source, if not properly controlled, is well illustrated by the estimate of 2.5 million shipments of such materials in 1977.

For purposes of the Nuclear Regulatory Commission study, the population groups being exposed were divided into commercial airline passengers, cabin attendants, aircraft crew, and ground crew. The estimated annual collective DE from this source is about 2,500 person-rems (see Table III-22). Comparable estimates of collective DE associated with the transportation of radioactive materials by trucks and vans, by rail, and by ships during 1975 were about 5,000, 25, and 10 person-rems, respectively.[19]

TABLE III-21   Annual DE from Cosmic Radiation to Aircraft Passengers and Crew, 1973[a]

| Population Group | No. Exposed | Dose Rates, mrems/yr | | Annual Population Dose, person-rems |
		Maximum	Average	
Passengers	35×10^{6b}	63^c	2.8^d	99,000
Cabin attendants	2.3×10^4	—	160^e	3,700
Aircraft crew	1.7×10^4	—	158^e	2,650
			TOTAL:	105,350

[a] Data from Wallace and Sondhaus.[23]

[b] About 25% of the adult population, or 35 million people, flew at least once in 1973.

[c] Based on assumption that a person made a maximum of 50 transcontinental flights (25 transcontinental round trips) during the year.

[d] Based on calculations that showed that average flight involved spending 1.4 h at altitude of 9.47 km with average DE rate of 0.20 mrem/h. On this basis, average DE per flight was about 0.28 mrem, and average number of flights taken by average passenger was about 10 per year.

[e] Dose-rate estimates and estimated number of cabin attendants and aircraft crew members based on assumed flying time of 720 h/yr.

TABLE III-22   Annual DE from Transport of Radioactive Materials in Passenger Aircraft, 1975[a]

| Population Group | No. Exposed | Dose Rates, mrems/yr | | Annual Approximate Collective Dose, person-rems |
		Maximum	Average	
Passengers	7×10^{6}[b]	108[c]	0.34	2,380
Cabin attendants	4×10^{4}[d]	13	3	120
Aircraft crew	3×10^{4}[d]	2.5	0.53	16
Ground crew (including bystanders)	$720/km^{2}$	85[e]		11
			TOTAL:	2,500

[a] Data from U.S. Nuclear Regulatory Commission.[19]
[b] Based on an average of 210 million revenue passengers per year with one of 30 flights transporting radioactive materials. Each of the 7 million people in this group is assumed to make only one trip per year on an aircraft transporting radioactive material.
[c] Based on a select group flying 500 h/yr between Knoxville, Tenn., and St. Louis, Mo.
[d] The numbers of cabin attendants and aircraft crew members listed differ from those given in Table III-21. Average flying time of 500 h/yr assumed here; in Table III-21, assumed flying time was 720 h/yr, requiring smaller number of people to handle these tasks.
[e] Applies only to most exposed member of ground crew. Calculated population dose based on assumed ground time per flight of 1 h.

SUMMARY

Annual dose rates from each of the important sources of radiation exposure in the United States are summarized in Table III-23.

TABLE III-23 Annual Dose Rates from Important Significant Sources of Radiation Exposure in United States

Source	Exposed Group		Body Portion Exposed	Average Dose Rate, mrems/yr	
	Description	No. Exposed		Exposed Group	Prorated over Total Population
Natural background					
Cosmic radiation	Total population	220×10^6	Whole body	28	28
Terrestrial radiation	Total population	220×10^6	Whole body	26	26
Internal Sources	Total population	220×10^6	Gonads	28	28
			Bone marrow	24	24
Medical x rays					
Medical diagnosis	Adult patients	105×10^6/yr	Bone marrow	103	77
Medical personnel	Occupational	195,000	Whole body	300–350[a]	0.3
Dental diagnosis	Adult patients	105×10^6/yr	Bone marrow	3	1.4
Dental personnel	Occupational	171,000	Whole body	50–125[a]	0.05
Radiopharmaceuticals					
Medical diagnosis	Patients	10×10^6 to 12×10^6/yr	Bone marrow	300	13.6
Medical personnel	Occupational	100,000	Whole body	260–350	0.1
Atmospheric weapons tests	Total population	220×10^6	Whole body	4–5	4–5
Nuclear industry					
Commercial nuclear power plants (effluent releases)	Population within 10 mi	$<10 \times 10^6$	Whole body	$\ll 10$	$\ll 1$
Commercial nuclear power plants (occupational)	Workers	67,000	Whole body	400[b]	0.1
Industrial radiography (occupational)	Workers	11,250	Whole body	320	0.02
Fuel processing and fabrication (occupational)	Workers	11,250	Whole body	160	0.01

Handling byproduct materials (occupational)	Workers	Whole body	3,500	350	0.01
Federal contractors (occupational)	Workers	Whole body	88,500	~250	0.1
Naval nuclear propulsion program (occupational)	Workers	Whole body	36,000	220	0.04
Research activities					
Particle accelerators (occupational)	Workers	Whole body	10,000	Unknown	<<1
X-ray diffraction units (occupational)	Workers	Extremities and whole body	10,000–20,000	Unknown	<<1
Electron microscopes (occupational)	Workers	Whole body	4,400	50–200	0.003
Neutron generators (occupational)	Workers	Whole body	1,000–2,000	Unknown	<<1
Consumer products					
Building materials	Population in brick and masonry buildings	Whole body	110 × 10^6	7	3–4
Television receivers	Viewing populations	Gonads	100 × 10^6	0.2–1.5	0.5
Miscellaneous					
Airline travel (cosmic radiation)	Passengers	Whole body	35 × 10^{6c}	3	0.5
	Crew members and flight attendants	Whole body	40,000	160	0.03
Airline transport of radioactive materials	Passengers	Whole body	7 × 10^{6d}	~0.3	0.01
	Crew members and flight attendants	Whole body	40,000	~3	<0.001

[a] Based on personnel dosimeter readings; because of relatively low energy of medical x rays, actual whole-body doses are probably less.

[b] Average dose rate to the approximately 40,000 workers who received measurable exposures was 600–800 mrems/yr.

[c] Total number of revenue passengers per year is 210 × 10^6; however, many of these are repeat airline travelers.

[d] About one in every 30 airline flights includes the transportation of radioactive materials; assuming 210 × 10^6 passengers per year (total), approximately 7 × 10^6 would be on flights carrying radioactive materials.

REFERENCES

1. Bogg, R. F., and T. M. Moore. A Summary Report on X-ray Diffraction Equipment. Report MORP 67-5. Rockville, Md.: U.S. Department of Health, Education, and Welfare, Bureau of Radiological Health, 1967.
2. Gesell, T. F., and H. M. Prichard. The Contribution of Radon in Tap Water to Indoor Radon Concentrations. Paper presented at the Third International Symposium on the Natural Radiation Environment, Houston, Tex., April 27, 1978.
3. Jacobson, A. P., P. A. Plato, and D. Toeroek. Contamination of the home environment by patient treated with iodine-131. Initial results. Am. J. Public Health 68:225-230, 1978.
4. Klement, A. W., Jr., C. R. Miller, R. P. Minx, and B. Shleien. Estimates of Ionizing Radiation Doses in the United States, 1960-2000. Report ORP/CSD 72-1. Washington, D.C.: U.S. Environmental Protection Agency, 1972.
5. McIntyre, A. B., D. R. Hamilton, and R. C. Grant. A Pilot Study of Nuclear Medicine Reporting through the Medically Oriented Data System. Publication (FDA) 76-8045. Rockville, Md.: Department of Health, Education, and Welfare, 1976.
6. Miles, M. E. Occupational Radiation Exposure from U.S. Naval Nuclear Propulsion Plants and Their Support Facilities. Report NT-78-2. Washington, D.C.: U.S. Department of the Navy, Naval Sea Systems Command, 1978.
7. Miller, L. A., C. I. Vassar, Jr., and R. A. Moats. Report of State and Local Radiological Health Programs, Fiscal Year 1978. HEW Publication (FDA) 80-8034. Rockville, Md.: U.S. Department of Health, Education, and Welfare, Food and Drug Administration, Bureau of Radiological Health, 1979.
8. National Council on Radiation Protection and Measurements. Natural Background Radiation in the United States. Report No. 45. Washington, D.C.: National Council on Radiation Protection and Measurements, 1975.
9. National Council on Radiation Protection and Measurements. Radiation Exposure from Consumer Products and Miscellaneous Sources. Report No. 56. Washington, D.C.: National Council on Radiation Protection and Measurements, 1977.
10. National Research Council, Advisory Committee on the Biological Effects of Ionizing Radiations. The Effects on Populations of Exposure to Low Levels of Ionizing Radiation. Washington, D.C.: National Academy of Sciences, 1972.
11. Nuclear Energy Policy Study Group (S. M. Keeny, Chairman). Nuclear Power Issues and Choices. Cambridge, Mass.: Ballinger Publishing Co., 1977.
12. Oakley, D. T. Natural Radiation Exposure in the United States. EPA Report ORP/SID 7201. Washington, D.C.: U.S. Environmental Protection Agency, 1972.
13. Shleien, B., T. T. Tucker, and D. W. Johnson. The Mean Active Bone Marrow Dose to the Adult Population of the United States from Diagnostic Radiology. Publication (FDA) 77-8013. Rockville, Md.: U.S. Department of Health, Education, and Welfare, Food and Drug Administration, Bureau of Radiological Health, 1977.
14. Smith, J. M., J. A. Broadway, and A. B. Strong. United States population dose estimates for iodine-131 in the thyroid after the Chinese atmospheric nuclear weapons test. Science 200:44-46, 1978. U.S. Office of Radiation Programs.
15. U.S. Bureau of Radiological Health. Gonad Doses and Genetically Significant Dose from Diagnostic Radiology—U.S., 1964 and 1970. Publication (FDA) 76-8034. Rockville, Md.: U.S. Department of Health, Education, and Welfare, Food and Drug Administration, Bureau of Radiological Health, 1976.
16. U.S. Bureau of Radiological Health. Population Exposure to X Rays, U.S., 1970.

Publication (FDA) 73-8047. Rockville, Md.: U.S. Department of Health, Education, and Welfare, Food and Drug Administration, Bureau of Radiological Health, 1973.

17. U.S. Comptroller General. Cleaning Up the Remains of Nuclear Facilities—A Multi-billion Dollar Problem. Report to the Congress. Report EMD-77-46. Washington, D.C.: U.S. Comptroller General, 1977.

18. U.S. Environmental Protection Agency. 40 CFR 190: Environmental Radiation Protection Requirements for Normal Operations of Activities in the Uranium Fuel Cycle. Report EPA 520/4-76-016. Volume I. Washington, D.C.: U.S. Environmental Protection Agency, 1976.

19. U.S. Nuclear Regulatory Commission. Final Environmental Statement on the Transportation of Radioactive Material by Air and Other Modes. Report NUREG-0170. Vol. 1. Ch. 4. Washington, D.C.: U.S. Nuclear Regulatory Commission, 1977.

20. U.S. Nuclear Regulatory Commission. Ninth Annual Occupational Radiation Exposure Report, 1976. Report NUREG-0322. Washington, D.C.: U.S. Nuclear Regulatory Commission, Office of Management Information and Program Control, 1977.

21. U.S. Nuclear Regulatory Commission. Numerical guides for design objectives and limiting conditions for operation to meet the criterion "as low as practicable" for radioactive materials in light-water-cooled nuclear power reactor effluents, pp. 277–345. In Rulemaking Hearing, Docket RM-50-2, Nuclear Regulatory Commission Issuances, Report NRCI-75/4, 1975.

22. U.S. Office of Radiation Programs. Radiological Quality of the Environment in the United States, 1977. Report EPA 520/1-77-009. Washington, D.C.: U.S. Environmental Protection Agency, Office of Radiation Programs, 1977.

23. Wallace, R. W., and C. A. Sondhaus. Cosmic radiation exposure in subsonic air transport. Aviat. Space Environ. Med. 610–623, 1978.

IV
Genetic Effects

INTRODUCTION AND BRIEF HISTORY

This chapter considers the health consequences of genetic damage that result when human populations are exposed to low levels of ionizing radiation in addition to natural background radiation. As in the 1972 review, *The Effects on Populations of Exposure to Low Levels of Ionizing Radiation* (BEIR I),[48] the main text is intended for the informed, nontechnical reader; further details are given in notes at the end of the chapter. This chapter constitutes an updating of Chapter V of BEIR I; our task would have been vastly more difficult had we not had that work to build on. (Indeed, where it is feasible, material from BEIR I is merely repeated here.) The recently completed review prepared by the United Nations Scientific Committee on the Effects of Atomic Radiation (UNSCEAR)[106] has also been extremely helpful.*

Since the publication of BEIR I, new data have been obtained, and perspectives have been modified to an extent that makes a new review desirable. The methods of BEIR I remain valid; however, new numbers have caused some changes in the estimates and some new methods of estimation have been added.

---

*UNSCEAR has issued a series of reports that collectively constitute a wealth of information on this subject.[101-106] In general, throughout this report, we shall not further document conclusions that are in the UNSCEAR reports, but instead simply refer to those reports. The bibliographies therein are very extensive, and the reader is referred to them for more detailed information.

HISTORICAL BASIS FOR RADIATION PROTECTION GUIDES FOR
THE GENERAL POPULATION

Concern over radiation effects on humans was limited at first to effects on radiation workers. Only later was this concern broadened to include non-occupational exposures and their genetic effects in later generations. In the 1920s, it was learned that ionizing radiation could produce a variety of genetic effects, but the interest of geneticists in radiation, before World War II, was related primarily to the use of radiation-induced variants to study genetic mechanisms, rather than to the measurement of health hazards. After the war, concern over radioactive fallout from nuclear detonations and increasing awareness of medical uses of radiation as sources of hazards led to some shifting of the focus of interest. A few years earlier, ionizing radiation was simply a laboratory tool for studying genetic principles; now, the methods of genetics provided the "tools" for studying the effects of radiation on human health. The exposure of entire human populations became the focus of concern, rather than the exposure only of occupational groups. This change was formalized in 1956 by the National Academy of Sciences–National Research Council Committee on the Biological Effects of Atomic Radiation (the BEAR Committee), which introduced the concept of regulation of the overall population dose.[49]

The BEAR Committee had the idea of using background radiation as a yardstick for setting standards. It was thought that the average background radiation was about 5 R over 30 yr, and that from medical exposure about 3 R. On the basis of these estimates and its appraisal of the genetic risk to future generations from population exposure, the BEAR Committee recommended that man-made radiation be so limited as to keep the average individual exposure less than 10 R before the mean age of reproduction, taken to be 30 yr. Specifically, it recommended:

That for the present it be accepted as a uniform national standard that X-ray installations (medical and nonmedical), power installations, disposal of radioactive wastes, experimental installations, testing of weapons, and all other humanly controllable sources of radiations be so restricted that members of our general population shall not receive from such sources an average of more than 10 roentgens, in addition to background, of ionizing radiation as a total accumulated dose to the reproductive cells from conception to age 30.

Simultaneously, a report with similar recommendations was issued by the British Medical Research Council,[8] which stated:

Those responsible for authorizing the development and use of sources of ionizing radiation should be advised that the upper limit, which future knowledge may set to the total dose of extra radiation which may be received by the population as a whole, is not likely to be more than twice the dose which is already received from the natural background: the recommended figure may indeed be appreciably less than this.

In 1957, an *ad hoc* committee of the National Council on Radiation Protection and Measurements (NCRP) addressed the question of limiting the *somatic* radiation exposure of the general population. It recommended that the general-population permissible dose of man-made radiation, excluding medical and dental sources, not be larger than that due to natural background. The *ad hoc* committee stressed, as had the BEAR Committee, that exposure to man-made radiation should be kept as far below the permissible levels as feasible. The present Radiation Protection Guides for the general population grew out of these recommendations (see Note 1).

The BEAR Committee divided the 10-rem recommended ceiling into 5 rems from medical procedures and 5 rems from exposure to nonmedical sources. The Federal Radiation Council excluded medical irradiation from its Radiation Protection Guides, but did take 5 rems as the 30-yr limit for the average population exposure to all nonmedical man-made radiation. The Radiation Protection Guides are stated in rems, rather than in roentgens, attempting to take into account biologic effectiveness and its dependence on radiation quality. The present Subcommittee on Genetic Effects recognizes that there is uncertainty as to the adequacy with which the rem does in fact take relative biologic effectiveness into account for genetic effects, particularly for low doses or dose rates. Nevertheless, for reasons of precedence, as well as in the conviction that solution of this problem is beyond its capabilities, the Subcommittee continues to use the rem in its estimates.

EARLIER ESTIMATES OF GENETIC RISK

The estimation of genetic risk to humans is based largely on animal studies, inasmuch as the few human data available were derived from limited observations and from dosimetry that was generally based more on estimate than on precise measurement. There has been no unequivocal demonstration of radiation-induced gene mutation in humans, and thus there are no data on induced mutation rate. However, human information is used as fully as possible, when it is pertinent to our problem.

The methods of the BEAR Committee were summarized in BEIR I (see Note 2). The BEAR Committee used two methods of calculating the amount of damage. It estimated the increase in incidence of mutation-maintained disease, and, although there was some doubt as to the validity of the effort, it attempted to estimate the total (per genome) mutation rate. There are many uncertainties inherent in this latter method and in its use in estimating health consequences. BEIR I declined to use it in 1972.

BEIR I used two methods of estimating genetic effects and presented the results in several contexts. Where possible, as with reciprocal translocation, direct estimates were made. Otherwise, an indirect approach was used, as in the case of gene mutation. A quantity, the relative mutation risk (the reciprocal of the "doubling dose"), which relates induced and spontaneous mutation rates, was used to estimate the increased incidence of genetic disease due to increased radiation exposure. This procedure requires knowledge of the mutation rates, of the incidence of genetic diseases in human populations, and of the extent to which the incidence depends on recurrent mutation. For translocations, etc., BEIR I derived equilibrium incidences from the "direct," first-generation expectations; for gene-mutation effects, the first-generation expression was derived from the "relative mutation risk" equilibrium expectations, by using the estimated rates of elimination to project the ratio of newly induced damage to transmitted damage.

ADVANCES IN KNOWLEDGE SINCE 1972

The period between the publication of the BEAR report in 1956 and that of BEIR I in 1972 saw rapid accumulation of new data and the emergence of new concepts. The dose-rate effect for specific-locus mutation was found, as were differences in radiation sensitivity between male and female germ cells and between different developmental germ-cell stages within each sex. By the time of BEIR I, chromosomal aberration rates in the mouse had been measured and the significance of chromosomal aberrations in humans had been recognized.

Since 1972, new data on the incidence of genetic disorders in human populations have been obtained that are useful in improving estimates made by the relative-mutation-risk method. New data on induced, transmissible genetic damage expressed in the first-generation progeny of irradiated male mice now allow direct estimation of first-generation consequences of gene mutation in humans. There has been a clearer delineation of the timing of oocyte development in the mouse, permitting better correlation of stages and changes in sensitivity and thereby providing a firmer basis for interpreting data derived from irradiation of immature oocytes. Although differences in viewpoint regarding the underlying mechanisms of dose-rate effects for specific-locus mutation persist, the alternative views fortunately lead to similar risk estimates for low levels of radiation exposure of humans.

As with BEIR I, a major obstacle continues to be the almost complete absence of information on radiation-induced genetic effects in humans. Hence, we still rely almost exclusively on experimental data, to the extent possible from studies involving mammalian species.

WHAT KIND OF GENETIC DAMAGE DOES RADIATION CAUSE?

"The genetic effect of radiation is to produce gene mutations and chromosome aberrations. Some of the ways in which radiation produces such effects are given in Note [3].* The effect of radiation on the well-being of the future population is a consequence of these changes. Because mutations and chromosome aberrations occur spontaneously, it follows that the consequences of radiation are not something new but rather an increase in frequency of various deleterious traits with which we are already beset. Since almost every aspect of the living organism is determined to some extent by its genes, the range of possible mutational effects encompasses virtually every aspect of our physical and mental well-being. The major exception is infectious disease, but even here inherited susceptibilities play a role.

"Some results of genetic change are conspicuous, others are invisible; some are tragic, others so mild as to be trivial; some occur in the first generation following the gene or chromosome change, others are postponed tens or hundreds of generations into the future. Furthermore, most of the effects that are produced by mutation are mimicked by others, of nongenetic origin.

"For all these reasons, radiation (or some other environmental agent) could be having an important effect on human well-being and yet this could go unnoticed. Even if the increase in mutation rate is large, the consequences are likely to be so heterogeneous in their nature, so diluted by space and time, and so obscured by similar conditions from other causes as to make it impossible to associate them with their cause. Only if all the affected persons in future generations could somehow be identified and brought together at one time and place could the total impact of the mutations be apparent.

*Changes from BEIR I are bracketed.

"One of the simplest categories of mutational damage includes those diseases and abnormalities that are caused by a single dominant mutation. The most recent compilation by McKusick[41, 5th ed.] lists [736] such conditions with an additional [753] that are less well established. The collective incidence is very roughly one percent of persons born. Some examples are polydactyly (extra fingers and toes), achondroplasia (short-limbed dwarfism), Huntington's chorea (progressive involuntary movements and mental deterioration), [two types] of muscular dystrophy, several kinds of anemia, and retinoblastoma (an eye cancer). . . .

"In contrast, recessive mutations, which require that the gene be present on both members of a pair of homologous chromosomes in order to produce the trait, may not be expressed for many generations. The trait will appear only when two mutant genes are inherited, one from each of the two parents. . . . However, this may not occur for a [great number of generations]. Indeed, the gene may be lost purely by chance in the Mendelian lottery, although this is balanced on the average by those mutants that increase in number by the same process. More important, there is good reason to think from animal experiments and from fragmentary human evidence that mutant genes are often lost from the population because of mild dominant effects on viability and fertility when the gene is heterozygous. Thus, there is a good chance that the gene will be eliminated from the population before it ever encounters another like itself.

"McKusick lists [521] recessive diseases, plus [596] that are less certain. Some examples are phenylketonuria (or PKU, a form of mental deficiency), Tay-Sachs disease (blindness and death in the first few years of life), sickle cell anemia and cystic fibrosis. These are fairly common and well known, but most recessive conditions listed in the book are very rare.

"Recessive mutations located on the X chromosome are characterized by being expressed almost exclusively in males. Well known examples are hemophilia (failure of blood clotting), color blindness, and a severe form of muscular dystrophy. McKusick lists [107] well established and [98] probable conditions of this sort. Because the gene can be expressed in a single dose in males, [who] have only a single X chromosome, X-chromosome-linked recessive mutations are somewhat like dominant mutations on other chromosomes in that they are expressed soon after occurrence instead of being spread out over an extended time span.

"Some of these dominant and recessive genes cause traits that we regard as normal, such as hair and eye color and blood groups. Others are not normal, but are so mild as to cause little concern. The great majority, however, cause diseases ranging from relatively mild to severe or even lethal. Most are so rare that they are known only to specialists. But, collectively, they are numerous enough that more than one percent of all children born will have a simply inherited disease causing an appreciable handicap.

"Another type of easily classified genetic damage is due to chromosome aberrations. Errors in chromosome distribution can lead to an individual whose cells contain too many or too few chromosomes. The well known disease [Down's syndrome] is caused by an extra representative of a specific chromosome (number 21). Most of the time, however, having too many or too few chromosomes leads to embryonic death; sometimes this is detected as a miscarriage, more often the death is so early as not to be detected at all. This kind of chromosome error is not thought to be strongly influenced by radiation, particularly at low doses.

"Another source of chromosome imbalance is chromosome breakage [leading to rearrangement of chromosomes]. This is less frequent than the type of distribution error mentioned above among spontaneous instances of severe human anomalies. But ionizing radiation is much more effective at [producing rearrangements] than in causing errors in chromosome distribution. The broken chromosomes may then reattach in various ways leading to rearranged gene orders, or they may be lost." (BEIR I, pp. 46–47)

Losses of small segments of chromosomes may have consequences quite similar to and often operationally indistinguishable from single-gene mutations. This class of chromosomal damage is expected to occur even at low doses and low dose rates, but is included among estimated single-gene effects. A kind of gross rearrangement frequently seen in human populations is the reciprocal translocation—the exchange of segments between two or more chromosomes. Some rearrangements have associated phenotypic effects (see Notes 13 and 14); otherwise, reciprocal translocations are not harmful as long as both rearranged chromosomes are present and have, among them, a normal gene content. However, gametes of persons having such "balanced" translocations frequently receive only one of the two parts of the rearrangement, and the zygotes that they produce are, as a result, genetically unbalanced. The nature and extent of the resulting developmental abnormality depend on the particular chromosomal regions that are duplicated or deficient, as well as on how large these regions are. Most such imbalance results in early embryonic death; when it does not, it often leads to physical abnormalities, usually accompanied by mental retardation.

"What is most severe in one sense may not be the most tragic from the standpoint of human welfare. A chromosome aberration that causes early embryonic death may cause very little trauma, whereas the 'milder' effect that permits the embryo to develop into a viable infant that is malformed and mentally retarded may be far more traumatic by any realistic measure of human suffering, both of the child and of his family." (BEIR I, p. 47)

Another common kind of gross rearrangement found among phenotypically normal humans is the Robertsonian translocation, which results from the "fusion" of two chromosomes (each originally having a spindle attachment near the end) to form a single chromosome whose spindle attachment is nearer the center. These constitute a special kind of reciprocal translocation in which the reciprocal product, the other rearranged chromosome, lacks significant genetic content and is likely eventually to be lost. Robertsonian translocations occur in the population with a frequency of about 8 per 10,000. The children of a carrier of such a translocation are usually normal because they inherit either the large translocated chromosome or the separate, normal chromosomes. However, carriers of these translocations produce some unbalanced gametes, which can lead to embryonic death or to congenital anomalies. Radiation does not appear to be a major cause of these translocations; instead, radiation-induced translocations are predominantly of the reciprocal type described earlier (see Note 14).

"In addition to these abnormalities and diseases that are caused by mutation of a single gene or by chromosome breakage, there are other diseases to which gene variation undoubtedly contributes but where the inheritance is more complex. There is abundant evidence that there are inherited predispositions for many common conditions—for example, diabetes, schizophrenia, cancer, and mental retardation.

"It is hard to assess the magnitude of the genetic component and it is even harder to assess what we want to know in the context of this report—the extent to which the disease incidence depends on the mutation rate. . . .

"There is an additional class of mutation whose importance we don't know how to assess—those whose effects are so mild that they are not detected individually. As mentioned before, it is known in *Drosophila* that the most frequent of all mutations belong to a group that causes effects so mild that they can only be detected statistically in experiments involving large numbers. For example, a mutation might cause a one-percent reduction in the probability of surviving from the egg to the adult stage. Such a mutation is clearly impossible to detect in man, and very few [if any] mouse experiments are of a size to reveal it. We don't know what the other manifestations of such a mutant would be. . . . Perhaps the human counterparts of these mutations, in addition to causing a slight reduction in life expectancy, are responsible for [slightly] greater susceptibility to disease, [slightly] impaired physical or mental vigor, or a slight malformation of some organ.

"We cannot ignore such mild mutations as unimportant, because (1) if *Drosophila* is any indication, they are by far the most frequent class of mutations; and (2) being mild, with less effect on viability and fertility, they are more likely to be transmitted to future generations and continue to have their effect over a longer time, thereby affecting more persons. Thus, their impact is multiplied by the number of generations through which they persist; and taken over the whole period, and in conjunction with other mutants, their effect may be far from negligible [see Note 4].

"Despite a concern for this effect, we shall not attempt to estimate it quantitatively, for reasons to be discussed below. It is worth noting ... that in *Drosophila* the evidence is now good that this class of mutation is relatively less frequent among radiation induced mutations than among spontaneous mutations.

"The contrast between genetic and somatic concerns is striking. The low-dose somatic effects that are most feared are cancer and leukemia. The evidence that high radiation doses have these effects is unequivocal. The evidence for low doses is less clear. For genetic effects of radiation, we have no direct evidence of human effects, even at high doses [except for reciprocal translocations cytologically detected in spermatocytes. (See Note 14 and 'Direct Estimation of First-Generation Incidence of Induced Disorders Resulting from Chromosomal Aberrations,' below.)] Nevertheless, the animal evidence is so overwhelming that we [can only assume] that humans are affected in much the same way. In contrast to somatic effects, where the concern is concentrated mainly on malignant disease, the genetic effects are on all kinds of conditions—for the spectrum of radiation-caused genetic disease is almost as wide as the spectrum from all other causes." (BEIR I, p. 48)

COULD AN INCREASED MUTATION RATE BE BENEFICIAL?

It has been suggested on occasion that increasing the mutation rate might be beneficial, in that mutations are "the raw material on which evolutionary progress depends." We see no merit in this view. Although the optimal rate of mutation is not known, there are strong reasons for believing that the rate of evolutionary adaptation is not limited by the rate of mutation. Furthermore, almost without exception, detectable mutations have been found to be deleterious—mildly or strongly—in their effects.

"We believe that a genetically diverse population is more to be desired than a uniform one, and this might be regarded as an argument for a high mutation rate. But the amount of genetic variability existing in the population is far greater than that which arises by mutation in a single generation. Furthermore, in some polymorphisms such as blood groups, hemoglobins, and serum proteins the entire variability may have arisen from a few mutant genes. If human mutation were to stop entirely, we should probably not notice any effect at all for many generations, except for some reduction in the incidence of severe dominant [diseases]. ... The mutant genes now in the population arose in the past and have been pre-tested to some extent, the worst ones having been eliminated by natural selection. What we are saying is that there is ample genetic variability in the population for any evolutionary progress that is likely to occur in the foreseeable future. Indeed, some geneticists argue that for a long time to come the closer we can come to a mutation rate of zero, the better off we will be. Whether this is correct or not (and in any case lowering the spontaneous mutation rate is not now possible) the Subcommittee is convinced that any increase in the mutation rate will be harmful to future generations." (BEIR I, p. 49)

RADIATION EXPOSURES OF GENETIC IMPORTANCE

The sources of population gonadal exposure are treated in detail in Chapter III and are summarized in Table IV-1. For estimation of genetic effects, additional physical and demographic factors must be considered. Also, concerns have been expressed regarding the induction of genetic effects by radiation from two particular sources: the transuranic actinide radionuclides resulting from nuclear power and weapons activities, and those radionuclides which can be directly incorporated into DNA, principally tritium and carbon-14.

Genetic disorders, by definition, do not occur in persons whose germ cells have been affected by radiation; rather, the effects are seen in their offspring and in later generations to which the altered genetic material is transmitted. Hence, these effects depend quantitatively on the portion of the dose that is received by the gonads of future parents, rather than on the total dose to the entire population. BEIR I followed the precedent of the BEAR Committee, taking 30 yr as the mean length of a human generation. Hence, it is the average 30-yr individual dose accumulated by all the parents of the new generation that concerns us in making estimates. The type of exposure regimen assumed in our calculations is of an entire population exposed uniformly over very long periods (many generations).

TABLE IV-1 Estimated Annual Average Genetically Significant Dose Equivalents

Source	Dose Equivalent Rate, mrems/yr
Natural radiation	
Cosmic radiation	28
Radionuclides in the body	28
External gamma radiation from terrestrial sources	26
SUBTOTAL	82
Man-made radiation	
Medical and dental x rays	
Patients	20
Occupational	< 0.4
Radiopharmaceuticals	
Patients	2–4
Occupational	< 0.15
Commercial nuclear power	
Environment	< 1
Occupational	< 0.15
National laboratories and contractors—occupational	< 0.2
Industrial applications—occupational	< 0.01
Military applications—occupational	< 0.04
Weapons-testing fallout	4–5
Consumer products	4–5
Air travel	< 0.5
SUBTOTAL	30–40 (approx.)

Where exposures are not uniformly or randomly delivered to the entire population, *the age distribution of the exposed population and the probability of having children for each age and sex need to be taken into account*, as was done, for example, in the Reactor Safety Study.[108] Also, calculation of the genetic consequences of occupational exposure requires special consideration of the distribution of dose in the population, as well as possible exposure to high-LET radiation.

Fuel reprocessing for the mixed-oxide reactor fuel cycle and the breeder reactor cycle will result in exposure that is primarily occupational, from plutonium-238 and other transuranic nuclides. There is concern over genetic hazards from this source, because of the high LET of the emitted alpha particles. However, very little of the plutonium to which the general population is exposed is deposited in the gonads, and this, because of the short range of the alpha particles, reduces the dose that can have genetic consequences. Although the fraction of the plutonium to which people are exposed occupationally that is deposited in their gonads may be larger,

the size of this work force is small, again minimizing the genetic consequences in the population. Furthermore, measured RBE values are either
lower than, or of the same magnitude as, those for other high-LET radia-
tion-like neutrons. Special consideration of the genetic hazards of
plutonium and other transuranics thus appears unnecessary (see Note 5).

In addition to direct effects of radiation, some genetic effects are to be
expected from the transmutation of radionuclides that have been incorporated into the genetic material, DNA. This includes the nuclides
hydrogen-3, carbon-14, and phosphorus-32. There are a number of positions in DNA bases in which hydrogen-3 transmutation leads to appreciable mutation in microorganisms, and some small effects in fruit flies
can be ascribed to transmutation of phosphorus-32. The yields from such
transmutations are small and, as pointed out in BEIR I, the risk from
transmutation is far smaller than that from the radiation emitted by the
decay of the same nuclides (see Note 6).

METHODS OF ESTIMATING THE GENETIC RISKS FROM RADIATION

In making our estimates, we have adhered to the principles enumerated in
BEIR I:

"1. Use relevant data from all sources, but emphasize human data
when feasible. In general, when data of comparable accuracy exist, place
greater emphasis on organisms closest to man.

"2. Use data from the lowest doses and dose-rates for which reliable
data exist, as being more relevant to the usual conditions of human exposure.

"3. Use simple linear interpolation between the lowest reliable dose
data and the spontaneous or zero dose rate. In order to get any kind of
precision from experiments of manageable size, it is necessary to use
dosages much higher than are expected for the human population. Some
mathematical assumption is necessary and the linear model, if not always
correct, is likely to err on the safe side (see Note 7).

"4. If cell stages differ in sensitivity, weight the data in accordance with
the duration of the stage.

"5. If the sexes differ in sensitivity, use the unweighted average of data
for the two sexes." (BEIR I, p. 51)

One way of looking at genetic risk is simply to compare the incremental
radiation exposure with that due to natural background radiation. BEIR I

did this, not as a risk estimate, but as a potentially useful policy guide. As mentioned in Chapter III, the gonadal dose from natural radiation averages about 80 mrems/yr.

Where BEIR I used four ways of estimating the genetic risk of added population exposures to ionizing radiation, we have used only two. The first is the indirect relative-mutation-risk method used in BEIR I. The second is a new, direct method of estimating total phenotypic damage induced in a single generation. As in BEIR I, abnormalities attributable to chromosomal aberrations are estimated from data derived by chromosomal cytology. BEIR I gave genetic-effects estimates for a population exposed to an added increment of radiation exposure (above natural background) of 5 rems/generation. Because the average human generation span was assumed to be 30 yr, this figure of 5 rems was a convenient approximation of exposure of the general population to the current maximal permissible dose of 170 mrems/yr from nonmedical sources. However, we found it more convenient to derive our estimates in terms of the risk per rem of added exposure per generation.

ESTIMATES BASED ON THE RELATIVE MUTATION RISK

In every organism that we study, an appreciable amount of spontaneous mutation occurs, in the absence of special causes. As discussed later in this section, background radiation probably accounts for only a small fraction of the spontaneous mutation frequency. Radiation (as well as a variety of chemical agents) can bring about an added amount of induced mutation. The rate of induced mutation can be stated in absolute terms (i.e., the probability of mutation per locus per rem), or it can be stated in relative terms, such as the ratio of the induced mutation rate to the spontaneous mutation rate. This ratio, the fraction by which each rem of added exposure would increase the mutation rate above the spontaneous level, is called the "relative mutation risk." Frequently, the reciprocal of this ratio, the "doubling dose," is given.

There is no single, simple way of relating the amount of genetic disorder in a population to the mutation rate. Each category of disorder must be dealt with in its own special way. If a single mutant gene has a simple dominant effect, the incidence of the trait will be proportional to the frequency (relative number) of the corresponding mutant gene in the population. The effect of mutation is to increase the frequency of mutant genes in the population. However, mutant genes are also being eliminated at a rate, for these simple dominant traits, proportional to their frequency. As the number of such genes in the population increases (because of the higher mutation rate), the number being eliminated will also increase.

Eventually, the rate of elimination will exactly balance the rate of increase through mutation; at this point, the new equilibrium frequency will bear the same numerical relationship to the old frequency as the new mutation rate bears to the old one. Hence, for disorders with simple, autosomal dominant expression that are maintained exclusively by recurrent mutation, the increase in incidence at equilibrium is proportional to the amount by which the mutation rate has been increased. If there were a 1% increase in the mutation rate—i.e., a relative mutation risk of 0.01—and if the higher mutation rate continued over a number of generations, the incidence of disorders that are maintained exclusively by mutation would eventually be 1% higher than it had been initially. If the incidence of some such specific condition before the onset of the radiation exposure had been 100 per million liveborn, the expected increase at equilibrium (perhaps 10 generations later) would be 1 per million.

Note 8 discusses the background information that went into our adoption of the range of 0.004–0.02 per rem (a doubling dose of 50–250 rem) for the value of the relative mutation risk. This range takes into account the mutation rates found in different oocyte stages in the mouse; this was done because there is still some uncertainty as to which stage is most representative of the human resting oocyte (see Note 9).

The numbers that we use as the current incidence of hereditary disorders are derived from epidemiologic studies in British Columbia and Northern Ireland, as well as from other studies.[94, 98, 106] In British Columbia, the incidence of autosomal dominant plus X-linked mutation-caused disorders is now thought to be about 10,000 per million, which is in close agreement with the corresponding incidence reported in the earlier study in Northern Ireland.

Estimates based on these incidences and on the relative mutation risk are given in Table IV-2, which shows the increase expected in the different classes of genetic disorders among 1 million liveborn people whose ancestors have received an increased radiation exposure of 1 rem per 30-yr reproductive generation (33 additional mrems/yr). The method of calculation for this table is given in Note 10.

The number of people affected by autosomal dominant and X-linked traits is expected to increase by 40–200 per million above the estimated present incidence of about 10,000. (X-linked disorders account for only about 400 of these 10,000; the incidence of disorders of this type is expected to follow a different pattern of change from that of the autosomal dominants, but for our purposes no great error will be introduced by lumping these two categories. For a brief discussion of this matter, see Note 11.)

The contribution of recessive disorders will be negligible, compared

TABLE IV-2   Genetic Effects of an Average Population Exposure of
1 Rem per 30-Yr Generation

Type of Genetic Disorder[a]	Current Incidence, per Million Liveborn Offspring	Effect per Million Liveborn Offspring, Rem per Generation	
		First Generation[b]	Equilibrium[c]
Autosomal dominant and X-linked	10,000	5–65[d]	40–200
Irregularly inherited	90,000		20–900[e]
Recessive	1,100	Very few; effects in heterozygotes accounted for in top row	Very slow increase
Chromosomal aberrations[f]	6,000	Fewer than 10[g]	Increases only slightly

[a] Includes disorders and traits that cause serious handicap at some time during lifetime.
[b] Estimated directly from measured phenotypic damage or from observed cytogenetic effects.
[c] Estimated by the relative-mutation-risk method.
[d] No first-generation estimate available for X-linked disorders; the expectation is that it would be relatively small.
[e] Some estimates have been rounded off to eliminate impression of considerable precision.
[f] Includes only aberrations expressed as congenital malformations, resulting from unbalanced segregation products of translocations and from numerical aberration.
[g] Majority of Subcommittee feels that it is considerably closer to zero, but one member feels that it could be as much as 20.

with that of the other classes of genetic disorders, especially in the early generations. When the disorder is not completely recessive, the equilibrium frequency is approximately proportional to the mutation rate. Whatever mechanisms of elimination operate, equilibrium is reached very slowly, and any effect of an increased mutation rate on the incidence of recessive traits would be spread over a very large number of generations.

The population survey in British Columbia reported that at least 9% of all liveborn humans will be seriously handicapped at some time during their lifetimes by genetic disorders of complex etiology, manifested as congenital malformations, anomalies expressed later, or constitutional and degenerative diseases. This, the largest category of genetic disorder listed in Table IV-2, we refer to as "irregularly inherited" disorders. The mutations responsible for the many hundreds of disorders in this category are thought to be maintained in the population by a variety of mechanisms, some of which would not be influenced by changes in mutation rate. It is felt that most cases of these conditions are caused by the cumulative effects of many different genes and environmental factors. It seems likely that some cases now classified as irregularly inherited disorders may turn out to be single-gene traits involving incomplete penetrance.

An estimate of the number of induced irregularly inherited disorders present at equilibrium must take into account the proportion of the incidence of these disorders that would vary directly with the mutation rate, a quantity that BEIR I called the "mutational component." More precisely, if the equilibrium incidence, I, of a disorder is a linear function of the mutation rate, m, i.e., $I = a + bm$, then we define the mutational component to be $MC = bm/(a + bm)$, in which case the relative increase of the disorder incidence after an increase in the mutation rate from m to, say, $m(1 + k)$ is $(I' - I)/I = (MC)k$. Each disorder may have its own mutational component, and a class of disorders, such as irregularly inherited disorders, its average mutational component.

Except in simple cases, the mutational component, however defined, is difficult to estimate, and there is no consensus among geneticists as to its most likely value. For it to be near its lowest value, zero, it would be necessary that all alleles responsible for the disorders be held in the population by balancing selection. That is, mutations capable of causing these irregularly inherited disorders would need to be of enough benefit in heterozygotes that their maintenance would be essentially independent of mutation.

However, balancing selection has not yet been proved to be operating for the maintenance of even one of the hundreds of different irregularly inherited human disorders. In this regard, it is of interest that, in empirical studies on mice exposed to large doses of radiation for many generations, the offspring showed no demonstrable effect on viability, fertility, or growth and no detected abnormalities attributable to the radiation (see Note 12). At first sight, this might suggest that the mutational component must be very small for irregularly inherited disorders. But that is not necessarily so. For example, if mice are like *Drosophila* in that the great majority of their mutations have small selection coefficients, the effects exerted by the few induced mutations per mouse that would have accumulated in the multigeneration experiments may have been too slight to show up in the presence of large amounts of nongenetic variability found in mouse strains for such characteristics as viability, fertility, and growth. Furthermore, the end points sought in these multigeneration experiments were those which would be considered important "components of fitness"—i.e., factors important in the extinction of mutant genes—and consequently less relevant to questions about irregularly inherited disorders. The suspicion that effects relevant to human genetic disorder must have occurred in those experiments is borne out by the recent one-generation empirical study, on mice, that has demonstrated clear-cut dominant damage to the skeletal system (see the following section and Note 13).

Uncertainties as to the relative roles of mutation and balancing selection in maintaining the current incidences of irregularly inherited disorders appear to center on three questions: (1) For any given clinically defined disorder, how many loci are there at which mutation can cause the disorder? (2) How small are the mean selection coefficients for mutations that cause handicaps in heterozygotes? (3) Are there loci involved in such disorders that have unusually high mutation rates? Firmer genetic analysis of disorders of this kind is needed before answers can be given as to how large the mutation rate would have to be for disorder incidence to depend primarily on recurrent mutation.

The uncertainties involved in relating the incidence of irregularly inherited disorders to recurrent mutation remain too great to permit any narrowing of the range of the mutational component used in BEIR I, 5–50%. These uncertainties enter into the calculated range of increase in the equilibrium incidence of these disorders resulting from an additional 1-rem/generation increase in population exposure (Table IV-2). The current incidence of irregularly inherited disorders is approximately 90,000/million liveborn offspring. The increase expected at equilibrium would be about 20–900 per million liveborn.

Radiation is only one of a number of environmental insults that can cause mutations. Our adoption of a range of 0.004–0.02 for the relative mutation risk implies that only 1–6% of the mutations responsible for disorders in the human population result from exposure to background radiation of approximately 3 rems/generation. Many spontaneous mutations may result purely from errors made during the replication of the genetic material. An alternative way of approximating the fraction of spontaneous mutations that is induced by radiation is to calculate the expected spontaneous mutation frequency with an assumption of a linear relationship of mutation to dose in the low-dose range. Such calculations for mouse spermatogonia indicate that the spontaneous-mutation rate is two to three orders of magnitude greater than would be expected; this suggests that the contribution of radiation-induced mutations to the spontaneous frequency is even smaller than was suggested by the earlier calculations based on the average relative mutation risk for both sexes.

DIRECT ESTIMATION OF FIRST-GENERATION EXPRESSION OF DAMAGE

BEIR I made a direct estimate of the first-generation incidence of induced chromosomal aberration, but did not do so for gene mutation. New data on the induction of chromosomal aberration in laboratory mammals, as well as some human data, permit refining the estimates of chromosomal mutation; and new data on transmissible skeletal damage, found in the first generation after the irradiation of male mice, make it possible to estimate nearly all the combined first-generation expression of damage from gene mutations.

Direct Estimation of First-Generation Incidence of Induced Disorders Resulting from Gene Mutations

One of the long-standing uncertainties about estimates of the incidence of induced genetic disorders resulting from gene mutations after increased human radiation exposure has been related to the lack of acceptable data on the amount of induced genetic disorder in any mammal. To bridge the gap between existing mutation-frequency data in mice and the expected incidence of induced genetic disorders in humans, it was necessary to apply to man estimates of the degree of reduced genetic fitness of induced mutations in *Drosophila* or, as in the relative-mutation-risk approach, to assume that induced and spontaneous mutations have an equal likelihood of causing genetic disorders. Now, however, new data permit the expected incidence of induced genetic disorders in humans to be estimated directly on the basis of the frequency of a type of radiation-induced genetic disorder—namely, skeletal abnormalities—in mice. This approach is thus termed a "direct estimate." It should be recognized that what we have continued to refer to as a direct estimate of chromosomal disorders is not a direct estimate in quite the same sense. However, we also call this latter method "direct" in the sense that it involves first-generation estimation that neither uses a relative-mutation-risk factor nor is derived from an equilibrium estimate.

Data on the amount of presumed mutational damage to the skeleton in the first generation after irradiation (in mice) have actually been available for some time.[19] However, they were not commonly used in making risk estimates, mainly because the animals in which the skeletal defects were observed had been killed for examination and therefore could not be used for genetic testing of the presumed mutations. A new group of similar mutants has now been obtained and tested for transmissibility, and the data from this recent study can serve as a basis for predicting genetic damage. For protracted exposure of spermatogonia, the estimated induced-mutation frequency per gamete per rem is 4×10^{-6} (see Note 13).

To expand this estimate of the induced-mutation frequency for all detected mutations, in this one body system, to the total number of effects, in all systems, that would cause a serious handicap if they occurred in humans, this mutation frequency, as explained in Note 13, is multiplied by 5–15 and by 0.25–0.75. To make the estimate apply to exposure of both sexes, the upper bound of the range thus obtained is multiplied by 1.44 and the lower bound is kept the same. This gives an estimate of 5–65 induced dominant disorders leading to serious handicaps at some time during life per million liveborn as the first-generation expression, after exposure of the entire population to 1 rem/generation (Table IV-2).

As shown in Table IV-2, this estimate for dominant disorders applies to the sum of effects of the categories of autosomal dominant and irregularly inherited disorders listed. The reason for this is that in humans the irregularly inherited disorders that would undergo an increase in incidence in the first generation after increased radiation exposure would be almost entirely dominants with penetrance low enough that geneticists would not be able to recognize them as autosomal dominants.

No direct estimate for X-linked disorders is available from the skeletal experiment, because male mice were irradiated and only their male offspring were examined for mutations. Compared with the risk estimate for dominantly inherited disorders given in Table IV-2, however, the relative risk from X-linked mutations seems certain to be very small (see Note 11). Any completely recessive autosomal mutations induced would have no effects in the first generation, except in the unlikely event that such a mutation came together with an independent mutant allele at the same locus. Heterozygous effects of recessive mutations would be accounted for among the dominantly inherited disorders.

For disorders due to gene mutation, there will be an increase in later generations, but the increase will be smaller and smaller with each passing generation until equilibrium is eventually reached. We have no direct evidence as to how many generations are required; this would depend on the rate of elimination of mutant genes.

Direct Estimation of First-Generation Incidence of Induced Disorders Resulting from Chromosomal Aberrations

Gross Rearrangement   We have made direct estimates of induced gross chromosomal aberration due to rearrangement or to error in assortment. Among rearrangements, Robertsonian translocations are a major cause of severe abnormalities due to secondary trisomy, but there is no evidence of induction of these in irradiated spermatogonia of mice. A few were found after the irradiation of male *Drosophila* in a very large experiment,[56,59] and a few have been induced by irradiating females of this insect,[57] but the rates are low; we feel that the risk of inducing these in humans is quite small in comparison with reciprocal translocations. However, the data that we use in estimating serious abnormalities due to gross rearrangement come from observations on all kinds of translocation heterozygotes currently found in human populations (see, e.g., Jacobs[28]). Inversions will be induced much less frequently than translocations, and there may also be some associated semisterility, the amount depending on the size and location of the chromosomal segment that is inverted. The BEIR I method of estimation, based on the rate of induced transmissible semisterility, did not distinguish between that due to translocations and that due to inversions or other causes. In view of the similarity of the effects and the infrequency of inversions relative to translocations, as well as the adequately broad range of uncertainty as to rearrangement damage, we feel that no separate treatment of inversion damage need be included. Small deficiencies (as well as duplications) will be produced by irradiation, and their incidence relative to gross rearrangements may be somewhat greater at lower doses, owing to the possibility that more may result from one-track events. Larger deficiencies and duplications are believed to be less likely to occur at low doses; furthermore, zygotes carrying these are likely to be lost because of genetic imbalance, as in the case of segmental aneuploidy in chromosomally unbalanced translocation carriers (discussed below). An exception to this may be certain imbalances affecting the sex chromosomes. Even in autosomes, duplications or deficiencies involving some chromosomal segments may be compatible with survival and sometimes with reproduction. Duplications involving up to one-third of a long chromosome where heterozygotes attain reproductive age and are fertile are known in mice.[68] All these long duplications and deficiencies have readily detectable adverse effects in heterozygotes, but shorter multilocus deficiencies are known for which adverse effects, if there are any, must be small.

Human and marmoset data are now available[6] for use in estimating the rate of induction of transmissible reciprocal translocations. Combined data from these species indicate that the rate of induction of reciprocal translocations in spermatogonia, scored as multivalents in primary spermatocytes, is 7×10^{-4} per rem per cell. In mice, the rate is known to be lowered by low-dose-rate exposure; furthermore, spermatocytes that are heterozygous for some translocations fail to complete the process of gametogenesis and hence give rise to no functional germ cells. Finally, in meiosis (maturation), the translocation multivalent may orient on the spindle in a number of ways, with the result that many of the gametes formed will not carry a balanced reciprocal translocation. The calculations given in Note 14 take each of these factors into account and yield an estimated incidence of recoverable balanced reciprocal translocations of $0.17-1.7 \times 10^{-4}$ per rem of chronic paternal exposure.

The incidence of zygotes with unbalanced segregation products (segmental aneuploids) will be slightly more than twice the incidence of newborns with a balanced translocation. Most of these segmental aneuploids will be eliminated in early development, usually too early to be recognized as spontaneous abortions. In the vast majority of cases, no effect will be seen on the reproductive history of the carriers of balanced translocations. On the average, half their children will be of normal karyotype, and half will carry the balanced rearrangement. The possibility of producing seriously affected children in this way exists for only a very few translocations. Another small fraction of carriers of balanced reciprocal translocations will be completely sterile (see Note 14).

We have no data on induced translocations in human oocytes. It is not clear that the oocyte rate would be higher than that in the spermatogonium, and there is some evidence that it might be lower. We have followed the approach of BEIR I and assumed that it will equal the rate in the spermatogonium.

On the basis of the known properties of transmissible translocations induced in mice and in other experimental organisms, we believe that considerably fewer than 5% of all transmissible induced translocations (most of which will be reciprocal) will be of such a nature that abnormal liveborn offspring could be produced, the majority of unbalanced zygotes being eliminated in very early development. In the few cases in which viable abnormal offspring could result, it appears unlikely that more than one of each of the possible types of unbalanced zygote would be able to survive, and we estimate that there would be at most about 10 such children per million liveborn per rem of parental exposure (Note 14). In view of an independent approach based on the litter-size reduction observed after acute irradiation of mouse germ cells (outlined in Note 14), most Subcommittee members felt that the estimate of 10 was too high and that the true value may be near zero.

Numerical Aberration The risk of induced numerical aberration is very small; there is no clear evidence of the induction of trisomy after the irradiation of mouse spermatogonia or mouse oocytes (see Note 15). The experimental analysis of induced trisomy in other species has shown it to be due, perhaps entirely, to damage incurred during the prophase preceding the first meiotic (maturation) division. This stage accounts for only a very small fraction of the total time that male germ cells are at risk, but is the stage of longest duration in the female—the primary oocyte. There have been conflicting claims concerning radiation-induced trisomy in humans: some epidemiologic studies have shown a relationship between maternal irradiation and trisomy (e.g., Down's syndrome), but there are strong reasons for doubting a cause-and-effect relationship. There is a large maternal-age component in cases in which a relationship has been claimed, as well as other complications, such as the pooling of heterogeneous data. In addition, high-dose exposures of human populations in Japan (Hiroshima and Nagasaki) failed to yield evidence of induced trisomy, as had also the mouse experiments,[65] using genetically marked X chromosomes. High doses given to oocytes of insects do result in a measurable increase in the amount of trisomy, and the analysis of these cases leads us to believe that it should also happen in humans. In insects, the risk of trisomy is extremely small, relative to that of other cytogenetic damage, and has not been measurable at doses below about 1,000 R.

Table IV-2 gives the direct estimates for the first-generation incidence of all abnormalities due to chromosomal aberrations, both structural and numerical, from an increase of 1 rem/generation in exposure of the general population. Because of the high rate of elimination, the incidence of abnormalities due to chromosomal aberrations would increase only slightly in later generations.

COMPARISON OF DIRECT AND INDIRECT METHODS OF ESTIMATION

There is some overlapping in the data used in the two different methods; for example, specific-locus information is used in estimating the total phenotypic damage that would be induced in females. Although the two methods measure different quantities, there is no major disagreement as to the incidence of the effects expected. One method estimates first-generation effects, the other equilibrium effects. If the rate of elimination of mutants were known with some precision, it would be possible to convert first-generation incidences into equilibrium incidences and vice versa; BEIR I made this conversion by using assumed rates of elimination. The important consideration, at this point, is that the rates of elimination needed to reconcile the two ranges do not appear unreasonable. In other words, the two methods give estimates that are in quite good agreement.

Our refusal to use the relative-mutation-risk method in estimating the risk associated with chromosomal aberrations may be questioned, but there were compelling reasons for this decision. Chromosomal disorders in human populations result largely from primary and secondary trisomy (resulting from nondisjunction and Robertsonian translocation, respectively), and these are not expected to be increased materially by low-level radiation exposures. A "doubling dose" determined for reciprocal translocations induced in spermatogonia, however accurate it might be, would have little relevance to the induction of these abnormalities. Furthermore, the degree of uncertainty involved in the direct estimate of reciprocal translocation rates seems much smaller than it would be if the estimate were based on a supposed doubling dose.

RISK ESTIMATES FOR SPECIAL SITUATIONS

Attention has already been given in this chapter to some situations in which the exposure to radiation may be quite different from that assumed in our calculations. It should be emphasized that, when the population of concern is a group that differs markedly from the general population, the differences must be taken into account. When the exposure of concern is relatively short, the projections called for may be short-term or intermediate-term.

If comparisons of the relative impact of genetic and somatic effects are made, it is imperative that differences in methods of calculating effects be kept in mind; otherwise, their interpretation may be unreliable. All estimates are of probabilities, giving the number of expected affected individuals as a fraction of some total population (e.g., number of affected individuals per million of population). However, the populations that supply the denominators are quite different in the two cases. On the one hand, genetic effects are seen in the *offspring* of exposed individuals and in later generations to which the damaged genetic material is transmitted. These effects may persist through some number of generations before a mutant gene or chromosome is eliminated from the population. Genetic effects are therefore usually expressed in terms of millions of liveborn. On the other hand, somatic effects, by definition, occur in and are limited to the *exposed* individuals. The exposed population is the general population (or some defined subdivision of it), and this supplies the denominator in the fraction that expresses rate.

Obviously, it is only the exposure of the germ cells that will actually become involved in reproduction that has genetic consequences. Thus, although it may be useful to speak of population doses of so many person-rems, this quantity is at best indirectly related to the average dose to germ cells that will later function in reproduction.

What counts is the dose, accumulated over one generation, to gametes (and, of course, to their precursor cells) that will function in conceptions. Thus, the quantity of importance is the number of effective gamete-rems received. Conversion from person-rems to gamete-rems requires demographic data on the population of interest. If age and sex distributions are known and reproductive patterns have been determined, it is possible to make detailed estimates (e.g., see U.S. Nuclear Regulatory Commission[108] and Note 16).

DISCUSSION

Much of the travail of this Subcommittee has stemmed from issues that were also of concern to the BEIR I Subcommittee. Few human data are—or are ever likely to be—available in the variety and depth needed to give us direct and simple answers to our questions. Hence, to estimate human genetic risks, we must find ways to apply information from other sources. Experimental data from laboratory organisms must be used, and this raises the inevitable question of how well chosen the sources of information were. This question is not new, but we have felt compelled to address it in somewhat greater detail than did BEIR I, usually in the course of defining the degrees of uncertainty encountered in our estimates.

We can draw one conclusion, not directly responsive to our charge: Improving the methods of estimating the health consequences of genetic damage should be given high research priority. We deplore having to use empirical approximations, rather than precise estimates based on a firm understanding of mechanisms. We believe that such a change in method can be made only when there is a better understanding of the organization, functions, and interactions of the genes of higher organisms. Such understanding is basic to the interpretation of the nature of the mutant gene and the analysis of the mechanisms of damage to the genetic material. Meanwhile, we have detailed our more obstinate uncertainties and have given our estimates correspondingly broad ranges of values. We are convinced that these estimates can be used wisely only if the sources of the uncertainties are understood by the user.

This report differs from BEIR I in having used a method of direct estimation of damage from gene mutations that was based on the incidence of skeletal mutations in mice. The assumptions used in this method overlap those used in the relative-mutation-risk method in only a few ways. The uncertainties of using mouse data to solve human problems are common to the two methods. There is an added uncertainty in extending the mutation-rate data on a single system to cover effects on all systems. This method and these data, alone, do not yield an estimate of the *equilibrium* incidence of these conditions, but only of first-generation effects.

However, there is uncertainty in the estimated value of the relative mutation risk, as well as in other numbers that enter into the projections derived by this method. It is not certain that the ratio of induced-mutation to spontaneous-mutation rates would be identical or even similar for all loci, nor that any of our methods would reveal the true effective doubling dose for the entire genome. We assume that spontaneously occurring and radiation-induced gene mutations will have a similar likelihood of producing disorders, but this expectation is based largely on studies of mutant genes not associated with dominant phenotypes. There are uncertainties, somewhat diminished since 1972, in the incidences of the different categories of human genetic disorders, and there are still uncertainties as to the role of recurrent mutation in maintaining "irregularly inherited disorders." Finally, our estimates of the equilibrium incidences can be regarded with greater confidence than could any for first-generation expression that could be derived from them.

For the reader faced with a choice between alternative estimates, we recommend that the direct method be used for first-generation estimates and the relative-mutation-risk method be used for equilibrium estimates. We have followed this practice in Table IV-2. By deriving first-generation estimates from the equilibrium values shown in Table IV-2 (assuming an average persistence of five generations for autosomal dominant and X-linked mutations and 10 generations for mutations causing irregularly inherited disorders), one obtains 10–130 per million per rem, compared with the direct estimate of 5–65. Going in the other direction, using the direct first-generation estimates and a mean persistence of 10 generations, one obtains an equilibrium estimate of 50–650, compared with the indirect estimate of 60–1,100 (sum of 40–200 and 20–900, from Table IV-2) based on the relative-mutation-risk method. We are reassured by the rather close agreement between the two estimates, given the reasonable assumptions of BEIR I regarding rates of elimination.

BEIR I addressed the question of placing an economic value on future genetic disorders; we were reluctant to approach this question. Presumably, a population would not purposely expose itself to increased radiation unless there were an associated benefit—a benefit usually to the population itself and only indirectly, if at all, to future generations. In the case of radiation-induced genetic damage, the major cost is felt in the future; i.e., the benefit accrues to one population, and the cost is borne by another. (We recognize that radiation may not be unique in this respect.)

SUMMARY

The genetic disorders that can result from radiation exposure are (1) those which depend on changes in individual genes (gene mutations or small deletions) and (2) those which depend on changes in chromosomes, either in total number or in gene arrangement (chromosomal aberrations). The former are expected to have greater consequences than the latter.

At low levels of exposure, the effects of radiation in producing either kind of genetic change will be proportional to dose, in that higher-order interactions (those involving more than one ionizing event) are extremely unlikely to occur. For reasons of prudence, and to the extent possible, estimates are based either on experimental findings at the lowest doses and dose rates for which reliable data have been obtained or on adjustment of the observed data obtained at high doses and dose rates by a dose-rate reduction factor deemed appropriate by the Subcommittee.

Two methods are used to estimate the changes in incidence of disorders caused by gene mutations. One method estimates the incidence of such disorders expected after the continuous exposure of the population over a large number of generations. The other method estimates the incidence of disorders expected to be seen in a single generation after the exposure of the parents.

By the first method, it is estimated that only about 1–6% of all spontaneous mutations that occur in humans can be ascribed to the effects of background radiation. Therefore, a small increase in radiation exposure above background will lead only to a correspondingly small relative increase in the rate of mutation. The numerical relationship of rates of induced and spontaneous mutation is shown as a relative-mutation-risk factor, which is the ratio of the rate of mutations induced per rem to the spontaneous rate. (The reciprocal of this is the "doubling dose," the amount of radiation required to produce as many more mutations as are already occurring spontaneously.) The estimated relative mutation risk for humans is 0.02–0.004 per rem (or a doubling dose of 50–250 rem). After many generations of increased exposure to radiation, it is expected that human hereditary disorders that are maintained in the population by recurrent gene mutation would show a similar increase in incidence. However, not all such human disorders have this simple relationship to mutation. It is estimated that the increase will be about 60–1,100 per million liveborn offspring per rem of parental exposure received in each generation before conception. The current incidence (resulting from causes other than the added radiation) of human genetic disorder is approximately 107,000 cases per million liveborn.

These expected incidences are reached only after a large number of generations of exposure, because, in any given generation, the disorders experienced result both from newly induced mutations and from mutations transmitted from an earlier generation. The number of generations required to reach an equilibrium between the induction of mutations and their elimination from the population depends on how long the induced damage persists before being eliminated.

In applying the second method of risk estimation, the incidence of induced, transmissible damage to one organ system (skeleton of the mouse) has been used to calculate the effects expected for all human organ systems. This estimate is for the effects in a single generation after exposure of the parents to radiation; it takes into account the proportion of all known human hereditary defects that affect the one system, and this is used to estimate the range of effects that is expected for all systems. An average parental exposure of 1 rem before conception is expected to produce 5–65 additional disorders per million liveborn offspring.

The estimates arrived at by the two different methods are in good agreement. One is for single-generation effects, and the other is for effects seen at equilibrium, after long-continued exposure of the population. Although no assumptions have been made in this report as to rates of elimination, the use of the estimates of persistence assumed in BEIR I (5 generations for autosomal dominants and 10 generations for irregularly inherited diseases) results in an agreement between the two sets of estimates that is quite good.

Disorders due to chromosomal aberrations, estimated from the aberration incidence seen in a late developmental stage of the germ cells (primary spermatocytes) after exposure of the immature germ cells (stem cell spermatogonia) to radiation, and assuming that the risk for oocytes is of equal size, will amount to fewer than 10 anomalies per million liveborn, and most Subcommittee members felt that the true value may be near zero.

NOTES TO CHAPTER IV

1. HISTORY OF RADIATION STANDARDS

In January 1957, the NCRP recommended that the population dose "not exceed 14 million man-rems per million of population over the period from conception up to age 30 and one-third that amount in each decade thereafter." This was based on the exposure practices and data of that period, and the contributions of the individual sources were estimated in man-rems per million population per 30 yr as:

Natural radiation	4,000,000
Medical irradiation	5,000,000
Occupational exposure	150,000
Radiation in plant environs	450,000
Fallout	200,000
TOTAL	9,800,000

The radiation exposures included medical, natural, and fallout radiation and that from all other man-made sources and allowed a cushion of over 4 million man-rems for future needs.

In April 1958, the concept of population dose of man-made radiation, exclusive of medical exposure, was made more specific in the statement that "the radiation ... shall be such that it is improbable that any individual will receive a dose of more than 0.5 rem in any 1 year from external radiation." It was also recommended, as in 1957, that the average body burden of radionuclides not exceed one-tenth that of radiation workers.

In September 1958, the International Commission on Radiological Protection (ICRP) suggested that "the genetic dose to the whole population from all sources, additional to the natural background, should not exceed 5 rems plus the lowest practicable contribution from medical exposure." Because the genetic dose is calculated for a 30-yr period, this would amount to an average of 170 mrems/yr.

The same value of 170 mrems/yr had been arrived at by a different route based on the 0.5 rem/yr recommended by the NCRP for an individual in the general population. It was reasoned that, to hold the dose to the individual to that level, the average for a population group would have to be approximately one-third of the maximum, or again 170 mrems/yr. On the basis of the published recommendations of the NCRP and ICRP, the population average of 170 mrems was adopted by the Federal Radiation Council in 1960.

The history of radiation protection standards is presented in further detail elsewhere (including staff reports of the Federal Radiation Council in 1960 and 1962).[26,27,45,46,96]

2. EARLY ESTIMATES OF GENETIC RISK

"The 1956 Genetics report relied mainly on data from *Drosophila* and the laboratory mouse, as there were almost no relevant human data. According to the BEAR report, the best one can do is to use the excellent information on such lower forms as fruit flies, the emerging information for mice, the few sparse data we have for man ... and then use the kind of biological judgment which has, after all, been so generally successful in interrelating the properties of forms of life which superficially appear so unlike but which turn out to be so remarkably similar in their basic aspects.

"The general principles that guided the committee at that time were: (1) Mutations, spontaneous or induced, are usually harmful; thus, the harm from an increased mutation rate greatly outweighs any possible benefit. (2) Any dose of radiation, however small, that reaches the reproductive cells entails some genetic risk. (3) The number of mutations produced is proportional to dose, so that linear extrapolation from high-dose data provides a valid estimate of the low-dose effects. (4) The effect is independent of the rate at which the radiation is delivered and of the spacing between exposures. The last of these principles has turned out to be incorrect, as will be discussed later.

"The BEAR Committee estimated that the amount of radiation required to produce a mutation rate equal to that which occurs spontaneously (a doubling dose) was almost surely between 5 R and 150 R and probably between 30 and 80 R. It also assumed that about 2 percent of all live-born children are or will be seriously affected by defects with a simple genetic origin. Under the assumption that for this fraction of human defects the incidence is proportional to the mutation rate, the effect at equilibrium after a continuing exposure to the recommended 10 R limit of radiation per generation was computed. Taking 40 R as a reasonable value for the doubling dose, the BEAR Committee calculated that 10 R per generation continued indefinitely would lead to about 5,000 new instances of tangible inherited defects per million births, with about one-tenth this number in the first generation after radiation begins.

"The BEAR Committee also estimated the total number of mutations which would be produced at all gene loci by 10 R of radiation. The principles listed above made these calculations relatively simple. The number of mutations produced is (the number of genes in the population) $\times$ (the dose) $\times$ (the mutation rate per gene per unit dose). For the last quantity, mouse data were available. But there was no evidence from any mammal as to the number of genes per cell. For this, the Committee used *Drosophila* data, dividing the total mutation rate by that for individual genes. So the estimates of the number of mutations induced were for a hypothetical organism whose mutation rate per gene is that of the mouse and whose gene number is that of *Drosophila*. ..." (BEIR I, pp. 42–43)

"Actually, this calculation does not assume that the number of genes is known, but rather it depends on the ratio of the overall mutation rate to that for a single locus. The ratio of the total lethal rate to that for a single locus was multiplied by 2 to 3 to allow for mutations with less than lethal effects. This led to an estimated ratio of about 10^4, subject to considerable uncertainty both as to accuracy of measurement and reliability of assumptions. The conclusion was reinforced by the fact that the number of bands in the salivary gland chromosomes in *Drosophila* is about 5000. There is recent evidence[5,29,35-37,62,91] that the number of genes (complementation units) in *Drosophila* is indeed equal to the number of salivary chromosome bands, which would be 5000 per gamete, or 10,000 in the diploid cell. The human number is probably larger, but there is no comparably reliable way to estimate it. We shall not use the gene number in any of our risk estimates." (BEIR I, p. 61) The present Subcommittee on Genetic Effects would prefer to substitute "agrees rather well with" for "is indeed equal to" in the third sentence before this one.

"The Committee then used the principle that each harmful mutant gene is eventually eliminated from the population and that this occurs by reduced viability or fertility. Thus, in a statistical sense each new mutant gene, in a population of stable size, must eventually be balanced by a gene extinction. This extinction occurs through prereproductive death or reduced fertility. The BEAR Committee was divided as to the usefulness of this kind of calculation. It was noted that the death of an early embryo is much less traumatic than the death of a child or adult and that the failure to reproduce cannot be equated to premature death in any tangible way. How is a single major defect to be judged in comparison with a number of minor risks?" As stated in the report: "This kind of estimate is not a meaningful one to certain geneticists. Their principal reservation is doubtless a feeling that, hard as it is to estimate numbers of mutants, it is much harder still, at the present state of knowledge, to translate this over into a recognizable statement of harm to individual persons. Also, they recognize that there is a risk involved in extrapolating from mouse and *Drosophila* to the human case. But the group concluded that in spite of all the difficulties and complications and ranges in numerical estimates, the result is nevertheless very sobering.

"Based on these estimates and other considerations which it regarded as germane, the BEAR Genetics Committee made two recommendations that are related to our present purposes:

"That for the present it be accepted as a uniform national standard that x-ray installations (medical and nonmedical), power installations, disposal of radioactive wastes, experimental installations, testing of weapons, and all other human controllable sources of radiation be so restricted that members of our general population shall not receive from such sources an average of more than 10 roentgens, in addition to background, of ionizing radiation as a total accumulated dose to the reproductive cells from conception to age 30.

"The previous recommendation should be reconsidered periodically with the view to keeping the reproductive cell dose at the lowest practicable level. If it is feasible to reduce medical exposures, industrial exposures, or both, the total should be reduced accordingly.

"The present subcommittee concurs with this recommendation for periodic review and it is in this spirit that the present study has been undertaken." (BEIR I, pp. 43-44)

Another way of looking at genetic risk is simply to compare the increased radiation exposure with that due to natural background radiation. BEIR I did this, not as a risk estimate, but as a potentially useful policy guide:

"As mentioned earlier, the natural level of radiation averages about 100 mrem per year. This varies considerably from one region to another, depending especially on the kinds of minerals present in the earth and on the altitude. A person who lives in a stone house may get more radiation than one who lives in a wooden house, because of the greater radioactivity of some rocks, such as granite. Likewise, a person who lives at a high altitude receives more radiation from cosmic rays. Exposure to man-made radiation near the level of background radiation will produce additional effects of a magnitude comparable to what man has experienced from this source throughout his entire history. Furthermore, since man-made radiations are not qualitatively different from natural radiation, they will not produce novel effects. These are particularly firm conclusions because they do not require any quantitative genetic information.

"Another way of stating this is to note that the annual difference in natural radiation between a location in Louisiana and one in Colorado might be [150 mrems or more—see Chapter III]. Even a person who knows this probably doesn't take this difference into account in deciding to change his residence. We can regard man-made radiation levels of this magnitude as comparable to other risks that are often accepted." (BEIR I, p. 52)

3. THE KINETICS OF MUTATION AND CHROMOSOMAL BREAKAGE BY RADIATION

The BEIR I Subcommittee on Genetic Effects stated:

"The genetic material is DNA which contains information in the sequence of its four nucleotides. [Sequences of three nucleotides (triplets) code for amino acids in proteins.] A gene is composed of many hundreds or more of nucleotides in a specific sequence. Not all DNA codes for proteins; probably the great majority has other functions, largely unknown. The DNA itself is organized into larger linear nucleoprotein structures, the chromosomes, found in the nucleus of the cell.

"Any change of a nucleotide such that a given [coding] triplet will now code for a different amino acid constitutes a mutation. Other changes in coding can also have mutagenic consequences. For instance, the addition, or deletion, of a nucleotide from DNA will shift the reading sequence of the code, since it is read 3 nucleotides at a time sequentially. Such frame shift mutants will change whole sequences of amino acids in the protein up to the point where a reverse shift can put the reading back into proper register. Thus, even a change, deletion, or addition of a single nucleotide in DNA can be a mutation.

"In addition, a larger class of mutational events arises from the breakage of the chromosome itself with subsequent deletion or rearrangement of the broken pieces. These changes are often large enough to be seen if the chromosomes are examined under the microscope. Their size distribution, however, forms a continuum from the very small deletion of a single nucleotide to the loss of a whole chromosome. At the bottom of the range, it is impossible to define just where a deletion should be considered a point mutation in the gene rather than a chromosome breakage type of mutation. For most of the chromosome rearrangements considered in this context, with low LET irradiation, the frequency of induced rearrangements is proportional to the dose over the dose range of interest. At higher doses, more complex kinetics are observed." (BEIR I, p. 64)

The present Subcommittee endorses this statement of the authors of BEIR I. We agree that many data for induced genetic damage describe a curvilinear relationship of yield to dose, with the slope increasing with dose over the dose range usually studied. In general, the data points for yield can be very satisfactorily fitted to a quadratic expression of the form,

$$y = \alpha D + \beta D^2 + C. \qquad \text{(IV-1)}$$

where y is yield, D is dose, and C is estimated zero-dose incidence. Disagreement arises, however, over the meaning of the coefficients, α and β; there are even some who harbor doubt that they have any real radiobiologic meaning, at least in the form in which they are determined by simple fitting of a quadratic equation to the experimental data points. The classical radiobiologic view is that these coefficients accurately measure the admixture of one- and two-track events. One modification of the classical theory would ascribe these values directly to the physical nature of radiation absorption, with the measured damage resulting from the interaction of two (or more) lesions, which may come about as an effect of either a single track or two separate tracks. In this view, α and β would vary according to the quality (LET) of the radiation.

The dose-effect relationship, based on physical microdosimetric considerations, can be expressed,

$$\epsilon = \kappa(\zeta D + D^2),\tag{IV-2}$$

where ϵ is effect, ζ is a physical quantity equal to the dose average of the specific energy deposited in the target volume by single ionizing events, and κ is a "sensitivity" coefficient. If the spontaneous rate is taken into account, Equation IV-2 reduces to Equation IV-1 if $\alpha = \kappa\zeta$ and $\beta = \kappa$. The virtue of either form of this classical formulation, as seen by its advocates, is that good data will yield good values of these coefficients, which, when accurately determined, will lead to precise estimation of the effects that would be produced at very low doses and very low dose rates.

Users of these equations verbally invoke an overlapping of lesion induction and lesion repair to account for the dose-rate effect, but do not include these formally in the calculations; time is not included as a variable. However, classical radiobiologic theory does include this in a seldom-used correction factor for two track events,[32]

$$G = 2(\tau/T)^2(T/\tau - 1 + e^{-T/\tau}),\tag{IV-3}$$

where G is a correction factor for yield of two-track events, τ is the average elapsed time between breakage and restitution (i.e., lesion induction and lesion repair), and T is duration of treatment. From the equation from which Equation IV-3 is derived, it is seen that the relation to yield for two-track events is $y \propto D^2G$.

The maximal value approached by G is unity, when T approaches zero. In the range where T and τ are approximately equal, the value of G is reduced to 0.736, which amounts to about a one-fourth reduction in yield below simple, two-track expectations. Although this correction factor is usually invoked only in relation to the use of the dose-rate effect to estimate the mean longevity of lesions, it is obvious that it can also result in different errors for each dose point in dose-response curves, where total dose is varied by varying time, rather than by varying the dose rate. It is important to note that *this correction factor is not dose-dependent.*

An alternative interpretation is that the end points in question—e.g., specific locus mutation—may depend on the operation of more than one mechanism. That is, there may be more than one biologic mechanism involved in addition to the presumed "dual-action" mechanism of physical absorption. There may be more than one class of event involved in "point mutation," as pointed out in the BEIR I paragraphs cited above. Furthermore, the end point, "mutation," may result from the operation both of repair and of damage mechanisms and may involve a variety of lesions. From this standpoint, it might be argued that, for reasons of prudent conservatism, the best estimate of damage at very low doses would be a linear extrapolation between the yield at the lowest dose point for which reliable data exist and the incidence at zero dose. Such an estimate would not differ appreciably from that based on the quadratic relationship, provided that the value of βD^2 at the lowest measured dose point is not appreciably different from zero.

There is yet another viewpoint, perhaps more pertinent to the kinetics of induction of two-break rearrangements than to gene mutation, but not strictly limited to the two-break rearrangements: the observed rates of damage may not reflect the rates of *induced* damage in any simple way, because of the nature of the screening process by which the end points were detected. In consequence, it can be argued that the values of α and β lack real biologic meaning; that is, they neither describe the real mechanisms of damage nor serve as useful indicators of the low-level effects that are to be expected. Statistical and sampling complexities are not properly taken into account by a direct fitting of data to a simple quadratic expression. As a result, the values of α and β obtained may differ markedly from their true values. Furthermore, because the estimations of α and β are not independent (the total number of observations predicted by the equation must equal the total number of observations made), the overestimation of one is accompanied by a compensatory underestimation of the other; this leads to an even greater error when it is their quotient, α/β, that is considered. Advocates of this point of view would also agree that it is prudent to use simple extrapolation between the lowest dose point for which good data exist and the zero-dose yield, when low-dose estimates are made.

Each of the views presented here may well contain some elements of truth. However, inasmuch as there is hardly any difference between them in the extent of effects to be expected at low doses, the Subcommittee feels no pressing need to adjudicate a difference of opinion that would be better resolved in the laboratory. Thus, we adhere to the words and methods of BEIR I in this matter.

---

4. THE HIGH FREQUENCY AND HETEROZYGOUS EXPRESSION OF MINOR MUTATIONS

"It has been known for many years that minor deleterious mutations in *Drosophila* are more numerous than those that produce a lethal or near-lethal effect. The first accurate quantitative assessment of the mutation rate of such minor genes was by Mukai,[43] who used the device of letting mutations accumulate on a chromosome that was protected from the effect of natural selection by being kept heterozygous generation after generation with careful precautions to minimize natural selection. From the mean and variance of the decline in viability when such chromosomes were later made homozygous, he inferred that the mutation rate is at least 15 times the lethal mutation rate. These results have ... been confirmed in three independent experiments.[44] [More recently, Ohnishi[55] found an increase of 12 times.] Further confirming evidence comes from microorganisms showing that mutations resulting from substituting one amino acid for another (missense mutations) are very much underrepresented relative to chain-terminating (nonsense) mutations among conditional lethals.[31,109] Presumably, the former are producing effects too small to be detected by the system employed.

"Although these mutants are found in very high frequency in natural populations of *Drosophila*, they are not as frequent as they would be if they were completely recessive. This means that they must be eliminated from the population through heterozygous effects[44,97] [and it appears from *Drosophila* data that, the smaller the effect of a mutant, the more nearly additive is its influence on viability, so lethals and mild detrimentals are likely to persist in the population for about the same period (30–50 generations[17,92])]. The high frequency of these mutants and their degree of heterozygous expression is such that they should have appreciable effects on the viability or fertility of the population. An increased mutation rate would, therefore, be expected to cause a general, non-specific reduction in the fitness of the individuals in the population through the production of such mutants.

"A mitigating factor is that these individually minor mutants are less frequent, relative to severe mutants, among radiation-induced than among spontaneous mutations.[21] [Another mitigating factor is the possibility that the elimination of such mutants will be to a large extent through fertility differences, rather than by differential viability.[42,92]] Radiation is known to produce genetic changes at all levels—single base replacements, insertions and deletions of nucleotides, changes involving several bases, and on up to gross chromosome rearrangements.[18] However, the ratio of deletions and chromosome rearrangements to single base effects is likely to be much higher for radiation-induced than for spontaneous changes." (BEIR I, pp. 63–64)

---

5. GENETIC HAZARDS OF PLUTONIUM

Autoradiographic studies have established that [^{239}Pu]plutonium citrate injected intravenously into mice is deposited quite nonrandomly in their testes;[24] this results in an alpha-particle dose to the peritubular spaces and tissue near the outsides of the seminiferous tubules that is about 2 or 2.5 times the average dose to the whole testicular mass. Because the stem cell spermatogonia are near the periphery of the tubules, the dose to them is larger than would be calculated simply from the testicular plutonium content per gram. Whether this is true for other isotopic or chemical species is unknown, and there is no information on the distribution of plutonium deposited in the mammalian ovary. The EPA, in a report in preparation on plutonium hazards, makes the reasonable assumption that the genetically significant doses resulting from a given blood plutonium content will be essentially the same in males and females, because the larger dose to spermatogonia will be essentially offset by the smaller per-gram plutonium content in testis than in ovary.

Although there is a good deal of information on genetic effects of other high-LET radiation, mainly fast neutrons (protons), experimental data that will allow the determination of plutonium alpha-particle RBE values for genetic effects in mammals are very sparse. Luning *et al.*[38] have reported experiments on dominant lethal induction in male mice that received intravenous injections of [^{239}Pu]plutonium citrate, but the information presented does not allow the calculation of an RBE value. (Interestingly, injected [^{239}Pu]plutonium nitrate appeared ineffective in the same experiments.) Searle *et al.*[83] have reported results for several genetic end points—including dominant lethals, reciprocal translocations, and chromosomal fragments—in male mice that received intravenous injections of [^{239}Pu]plutonium citrate or were subjected to chronic gamma irradiation. RBE values of 22–24 were calculated. Allowance for the nonuniform distribution of plutonium alpha-particle dose across the seminiferous tubules was deemed inappropriate, because the effects were induced in cells of various types, rather than in spermatogonial stem cells. Grahn (personal communication), who is also determining dominant-lethal and translocation frequencies in meiotic and postmeiotic stages in male mice that were given injections of [^{239}Pu]plutonium citrate, reported RBEs, compared with chronic gamma radiation, of 13 for dominant lethals and 40–50 for translocations. These values are about half and twice the respective values observed by Searle *et al.*[83] Preliminary results reported by Russell *et al.* (Russell *et al.*[78] and personal communication) on specific-locus mutations induced in mouse spermatogonia after injection of [^{239}Pu]plutonium citrate indicate an RBE of only 4, compared with chronic gamma irradiation. This is considerably lower than the RBE obtained with fission neutrons.

---

6. TRANSMUTATION EFFECTS

Three radioactive isotopes—hydrogen-3 (tritium), carbon-14, and phosphorus-32—may be incorporated directly into the DNA of organisms encountering them in the environment. When such incorporated atoms decay, the resulting change in atomic number, recoil, or excitation—often collectively referred to as transmutation—may give rise to biologic effects, including mutation, beyond those induced by the attendant ionizing radiation. In consequence, concern has arisen that the genetic hazard presented by these radionuclides might be seriously underestimated. The problem is compounded by the fact that all three isotopes decay by emission of a low-energy beta particle that, especially in the case of hydrogen-3, limits energy deposition to the vicinity of the decaying atom and greatly complicates the design of experiments to detect any effects of transmutation.

BEIR I concluded (in its Note 7) that the genetic effects of decays of hydrogen-3, carbon-14, and phosphorus-32 can in fact be attributed almost entirely to their beta radiation and that the contribution from transmutation is so small in comparison that it is "justified to consider the main effect to come from the radiation emitted when the isotope disintegrates." However, when BEIR I was being prepared there was evidence of a transmutation effect on mutation caused by decay of tritium in only one specific site in DNA (the number 5 ring position of cytosine) and some suggestive evidence of a slight transmutation effect of incorporated carbon-14 in *Drosophila*. A good deal of evidence has since been accumulated, and it seems appropriate to reevaluate the question of genetic effects of transmutation.

The evidence has been extensively documented[4] and will be only briefly summarized here.

Tritium has now been demonstrated to produce measurable transmutation effects in microorganisms when in the 6 ring position of thymidine, the 2 position of adenine,[61] and the 5 position of cytosine. The last one of the three has a much higher efficiency of transmutation than the other two. Tests for transmutation effects on mutation[61] for decay of tritium in the other stable DNA base positions—the methyl group of thymidine, the 6 ring position of cytosine, the 8 position of adenine, and the 8 position of guanosine—are negative, or nearly so. The three "sensitive" positions together constitute only about 6% of all DNA hydrogen and only about 0.1% of all nuclear hydrogen. Furthermore, data are now becoming available from mouse experiments on both dominant lethal and specific-locus mutations that suggest that any contribution of transmutation is too small to be detected when the tritium is randomly incorporated (see Carsten and Commerford;[11] Carsten and Commerford, personal communication; and W. L. Russell, personal communication). Thus, in spite of the demonstration of new positions in DNA at which tritium transmutation effects can occur, it still seems unlikely that they contribute importantly to mutation.

Earlier experiments had suggested that significant transmutation effects might be associated with the decay of carbon-14 incorporated into DNA in *Drosophila*, but more recent mutation experiments with substantial carbon-14 incorporation in this organism have failed to detect any mutations not attributable to the beta-particle dose alone.[34] It thus still seems unlikely that the genetic hazards from the decay of carbon-14 are significantly underestimated by considering only the ionizing-radiation dose accumulated by germ-line cells.

Mutation experiments in *Drosophila* have clearly demonstrated that the transmutation of phosphorus-32 incorporated into DNA does cause sex-linked lethal mutations in addition to those caused by the attendant beta-particle dose; interestingly, however, they are detected only in the F_3, suggesting that the F_2 flies are mosaics for the transmutation-induced mutations.[33] The efficiency is very low, and the yield of mutations from transmutation is thus very small, in comparison with that from the ionizing-radiation dose. It therefore still appears justifiable to ignore the small contributions of transmutation to genetic hazards associated with phosphorus-32.

---

7. THE LINEARITY, NO-THRESHOLD ASSUMPTION

"There is strong evidence that, for single locus mutations in *Drosophila*, the dose-response relationship is linear down to the lowest doses that have been adequately tested. There is no evidence for any threshold. If there is none, then the curve, when extrapolated to lower doses, should intersect the zero-dose ordinate at a value equal to the spontaneous rate. The observations are compatible with this, but the statistical error is too large for this expectation to be tested with any rigor.

"As mentioned in [Note 3], another reason to expect a linear relationship is that for very low doses there is very little opportunity for ionizations from independent ion tracks to occur in the same cell locality. Any effect following exponential kinetics with an exponent larger than one is bound to disappear at sufficiently low doses." (BEIR I, p. 64)

"In the mouse, two opposite types of departure from linearity have been found for acute irradiation of spermatogonia. One of these has been explained by differential cell killing, and the other by repair of premutational damage.

"The first departure consists of an upward convexity of the dose-effect curve at high doses: an x-ray dose of 1000 R actually produced fewer mutations than did a dose of 600 R.[71,79] Russell's hypothesis to account for this result is that in the heterogeneous population of spermatogonial cells some cells are more sensitive to both killing and mutation. Thus, at high doses, the sensitive cells are destroyed, leaving only those cell types that produce fewer mutations. If this effect were to extend down to lower dose levels, then the mutation rate at these levels would be higher than predicted from a linear interpolation between 600 R and 0 R. However, at 300 R, no significant departure from linearity was observed. Recent work by Oakberg[53] indicates that the true stem cells in the mouse testis are not as easily killed by radiation as are the rest of the spermatogonia, and that differential killing among these stem cells is not, in fact, likely to have any humping effect on the dose curve in the range below 500 R. Furthermore, mutation-rate studies in the low dose range indicate that if there is any tendency toward such a humping it is more than counterbalanced by the opposite departure from linearity, to be described below." (BEIR I, p. 65)

"The second type of departure from linearity observed in the mouse consists of an upward concavity of the dose-effect curve at low doses.[75] This non-linear relation for mutations that seem to be mainly the result of single-track ionization events[2,64,74,75] is explained on the hypothesis that there is repair of mutational or premutational damage, but that the repair process is either damaged or saturated at high doses and high dose rates. This hypothesis, which was originally derived from the discovery of a dose-rate effect in mouse spermatogonia and oocytes,[76,81] predicts that repair could operate even at high dose rates, provided that the total dose were small or given in small fractions at intervals long enough for the repair process to recover. As shown above, this prediction was met for small total doses. It has also proved true for fractionation. [As pointed out in Note 3, it is possible for events scored as point mutations to arise from either one- or two-track events. At low dose rates, two-track products are expected to decline, owing to a decreased likelihood of the simultaneous presence of two potentially interacting lesions. Because this model gives virtually the same hazard estimation as does the repair model, the Subcommittee leaves the evaluation of the models to those who would test them in the laboratory.]

"The finding of a dose-rate effect for mutation induction in mouse spermatogonia and oocytes raised anew the question of whether there might be a threshold dose or dose rate below which all mutational damage would be repaired. Exploration of a range of dose rates provides no evidence of a threshold dose rate for mutation induction in mouse spermatogonia.[72,75,77] Mutation frequency drops as the dose rate is lowered from 90 R/min through 9 R/min to 0.8 R/min; but below that level, to 0.009 R/min and even 0.001 R/min [and now 0.0006 R/min], there is no further reduction in mutation frequency. Therefore, we shall make the prudent assumptions that there is no threshold dose rate in the male and that the dose response at low dose rates is linear." (BEIR I, p. 65)

8. ESTIMATION OF THE RELATIVE MUTATION RISK

The rate of induced mutation is measured as an absolute rate, but its significance is best seen when it is related to the rate of spontaneous mutation. This relationship is commonly given as a "doubling dose," the amount of absorbed radiation that would be required to produce as many mutations as are already occurring spontaneously—i.e., to double the existing mutation rate. This quantity is expressed in rems and in the simplest case is determined by dividing the spontaneous-mutation rate by the induced-mutation rate per rem. Actually, we use its reciprocal, which BEIR I called the "relative mutation risk." This is the quotient that results from dividing the induced-mutation rate per rem by the spontaneous-mutation rate, and it gives the risk of induced mutation per rem, expressed as a fraction of the current risk of spontaneous mutation.

It is the average mutation rate for the two sexes that is used in the calculation of the relative mutation risk. For much of the reproductive cycle of the male, the germ cells are present as stem cell spermatogonia; mutation-rate data for use in risk estimates are taken from this stage in mice. The stage of longest duration in the female is the oocyte, which is formed in the ovary during fetal development and remains without further division until about ready for ovulation. The induced-mutation rate for chronic exposure in female mice is either a very small fraction or some fraction up to about 40% of the rate in spermatogonia, depending on the stage of oocyte development being treated (see Note 9). The average mutation rate that will be used in calculating relative mutation risk is thus from 0.5 to about 0.7 times the male rate. The average rate is expressed in this manner, rather than as an absolute per-rem rate, because it must be used with mutation-rate data based on different sets of loci.

A precise calculation of the relative mutation risk would require that measurements of rates of spontaneous and induced mutation be based on the same loci in the same species. Ideally, the loci chosen would be reasonably representative of the entire genome, and the human would be the species of choice for study. Because there are no per-locus induced-mutation rates for humans, BEIR I used the mouse induced-mutation rate measured for some 12 specific loci and the human spontaneous-mutation rate estimated from population surveys.

BEIR I took the average induced-mutation rate for the two sexes to be half the male rate and divided this number (0.25×10^{-7} per locus per rem) by the order-of-magnitude range of the estimated human spontaneous-mutation rate (0.5×10^{-6} to 0.5×10^{-5}) to obtain a relative mutation risk ranging from 0.005 to 0.05 per rem (corresponding to a doubling dose of 200–20 rems). If our factors are substituted for the average induced-mutation rate, this range is extended to become 0.005–0.063 per rem (or a doubling dose of 200–16 rems).

Some believe the BEIR I method to be flawed, in that the degrees of diligence exercised in choosing each of the sets of data that are used in the comparison are not the same. Specifically, which human loci will supply the data for the spontaneous-mutation rate, and are these loci comparable with the mouse loci used for determining the induced-mutation rate? Cavalli-Sforza and Bodmer,[13] the source of the spontaneous-mutation-rate estimates used by BEIR I, pointed out that, "if we want to study mutation rates and if we choose for this purpose to observe loci at which we already know mutations have occurred, we will be working with a biased sample." This, of necessity, is the case with the loci chosen for study in mice, and these authors so stated: "The average spontaneous mutation rate obtained for the seven loci was 8.4×10^{-6} in males and 1.4×10^{-6} in females.[76] The average for the two sexes is 4.9×10^{-6}, which must be biased upwards, for the reasons already discussed. . . ."

Cavalli-Sforza and Bodmer made the point that spontaneous-mutation rates are quite similar in a variety of animals, including humans, and the order of magnitude is 10^{-6} "or less." The major source of information that they used for human rates was a population survey giving an estimated mean rate for spontaneous X-linked mutations of about 4×10^{-6}, but with an estimated median rate about an order of magnitude lower. The BEIR I choice of a range of 0.5×10^{-6} to 0.5×10^{-5} conformed closely to the median and mean rates estimated from this survey.

Thus, the difficulty that some have in accepting the BEIR I estimate is that they have chosen to compare induced-mutation rates for loci that, from the standpoint of spontaneous mutability, are made up of "a biased sample" with spontaneous-mutation rates that include loci with much more typical spontaneous-mutation rates. It would not seem imprudent to choose, for comparison, loci that have similar spontaneous mutabilities. This difficulty can be avoided in either of two ways: by using the more biased mean human spontaneous-mutation rate, or by using the mouse data for spontaneous mutation at the same loci that are used to measure the induced-mutation rate. Exercise of either option would give quite similar results.

The estimates of the relative mutation risk that follow are based on specific-locus mutations in the mouse, both spontaneous and induced. We use data from fewer loci than did BEIR I: for BEIR I, it was necessary to base the mouse induced-mutation rate on as many loci as possible, to minimize the effects of large differences in mutability of different loci. To compare mutations at the same loci requires good data on rates of both induced and spontaneous mutation. For this reason, our consideration is limited to data from the seven most commonly used loci.

The point estimate of the spontaneous-mutation frequency in the male is 7.5×10^{-6} mutation per locus, and in the female it is either 2.1×10^{-6} or 5.6×10^{-6} mutation per locus, depending on how a cluster of mutations is dealt with in the calculation. Thus, the average for the two sexes is either 4.8×10^{-6} or 6.6×10^{-6} mutation per locus per generation.

The induced-mutation rate in mouse spermatogonia irradiated at 0.009 rem/min and below is 6.6×10^{-8} mutation per locus per rem. As discussed above, the average induced-mutation frequency for the two sexes depends on the oocyte stage in the mouse that is chosen as being more comparable with resting oocytes in humans and will be either 0.5 or 0.7 times the spermatogonial rate. Multiplying by these factors and dividing by the spontaneous-mutation rates yields an array of values of the relative mutation risk ranging from a high of 0.01 to a low of 0.005 per rem (corresponding to doubling doses of 100 and 200 rems).

The maximal relative mutation risk has also been estimated from data on off-spring of survivors of the Hiroshima and Nagasaki exposures.[50] For the types of genetic damage resulting in death during the first 17 yr after live birth, the max-imal value is not greater than 0.00725 (minimal doubling dose, 138 rads) for males and not greater than 0.001 (minimal doubling dose, 1,000 rads) for females. This gives an average value of 0.00412 (minimal doubling dose, about 240 rads) for the two sexes. These data, collected on humans, suggest that the experimentally derived range of values of the relative mutation risk may overestimate that risk. However, we feel that it is better to use the more cautious approach and adopt a range that takes experimental animal data into account.

We have adopted for our calculations a range for the relative mutation risk of 0.02–0.004 per rem (doubling dose, 50–250 rems). This is based mainly on our best substantiated estimate of the doubling dose—namely, 114 R for mouse sper-matogonia. (For x and gamma radiation, the roentgen, R, and the rem are virtu-ally equal.) We approximately halve and double this to get our range of 50–250 R, which we believe overlaps the true value. Further reason for thinking that this range is broad enough comes from the estimates of 100–200 R obtained when data from both sexes are combined. The approach used by BEIR I yields values of 16–200 rems. The value of 16 rems, as discussed above, seems unreasonably low. If the BEIR I approach is used, with the modification suggested above (using the mean human spontaneous-mutation rate), the estimated doubling dose is 200 rems. The few human data suggest that humans are not notably more sensitive, and are prob-ably less sensitive, than mice.

9. OOCYTE SENSITIVITY

"The reproductive cells of the female [mouse], for most of their lifetime, are very much less mutable than those in the male, even from acute irradiation. Further-more, the germ cell stages in the female that have a high mutational sensitivity to acute irradiation, namely, the mature oocytes, give a very low mutation rate with chronic irradiation." (BEIR I, p. 52)

"Mature oocytes in the mouse are relatively susceptible to radiation effects. The rate of production of point mutations is about 5×10^{-7} per locus per rem with acute radiation. However, there is a reduction to about 1/20 of this amount for chronic radiation. The stages prior to the mature oocyte are very resistant to muta-tion; hardly any mutations are produced. In the mouse the duration of the mature oocyte is about 7 weeks. It is reasonable to assume that in humans the stage of sen-sitivity is short relative to the total pre-reproductive life cycle, as it is in the mouse, but there is no direct evidence for this." (BEIR I, pp. 65–66)

BEIR I concluded that the data from irradiated female mice showed mutation at low doses and low dose rates to be so much less frequent than that in the spermatogonia that the average mutation rate for the two sexes effectively was half the male rate. This value was then used in its estimation of human risk. The present Subcommittee has reexamined the BEIR I conclusion in the light of new data and some published contrasting reassessments.[1,73]

The female germ cell stage of primary importance in radiation genetic hazards is the immature, arrested oocyte. In mice, this stage has zero or near-zero sensitivity to mutation induction by radiation.[69,77] However, two major questions must be considered in the application of these mouse results to women. One arises from a possible relationship between sensitivity to cell-killing and mutation induction, the other from differences in nuclear morphology.

Although immature oocytes of adult mice are resistant to mutation induction, they are highly sensitive to killing by radiation, whereas immature oocytes in adult humans are resistant to killing. In mice, maturing oocytes are resistant to killing, but show high mutability, at least with sufficiently high doses of acutely delivered radiation. If this is taken as evidence of a consistent inverse relationship, or negative correlation, between oocyte-killing and mutational response, then it can be argued that the resistance of human oocytes to direct killing implies a greater sensitivity to mutation induction.

In connection with this first question, however, there are two new pieces of evidence that fail to support such a negative correlation. It has been shown that the fully mature oocytes of mice are less sensitive to both killing and mutation induction by radiation than the slightly less mature stages.[73] Thus, here there is a positive correlation, rather than a negative one, between killing and mutational sensitivity. The author concluded: "It appears, from the lack of consistent correlation, that mutation induction and killing are independent events." For another type of genetic damage, an inverse relationship between cell-killing and mutation was denied by a recent study in which guinea pigs showed less killing of immature than of mature oocytes, and golden hamsters showed the reverse, whereas in both species lower amounts of dominant-lethal genetic damage occurred in immature than in mature oocytes.[16] On the basis of this finding, the authors stated: "Thus no general pattern has emerged from this work of correlation, either positive or negative, in the sensitivities of oocytes to killing and to dominant lethal induction."

The second question with regard to using the mutational insensitivity of immature, arrested oocytes of mice in risk estimation for arrested oocytes of women arises from the fact that the nuclear morphology of this stage in mice, the so-called dictyate, is not like the typical diplotene found in humans and many other species. Again, new evidence apparently diminishes this objection. Recent oocyte-maturation timing studies[54] showed that the shift in mutational sensitivity from low to high (at about 6 wk before ovulation) appears to coincide with the beginnings of zona pellucida formation, thereby confirming an earlier, independent report[4] of a change in oocyte nuclear morphology from the dictyate to a rather typical diplotene before this time. Furthermore, "resting" oocytes from stage 2 until about the time of zona pellucida formation apparently retain the low sensitivity to mutation induction found in resting oocytes in the dictyate stage (stage 1). Thus, there are oocytes in mice that show exceedingly low sensitivity to mutation induction and whose nuclear morphology is similar to that of arrested human oocytes.

It is concluded that there is less reason now than there was at the time of BEIR I for considering differences in sensitivity to cell-killing and nuclear morphology as grounds for preventing the use of the mutational response of arrested mouse oocytes as a guide for risk estimation in humans. The problem is not, however, fully solved, and the Subcommittee has reexamined the results of another approach used by experimenters—the determination of mutation rates in other mouse oocyte stages, particularly the maturing and mature ones, which are resistant to killing by radiation.

It was first determined many years ago that, although these stages are mutationally sensitive to high doses of acute irradiation, they have low mutational response to low-dose-rate irradiation.[80] It was recognized at that time that one of the difficulties in measuring the effect of low-dose-rate irradiation on mutation frequency in maturing oocytes is that the duration of radiation exposure necessary to accumulate a sizable dose may approach the duration of the oocyte stage under measurement. This was circumvented in the publication cited above by showing that a large dose-rate effect persisted when comparison was restricted to conceptions that occurred within 2 wk after the 3 wk required for accumulation of the dose. A few years later, it was discovered that the mutation rate resulting from acute irradiation decreases sharply to zero or near-zero in ovulations that take place 7 wk or more after exposure;[77] this afforded a firmer basis for determining the interval over which data from chronic irradiation could be collected for comparison with the results from acute irradiation. This was done in the computations used to arrive at the figure of one-twentieth, which, as quoted at the beginning of this note, was accepted in BEIR I as the ratio of effects of chronic to acute irradiation in maturing oocytes.

The "effective dose"—i.e., the portion of the dose of chronic irradiation received when the oocytes are in mutationally sensitive maturing stages—was later computed in a different way in a theoretical paper[1] that concluded that the dose-rate effect was less than had been estimated earlier; in other words, the mutation frequency from chronic irradiation of maturing oocytes was greater than had been calculated in BEIR I. The calculation of effective doses made use of follicle kinetics derived from data obtained from the labeling of granulosa cells with tritiated thymidine.[60] This method depends on an estimated doubling time for granulosa cells, which is calculated by using the labeling index together with the lengths of the S-phase and the G_2-phase plus half the mitotic phase. The latter quantities were estimated from the "percent labeled mitoses" curve. The transit time for each type of follicle could then be estimated by taking the doubling time in conjunction with the minimum and maximum of granulosa cells for that particular follicle type.

More recent timing studies have been based on the labeling of the zona pellucida;[16,54] these have shown a longer interval between the beginning and the completion of follicle development, with the period between the appearance of the zona pellucida and ovulation being about 6 wk.

There is no disagreement that effective doses should be calculated on the basis of the duration of the more sensitive stages of oocyte development. In the computation made for BEIR I and in a recent reevaluation[73] of all the chronic-irradiation results (including new data) on maturing mouse oocytes, the approach used depended not on either set of timing studies, but on the actual pattern of mutation recovery from acutely irradiated females. The data reaffirmed that the oocytes that are highly sensitive to mutation induction are not exhausted until 6 wk after irradiation. Hence, the estimated effective dose is based on the dose received during the 6 wk before ovulation.

The reality of a pronounced dose-rate effect on maturing oocytes arrived at in this evaluation was given further support by the reduction in mutation yield that was found when the total dose was fractionated.[39] High-dose-rate administration of a total dose of 200 rads in 20 fractions of 10 rads each, over a period of either 4 wk or 5 d, led to a much lower yield of mutation than administration of a single 200-rad exposure. Because of the manner in which the single, acute treatments were given, germ cell attrition can be ruled out as a contributing factor in the lower yields that followed fractionated treatment. This argument carries special weight in the series in which the fractionated treatment was given over a 5-d period.

There is disagreement over whether the dose-rate effect is due to two-track mutational events or mainly to single-track events combined with damage or saturation of the repair process at high doses and dose rates. The Subcommittee finds it unnecessary to discuss either view, because both agree that the results at low doses and dose rates are best fitted by the simple linear equation $y = C + \alpha D$, where y is mutation frequency per locus, C is control rate, α is induced-mutation frequency per rem, and D is dose.

In the latest reevaluation of data on maturing mouse oocytes,[73] weighted least-square regression lines were fitted to all the available low-dose-rate results and to the fractionation data described above, making use of the calculated "effective doses." Four values of α were obtained; they ranged from 0.113×10^{-7} to 0.296×10^{-7}, depending on which data and which control rate were used. It is noteworthy that only the highest of these four values is significantly greater than the control rate. The advantage of using these more recently estimated values is that they were based on effective doses determined from the actual pattern of mutation recovery, rather than on expectations as to the rate of follicle development and ovulation derived from the timing of oogenesis.

In summary, there seems to be more justification now than at the time of BEIR I for using data on immature, arrested mouse oocytes to estimate the risk to immature, arrested human ooctyes. If, on the side of caution, one continues to consider the possibility that immature, arrested human oocytes might be mutationally as sensitive as the most sensitive of all oocyte stages in mice—maturing oocytes—then the values given here can be used. These translate into estimated mutation frequencies of 0.17–0.44 times that in spermatogonia, but again it should be remembered that in three of the four estimates the frequencies are not significantly above control values.

The estimate of relative mutation risk discussed in Note 8 is given as a range of values that takes into account the degrees of uncertainty that have been encountered in our efforts to make use of the data on female mice.

10. RELATIVE-RISK CALCULATION FOR AUTOSOMAL TRAITS

If the "mutational component" of a deleterious trait is near unity—i.e., if there is strong reason to believe that the incidence of the trait is maintained in the population exclusively by recurrent mutation—then an increase in exposure of 1 rem per generation will lead to an equilibrium increase equal in amount to the calculated relative mutation risk. The value range that we have chosen, 0.004–0.02 (corresponding to a doubling-dose range of 250–50 rems), when multiplied by the current incidence of autosomal dominant traits (approximately 10,000 per million liveborn), yields the range of values found in the column of Table IV-2 that shows equilibrium expectations, 40–200 per million liveborn. That is, the incidence after a number of generations will have increased from an initial 10,000 per million liveborn (the incidence without any added radiation) to 10,040–10,200 per million liveborn.

The number of generations required to reach equilibrium will depend on the rate of elimination of these added mutants from the population. If, for autosomal dominants, we were to take the mean persistence to be about 5 generations (as was done in BEIR I), there would be about a 20% probability that the mutant would be eliminated in any given generation. Equilibrium would be reached when the rate of elimination was exactly balanced by the rate of addition of new mutants to the population. For all practical purposes, this would be achieved in some 10–20 generations in the example chosen. If the persistence is 5 generations, then the amount of first-generation expression would be one-fifth of the equilibrium expression; if it were 10 generations, the first-generation expression would be one-tenth of the equilibrium expression, etc. If we were to use the BEIR I method of estimating first-generation expression from equilibrium estimates, then a mean persistence of 5 generations would imply first-generation expression in the range of (40/5) to (200/5), or 8–40 per million liveborn per rem of parental exposure. These, of course, would be in addition to the 10,000 per million that would be expected in the absence of added radiation exposure.

For the more complex situation involving irregularly inherited diseases, we must also introduce a factor for the mutational component that is somewhat less than unity. We have rather arbitrarily chosen the range of 5–50% for this value. If we multiply the current incidence by these factors, as well as the relative risk factors, we arrive at the range of values listed in Table IV-2. The maximum is obtained by multiplying, $90,000 \times 0.02 \times 0.5 = 900$; the minimum, by multiplying, $90,000 \times 0.004 \times 0.05 = 18$. These mutants would be expected to persist for longer periods than would the simple, autosomal dominants. BEIR I assumed a mean persistence of 10 generations, which would lead to an expectation of a first-generation expression of about one-tenth the equilibrium expression.

Thus, at equilibrium many generations later, the incidence of diseases due to gene mutation would have increased from an initial approximately 107,100 per million liveborn to around 107,160–108,200 per million liveborn, if there were an average exposure of the general population amounting to 1 rem per person per 30-yr generation *in each intervening generation.*

11. COMPARISON OF EQUILIBRIUM INCIDENCE AND FIRST-GENERATION EXPRESSION OF X-LINKED DISEASES

We assume that X-linked mutations have an average fitness of roughly half the normal. (Both X-linked and autosomal recessive mutations, on the average, cause a greater reduction in viability than do dominants.) At equilibrium, the incidence of affected people, almost all of whom would be males, would be approximately 3 times the mutation rate. The equilibrium gene frequency would be $(3u)/s$, where u is the mutation rate and s is the selection coefficient, in this case 0.5. Thus, the affected proportion of males would be $6u$. Because almost all those affected are males, the affected proportion among both sexes would be $3u$. The incidence of persons affected by a new mutation is half the mutation rate (in this case, the mutation rate in females, inasmuch as an affected male gets the mutant gene from his mother.) Thus, we would expect the number of persons affected in the first generation to be about one-sixth of the number affected at equilibrium. If the female mutation rate were less than the male rate in humans, as it is in mice, the expected number affected by new mutations would be less than one-sixth of the equilibrium number.

Because the current incidence of such diseases is 400 per million liveborn, the incidence at genetic equilibrium, after an exposure increase of 1 rem per generation, is estimated to be increased by 1.6–8 per million (i.e., from 400×0.004 to 400×0.02 per million). Thus, under these exposure conditions, the number of induced serious genetic diseases of this type in the first generation should be less than 1.3 per million (i.e., 8 divided by >6).

12. EMPIRICAL STUDIES ON MOUSE POPULATIONS

"Although the simplest approach to assessing radiation risks would seem to be direct observation of harmful changes in offspring and later descendants of irradiated mammals, such studies are generally believed to reveal only part of the total genetic damage. Recessive lethal changes in particular tend to escape detection unless special stocks and special breeding systems are employed, and the same may be said of recessive detrimental changes and mutations associated with small dominant effects. Nevertheless, induced hereditary changes leading to skeletal anomalies,[19,88] loss of learning ability, and changes in such quantitative characteristics as body weight,[51,52] have been detected by this method.

"Where the irradiations have been repeated over many generations, such mammalian studies have posed a curious problem. If, as is generally believed, most induced mutations have slight deleterious effects in the heterozygous state, the continued accumulation of such change without apparent eliminations through deaths and failures to reproduce would be expected to cause eventually some obvious and substantial effects on the members of the population. This has not yet happened in any of the large-scale studies.

"Results obtained by Spalding and his co-workers[93] are of special relevance in that the exposures, in this case 200 rems per generation to the male line, were continued over a total of 45 generations. It was reasoned that, if mutations with individually small effects do, in fact, occur with much greater frequency than mutations with major effects, and can accumulate to constitute a damaging genetic load, the presumed effects would eventually be reflected in measurable alterations of the growth and death rates. The experiment was carried out with a highly inbred strain of mice to minimize initial chance differences in the irradiated and unirradiated lines. There were no significant differences between the irradiated and control strains in growth rate or in mortality; the lifetime survival curves are almost identical in the two groups. Other such studies of mammals have shown changes in growth rates, but not in any consistent direction.

"As summarized by Green,[25] these negative results may be due to the non-existence of induced mutations having only moderate individual effects on heterozygotes, to the failure to find the right indicator trait, or to the relatively small sizes of the experiments so far conducted and their relative lack of power for discriminating small genetic differences in the presence of large amounts of nongenetic variability." (BEIR I, pp. 61–62)

It is worth noting that, if mice are similar to *Drosophila* in that most mutations have small selection coefficients, the effects exerted by the few induced mutations per mouse that would have accumulated in these experiments may have easily been obscured by this nongenetic variability.

A recent experiment,[88] discussed in detail in this report, has shown that there is a fairly high frequency of induced dominant mutations that cause extensive skeletal anomalies. Most of these mutations can be maintained easily for many generations in the laboratory. Thus, there is no reason to assume that such mutations would not have accumulated in the multigeneration experiments. It now seems certain that the negative or equivocal results of the multigeneration experiments occurred because the traits studied had lower heritability or were associated with greater background noise than the traits studied in the skeleton experiments.

13. DIRECT ESTIMATION OF TOTAL PHENOTYPIC DAMAGE IN THE FIRST GENERATION

In the new skeleton study,[88] mouse spermatogonia were exposed to a fractionated dose of gamma radiation of 100 R + 500 R, delivered at 60 R/min with a 24-h interval between fractions. This procedure was used because it causes a high mutation frequency, which made it possible to subject as many suspected skeletal mutants as possible to breeding tests, to confirm (by their transmission of effects) that they were indeed mutants. Thirty-seven dominant mutations were found in the sample of 2,646 F_1 male progeny, for a mutation frequency of 1.4% per gamete. Thirty-one of the mutations were confirmed by breeding tests;[88,89] the remaining six are included on the basis of presumed-mutation criteria supported by the data,[88,90] even though they had no progeny. In the absence of a contemporary control, the mutation frequency of 1.4% is assumed to be the induced-mutation frequency. The reason for making this assumption is that the earlier skeleton studies indicated that the spontaneous-mutation frequency is very low, and the new study indicated that some induced mutations would almost certainly have been overlooked in the experimental approach used, owing to incomplete penetrance or viability effects. It was thought that these overlooked mutations counterbalanced any spontaneous mutations included in the 37 mutations reported.

Almost every region of the skeleton was affected by at least one of the 37 mutant genes. The abnormalities were easily seen in cleared and alizarin-stained skeletal preparations observed through a dissecting microscope. The effects found consisted mostly of the following changes: too few or too many bones, major changes in the shapes of bones caused by too little or too much bone growth, fusions of bones, and changes in the relative positions of bones. For three mutations, the only skeletal abnormality was a pronounced decrease in general body size. Many of the abnormalities are similar to malformations found in humans. Essentially all the mutations had incomplete penetrance for some or all of their effects. (At least nine had incomplete penetrance for all effects.) Very few of the mutations caused externally visible effects, most of which were manifest in only some of the carriers.

It is now known that some of the mutations, and very likely as many as four of them, are inseparable from reciprocal translocations, as though there were a dominant mutation at one of the breakpoints (P. B. Selby, personal communication). Such a class of translocation is not known to exist in humans, although it has been suspected in a few pedigrees. It is important to recognize that, in the direct estimation of abnormalities caused by chromosomal aberrations (made elsewhere in this chapter), reciprocal translocations themselves are not considered to be harmful in translocation heterozygotes. Thus, this small fraction (perhaps four of 37) of the skeletal mutations may represent a category of genetic disorders actually resulting from chromosomal aberration, which in humans might at present be confused with autosomal dominant and irregularly inherited disorders. That is, the genetic disorders caused by such translocations would be grouped with the non-chromosomal-aberration disorders in both the current-incidence figures and the risk estimates. It should be pointed out that translocations that have this phenotypic expression would also be scored in the direct measurement of the rate of induced translocation. However, their significance is far greater for their associated phenotypes than as ordinary rearrangements. In any case, genetic analysis of the skeletal dominants will reveal occurrences of this kind. The skeleton data suggest that very few if any of the remaining 33 dominant mutations are associated with rearrangements. Carriers of the great majority of dominant skeletal mutations have normal fertility.

To convert data from this experiment, which used acute fractionated exposures, to the expected rate of induction for continuous, low-dose-rate exposure, we divide by 1.9 to correct for the fractionation effect and by 3 to correct for the dose-rate effect; both corrections are based on results of specific-locus experiments. Because about three-fourths of specific-locus mutations are homozygous lethal, the assumption that results from specific-locus mutation experiments can be used for this correction is strengthened by the finding that the first four skeletal mutations tested were all homozygous lethal.[87] The application of these corrections to dominant skeletal mutations yields an estimated induced-mutation rate under protracted exposure of $(37/2,646) \times (1/600) \times (1/1.9) \times (1/3) = 4 \times 10^{-6}$ mutation per gamete per rem. In the earlier experiments,[19] a single acute x-ray dose of 600 R to spermatogonia produced five presumed skeletal mutations in 754 offspring, in comparison with the control observation of one presumed mutation in 1,739 offspring. After correction for the control observation and for the effect of low dose rate, the rate of induction of presumptive skeletal mutants in that experiment was 3.4×10^{-6} per gamete per rem for chronic exposure—in good agreement with the results of the recent experiment.

The proportion of dominant conditions in humans for which the main effect is on the skeleton can be used in conjunction with the mouse mutation rate to estimate total effects on all systems. A tabulation of monogenic disorders in man (see McKusick,[41] 4th ed.) showed that, of 583 "proven" autosomal dominants, 328 were clinically important, and about 20% of the latter (74 disease entities) involved at least one part of the skeleton to some extent. This figure is likely to be high, because of the ease of clinical diagnosis of such abnormalities, so we concur with the recent UNSCEAR estimate of 10%.[106] Because it is known that such dominant mutations may affect other systems in addition to the skeletal system, in both mouse and man, our estimate makes allowance for such pleiotropy. If mutations that affect the skeletal system constitute about 10% of mutations that affect any body system, then the total mutation rate must be some 10 times the rate of skeletal mutations alone, and we take this factor to be within the range of 5–15.

Many skeletal abnormalities caused by mutations in mice have effects that would undoubtedly impose no real harm if they occurred in humans, but would simply contribute to what is considered normal variation. Of the 37 dominant skeletal mutations in mice found in the experiment just described, about half were in this category; we thus reduce the estimated mutation rate by a factor of 0.25–0.75, to exclude mutations whose effects are slight.

Thus, after 1 rem of paternal (spermatogonial) irradiation, the probable increase in incidence of dominant genetic disorders that lead to serious handicaps at some time during life amounts to 5–45 per million liveborn.

No data are yet available on skeletal mutations resulting from maternal irradiation, but we can estimate the rate of such mutations if we assume that the relative sensitivities of oocytes and stem cell spermatogonia will be similar for different methods of detection of gene mutation. The mutational response of resting oocytes in mice is negligible, compared with that of spermatogonia, and mature and maturing oocytes in mice have a mutation rate no greater than 0.44 times that found in spermatogonia. We do not know which of these two classes of oocytes would have a mutational response more similar to that of arrested oocytes in women. To incorporate this range of uncertainty into our risk estimate for the combined effect of irradiation of both sexes, we have simply kept the lower limit of our estimate the same as it was (assuming a negligible mutation frequency in resting oocytes) and multiplied the upper limit by 1.44 (assuming the maximal estimate of the mutation frequency in mature and maturing oocytes). This gives an estimate of 5–65 induced serious dominant disorders per million liveborn as the first-generation expression, after exposure of the entire population to 1 rem per generation (Table IV-2).

14. CHROMOSOMAL REARRANGEMENTS

Translocations can be detected in a variety of organisms by testing the fertility of F_1 offspring and looking specifically for partial sterility of heterozygotes, or by examining cytologically the primary spermatocytes of irradiated males for multivalents at diakinesis or metaphase. Data obtained by the latter method have indicated a rate of induction about twice that determined by the partial-sterility method.[40] This discrepancy remains after one takes into account dominant lethality, which occurs in about the same ratio to transmissible partial sterility as would be expected from the usual frequencies of adjacent, as opposed to alternate, segregation products from translocation heterozygotes. No assumption other than selective elimination of some translocation-bearing cells can account for the discrepancy in the two methods of screening. Studies on postmeiotically induced translocations, recovered in partially sterile females,[40] have shown clearly that an appreciable fraction of these translocations cannot be transmitted through the male. Other studies on sons of males irradiated in postmeiotic stages have indicated that a considerable fraction (up to one-third) of translocation carriers are totally sterile, owing to a block in spermatogenesis.[9] Most of the translocations responsible for this male sterility are exchanges between two autosomes in which at least one of the breakpoints is near a centromere or telomere; and some are Y-autosome translocations. Independent work (see Russell and Montgomery[66] and L. B. Russell, personal communication) has also shown that all balanced segregants of balanced X-autosome translocations in the mouse are associated with this type of male sterility. Because most of the genetic effects of radiation in human populations would, however, result from exposure during spermatogonial—rather than postmeiotic—stages, male sterility resulting from spermatogenic blocks would probably be only a very rare consequence of translocations, inasmuch as it would be filtered out in spermatogenesis of the exposed male himself.

In BEIR I, it was not necessary to take nontransmissible rearrangements into account; they did not contribute to the incidence of transmissible partial sterility, and it was this quantity that was used to measure the rate of induction of translocations. (Although the occurrence of nontransmissible rearrangements is a matter of considerable theoretical interest, no appreciable hazard to human reproduction, survival, or health is to be expected if a very small fraction of human germ cells are unable to continue through their development and form functional sperm.) However, estimates for transmissible rearrangements can now be made on the basis of newer data, derived from human sources and from marmosets.[6] The disadvantage of having to estimate the fraction of new rearrangements that could not, under any circumstances, be transmitted to offspring is, we feel, more than offset by having good data derived from humans and from another primate. The frequency of translocation multivalents in the primary spermatocytes in humans and marmosets is approximately 7×10^{-4} per rem. In mice, the ratio of the observed incidence of partial sterility to that calculated on the basis of incidence of multivalents in primary spermatocytes was approximately 1:2 for several dose points, including 300 R and higher doses.[7,20,22] However, at the lowest dose tested, 150 R, the ratio was 1:1, for reasons not yet explained.[7,22] We have assumed an overall ratio of 1:1.5 for our calculations, so that we shall err in the direction of conservatism, if the true relation is 1:2, as indicated by the bulk of the data. We believe that it is reasonable to expect about the same ratio in other mammalian species, so we divide the observed incidence by 1.5, and this yields an incidence of potentially transmissible rearrangements in spermatocytes of about 4.7×10^{-4} per rem.

The new human data resulted from acute exposures at 100 R or less. Because lowering the dose rate or dividing the dose into small fractions has been found to decrease the total yield of aberrations, it is necessary to adjust this value to take into account our concern over low-dose, low-dose-rate exposures of humans. In mice, irradiation with a total gamma-ray dose of 600 R, delivered at a variety of dose rates from 83 R/min down to 0.02 R/min, showed a consistent lowering of yield as dose rate was lowered; the yield of the lowest dose rate was about one-ninth that of the highest dose rate.[84] To make our estimate adequately conservative, we have attempted to estimate the reduction factor, on the basis of a quadratic model, where the yield, y, is $\alpha D + \beta D^2 + C$ (see Note 3). This model may be unsatisfactory, especially at high doses (where considerable curve saturation may occur), but it will not result in overestimation of the reduction factor to be used for low dose rates at doses of around 100 R. On this basis, we assume that only one-track events are able to occur at the lowest dose rates; for mice, the values of α would be 2×10^{-5} ($\beta = 2.9 \times 10^{-7}$) for the 0.02-R/min data and 5×10^{-5} ($\beta = 2.5 \times 10^{-7}$) for the 0.09-R/min data. (The β values are those required to give, in conjunction with the one-track contribution, the total yield observed at 600 R, delivered at 83 R/min.) From these, it is simple to calculate an expected reduction in yield for lower total doses. The estimated factors at 100 R are 2.5 and 1.5, respectively, for the two pairs of estimates of α and β given.

If we assume that the dose-rate reduction factors in humans will be much the same as in mice, then it will be appropriate to reduce the incidence of expected multivalent configurations, at low doses and low dose rates, by a factor of 2. Accordingly, the incidence of newly induced translocation multivalents, where the spermatocytes carrying them would be capable of undergoing meiosis and could give rise later to functional sperm, would be 2.3×10^{-4} per rem.

Not all the products of such spermatocytes would carry reciprocal translocations. Transmission of a translocation requires that, in meiosis, alternate segregation or its equivalent occur. A translocation multivalent (usually quadrivalent) may orient on the division I spindle in a number of ways, influenced to some extent by the occurrence and placement of chiasmata. In turn, depending on this orientation and on the locations of chiasmata, the segregation products going into functional sperm will be of balanced or of unbalanced chromosomal constitution; i.e., segregation may be alternate (or its equivalent) or adjacent, of which there are two kinds. The probability of recovering a reciprocal translocation from such a quadrivalent therefore depends on the ratio of alternate to adjacent segregation. This ratio is not identical for all translocations; it depends on the locations of breakpoints. In no case will the ratio exceed unity; to the extent that adjacent-2 segregation occurs, the incidence of alternate segregation will be below 50%. Only if there is no adjacent-2 segregation, if there is at least one interstitial chiasma, and if centromere orientation is such that two always proceed to each pole of the spindle at division I, will the recovery of alternate segregation products reach the maximum of 50%. However, even in this case, half the recoveries will be of normal-sequence (nontranslocated) chromosomes. Hence, no more than 25% of the sperm produced by spermatocytes carrying a heterozygous translocation will themselves carry a balanced reciprocal translocation.

We have assumed that, on the average, alternate segregation will occur in 45% of these spermatocytes—only slightly less than the maximal rate. Application of these factors yields the expectation that the probability of transmitting a newly induced translocation to an offspring will be 5.2×10^{-5} per rem. To accommodate the uncertainties as to reduction because of low dose rate, we prefer to present the probability as a range: it almost certainly will not exceed 1.7×10^{-4} per rem, and it is rather unlikely to be less than 1.7×10^{-5} per rem.

Judging from data on partial sterility, the induction of reciprocal translocations in mature and maturing oocytes of mice is considerably lower than in spermatogonia, and it appears that immature oocyte stages also have a low incidence. This is expected from the finding that interchanges induced in dictyate oocytes are between chromatids,[10] which suggests that balanced reciprocal translocations should be poorly recoverable when induced in immature oocytes. That is the case in *Drosophila*, in which it is found that chromatid interchange, at least in the simpler cases, makes it highly likely that the halves of a reciprocal translocation will assort apart during meiosis.[58] Thus, a low yield of reciprocal translocations from irradiated immature oocytes can be expected on the basis of meiotic mechanics alone, regardless of differences in mechanisms of induction and repair of damage in different species.

Robertsonian translocation is a quite different matter, in that it is immaterial whether the reciprocal product of the translocated (metacentric) product is recovered. The reciprocal product is a small, centric fragment, almost or entirely lacking in significant genetic content. Robertsonian translocations can be induced in females of *Drosophila*, although at low frequencies. These, in their recovery, show a remarkable parallel to new recoveries of Robertsonian translocations in humans. In both cases, most recoveries are in zygotes of unbalanced (or, in the case of *Drosophila*, potentially unbalanced) chromosomal constitution, with eight of 10 recoveries in each species having involved an aneuploid gamete.[28,57] However, in both species, transmitted Robertsonian translocations show a much lower incidence of aneuploidy in the translocation carriers. This rather strongly suggests that these interchanges in humans occurred in meiotic prophase (as they are known to have occurred in the insect). It is likely that meiotic mechanics, at least in this respect, are similar in the two species.

If the probability that a chromosome will undergo breakage and rearrangement is a function of its size, then we should expect only a small fraction of all breakages to be in the short arms of the five acrocentrics of humans, and it would require an additional breakage in the proximal region of the long arm of another acrocentric for a Robertsonian metacentric to be formed. Of the possibilities for all types of interchange, four remaining acrocentric arms and 36 metacentric arms (in a haploid set of chromosomes) would be available to form the desired product. A nonrandom association of nonhomologous chromosomes is possible, but the expectation remains that most interchanges involving the short arm of an acrocentric would be with something other than the base of the long arm of another acrocentric.

In view of all the above, we expect that the production and transmission of new translocations from oocytes would be quite small, relative to those generated in spermatogonia.

Except as pointed out in Note 13 and except for rearrangements that cause male sterility, the deleterious consequences of reciprocal translocation will result only from the production of chromosomally unbalanced gametes. The incidence of zygotes with unbalanced segregation products (segmental aneuploids) will be slightly more than twice the incidence of zygotes carrying reciprocal translocations. However, most of these chromosomally unbalanced zygotes would be eliminated in early development—many too early to be recognized as spontaneous abortions, because the loss would occur before implantation.

Our confidence in making this assessment originates in the findings of an extensive cytogenetic survey of a human population in the United Kingdom;[28] it was found that, in spite of the substantial numbers of people studied, none carried an unbalanced form of the rearrangement, when ascertainment had been through a proband with a segregating balanced translocation. Furthermore, there had been no detectable reduction in the reproductive fitness of carriers of such balanced rearrangements. In contrast, when ascertainment was through an aneuploid person, the risk of having a child with an aneuploid form of the rearrangement was about 10-20% when it was the mother who carried the balanced rearrangement and about 2-5% when the father was the carrier. Although it was not explicitly stated, examination of the data strongly suggest that most of the rearrangements ascertained through aneuploid probands were of the Robertsonian type.

We assume that randomly induced translocations will not differ greatly in nature or behavior from randomly ascertained translocations in human populations, especially because about one-fifth of the latter are new, spontaneous occurrences of translocations. Certainly, the vast majority of translocations induced in mouse spermatogonia are not associated with the production of viable segmental aneuploids. Therefore, it is likely that very few of the translocations induced in humans would be capable of producing a class of viable segmental aneuploids.

If no more than 5% of all translocations are capable of producing viable aneuploids, then we can derive an expectation for the production of these from the figures given earlier in this note. We had concluded that the frequency of newly induced translocation multivalents in which the spermatocytes exhibiting them would be capable of undergoing meiosis to form functional sperm would be about 2.3×10^{-4} per rem. Because we assumed 45% alternate segregation, the combined expectation for all adjacent segregations would be 55% of 2.3×10^{-4}, or 1.3×10^{-4}. Five percent of this quantity is 6.3×10^{-6}. However, we would expect only one of the four kinds of aneuploid segregation products to be capable of giving rise to viable zygotes. Taking into account that this might not always be one of the less frequent or more frequent products, dividing by 4 to accommodate the one-out-of-four expectation gives the figure 1.6×10^{-6}. Using the order-of-magnitude range of uncertainty we adopted earlier gives the range 0.5×10^{-6} to 5×10^{-6} per rem.

We have no data on induced translocations in human oocytes. It is not clear that the female rate would be higher than the male rate, and there is some evidence that it may be lower; we therefore follow the approach taken in BEIR I and assume that it will be equal to the male rate. Thus, the expected frequency of viable aneuploids for both sexes is assumed to range from 1×10^{-6} to 10×10^{-6} per rem.

In keeping with the human population studies, we believe that in the vast majority of cases no adverse effects of segmental aneuploidy would be detectable, either by reduced reproductive performance or by increased fetal or infant deaths, except for 10 (or fewer) severely affected persons (accounted for in the preceding paragraph) per million liveborn per rem.

An alternative and independent approach based on the litter-size reduction observed after acute irradiation of mouse germ cells suggests that this upper limit of 10×10^{-6} for both sexes combined may be an overestimate and that the true value could indeed be near zero.

In an experiment in which spermatogonia were given an acute exposure of 600 R at 90 R/min, the mean litter size 3 wk after birth was 5.58 for a sample of 12,986 litters.[70] In contrast, it was 5.75 for an unirradiated control sample of 9,710 litters. Thus, the decrease attributable to irradiation was 2.96% [i.e., $(5.75-5.58)/5.75 = 0.0296$]. This includes dominant lethality from all causes (segmental aneuploidy, monosomy, trisomy, gene mutation, and so on, as well as dominant subvital mutations resulting in death during the first 3 wk of life), but as a "worst case" we will assume that all the decrease in litter size resulted from segmental aneuploidy. We will further assume that 6% of all human conceptions with a structurally unbalanced chromosome complement survive birth and are seriously handicapped[105,106] (BEIR I adopted a figure of 5%.). In the mouse, it is known that the unbalanced products of balanced translocations arise in meiosis at expected frequencies and show normal transmission. Most, however, cause lethality around the time of implantation, and only a very small fraction of the embryos carrying such products survive to produce viable offspring.[40,68] If we assume, on the basis of the human estimate, that 94% of the segmental aneuploids die before birth, then the maximal percentage of all zygotes that were segmental aneuploids in the mouse experiment is 3.15% [i.e., $(2.96)(100)/94 = 3.15$], and the percentage of such zygotes surviving birth is 0.19% (i.e., $3.15\% - 2.96\% = 0.19\%$). Dividing by 600 R, the maximal estimate of the frequency of all liveborn mice with segmental aneuploidy is 3.17×10^{-6} per roentgen of acute x irradiation. Allowing a dose-rate reduction factor of 9,[84] this yields an upper-bound estimate of 0.35×10^{-6} individuals seriously handicapped by segmental aneuploidy per roentgen of chronic gamma-ray exposure.

A similar calculation for mature oocytes that uses a dose-rate reduction factor of 12, a conservative value based on the consideration that the female is about twice as sensitive to specific-locus mutation induction as the male for acute irradiation but only about half (0.44) as sensitive for chronic irradiation,[73] yields a maximal estimate of 2.3×10^{-6} individuals seriously handicapped per roentgen of chronic gamma irradiation. (Dominant lethality in mature oocytes, after 400 roentgen of acute x irradiation, was 17%.[67])

Because these estimates represent extreme maximums, owing to the conservative assumptions involved, most Subcommittee members felt that our estimate of 10 seriously handicapped individuals per rem per million given in Table IV-2 is likely to be too high and that the true value may be near zero; one Subcommittee member, however, felt that the value of 10 per million live births, based as it is on human observations, is more likely to represent the true value.

---

15. IS NONDISJUNCTION INDUCED BY RADIATION?

Nondisjunction of a given chromosome produces trisomy and corresponding monosomy. The latter is less important in evaluation of human risk, because it is almost certainly lethal at a very early embryonic stage (probably before a woman even knows that she is pregnant), except when the sex chromosomes are involved. X-chromosome monosomy (X/0) leads to the Turner syndrome of amenorrhea and various morphologic anomalies. Monosomy can be produced by several mechanisms in addition to nondisjunction, so determinations of the frequency of monosomy give no clue as to frequencies of trisomy.

In man, trisomies are viable not only when the sex chromosomes are affected, but also in the case of a number of autosomes. Trisomies can cause severe abnormalities, some of them resulting in early death and others (such as Down's syndrome) afflicting people throughout life.

There have been conflicting results concerning radiation induction of trisomies in humans. Uchida[99] has reviewed 11 epidemiologic studies on the association between maternal preconception irradiation and subsequent birth of a child with Down's syndrome. Of these, four (including two by Uchida) showed a significant positive association; five, a nonsignificant positive association; and two, a nonsignificant negative association. A population exposed to high natural background radiation (1.5–3 R/yr) in Kerala, India, has recently been reported[30] to have a higher frequency of Down's syndrome than that found in nearby controls; but, because the control frequency seems unusually low and the Kerala frequency is not out of line with data from unexposed populations elsewhere, a further investigation seems warranted.

Among studies that failed to show an effect of preconception irradiation of either parent on the frequency of Down's syndrome is a recent Baltimore series[14] involving 150 cases and 150 controls. An earlier Baltimore study (one of the four positive studies just referred to) by some of the same authors[15] showed a positive association; and the difference between the two sets of results was ascribed to the more careful radiologic techniques (and thus smaller doses) used in recent years.

Another negative body of data came from the large series of children examined for Down's syndrome in Hiroshima and Nagasaki (three of 5,582 of exposed parents versus 12 of 9,452 of control parents).[82] A later cytogenetic study[3] of roughly the same populations found no autosomal trisomies in 2,885 children of exposed parents and 1,090 controls and a nonsignificant increase (eight of 2,885 versus one of 1,090) in the frequency of cases with supernumerary sex chromosomes (XXY, XYY, XXX).

Studies on the induction of nondisjunction in experimental mammals have involved either the sex chromosomes (diagnosis after birth) or all chromosomes (cytologic diagnosis in embryonic stages or at meiosis). The former studies,[65] which have involved irradiation of virtually the gamut of germ-cell stages in both sexes, have failed to demonstrate radiation-induced nondisjunction (i.e., supernumerary sex chromosomes), although X-monosomy was readily induced.

Because human data indicated a higher incidence of Down's syndrome in children of older mothers, recent experimental series have attempted to determine whether there is an age effect on induction of nondisjunction. Of five sets of data, one was positive and four were equivocal or negative (although one was claimed to be positive). The positive results were reported by Uchida and Freeman[100] in a study of meiotic (metaphase II) chromosomes after irradiation of oocytes with 10, 20, or 30 R. In old females, the frequency of hyperhaploid metaphases was significantly greater in the combined radiation groups than in controls. In young females, this difference was on the borderline of significance. There was no significant effect of age within the irradiated or control groups. Using a similar method, Reichert et al.[63] found a nonsignificant increase in hyperhaploids after 22–200 R irradiation of the "preovulatory" stage (probably diakinesis) of young females (six of 204 in irradiated groups; none of 143 in controls). In two experiments the cytologic scoring was done in 10.5-d embryos of mothers whose oocytes had been irradiated in dictyate. Yamamoto et al.[110] claimed an increase in nondisjunction from low-dose irradiation (5 R of x rays) of a rather small group of aged mice, but they included mosaics and monosomics in their calculations. A statistical evaluation of their data by Gosden and Walters[23] yielded no evidence of an interaction of maternal age and radiation treatment. Strausmanis et al.[95] failed to detect any nondisjunctional effect of 4, 8, or 16 R given to aged female mice. Similarly, L. B. Russell (personal communication), studying sex-chromosome nondisjunction with genetic markers, obtained negative results for both young and old female mice that had received 200 R. Thus, only the data of Uchida and Freeman[100] indicate a significant effect of radiation on the induction of trisomy.

On the assumption that all the autosomal monosomies resulting from nondisjunction, and virtually all the trisomies as well, would lead to prenatal death, dominant-lethal projections may be made from the data of Uchida and Freeman.[100] Such a calculation for 400 R, for example, results in frequencies that are not accommodated by the observed dominant-lethal incidence (Note 14). Indeed, each embryo would have to die more than once. Because other factors are known to contribute to dominant lethality, it may be concluded from the mouse data that any hazard from radiation-induced nondisjunction is probably quite small.

Studies on insects showed conclusively that trisomy can be induced by the irradiation of prophase I in oocytes, and there is a considerable volume of evidence that this induction results largely, if not exclusively, from a kind of direct damage to the chromosomes.[58] There is no longer any support for an earlier belief that there might be a threshold for the induction of trisomy. The implication of the insect work is that, as a result of chromatid interchanges and of the operation of simple meiotic mechanics, similar mechanisms might also operate in meiosis in any other species that have discrete centromeres (such as humans and mice). Hence, it is prudent to believe that the exposure of appropriate stages of the oocytes of humans can lead to trisomies. However, even assuming much higher sensitivity in the human female germ cells than in those of insects, the incidence expected would be quite low, because doses as high as 1,000 R produced only a barely detectable increase in insects, even in experiments in which large numbers of offspring were examined.

16. CALCULATIONS OF GENETIC RISK FOR OCCUPATIONAL EXPOSURES AND OTHER LIMITED EXPOSURES

The genetic-risk estimates in BEIR I and those in Table IV-2 are for a model population in which the germ cells involved in each conception have accumulated an average radiation dose essentially chronically over an average interval of 30 yr. Specifying the population dose in terms of that accumulated by gametes before conception offers the not inconsiderable advantage that one need not be concerned with the precise makeup of the parent population, nor worry about the distribution of dose, at least within reasonable limits, among its members. However, one of the uses to which we presume our estimates will be put is the calculation of genetic effects to be anticipated as a consequence of an increase in the radiation exposure of a real population. The exposure is likely to be expressed in person-rems per year, making conversion to gamete- or zygote-rems necessary. The exposure will often be markedly inhomogeneously distributed among the population, with a substantial fraction contributed by occupational or other exceptional exposure of a relatively small group. The added radiation exposure is often of short duration, so genetic equilibrium is never established, and some of the dose may even be delivered acutely. The estimation of genetic effects for such cases requires consideration of a number of factors.

For convenience, we have taken the mean parental age at the birth of a child to be 30 yr (as did BEIR I), although it is currently several years less in the U.S. population. Therefore, the preconception dose per year accumulated by the gametes contributing to the million live births for which estimates are given in Table IV-2 is 1 rem divided by 30 yr times 2×10^6 gametes, or 66,667 gamete-rems/yr, but this is clearly not necessarily the total person-rem exposure of the total population within which the million live births occur. For example, if the parent population for which the genetic-effects estimates are given in Table IV-1 is assumed to have the same age, sex, and reproductive characteristics as those of the U.S. population as estimated for 1974,[107] then this gamete-rem dose corresponds to about 66,667 gamete-rems per 3.1 children per mother, which is about 21,500 person-rems/yr to the parents. Because about half the population is over the age of 30, the total person-rem dose to the entire population would be about twice this figure, or about 43,000 person-rems/yr. Obviously, differences in the average family size or in the age distribution of the population would change the population doses correspondingly.

The practice of expressing population dose in person-rems implies that the distribution of dose among population members makes no difference; but it plainly does make a difference for estimates of genetic effects in real situations. Genetic consequences of person-rems received by women over 50 are clearly much less likely than those of person-rems received by adolescent males. This factor can become especially important if the bulk of the population dose is received by a small group, such as diagnostic x-ray technicians or nuclear-power-plant workers, whose age and sex distribution is very different from that of the general population. In cases involving substantial occupational or other limited exposures, then, it is particularly important to take into account the age, sex, and dose distribution and the probability that people at each age and of each sex will have further children, as well as the mutational-sensitivity difference between males and females. This may be done by calculating a gonadal-dose estimate for separate age cohorts of each sex, weighting this by the probable number of additional children for each cohort, and then multiplying by the appropriate factor to allow for the different mutational sensitivities of spermatogonia and immature oocytes (this assumes, of course, that procreation will be delayed long enough after any substantial acute exposures for these germ-cell stages to be applicable). Such a procedure gives doses in gamete-rems, so the estimates of Table IV-2 may be used directly. Because, in making these estimates, the Subcommittee assumed the mutational sensitivity of immature oocytes to be between 0 and 0.44 (an average of 0.22) times that of spermatogonia (see Notes 8, 9, and 13), appropriate adjustment factors are roughly 0.82 and 0.18, respectively, for male and female exposures.

Occupational radiation exposures may include an irregular series of acute doses, any one of which can, at least in theory, be as large as the 3-rem quarterly maximal permissible dose. It may thus be questioned whether the estimates in Table IV-2, for which a low dose rate was assumed, are entirely appropriate for calculating the genetic effects expected as a consequence of occupational exposure. However, both radiobiologic theory (see Note 3) and empirical data from mouse mutation studies with fractionated doses strongly suggest that even doses as large as 3 rems are small enough to be properly treated as though delivered at low dose rates, even if they are acute. Nevertheless, other special cases for which genetic-effects estimates might be wanted, such as accidental or therapeutic radiation exposures, may involve acute gonadal doses of tens of rems or more. In such cases, it is appropriate to adjust the estimates of Table IV-2 upward by a factor of 3 to allow for the high dose rate (again, assuming that procreation is delayed long enough for the estimates to be appropriate).

Many occupational exposures include a substantial high-LET component—e.g., the fast-neutron exposures of high-energy-particle accelerator workers. Accidental and therapeutic exposures may also involve high-LET radiation. Not only have questions been raised as to whether the RBE values currently used in the conversion of rads to rems for high-LET radiation are entirely appropriate for low doses (see

Note 5), but it is well established that there is little if any dose-rate effect of such radiation. Because the RBE factor incorporated into the calculation of rems is usually based on acute-exposure data, the dose-rate reduction factor incorporated in the genetic-effects estimates in Table IV-2 is inappropriate for high-LET exposure, and upward correction by a factor of 3 is in order for such exposures.

The estimation of genetic effects to be expected from exposure of limited populations has a further complication: the size of the population in which the genetic effects will be expressed is often not clear. This becomes particularly troublesome when, as is not infrequently the case, what is wanted is an estimate of the increase in relation to the "current incidence" in the absence of the added radiation exposure. First-generation risk may be calculated on a per-live-birth basis for any offspring of exposed individuals or groups. If genetic equilibrium is eventually established, however, the effects ascribable to exposure of the prior generations will have been distributed among the members of an undefined larger population, and no estimate of individual risk will be possible.

When an added radiation exposure is limited to one individual, or to a group in which all members are of one or a few generations, no genetic equilibrium will be established. An alternative to the equilibrium estimate that is appropriate for such instances is an absolute "effects-over-all-time" estimate. At genetic equilibrium, exactly as many future genetic effects are induced as are eliminated in any one generation. It follows that the total of all genetic effects that will be expressed over all future generations as a consequence of exposure limited to a single generation is numerically equal to the total for each generation in the equilibrium situation. Thus, an absolute "all-time" estimate may be derived directly from the equilibrium estimates in Table IV-2.

Another type of exposure requiring special consideration is that delivered prenatally. Although the mutational sensitivity of the early germ-line cells in the male seems unlikely to be very different from that of spermatogonia, the sensitivity of the early germ-line cells and the oogonia present during female fetal development could well be somewhat higher than that of the immature oocytes, which are at risk after birth. Unfortunately, no experimental data are available; but it seems possible that the sensitivity of these cells to chronic irradiation could even be as high as the sensitivity of spermatogonia. The estimates in Table IV-2 assume a female sensitivity ranging from negligible to 0.44 that of the male, so it seems conservative to adjust them upward by a factor of about 1.6 (thus making female equal to male in mutational sensitivity) if they are to be used to estimate genetic effects that might result from chronic fetal irradiation. For acute exposures of immature cells in the male and female, published data on the mutational sensitivity of mouse gonocytes[85] and of those oocytes present in mouse fetuses at 17.5 d after conception[12] and at birth[86] make it seem likely that the genetic risk would not be much, if any, greater than the range calculated for chronic exposure of these stages. No mutation-induction data exist for mouse oogonia exposed to low-LET radiation. However, because at a few months after conception human ovaries already contain many oocytes that are probably comparable with those studied in mutation studies on newborn mice, it seems reasonable to think that the mouse studies mentioned may relate to hazard estimates made for the greater part of human pregnancy.

REFERENCES

1. Abrahamson, S., and S. Wolff. Reanalysis of radiation-induced specific locus mutations in the mouse. Nature 264:715–719, 1976.
2. Auerbach, C., and B. J. Kilbey. Mutation in eukaryotes. In H. Rowan, Ed. Annual Review of Genetics 4:163–218. Palo Alto, Calif.: Academic Annual Review, Inc., 1971.
3. Awa, A. A. Cytogenetic study. In S. Okada *et al.,* Eds. Thirty Years Study of Hiroshima and Nagasaki Atomic Bomb Survivors. Radiat. Res. 16(Suppl.):75–81, 1975.
4. Baker, T. G. The effects of ionizing radiation on the mammalian ovary with particular reference to oogenesis, pp. 349–361. In R. O. Greep, E. B. Astwood, and S. R. Geiger, Eds. Handbook of Physiology. Section I. Endocrinology. Vol. II. Female Reproductive System. Part 1. Washington, D.C.: American Physiological Society, 1973.
5. Berendes, H. D. Polytene chromosome structure at the submicroscopic level. A map of region X, 1-4E of *Drosophila melanogaster.* Chromosoma (Berlin) 29:118–130, 1970.
6. Brewen, J. G., and R. J. Preston. Analysis of x-ray induced chromosomal translocations in human and marmoset stem cells. Nature 253:468–470, 1975.
7. Brewen, J. G., R. J. Preston, and W. M. Generoso. X-ray-induced translocations. Comparison between cytologically observed and genetically recovered frequencies, pp. 74–75. In Biology Division Annual Report for period ending June 30, 1974. Oak Ridge National Laboratory Report ORNL-4993. 1974.
8. British Medical Research Council. The Hazards to Man of Nuclear and Allied Radiations. London: Her Majesty's Stationery Office, 1956, 1960.
9. Cacheiro, N. L. A., L. B. Russell, and M. S. Swartout. Translocations, the predominant cause of total sterility in sons of mice treated with mutagens. Genetics 76:73–91, 1974.
10. Caine, A., and M. F. Lyon. The induction of chromosome aberrations in mouse dictyate oocytes by X-rays and chemical mutagens. Mutat. Res. 45:325–331, 1977.
11. Carsten, A. L., and S. L. Commerford. Dominant lethal mutations in mice resulting from chronic tritiated water (HTO) ingestion. Radiat. Res. 66:609–614, 1976.
12. Carter, T. C., M. F. Lyon, and R. J. S. Phillips. The genetic sensitivity to x-rays of mouse foetal gonads. Genet. Res. Camb. 1:351–355, 1960.
13. Cavalli-Sforza, L. L., and W. F. Bodmer. The Genetics of Human Populations. San Francisco: W. H. Freeman, 1971.
14. Cohen, B. H., A. M. Lilienfeld, S. Kramer, and L. C. Hyman. Parental factors in Down's syndrome. Results of the second Baltimore case-control study. In E. B. Hook and I. H. Porter, Eds. Population Cytogenetics. New York: Academic Press, 1977.
15. Cohen, B. H., A. M. Lilienfeld, and A. T. Sigler. The epidemiological study of mongolism in Baltimore. Ann. N.Y. Acad. Sci. 171:320–327, 1970.
16. Cox, B. D., and M. F. Lyon. X-ray induced dominant lethal mutations in mature and immature oocytes of guinea pig and golden hamsters. Mutat. Res. 28:421–436, 1975.
17. Crow, J. F. Minor viability mutants in Drosophila. Genetics. (in press)
18. de Serres, F. J., and H. V. Malling. Identification of the genetic alterations in specific locus mutants at the molecular level. Jap. J. Genet. 44(Suppl. 1):106–113, 1969.
19. Ehling, V. H. Dominant mutations affecting the skeleton in offspring of x-irradiated male mice. Genetics 54:1381–1389, 1966.
20. Ford, C. E., A. G. Searle, E. P. Evans, and B. J. West. Differential transmission of translocations induced in spermatogonia of mice by x-irradiation. Cytogenetics 8:447–470, 1969.
21. Friedman, L. X-ray induced sex-linked lethal and detrimental mutations and their effect on the viability of *Drosophila melanogaster.* Genetics 49:689–699, 1964.

22. Generoso, W. M., K. J. Cain, and S. W. Huff. Dose effects of acute x-rays on induction of heritable reciprocal translocations in mouse spermatogonia, pp. 136–138. In Biology Division Annual Report for period ending June 30, 1974. Oak Ridge National Laboratory Report ORNL-4993. 1974.

23. Gosden, R. G., and D. E. Walters. Effects of low-dose x-irradiation on chromosomal nondisjunction in aged mice. Nature 248:54–55, 1974.

24. Green, D., G. R. Howells, E. R. Humphreys, and J. Vennart. Localization of plutonium in mouse testis. Nature (London) 255:77, 1975.

25. Green, E. L. Genetic effects of radiation on mammalian populations. Annu. Rev. Genet. 2:87–120, 1968.

26. Recommendations of the International Commission on Radiological Protection (adopted September, 1958). ICRP Publ. 1. Oxford: Pergamon Press, 1959.

27. Recommendations of the International Commission on Radiological Protection, as amended 1959 and revised 1962. ICRP Publ. 6. Oxford: Pergamon Press, 1964.

28. Jacobs, P. A. Human population cytogenetics, pp. 232–242. In J. de Grouchy, F. J. G. Ebling, and I. W. Henderson, Eds. Human Genetics. Proc. IV Int. Congr. Hum. Genet. Amsterdam: Excerpta Medica, 1972.

29. Judd, B. H., M. W. Shen, and T. C. Kaufman. The anatomy and function of a segment of the x chromosome of *Drosophila melanogaster*. Genetics 71:139–156, 1972.

30. Kuchupillai, N., I. C. Verma, M. S. Grewal, and V. Ramalingas. Down's syndrome and related abnormalities in an area of high background radiation in coastal Kerala. Nature 262:60, 1976.

31. Langridge, J., and J. H. Campbell. Classification and intragenic position of mutations in beta-galactosidase gene of *Escherichia coli*. Mol. Gen. Genet. 103:339–347, 1969.

32. Lea, D. E. Action of Radiation on Living Cells. 2nd ed. Cambridge: Cambridge Univ. Press, 1955.

33. Lee, W. R., G. A. Sega, and C. F. Alford. Mutations produced by transmutation of phosphorus-32 to sulphur-32 within Drosophila DNA. Proc. Natl. Acad. Sci. U.S.A. 58:1472–1479, 1967.

34. Lee, W. R., G. A. Sega, and E. S. Benson. Transmutation of carbon-14 within DNA of *Drosophila melanogaster* spermatozoa. Mutat. Res. 16:195–201, 1972.

35. Lefevre, G. Salivary chromosome bands and the frequency of crossing over in *Drosophila melanogaster*. Genetics 67:497–515, 1971.

36. Lifschytz, E., and R. Falk. Fine structure analysis of a chromosome segment in Drosophila melanogaster. Analysis of ethyl methane sulphonate-induced lethals. Mutat. Res. 8:147–155, 1969.

37. Lifschytz, E., and R. Falk. Fine structure analysis of a chromosome segment in Drosophila melanogaster. Analysis of x-ray induced lethals. Mutat. Res. 6:235–244, 1968.

38. Luning, K. G., H. Frolen, and A. Nilsson. Genetic effects of ^{239}Pu salt injections in male mice. Mutat. Res. 34:539–542, 1976.

39. Lyon, M. F., and R. J. S. Phillips. Specific locus mutation rates after repeated small radiation doses to mouse oocytes. Mutat. Res. 30:375–382, 1975.

40. Lyon, M. S., and R. Meredith. Autosomal translocations causing male sterility and viable aneuploidy in the mouse. Cytogenetics 5:335–354, 1966.

41. McKusick, V. A. Mendelian Inheritance in Man: Catalogs of Autosomal Dominant, Autosomal Recessive, and X-linked Phenotypes. Baltimore: The Johns Hopkins University Press, 1975 (4th ed.), 1978 (5th ed.).

42. Mitchell, J. Fitness effects of EMS-induced mutations on the X-chromosome of *Drosophila melanogaster*. Viability effects and heterozygous fitness effects. Genetics 87:763–774, 1977.

43. Mukai, T. The genetic structure of natural populations. I. Spontaneous mutation rate of polygenes controlling viability. Genetics 40:1–19, 1924.
44. Mukai, T., S. I. Chigusa, L. E. Mettler, and J. F. Crow. Mutation rate and dominance of genes affecting viability in *Drosophila melanogaster*. Genetics 72:335–355, 1972.
45. National Council on Radiation Protection and Measurements. Maximum permissible radiation exposures to man. Amer. J. Roent. 77:910–913, 1957.
46. National Council on Radiation Protection and Measurements. Maximum permissible radiation exposures to man. Radiology 71:263–266, 1968.
47. National Council on Radiation Protection and Measurements, Scientific Committee 24. Radionuclides and labelled organic compounds incorporated in genetic material. (in press)
48. National Research Council, Advisory Committee on the Biological Effects of Ionizing Radiations. The Effects on Populations of Exposure to Low Levels of Ionizing Radiation. Washington, D.C.: National Academy of Sciences, 1972.
49. National Research Council, Committee on Genetic Effects of Atomic Radiation. The Biological Effects of Atomic Radiation, pp. 3–31. Washington, D.C.: National Academy of Sciences, 1956.
50. Neel, J. V., H. Kato, and W. L. Schull. Mortality in the children of atomic bomb survivors and controls. Genetics 76:311–326, 1974.
51. Newcombe, H. B., and J. F. McGregor. Heritable effects of radiation on body weight in rats. Genetics 52:851–860, 1965.
52. Newcombe, H. B., and J. F. McGregor. Learning ability and physical well-being in offspring from rat populations irradiated over many generations. Genetics 40:1065–1081, 1964.
53. Oakberg, E. F. A new concept of spermatogonial stem cell renewal in the mouse and its relationship to genetic effects. Mutat. Res. 11:1–7, 1971.
54. Oakberg, E. F. Timing of oocyte maturation in the mouse and its relevance to radiation-induced cell killing and mutational sensitivity. Mutat. Res. 59:39–48, 1979.
55. Ohnishi, O. Spontaneous and ethyl methanesulfonate-induced mutations controlling viability in *Drosophila melanogaster*. I. Recessive lethal mutations. II. Homozygous effects of polygenic mutations. III. Heterozygous effects of polygenic mutations. Genetics 87:519–556, 1977.
56. Painter, T. S., and W. Stone. Chromosome fusion and speciation in Drosophila. Genetics 20:327–341, 1935.
57. Parker, D. R. Radiation-induced nondisjunction and Robertsonian translocation in Drosophila. Mutat. Res. 24:149–162, 1974.
58. Parker, D. R., and J. H. Williamson. Aberration induction and segregation in oocytes, pp. 1251–1268. In M. Ashburner and E. Novitski, Eds. The Genetics and Biology of Drosophila. London: Academic Press, 1976.
59. Patterson, J. T., W. Stone, S. Bedichek, and M. Suche. The production of translocations in Drosophila. Amer. Nat. 68:359–369, 1934.
60. Pedersen, T. Follicle growth in the mouse ovary, pp. 301–376. In J. D. Biggers and A. W. Schuetz, Eds. Oogenesis. Baltimore: University Park Press, 1972.
61. Person, S., W. Snipes, and F. Krasin. Mutation production from tritium decay. A local effect for [^{3}H]2-adenosine and [^{3}H]6-thymine decays. Mutat. Res. 34:327–332, 1976.
62. Rayle, E. E., and M. M. Green. A contribution to the genetic fine structure of the region adjacent to white in *Drosophila melanogaster*. Genetics 39:497–507, 1968.
63. Reichert, W., I. Hansmann, and G. Rohrborn. Chromosome anomalies in mouse oocytes after irradiation. Humangenetik 28:25–38, 1975.
64. Russell, L. B. Definition of functional units in a small chromosomal segment of the

mouse and its use in interpreting the nature of radiation-induced mutations. Mutat. Res. 11:107–123, 1971.

65. Russell, L. B. Numerical sex-chromosome anomalies in mammals. Their spontaneous occurrence and use in mutagenesis studies, pp. 55–91. In A. Hollaender, Ed. Chemical Mutagens. Principles and Methods for Their Detection. New York: Plenum Press, 1976.

66. Russell, L. B., and C. S. Montgomery. Comparative studies on X-autosome translocations in the mouse. I. Origin, viability, fertility, and weight of five T(X; 1)'s. Genetics 63:103–120, 1969.

67. Russell, L. B., and W. L. Russell. The sensitivity of different stages in oogenesis to the radiation induction of dominant lethals and other changes in the mouse, pp. 187–192. In J. S. Mitchell, B. E. Holmes, and C. L. Smith, Eds. Progress in Radiobiology. Edinburgh: Oliver and Boyd, 1956.

68. Russell, L. B., W. L. Russell, N. L. A. Cacheiro, C. M. Vaughn, R. A. Popp, and K. B. Jacobson. A tandem duplication in the mouse. Genetics 80:s71, 1975.

69. Russell, W. L. Evidence from mice concerning the nature of the mutation process, pp. 257–264. In Genetics Today. Vol. 2. Proc. VI Int. Congr. Genet. (The Hague). Oxford: Pergamon Press, 1965.

70. Russell, W. L. Genetic effects of radiation in mammals, pp. 825–859. In A. Hollaender, Ed. Radiation Biology. Vol. 1. New York: McGraw-Hill, 1954.

71. Russell, W. L. Lack of linearity between mutation rate and dose for x-ray-induced mutations in mice. Genetics 41:658–659, 1956.

72. Russell, W. L. Mutagenesis in the mouse and its applicability to the estimation of the genetic hazards of radiation. In Proceedings of IV International Congress of Radiation Research, Evian, France, June 29–July 4, 1970.

73. Russell, W. L. Mutation frequencies in female mice and the estimation of genetic hazards of radiation in women. Proc. Natl. Acad. Sci. U.S.A. 74:3523–3527, 1977.

74. Russell, W. L. Observed mutation frequency in mice and the chain of processes affecting it, pp. 216–228. In G. E. W. Wolstenholme and Maeve O'Connor, Eds. Mutation as Cellular Process. Ciba Foundation Symposium. London: J. & A. Churchill, 1969.

75. Russell, W. L. Repair mechanisms in radiation mutation induction in the mouse. In Recovery and Repair Mechanisms in Radiology. Brookhaven Symp. Biol. 20:179–189, 1968.

76. Russell, W. L. The effect of radiation dose rate and fractionation on mutation in mice, pp. 205–217, 231–235. In F. Sobels, Repair from Genetic Radiation Damage. Oxford: Pergamon Press, 1963.

77. Russell, W. L. The genetic effects of radiation, pp. 487–500. In Peaceful Uses of Atomic Energy. Vol. 13. Vienna: IAEA, 1972.

78. Russell, W. L., R. B. Cumming, E. M. Kelly, and A. Lindenbaum. Plutonium-induced specific locus mutations in mice. Genetics 88(Suppl.):585, 1978.

79. Russell, W. L., and L. B. Russell. Radiation-Induced Genetic Damage in Mice, pp. 179–188. In J. C. Bugher, J. Coursaget, and J. F. Loutit, Eds. Progress in Nuclear Energy. Series VI. Vol. 2. Biological Sciences. London: Pergamon Press, 1959.

80. Russell, W. L., L. B. Russell, and M. B. Cupp. Dependence of mutation frequency on radiation dose rate in female mice. Proc. Natl. Acad. Sci. U.S.A. 45:18–23, 1959.

81. Russell, W. L., L. B. Russell, and E. M. Kelly. Radiation dose rate and mutation frequency. Science 128:1546–1550, 1958.

82. Schull, W. J., and J. V. Neel. Maternal radiation and mongolism. Lancet 1:537, 1962.

83. Searle, A. G., C. V. Beechey, D. Green, and E. R. Humphreys. Cytogenetic effects of protracted exposures to alpha particles from plutonium-239 and to gamma rays from cobalt-60 compared in male mice. Mutat. Res. 41:297–310, 1976.

84. Searle, A. G., E. P. Evans, C. E. Ford, and B. J. West. Studies on the induction of translocations in mouse spermatogonia. I. The effect of dose-rate. Mutat. Res. 6:427–436, 1968.

85. Selby, P. B. X-ray-induced specific-locus mutation rate in newborn male mice. Mutat. Res. 18:63–75, 1973.

86. Selby, P. B. X-ray induction of specific-locus mutations in newborn female mice, p. 92. In Biology Division Annual Progress Report for period ending June 30, 1971. Oak Ridge National Laboratory Report ORNL 4740. 1971.

87. Selby, P. B., and V. S. Mierzejewski. Studies on mice suggest that the dominant deleterious effects of recessive lethals may account for an important fraction of human genetic disease. Genetics 86:s57, 1977.

88. Selby, P. B., and P. R. Selby. Gamma-ray-induced dominant mutations that cause skeletal abnormalities in mice. I. Plan, summary of results and discussion. Mutat. Res. 43:357–375, 1977.

89. Selby, P. B., and P. R. Selby. Gamma-ray-induced dominant mutations that cause skeletal abnormalities in mice. II. Description of proved mutations. Mutat. Res. 51:199–236, 1978.

90. Selby, P. B., and P. R. Selby. Gamma-ray-induced dominant mutations that cause skeletal abnormalities in mice. III. Description of presumed mutations. Mutat. Res. 50:341–351, 1978.

91. Shannon, M. P., T. C. Kaufman, M. W. Shen, and B. H. Judd. Lethality patterns of selected lethal and semi-lethal mutations in the *zeste-white* region of *Drosophila melanogaster*. Genetics 72:615–638, 1972.

92. Simmons, M. J., and J. F. Crow. Mutations affecting fitness in Drosophila populations. Annu. Rev. Genet. 11:49–78, 1977.

93. Spalding, J. F., M. R. Brooks, and G. L. Tietjen. Lifetime body weights and mortality distribution in mice with 10 to 35 generations of ancestral x-ray exposure. Genetics 63:897–906, 1969.

94. Stevenson, A. The load of hereditary defects in human populations. Radiat. Res. Suppl. 1:306–325, 1959.

95. Strausmanis, R., I.-B. Henrikson, M. Holmberg, and C. Rönnbäck. Lack of effect on the chromosomal non-disjunction in aged female mice after low dose x-irradiation. Mutat. Res. 49:269–274, 1978.

96. Taylor, L. S. Radiation Protection Standards. Cleveland: CRC Press, 1971. 110 pp.

97. Temin, R. G., H. U. Meyer, P. S. Dawson, and J. F. Crow. The influence of epistasis on homozygous viability depression in *Drosophila melanogaster*. Genetics 61:497–519, 1969.

98. Trimble, B. K., and J. H. Doughty. The amount of hereditary disease in human populations. Ann. Hum. Genet. (Lond.) 38:199–223, 1974.

99. Uchida, I. A. Maternal radiation and trisomy 21. In E. B. Hook and I. H. Porter, Eds. Population Cytogenetics. New York: Academic Press, 1977.

100. Uchida, I., and C. P. V. Freeman. Radiation-induced non-disjunction in oocytes of aged mice. Nature 265:186–187, 1977.

101. United Nations Scientific Committee on the Effects of Atomic Radiation. Report A.3838. General Assembly Official Records. 13th Sess. Suppl. No. 17. New York: United Nations, 1958.

102. United Nations Scientific Committee on the Effects of Atomic Radiation. Report A/5216. General Assembly Official Records. 17th Sess. Suppl. No. 16. New York: United Nations, 1962.

103. United Nations Scientific Committee on the Effects of Atomic Radiation. Report A/8314. General Assembly Official Records. 21st Sess. Suppl. No. 14. New York: United Nations, 1966.

104. United Nations Scientific Committee on the Effects of Atomic Radiation. Report A/6314. General Assembly Official Records. 24th Sess. Suppl. No. 13. New York: United Nations, 1969.

105. United Nations Scientific Committee on the Effects of Atomic Radiation. Report A/8725. General Assembly Official Records. 27th Sess. Suppl. No. 25. New York: United Nations, 1972.

106. United Nations Scientific Committee on the Effects of Atomic Radiation. Report A/32/40. General Assembly Official Records. 32nd Sess. Suppl. No. 40. New York: United Nations, 1977.

107. U.S. Department of Commerce, Bureau of the Census. Statistical Abstract of the United States: 1976. 97th ed. Washington, D.C.: U.S. Government Printing Office, 1976.

108. U.S. Nuclear Regulatory Commission. Genetic effects, pp. I-1-I-15. In Reactor Safety Study. An Assessment of Accident Risks in U.S. Commercial Nuclear Power Plants. Appendix VI. Calculation of Reactor Accident Consequences. WASH-1400. (NUREG 75/014.) Washington, D.C.: U.S. Nuclear Regulatory Commission, 1975.

109. Whitfield, J. J., R. G. Martin, and B. N. Ames. Classification of amino-transferase (c gene) mutants in the histidine operon. J. Mol. Biol. 21:335–355, 1966.

110. Yamamoto, M., T. Shimada, A. Endo, and G. Watanabe. Effects of low-dose x irradiation on the chromosomal nondisjunction in aged mice. Nature New Biol. 244:206–208, 1973.

V
Somatic Effects: Cancer

This chapter deals with cancer induction, which the Committee considers the most important somatic effect of low-dose ionizing radiation. The chapter reviews the extensive epidemiologic and laboratory-animal literature and describes the train of logic that leads to the estimation of cancer risk coefficients; it is necessarily long, and to some it will appear complex. The following may guide the reader.

Chapter II, "Scientific Principles in Analysis of Radiation Effects," discusses the basic information that Chapter V applies to the subject of cancer induction due to low-dose ionizing radiation.

For the reader not intimately knowledgeable in the biologic effects of ionizing radiation, the first section of the present chapter, "Summary and Conclusions," may provide sufficient information on the Committee's major conclusions. The section includes an abbreviated set of tables of lifetime cancer risk estimates for various situations of exposure to low-dose, low-LET, whole-body radiation.

The remainder of the chapter is directed primarily to those who are knowledgeable in the subject. The second section, "General Considerations," describes both the major data sources and the major assumptions used in interpreting the data on the carcinogenic effects of ionizing radiation.

Two members of the Committee, Dr. Radford and Dr. Rossi, dissent from the report. Their statements appear immediately after the section, "Estimating the Total Cancer Risk of Low-Dose, Low-LET, Whole-Body Radiation," to which their dissent is primarily related. Comments by

Dr. Webster subscribed to by two other members of the Committee also follow this section.

Appendix A presents and evaluates the results of studies of radiation effects on specific organ sites and thus provides a perspective on the assessment of total risk from radiation by discussing the relative importance of radiation induction of cancers at various individual sites.

The information in the "General Considerations" section and the literature review in Appendix A is used to develop the third section, "Estimating the Total Cancer Risk of Low-Dose, Low-LET, Whole-Body Radiation." This necessarily detailed section concludes with risk estimates based on the Committee's extensive deliberations. Those deliberations dealt largely with the question of which method to use in estimating carcinogenic effects of low doses of ionizing radiation, in light of the lack of definitive data on the effects of such exposure and the lack of agreement on how to extrapolate data on high doses to estimate the effects of low doses. The Committee chose to explain, in detail, its process of accepting or discarding various lines of reasoning, so that those who must decide on radiation-protection policy and those who in the future will be able to refine the estimates with the benefit of additional data can trace the Committee's steps.

Appendix B contains the Committee's evaluation of specific studies that have attracted much public attention. These studies have provoked public controversy, because their authors have attached a greater risk of cancer to exposure to low-dose ionizing radiation than that identified by most other investigators in this field or predicted by the various models used to estimate such effects.

SUMMARY AND CONCLUSIONS

The Committee considers cancer induction to be the most important somatic effect of low-dose ionizing radiation. The induction of cancer by radiation is detectable only by statistical means; that is, the cancer of any given person cannot be attributed with certainty to radiation, as opposed to some other cause. In general, the smaller the dose of radiation, the smaller the likelihood that radiation was the cause.

There are good observational data relative to cancer induction in humans over a range of higher doses, but little direct evidence is available for doses of a few rads. Estimation of the excess risks at these low doses usually involves extrapolation from observations at higher doses on the basis of assumptions about the nature of the dose-response relationship. Unfortunately, too little is known about the mechanisms of radiation carcinogenesis for dose-response models to be specified with any certainty, except as a general parametric family of functions.

CANCER INDUCTION

In considering the cancers attributable to radiation exposure, the following comments are pertinent:

• Cancers induced by radiation are indistinguishable from those occurring naturally; hence, their existence can be inferred only on the basis of a statistical excess above the natural incidence.
• Cancer may be induced by radiation in nearly all the tissues of the human body.
• Tissues and organs vary considerably in their sensitivity to the induction of cancer by radiation.
• The natural incidence of cancer varies over several orders of magnitude, depending on the type and site of origin of the neoplasm, age, sex, and other factors.
• With respect to excess risk of cancer from whole-body exposure to radiation, solid tumors are now known to be of greater numerical significance than leukemia. Solid cancers characteristically have long latent periods; they seldom appear before 10 yr after radiation exposure and may continue to appear for 30 yr or more after radiation exposure. In contrast, the excess risk of leukemia appears within a few years after radiation exposure and largely disappears within 30 yr after exposure.
• The major sites of solid cancers induced by whole-body radiation are the breast in women, the thyroid, the lung, and some digestive organs.
• The incidence of radiation-induced human breast and thyroid cancer is such that the total cancer risk is greater for women than for men. Breast cancer occurs almost exclusively in women, and absolute-risk estimates for thyroid-cancer induction by radiation are higher for women than for men (as is the case with the natural incidence). With respect to other cancers, the radiation risks in the two sexes are approximately equal.
• There is now considerable evidence from human studies that age is a major factor in the risk of cancer from exposure to ionizing radiation. Both age at exposure and age at cancer diagnosis are important for interpretation of human data. If risks are given in absolute form—i.e., numbers of cancers induced per unit of population and per unit of radiation exposure—then a single value independent of age may be

inappropriate. The 1972 BEIR report[56] concluded that the risk of some kinds of cancer was greater after irradiation in childhood and *in utero* than in adult life. It is now apparent that other age groups may also have risks that differ from the average for all ages; e.g., women exposed during the second decade of life have the highest risk of radiation-induced breast cancer.

• Various host or environmental factors may interact with radiation to affect cancer incidence in different tissues. These may include hormonal influences, immunologic status, exposure to various oncogenic agents, and nonspecific stimuli to cell proliferation in tissues sensitive to cancer induction by radiation.

• The time elapsing between irradiation and the appearance of a detectable neoplasm is characteristically long, i.e., years or even decades. This long latent period must be taken into consideration in all risk calculations, whether these are estimates of the risk experienced by populations under study or projections into the future.

• The variety of possible biologic mechanisms responsible for human cancer suggests that the dose-response relationship may not be the same for all types of radiation-induced cancer. The fact, however, that epidemiologic studies of widely differing human populations exposed to radiation have given reasonably concordant results for some cancer sites and for a broad range of radiation dose adds considerable strength to the dose-response information now available.

• Some of the existing human and animal data on radiation-induced cancers are derived from populations exposed to internally deposited radionuclides for which dose-incidence relationships are influenced by marked nonuniformities in the temporal and spatial distribution of radiation within the body.

• Some of the human data concern cancer mortality; others, cancer incidence. It is appropriate to distinguish radiation-induced cancers that may not greatly alter the death rate (e.g., skin and thyroid cancer) from others that are generally fatal (e.g., leukemia and lung cancer).

• It is not yet possible to estimate precisely the risk of cancer induction by low-dose radiation, because the degree of risk is so low that it cannot be observed directly and there is great uncertainty as to the dose-response function most appropriate for extrapolating in the low-dose region.

Studies by a number of scientists who have claimed a greater carcinogenic effect due to exposure to low-dose ionizing radiation than generally accepted are reviewed in Appendix B. None of these studies was considered by the Committee to constitute reliable evidence at present for use in risk estimation, for various reasons, including inadequate sample size in some instances, inadequate statistical analysis, and unconfirmed results.

Despite the difficulties and uncertainties, a clear-cut increase in incidence or mortality with increasing radiation dose has been demonstrated for many types of cancer in human populations, as well as in laboratory animals. At the time of the 1972 BEIR report,[56] almost all evidence of radiation-induced cancer was from observations at relatively high doses and high dose rates. This is still the case, although there is now somewhat more dose-response information related to lower doses. Most of the information now available is reasonably consistent from one irradiated human population to another; this suggests that it can be applied to the general population for purposes of risk estimation.

PROBABILITY OF CANCER INDUCTION AT LOW DOSES AND LOW DOSE RATES

There are two questions of major interest: (1) Will such effects as may be calculated with the use of the available risk factors actually occur in a general population exposed to tens or a few hundreds of millirads of low-LET radiation per year in addition to the natural background of about 100 mrems/yr? (2) Will such effects actually occur in an occupational population exposed to about 0.5–5 rems/yr in addition to the natural background and medical exposures? With respect to question 1, the 1972 BEIR report[56] stated that

expectations based on linear extrapolation from the known effects in man of larger doses delivered at high dose rates in the range of rising dose-incidence relationship may well overestimate the risks of low-LET radiation at low dose rates and may, therefore, be regarded as upper limits of risk for low-level low-LET irradiation. The lower limit, depending on the shape of the dose-incidence curve for low-LET radiation and the efficiency of repair processes in counteracting carcinogenic effects, could be appreciably smaller (the possibility of zero is not excluded by the data). On the other hand, because there is greater killing of susceptible cells at high doses and high dose rates, extrapolation based on effects observed under these exposure conditions may be postulated to underestimate the risks of irradiation at low doses and low dose rates. (p. 88)

The present Committee endorses this view. It is by no means clear whether dose rates of gamma or x radiation of about 100 mrads/yr are in any way detrimental to exposed people; any somatic effects would be masked by environmental or other factors that produce the same types of health effects as does ionizing radiation. It is unlikely that carcinogenic effects of low-LET radiation administered at this dose rate will be demonstrated in the foreseeable future. Notwithstanding these limitations, the Committee recognizes the need to estimate the effects on human populations exposed to radiation at very low doses. In most cases, the linear hypothesis, as the 1972 BEIR report indicated, probably overestimates, rather than underestimates, the risk from low-LET radiation. For high-LET radiation, such as from internally deposited alpha-emitting radionuclides, the application of the linear hypothesis is less likely to lead to overestimates of risk and may, in fact, lead to underestimates. (See Figure II-2, Chapter II, for the equations describing the linear, the quadratic, and the linear-quadratic functions.)

In studies of animal or human populations, the shape of a dose-response relationship at low doses may be practically impossible to ascertain statistically. This is because the sample sizes required to estimate or test a small absolute cancer excess are extremely large; specifically, the required sample sizes are approximately inversely proportional to the square of the excess. For example, if the excess is truly proportional to dose and if 1,000 exposed and 1,000 control subjects are required to test the cancer excess adequately for 100 rads, then about 100,000 in each group are required for 10 rads; and about 10,000,000 in each group are required for 1 rad. Experimental evidence and theoretical considerations are more likely than empirical data to guide the choice of a function for radiation carcinogenesis in humans.

In regard to question 2, the Committee believes that a distinct carcinogenic effect could be discernible for the large doses that may be associated with lifetime occupational exposure.

RELATIVE BIOLOGIC EFFECTIVENESS

There is substantial evidence, from both epidemiologic and experimental data, of wide variation in relative biologic effectiveness (RBE) for different types of ionizing radiation. This variation, which is related to differences in the microdistribution of radiation energy deposited in the tissues and of linear energy transfer (LET), may cause a given absorbed dose to differ in its biologic effect by a factor of 20 or more, depending on the type of radiation. The RBE for a given type of radiation is defined as the ratio between the doses of a reference radiation and the type in question that produce the same biologic effect. The wide variations in RBE pertain directly to the interpretation of epidemiologic data from several of the important available sources—atomic-bomb survivors of Hiroshima, underground miners exposed to radon gas and its radioactive decay products, and a number of populations with relatively high body burdens of alpha-emitting radionuclides.

Many radiobiologic experiments indicate that the risk per rad for low-LET radiation, such as x rays and gamma rays, decreases to a greater degree with decrease in the dose and dose rate than does the risk for high-LET radiation. Hence, the RBE of high-LET radiation can be expected to increase with decreasing dose and dose rate. For radiation-protection purposes, the RBE for fast neutrons relative to gamma radiation has been fixed at 10 by standard-setting organizations (e.g., ICRP, NCRP). However, this Committee has chosen not to use an arbitrary average RBE for fission neutrons in its calculations, but to derive RBE estimates from the Hiroshima and Nagasaki data.

The data available on human populations exposed to alpha-emitters (e.g., underground miners, Thorotrast- or radium-treated patients, and radium-dial painters) indicate that, for cancer production, alpha radiation is many times more effective per rad of average tissue dose than are x rays or gamma rays delivered at high dose rates. Epidemiologic and experimental data suggest that the effect per dose of alpha radiation at low dose rates (i.e., because of protraction or fractionation) is greater than that at high dose rates.

DOSE-RESPONSE MODELS

The sampling requirements for direct observations of cancer risk at low doses are so formidable that estimation can be done only by applying one or another dose-response model to exposure data that include observations at doses high enough to give fairly stable risk estimates.

The cancer-risk estimates presented in the 1972 BEIR report[56] for whole-body exposure were derived from observations at doses generally of a hundred or more rads. These linear-model estimates have been criticized on the grounds that the increment in cancer risk per rad may well depend on dose and that the true risk at low doses may therefore be lower or higher than the linear model predicts.

In animal experiments, it has been shown, often with considerable statistical precision, that the dose-effect curve for radiogenic cancer can have a variety of shapes (sometimes including even a negative initial slope). As a rule, the curve has positive curvature for low-LET radiation, i.e., the slope of the curve increases with increasing dose. However, at higher doses (around 100 rads or more), the slope often decreases and may even become negative. Dose-effect curves may also vary with the kind of cancer, with species, and with dose rate. On the basis of extensive experimental work and current microdosimetric theory, the Committee has adopted a parametric family of functions as a general dose-response model for low-LET radiation. This family is the product of a general quadratic ("linear-quadratic") form representing carcinogenesis and an exponential form representing the competing effect of cell-killing often suggested by experimental data in which the observed dose response has declined at high doses. The cell-killing phenomenon has been less often suggested by epidemiologic data, particularly those involving whole-body exposure; in this report, only the linear-quadratic form representing carcinogenesis and its limiting forms having either only a linear term in dose ("linear") or only a dose-squared term ("pure quadratic") have been fitted to human data.

Human populations are genetically more diverse than the inbred animal strains used in most experimental studies. The existence in man of subsets of high or low susceptibility to radiation carcinogenesis could very well influence the dose-effect curve. The most likely effect of such diversity is probably a tendency toward linearity, although the existence of exquisitely sensitive subgroups of suitable size conceivably would produce a dose-response curve that showed a greater effect per rad at very low doses than at high. The hypothesis of sensitive subgroups does not itself suggest a particular shape for the dose-effect curve.

For the most part, the available human data fail to suggest any specific dose-response model and are not robust enough to discriminate among *a priori* models suggested by theoretical and experimental work. However, there are exceptions. For example, cancer of the skin is not observed at low doses, and dose-response relationships observed in the Nagasaki leukemia data appear to have positive curvature. The incidence of breast cancer seems to be adequately described by a linear dose-response model.

The Committee was in general agreement that, for most radiation-induced solid cancers, the dose-response relationship for low to intermediate doses of low-LET radiation is best described by a linear-quadratic function of dose with nonnegative curvature. Nevertheless, there are arguments in favor of other models, especially the linear and the quadratic, which lead to widely divergent estimates. For these reasons, and because of the basic uncertainty associated with the choice of a single model, the Committee decided to present an envelope of estimates bounded by the linear and the pure quadratic models, with the linear-quadratic providing intermediate values.

RISK ESTIMATION

The quantitative estimation of the carcinogenic risk of low-dose, low-LET radiation is subject to numerous uncertainties. The greatest of these concerns the shape of the dose-response curve. Others pertain to the length of the latent period, the RBE for fast neutrons and alpha radiation relative to gamma and x radiation, the period during which the radiation risk is expressed, the model used in projecting risk beyond the period of observation, the effect of dose rate or dose fractionation, and the influence of differences in the natural incidence of specific forms of cancer. In addition, uncertainties are introduced by the characteristics of the human experience drawn on for the basic risk factors, e.g., the effect of age at irradiation, the influence of any disease for which the radiation was given therapeutically, and the influence of length of followup. Moreover, these uncertainties, unlike sampling variation, cannot be summarized in probabilistic terms; their collective influence is such as to deny great credibility to any estimates that can now be made for low-dose, low-LET radiation. Therefore, the Committee has placed more emphasis on methods of estimation than on any numerical estimates derived thereby.

The chief sources of data used in this report are the populations exposed to whole-body irradiation in Hiroshima and Nagasaki, patients with ankylosing spondylitis and other patients who were exposed to partial-body irradiation therapeutically, and various occupationally exposed populations, such as uranium miners and radium-dial painters. Most epidemiologic data do not systematically cover the range of low to moderate doses for which the Japanese data appear to be fairly strong. Analysis in terms of dose response must therefore rely heavily on the Japanese atomic-bomb survivor data. The substantial neutron component of the dose in Hiroshima and its strong correlation with gamma dose severely limit the relevance of the more numerous Hiroshima data for the estimation of risk from low-LET radiation. The Nagasaki data, for which the neutron component of dose is small, are weaker for doses below 100 rads; it is necessary, therefore, to obtain the maximal benefit from the Hiroshima data. In any analysis of the Japanese data that attempts to separate the effects of neutrons and gamma rays, however, the gamma-ray coefficients are determined mainly by the Nagasaki data. With respect to incidence, the Hiroshima data are known to be incomplete, and some estimates have been computed from the Nagasaki Tumor-Registry data alone.

For its illustrative computations of the lifetime risk from whole-body exposure, the Committee chose three situations:

- A single exposure of a representative (life-table) population to 10 rads.
- A continuous, lifetime exposure of a representative (life-table) population to 1 rad/yr.

• An exposure to 1 rad/yr over several age intervals exemplifying conditions of occupational exposure.

The three exposure situations do not reflect any circumstances that would normally occur, but embrace the areas of concern—general population and occupational exposure and single and continuous exposure. Below these doses, the uncertainties of extrapolation of risk were believed by some members of the Committee to be too great to justify calculation. The selected annual exposure, although only one-fifth the maximal permissible dose for occupational exposure, is consistent with average occupational exposures in the nuclear industry. The U.S. 1969–1971 life tables[54] were used as the basis for the calculations, and all results were expressed in terms of excess cancers per million persons throughout their lifetime after exposure. Although in the 1972 BEIR report[56] estimates were made for an expression time of 25 yr for leukemia and either 30 yr or a full lifetime after exposure for other forms of cancer, in the present report the expression time is taken as 25 yr for leukemia and the remaining years of life for other cancers. Separate estimates were made for mortality and incidence.

The resulting mortality estimates for all forms of cancer differ by as much as an order of magnitude; and, clearly, some are more plausible than others. The uncertainty derives chiefly from the range of dose-response models used, from the alternative projection models, and from the sampling variation in the source data. The lowest estimates are derived from the pure quadratic model; the highest, from the linear model. The constrained linear-quadratic model provides estimates intermediate between these two extremes.

In the absence of any increased radiation exposure, among one million persons of life-table age and sex composition in the United States, about 164,000 would be expected to die from cancer, according to present mortality rates. For a situation in which these one million persons are exposed to a single dose increment of 10 rads of low-LET radiation, the constrained linear-quadratic model predicts about 766 additional deaths from all forms of cancer according to one projection model and about 2,255 according to the other—increases of about 0.5% and 1.4% over the normal expectation of cancer mortality, respectively (see Table V-1 and similar tables, which demonstrate the range of estimates that results from use of alternative projection models).

Table V-2 shows the variation in mortality risk estimates associated with the choice of dose-response model and projection model for a single exposure of the general population to 10 rads of low-LET radiation. In that exposure situation, the choice of dose-response model involves dif-

TABLE V-1 Estimated Excess Mortality per Million Persons from All Forms of Cancer, Linear-Quadratic Dose-Response Model for Low-LET Radiation[a]

	Absolute-Risk Projection Model	Relative-Risk Projection Model
Single exposure to 10 rads		
Normal expectation	163,800	163,800
Excess cases: number	766	2,255
% of normal	0.47	1.4
Continuous exposure to 1 rad/yr, lifetime		
Normal expectation	167,300	167,300
Excess cases: number	4,751	11,970
% of normal	2.8	7.2

[a] Intermediate results in this table and throughout the report, including estimated risk coefficients and numbers of normally expected and excess cancer deaths and cases, are given to four significant digits. The intention is to facilitate the reconstruction of the final results by readers who may wish to reconstruct them, rather than to suggest an unwarranted accuracy of the estimates.

TABLE V-2 Estimated Excess Mortality per Million Persons from All Forms of Cancer, Single Exposure to 10 Rads of Low-LET Radiation, by Dose-Response Model

Dose-Response Model			Absolute-Risk Projection Model	Relative-Risk Projection Model
Leukemia and Bone	Other Cancer			
		Normal expectation of cancer deaths	163,800	163,800
LQ-L	$\overline{LQ\text{-}L}$	Excess deaths: number	766	2,255
		% of normal	0.47	1.4
L-L	$\overline{L\text{-}L}$	Excess deaths: number	1,671	5,014
		% of normal	1.0	3.1
Q-L	$\overline{Q\text{-}L}$	Excess deaths: number	95	276
		% of normal	0.058	0.17

ferences of about 20 to 1, and the choice of projection model, differences of 3 to 1.

Table V-3 shows that, for continuous lifetime irradiation at 1 rad/yr beginning at birth, risk estimates increase roughly in proportion to the increase in duration of exposure and, therefore, total dose (from 10 rads to a lifetime dose of about 75 rads). Here, the normal lifetime expectation of dying of cancer is 167,000 per million births.

For continuous lifetime exposure to 1 rad/yr, the increase in cancer mortality, according to the linear-quadratic model, ranges from about 3% to 8% over the normal expectation (Table V-1).

To compare these estimates with those of the 1972 BEIR report[56] and the 1977 UNSCEAR report,[82] it is convenient to express them as cancer deaths per million persons (including both sexes) per rad of continuous lifetime exposure. The age-specific risk factors in the 1972 BEIR report were used with 1969–1971 life tables and with the computation procedures developed for the present report. In this form, the risks are based on average values per rad received over a lifetime and should not be taken as estimates of the excess for a single dose of 1 rad. The linear-quadratic dose-response model for low-LET radiation yielded estimates below the comparable linear estimates given in the 1972 BEIR report, especially for the relative-risk model (Table V-4). For continuous lifetime exposure to 1 rad/yr, the relative-risk projection is 169 excess

TABLE V-3   Estimated Excess Mortality per Million Persons from All Forms of Cancer, Continuous Lifetime Exposure to 1 Rad/Yr of Low-LET Radiation, by Dose-Response Model

Dose-Response Model			Absolute-Risk Projection Model	Relative-Risk Projection Model
Leukemia and Bone	Other Cancer			
		Normal expectation of cancer deaths	167,300	167,300
LQ-L	L̄Q̄-L̄	Excess deaths: number	4,751	11,970
		% of normal	2.8	7.2
L-L	L̄-L̄	Excess deaths: number	11,250	28,690
		% of normal	6.7	17.2
Q-L	Q̄-L̄	Excess deaths	a	a

[a] Estimates were not calculated, because Q-L coefficients were very small—0.014 excess leukemia and bone cancer per million per year per rad[2] and 0.018 excess fatal cancer other than leukemia and bone cancer.

TABLE V-4 Comparative Estimates of the Lifetime Risk of Cancer
Mortality Induced by Low-LET Radiation—Excess Deaths per Million,
Average Value per Rad by Projection Model, Dose-Response Model,
and Type of Exposure

| | | Projection Model | | | |
| | | Single Exposure to 10 Rads | | Continuous Lifetime Exposure to 1 Rad/Yr | |
Source of Estimate	Dose-Response Models[a]	Absolute	Relative	Absolute	Relative
BEIR, 1980[b]	$LQ\text{-}L, \overline{LQ\text{-}L}$	77	226	67	169
1972 BEIR report factors[c]	Linear	117	621	115	568
UNSCEAR 1977[d]	Linear			75–175	

[a] For BEIR 1980, the first model is used for leukemia, the second for other forms of cancer. The corresponding estimates when the other models are used (thereby providing an envelope of risk estimates) are:

	$L\text{-}L, \overline{L\text{-}L}$	167	501	158	403
	$Q\text{-}L, \overline{Q\text{-}L}$	10	28		

[b] The values are average values per rad, and are not to be taken as estimates at only 1 rad of dose.

[c] 1972 BEIR[56] postnatal, age-specific risk factors used with 1969–1971 life tables, with plateau extending throughout the lifetime remaining after irradiation, estimate (b) in the 1972 report. The average age of the 1969–1971 life-table population exceeds that of the 1967 U.S. population used for the 1972 BEIR report. For this reason, the numbers shown here for continuous exposure are larger, on a per-rad basis, than those obtainable from Tables 3-3 and 3-4 of the 1972 BEIR report.

[d] UNSCEAR range of estimates for low-dose, low-LET radiation (p. 414, para. 318).[82]

cancer deaths per million persons exposed per rad, compared with 568
in the 1972 BEIR report, and the absolute-risk projection is 67 deaths,
compared with 115. Although the present report uses much information
not available for the earlier report, the differences mainly reflect differences in the assumptions made by the two Committees. The present
Committee preferred a constrained linear-quadratic, rather than the
linear, dose-response model for low-LET radiation and preferred not to
assume a fixed relationship between the effects of high- and low-LET
radiation. There is an additional difference between the two reports
with respect to the relative-risk projections: the present estimates do not,
as in the 1972 report, carry through to the end of life very high relative-
risk coefficients obtained with respect to childhood cancers induced by
radiation; this accounts for about half the magnitude of the 1972 rela-

tive-risk estimates. The present linear-quadratic estimates do not differ appreciably from those in the 1977 UNSCEAR report.[82] The comparison differs when made on the basis of the alternative BEIR 1980 models, because the linear model estimates in the first footnote of Table V-4 are about twice as large as the linear-quadratic model estimates, and the pure-quadratic model estimates are substantially smaller than the linear-quadratic.

Cancer-incidence risk estimates are less firm than mortality estimates. The present Committee used a variety of dose-response models and several data sources. The dose-response models produced estimates that differed by more than an order of magnitude. The different data sources gave broadly similar results. In particular, analysis of the Nagasaki Tumor Registry data agreed well with sex-specific expansions of estimates of excess cancer mortality. For the linear-quadratic model and for continuous lifetime exposure to 1 rad/yr, the increased risks, based on the Nagasaki Tumor Registry and on the leukemia registries in Hiroshima and Nagasaki and expressed as percent of the normal incidence of cancer in males, were: (a) relative-risk projection model, 3.1%, and (b) absolute-risk projection model, 2.0%. As previously mentioned, risks for females are substantially higher than those for males, owing primarily to the relative importance of radiation-induced thyroid and breast neoplasia.

Estimates of excess risk for individual organs and tissues depend in large part on partial-body irradiation and use a wider variety of data sources. Except for leukemia and bone cancer, estimates for individual sites of cancer were made only on the basis of the *linear* model (cf. Appendix A) and are stated in terms of excess cases per year per million persons exposed per rad. For leukemia, the *linear-quadratic* model yielded: for males, excess leukemia cases (or deaths) per million persons per year = 1.37(gamma dose) + 0.0117(gamma dose)^2; for females, excess leukemia cases (or deaths) per million persons exposed per year = 0.904(gamma dose) + 0.00772(gamma dose)^2. For solid tumors, parallel *linear*-model estimates, expressed as excess cancer cases per year per million persons exposed per rad, were, for example: for thyroid, male, 2.2, and female, 5.8; for female breast, 5.8; and for lung, male, 3.6, and female, 3.9. These coefficients of risk derive largely from epidemiologic data in which exposure was at high doses, and these risk values may, in some cases, overestimate risk at low doses. They yield final estimates of excess solid tumors that are 3–6 times those derived by expanding the mortality estimates on the basis of the linear-quadratic model.

GENERAL CONSIDERATIONS

In this chapter, we consider the effects of radiation in those who are exposed (somatic effects), as distinct from effects transmitted to their offspring genetically. Effects on the developing embryo and fetus irradiated *in utero* and other noncancer somatic effects are described in Chapter VI.

This report does not address pathologic changes caused by high doses of radiation delivered at high dose rates. The primary emphasis is on effects of relatively low doses of radiation—that is, below 100 rads—because we anticipate that most future exposures of radiation workers and the general public should be well below such doses. In addition, we consider only effects of radiation that lead to disease or disability; thus, we do not deal with cytologic or cytogenetic abnormalities, which may be detected in cells of persons exposed to relatively low doses, but whose significance with respect to later disease or disability remains unclear (see discussion of cytogenetic effects in the 1977 UNSCEAR report[82]).

The effects considered here are expressed in defects or in disease states that usually arise many years after radiation exposure. These diseases are, however, similar to those occurring spontaneously in man. Thus, detection of the contribution of radiation exposure is often difficult and depends heavily on an adequate duration of followup of irradiated groups. An important justification for this updating of the 1972 BEIR report is therefore the need to assess the extent to which our understanding of delayed or long-term effects of radiation exposure has been altered by new evidence, particularly from further followup of a number of human groups irradiated under a variety of circumstances. Especially important have been data obtained from such populations exposed to doses well below 100 rads; effects of these low doses are usually difficult to distinguish from spontaneous disease, so adequate followup is crucial. Regrettably, there is still a paucity of information at these low doses, but additional studies have become available since the 1972 BEIR report.[56]

In the 1972 BEIR report, the Committee tried to estimate the risk of somatic effects in an entire population that received a cumulative radiation exposure at a low dose rate. The principal risk was from the major somatic effect produced—cancer. The present Subcommittee on Somatic Effects has also made quantitative risk estimates, especially for cancer, albeit with some modifications of the methods used in the 1972 BEIR report.

This chapter emphasizes risk estimates derived from epidemiologic data on human populations, as was the case in the 1972 report. Chapter II gives some of the scientific background on how risks are assessed by these methods. We include some discussion of laboratory animal data where relevant, to emphasize the importance of laboratory animal studies to an understanding of effects on man (see Appendix A).

Two long-term studies of irradiated human populations are of special interest, because of the numbers of persons involved, the fraction of the body exposed to radiation, and the duration of followup: the Japanese survivors of the atomic bombings of Hiroshima and Nagasaki and a group of British patients with ankylosing spondylitis who were treated with deep x-ray therapy to the spine and pelvis. Because of their importance and because new information available from them since the 1972 BEIR report has added to our understanding of cancer induction by radiation, these two studies are described in some detail in this chapter. Important new dosimetric data from these two studies are also included.

At the time of the 1972 BEIR report, the Subcommittee on Somatic Effects concluded that evidence on man was insufficient to determine whether the absolute-risk or the relative-risk projection model (see Chapter II) of radiation-induced cancer was more applicable to the determination of lifetime risks on the basis of incomplete followup data. Review of the current data has led the present Subcommittee to conclude that the relative-risk model does not apply generally, but is applicable to the effect of age on cancer incidence for many sites at which cancer is induced by radiation. Thus, age at exposure and at cancer development has emerged as a major determinant of cancer risk from radiation. For this reason, this subject is also considered in some detail; both projection models have been used.

THE EXPERIENCE OF THE ATOMIC-BOMB SURVIVORS AS A SOURCE OF DATA ON THE LATE SOMATIC EFFECTS OF IONIZING RADIATION

The experience of the survivors of the atomic bombings of Hiroshima and Nagasaki constitutes a major source of information on the late somatic effects of ionizing radiation.[58] Some of the characteristics that make it extremely useful for radiobiologic research are as follows: the samples available for study are generally the largest of their kind; the population was relatively unselected with respect to disease or working status; a formal program of intensive study has been under way for over 30 yr; an elaborate dosimetry program has yielded individual dose estimates for the major samples, initially in terms of kerma, but recently for the tissues of major organs; and the family registration system of Japan guarantees virtually 100% mortality followup. There are, however, disadvantages and limitations: the samples available are not large enough to detect with high probability some of the presumably smaller risks, e.g., in the very-low-dose region; there was a single exposure at a very high dose rate; the radiation was a mixture of neutrons and gamma rays; the radiation dose of a survivor depended on his location and shielding situation, i.e., there was no random assignment; the energy released by the bombs was in the form of heat and blast, as well as radiation; each city was so devastated that living patterns were profoundly disrupted; and the fact that tens of thousands were killed by the bomb in each city raises the possibility that the survivors were, in some respects, fitter than those who died.

At the time of the 1950 national census in Japan, 284,000 survivors were enumerated by means of a supplementary schedule,[37] and this is the source that was used in selecting survivors for the life-span study (LSS) sample of which the adult health study (AHS) sample is a part. Table V-5 lists the major fixed samples from which data have been derived on late somatic effects; information on cataractogenesis was developed from *ad hoc* samples before the Atomic Bomb Casualty Commission (ABCC) adopted its relatively strict fixed-sample approach. The LSS sample, is the primary source of information on mortality, but is also used in studies on the incidence of cancer[36,50] and heart disease. For this sample, observations began October 1, 1950. The AHS sample is the vehicle for biennial physical examinations and history-taking[13] and also for much of the cytogenetics work.[3] The two *in utero* samples are based on birth certificates and early special censuses,[12] and their functions parallel those of the LSS and AHS samples.

TABLE V-5   Major Fixed Samples[a] Studied at the Radiation Effects Research Foundation (Formerly Atomic Bomb Casualty Commission)

Sample	Year Observations Begun	Size
Life-span study (LSS) sample (extended)	1950	109,000
Adult health study (AHS) sample[b]	1958	20,000
In utero mortality sample	1945	2,800
In utero morbidity sample[c]	1950	1,600

[a] All samples include some representation of those who were not in the city at the time of the bombing, i.e., not directly exposed.

[b] A subsample of the life-span study sample.

[c] Progressively enlarged from 1950 to 1959.

The dosimetry of atomic-bomb survivors rests on the exposure histories of survivors obtained by ABCC technicians[57] and the work of the Ichiban group of the Health Physics Division, Oak Ridge National Laboratory (ORNL),[2,41] supplemented in recent years by parallel research at the National Institute of Radiological Sciences (NIRS), Chiba, Japan.[32] Individual (T-65) dose estimates are available for all but 3% of the survivors in the LSS sample. Gamma and neutron components are estimated separately and expressed in terms of rads of tissue kerma* in air. The potential contribution of any fallout or residual radiation is believed to have been negligible.[57] Correspondingly, early entrants into the cities within the first few days after the bombings are not thought to have been exposed to substantial radiation. For selected organs of interest, the Oak Ridge group[41] and the NIRS group[32] have provided the basis for a preliminary conversion of the T-65 kerma dose into a tissue dose. Ratios (R) of absorbed tissue dose to kerma have been estimated for external gamma radiation as $R_\gamma = D_\gamma/K_\gamma$, for neutrons as $R_n = D_n/K_n$, and for n-capture gamma radiation as $R_c = D_c/K_n$,

TABLE V-6 Ratio of Organ Dose to Kerma Dose for Atomic-Bomb Survivors[a]

Organ or Tissue	Ratio, Organ Dose to Kerma Dose			
	D_γ/K_γ (R_γ)	D_n/K_n (R_n)	D_γ/K_n (R_c)	Average $H + N$[b] RBE = 1
Bone marrow	0.56	0.28	0.067	0.53
Bladder	0.45	0.18	0.072	0.42
Breast	0.80	0.55	0.045	0.77
Fetus	0.42	0.14	0.077	0.39
Intestinal tract	0.40	0.14	0.077	0.38
Kidney	0.52	0.24	0.065	0.49
Liver	0.47	0.18	0.075	0.44
Lung	0.50	0.22	0.070	0.47
Ovary	0.40	0.12	0.080	0.37
Pancreas	0.40	0.12	0.080	0.37
Stomach	0.47	0.18	0.072	0.44
Thyroid	0.70	0.45	0.035	0.67

[a] Data from Kerr.[41]

[b] For survivors exposed to 10+ rads, combined Hiroshima and Nagasaki average.

*Kerma (kinetic energy released in material) = a unit of quantity that represents the kinetic energy transferred to charged particles by the uncharged particles per unit mass of the irradiated medium.

for various organs, where D_γ, D_n, and D_c represent absorbed dose to tissue for gamma, neutron, and n-capture gamma radiation, respectively, and K_γ and K_n are the kerma doses for gamma radiation and neutrons, respectively.

The recent ORNL ratios[41] are given in Table V-6. They can be used to estimate the ratios of organ dose to average kerma dose for survivors exposed to 10$^+$ rads. Column 4 in Table V-6 shows the values for specific organs and tissues, on the assumption of an RBE of 1. The calculation requires only the ratios (R), the fraction of the kerma dose derived from gamma radiation (F_γ) and that derived from neutrons (F_n), and an RBE factor (Q) for neutrons. Then the conversion ratio for a kerma of 1 rad is equal to

$$F_\gamma R_\gamma + F_n R_c + Q F_n R_n.$$

The NIRS values of R for selected tissues are similar to those of ORNL, but not precisely the same, especially for R_γ. The effect of varying RBE may be seen in the following comparison by city, for several organs:

Organ	RBE	Organ dose (rems)/air kerma (rads)		
		H + N	H	N
Marrow	1	0.53	0.51	0.56
	5	0.69	0.77	0.57
	10	0.88	1.09	0.59
	15	1.07	1.41	0.61
Breast	1	0.77	0.75	0.80
	5	1.08	1.26	0.82
	10	1.46	1.89	0.86
	15	1.83	2.52	0.89
Lung	1	0.47	0.45	0.50
	5	0.59	0.65	0.51
	10	0.75	0.90	0.53
	15	0.90	1.16	0.54

Ratios like those in Table V-6 may be used to adjust any linear coefficient of the combined gamma and neutron risk expressed in rads kerma to an approximate rad (or rem) organ-dose risk estimate by dividing the estimate expressed in rads kerma by the appropriate ratio. For example, McGregor *et al.* gave absolute-risk estimates of 1.8 excess breast cancers per million women per year per rad kerma for Hiroshima, and 2.0 for Nagasaki.[50] Division by the ratios 0.75 and 0.80 gives approximate organ-dose risk estimates of 2.4 and 2.5, respectively. These values are very close to those obtained by transforming average kerma values for the dose groups used in the regression calculation.

The completeness of ascertainment of disease states, impairments, laboratory abnormalities, and deaths is variable, as is the quality of the observations made. By virtue of the family registration system of Japan and restriction of samples to those of Japanese citizenship and known place of family registration, the ascertainment of mortality is virtually complete.[10] Clinical and laboratory observations have been less complete, but participation has been consistently above 80% for those living within the range of contacting, and essentially independent of dose or exposure status.[13] Death-certificate diagnoses differ greatly in their accuracy, even within the set of neoplastic diseases; but for the atomic-bomb survivors, an active autopsy program in the period 1961–1969 has provided unusually good information on errors in death-certificate diagnoses.[75] The quality and uniformity of laboratory observations and the cytogenetics have been controlled at a high level, whereas the clinical observations have been more variable in quality.

Special disease registries are of particular importance. A registry was set up for leukemia very early, and efforts to ensure its completeness, combined with binational reviews of the series, have made the leukemia registry invaluable for epidemiologic studies.[36] In 1957–1958, tumor registries were established in both cities, and these were supplemented in the early 1970s by the so-called "tissue registries" for neoplasms that include permanent slide collections and review diagnoses. The tumor-registry data for 1959–1970 have been used in parallel with the analyses of mortality in the LSS series for 1950–1974;[9] but, because of their incompleteness, especially in Hiroshima, less confidence can be placed in them than in the death-certificate data. The tumor-registry data do, however, have some value for selected sites of cancer often not well recognized by the certifying physicians as primary causes of death, e.g., lung, liver, and pancreas. In Nagasaki, where reporting was more complete, there is no reason to suppose that ascertainment depended in any way on radiation dose, but these data have not been critically examined. Estimates of excess risk for breast and lung cancer based on tumor-registry data for Hiroshima, where the opportunity for ascertainment bias was greater than in Nagasaki, are consistent with estimates based on the mortality ascertainment, which is unbiased as to dose. When tumor-registry data are used here, they are limited to cases within the fixed LSS cohort, as is true with death-certificate data. Because local registries do not cover migrants from the areas, the registry counts are necessarily incomplete and absolute-risk estimates correspondingly low; but migration from the cities has been carefully monitored and is known to be relatively independent of radiation dose and to be appreciable for only the youngest members of the cohort.

There are no true "controls" in the experimental sense, but the not-in-city (NIC) components of the samples have some value as a comparison group and are often used in conjunction with low-dose groups in making high-dose–low-dose comparisons. The zero-dose group is believed to be superior to the NIC group as a control group, however, especially in the early years of followup, largely because the NIC group has the characteristics of an immigrant population with a different medical history before 1950.[7] Unfortunately, although the zero-rads group amounts to 48% of the Hiroshima sample, it is only 23% of the smaller Nagasaki sample and often seems too small to stand alone, especially in the examination of end results of low frequency. For this reason, the 0–9-rads group is often used as the basis for statistical comparisons with higher-dose groups. The 0–9 component represents 22% of the Hiroshima survivors and 33% of the Nagasaki survivors; their average doses were 3.7 and 3.9 rads (kerma), respectively. Increasingly, however, the need for information about dose response requires that various parametric functions be fitted to data arrayed by size of dose, and simple case-control comparisons are replaced by regression analyses. In this context, the NIC group may or may not be used. The use of all-Japan age-, sex-, and time-specific death rates as a basis for calculating expected deaths among atomic-bomb survivors, to be compared with observed deaths, is valid only when the national mortality rates coincide with those for the cities of Hiroshima and Nagasaki, and often this is not the case. Investigators at ABCC have generally preferred not to draw inferences about the effects of radiation from such comparisons and have used the national rates only as a device to standardize dose-specific mortality-risk estimates for age and sex, preparatory to comparing the dose groups directly.[40]

The general applicability of risk estimates based on the experience of the atomic-bomb survivors has been questioned by Rotblat[67] and by Kneale and Stewart.[42] Impressed by the failure of investigators to derive positive evidence of genetic effects of the atomic radiation, by the absence of evidence of a general life-shortening proportional to dose, and by the absence of evidence of a carcinogenic effect of fetal irradiation, Rotblat[67] argued that the survivors of the bombings might have been genetically selected for a lower sensitivity to the late effects of ionizing radiation. He also cited published estimates of leukemia incidence among early entrants into the cities after the bombings that seemed to show that their leukemogenic response was considerably stronger than that of survivors directly exposed to comparable doses. Kneale and Stewart[42] argued that, when cancers originate in the reticuloendothelial system (RES tumors), they may cause loss of immunologic competence before they are clinically recognizable and thus pave the way for lethal infections. Because the atomic-bomb survivors had high mortality rates from infectious diseases for several years after the bombings, their argument continues, atomic-bomb victims with early radiogenic cancers may have succumbed to fatal infections to such an extent that estimates of carcinogenic risk based on the atomic-bomb experience would not be generally applicable to populations for which radiation protection guidelines are written.

Whether the risk estimates derived from the experience of the Japanese atomic-bomb survivors are generally applicable is best determined empirically, by applying the test of consistency with other human data. When this is done with attention to age at irradiation, quality of radiation, attenuation of external whole-body radiation by the various tissues of the body itself, and length of followup, risk estimates derived from the atomic-bomb experience are seen to be generally consistent with those based on other human exposure. The only very marked exception is the absence of a carcinogenic effect among those exposed *in utero* at Nagasaki and Hiroshima,[39] in contrast with the significant relative-risk estimates of Stewart and Kneale[76] and of MacMahon[47] for prenatal x-ray exposure. Other apparent differences in carcinogenic effect—e.g., on tissues of the lung and stomach—apply only to the Nagasaki experience and may ultimately be better understood in terms of the shape of the dose-response curve for low-LET radiation.

That genetic effects have not thus far been found does not necessarily argue against the general applicability of the atomic-bomb experience: no direct evidence of a genetic effect has been forthcoming for man, and presumptions as to the order of magnitude of any such effect (see Chapter IV) suggest that it is too small to be easily seen in samples of the size available to investigators in Japan. Nor is the absence of a general life-shortening effect any indication that the atomic-bomb experience is a dangerous basis for generalization: testing the hypothesis of radiation-accelerated aging is no longer a promising line of experimental investigation and has been replaced by the view that radiation at moderate-to-low doses (under 300 rads of low-LET radiation) shortens life principally, and perhaps exclusively, by induction or acceleration of neoplastic disease.[87] The incidence of leukemia among "early entrants," although the subject of several publications, remains essentially unknown. The published data are inconsistent with what we know of the leukemogenic response from other human data and are not of a quality to challenge estimates of the leukemogenic effect of ionizing radiation derived from the experience of those directly exposed to the atomic bombings.[11] In any event, even for Nagasaki, the latter estimates are no lower than those obtained from other human observations.[82]

The argument of Kneale and Stewart[42] rests on observations, made in the Oxford Survey of Childhood Cancers, that many childhood illnesses and injuries are more frequent before death from cancer than in live controls, and progressively so as death approaches. The argument that early-occurring radiation-induced cancers would not come to light because of an "exceptionally high infection death rate for several years after the event" is not borne out. There were no major epidemics in Hiroshima and Nagasaki, perhaps because all who could do so fled the bombed areas.[59] Whether the excess mortality that continued for perhaps 3 yr among the more proximally exposed[77] differentially removed from observation those in whom a radiation-induced carcinogenic process had already started seems doubtful, for the following reasons: the leukemogenic effect began in 1948, peaked in 1951–1952, and then fell to nearly zero by 1974, in a temporal pattern quite like that seen in the ankylosing-spondylitis patients;[73] for solid tumors, minimal latent periods are much longer, and one would have to suppose that a carcinogenic process normally latent for many years would have such profound effects in 1945–1948 as to have influenced survival; and, in most instances, where close comparisons can be made, the atomic-bomb experience is not out of line with other human data.

THE STUDY OF LATE EFFECTS OF X-RAY TREATMENT OF BRITISH PATIENTS WITH ANKYLOSING SPONDYLITIS

In December 1954, the British Medical Research Council reviewed the evidence that leukemia mortality rates were rising sharply in the United Kingdom, as well as in many other countries, including the United States. Although some of the rise could be ascribed to better diagnostic criteria of the disease, recognized as a malignant neoplasm only in the 1930s, the impression in 1954 was that at least some of the more recent increase was real and reflected exposure to an environmental stimulus. In view of the mounting evidence that exposure to ionizing radiation could lead to increased risk of leukemia, especially the evidence from the Japanese atomic-bomb survivors,[44,53] the Council decided to initiate a study of late effects of exposure to radiation. Preliminary evidence had already been obtained by Court Brown and Windeyer that patients with ankylosing spondylitis given deep x-ray treatment to the spine and sacroiliac region had an increased risk of developing leukemia. This preliminary study was published in June 1955 by Court Brown and Abbatt. [21]

Accordingly, an appeal was made in January 1955 through the British medical press to locate patients with the disease who later were found to have leukemia. It was recognized, however, that case-finding alone could not settle the questions of the relationship of radiation dose to the probability of developing leukemia or other bone-marrow disease; therefore, an extensive survey of patients with ankylosing spondylitis was commissioned in 1955. This extensive epidemiologic investigation was put under the direction of Court Brown and Doll. With the aid especially of the directors of the radiotherapy clinics, within 9 mo the followup of over 13,000 patients was completed, radiation doses to the spinal marrow calculated, and dose-response data presented. [79] The full report of this phase of the study was published in 1957. [22]

The patients enrolled in the study were identified from records of 81 radiotherapy centers throughout the United Kingdom. All the patients had received x-ray therapy during the period 1935–1954. The treatment was given in "courses" lasting usually for 2 wk to 2 mo, generally with one to five treatments per week. On the average, about 10 individual radiation exposures over a period of a month constituted a single course, although there was considerable variation from patient to patient. Elaborate efforts were made in 1956 to determine the exposure dose to the spinal marrow. A 16% random sample of the entire study group was drawn from each of the clinics (stratified by number of treatment courses), and the mean spinal-marrow dose was calculated from the treatment records of each person in the sample.

By 1960, the study had been extended in two ways. First, the group treated with x rays before 1955 was somewhat increased by inclusion of patients from an additional six clinics, and the decision was made to investigate other causes of mortality by further followup. Second, records of a group of patients who were diagnosed as having ankylosing spondylitis at the same clinics, but whose records indicated that they had not been given x-ray therapy, were also collected to permit this "untreated" group to serve as a control for the irradiated group. The untreated patients were enrolled during the period 1935–1957.

In 1965, a further followup, to December 31, 1962, of the x-ray-treated group, now composed of more than 12,000 men and 2,300 women, was reported. [23] The results of this followup included an evaluation of the risk of cancers other than leukemia, as well as of all causes of death. This study, which was reviewed in the 1972 BEIR report, showed that excess mortality in this large patient population was present for many causes of death besides cancer; not only leukemia, but also cancers of organs in the heavily irradiated areas of the body were significantly greater than expected, whereas the excess was not significant for cancers at sites likely to be lightly irradiated.

Investigation of mortality in the patients not given x-ray therapy was carried out through 1967.[64,74] Among the untreated men, deaths from all causes were significantly greater than expected; and, for all causes except cancer, their mortality, compared with national statistics, was slightly greater than that of the group given x rays. On this basis, there is no indication that x-ray treatment caused vascular or other degenerative diseases that could be associated with accelerated aging. The evidence on this point is not strong, however; there may have been a difference in the severity of spondylitis between the untreated men and those given x rays. The untreated women clearly had less severe spondylitis than the treated group as a whole.

An important observation in the men was a high mortality rate in the period from their first clinic visit and enrollment in the study through the next year. The excess mortality arose from a small excess of cancer deaths, as well as from causes associated with spondylitis. A similar phenomenon was observed in the group given x-ray treatment;[23] the interpretation is that on their first visit a small proportion of the patients were terminally ill with spondylitis or its complications or had cancer that already involved the spine. In the latter case, because their symptoms were ascribed to reactivated spondylitis, they were included in the study, but they died of metastatic disease soon after enrollment.

Analysis of mortality after this initial period showed that the untreated group had no increased risk of death from cancer: through 1967, 21 cancer deaths were observed, compared with 21.51 expected, and there were no deaths from leukemia. Cancer was the only major category of cause of death that was not increased in the untreated group. Thus, the increased cancer mortality, observed 2 yr or more after x-ray treatment was begun in the group given radiation therapy, can reasonably be ascribed to the x-ray exposure.

A preliminary report dealing with followup of the x-ray-treated patients to January 1, 1970, has been published recently.[73] This was a study of patients who had received only one course of x-ray treatment; about half the original 14,000 patients were later given a second x-ray course, and followup was included only through 18 mo after the second course was begun. Thus, the period of study for this re-treated group was generally only a few years, and the primary long-term followup was for the group given only a single course. This makes the study especially valuable, because the time of the split-dose x-ray exposure can be clearly specified.

On January 1, 1970, the group not re-treated with a second course numbered 4,420 patients still alive and 1,759 who had died (477 had emigrated or were otherwise lost to followup). A more extensive report of this followup has been made available to the Committee (R. Doll and P. G. Smith, personal communication) and was included in the 1977 UNSCEAR report. [82]

Unfortunately, no radiation-exposure data for these patients have been published since the original very detailed analysis in 1957. [22] In the following section, the doses delivered to various tissues in this group are estimated. We believe that these estimates are valid at least to within 50%, and they have been used to calculate risk estimates from the epidemiologic evidence of excess cancers (by site) observed in these patients.

ESTIMATES OF RADIATION DOSES IN TISSUES AND ORGANS
IN THE SINGLE-COURSE RADIOTHERAPY PATIENTS TREATED
FOR ANKYLOSING SPONDYLITIS IN ENGLAND AND WALES

The available data on patients with ankylosing spondylitis who received a single treatment course with x rays in the Doll and Smith study[26,73] were reviewed, to estimate average radiation doses in tissues and organs giving rise to excess leukemias and cancers of heavily irradiated sites. It was not possible to review the radiotherapy charts of each patient, and it was therefore necessary to make assumptions on, for example, the selection of patients, the extent and severity of disease, the method of therapy, the radiotherapy dosimetry and exposures, the tissues irradiated, and the doses absorbed. The number of assumptions was kept to a minimum; it was recognized that the selection of patients and the clinical courses of treatment chosen by the radiotherapist in individual cases were, understandably, extremely variable. In spite of these limitations, however, it has been possible to make reasonable assumptions and to develop a model of what probably occurred in the radiotherapy planning and treatment of these patients, on the basis of conventional orthovoltage radiotherapy of the 1930s–1950s. The estimates of radiation doses absorbed thus derived, however, are imprecise and must be further corrected as new information becomes available, not only with respect to the assumptions made above, but also with respect to subtle information still lacking—for example, on the location of the organ during treatment and on the fraction of the organ or tissue that was irradiated.

Patients Studied

Of the original 14,558 patients in the ankylosing-spondylitis study group, 4,420 (30.4%) patients who had received only one course of treatment were studied by Doll and Smith.[26,73] The average period of followup for patients who had only one course of treatment was 16.2 yr. In addition, 52 patient histories and clinical courses of disease were carefully detailed in the Medical Research Council (MRC) report;[22] these patients had all developed leukemia after either single or multiple courses of x-ray therapy.

Radiotherapy Dosimetry and Treatment Planning

The radiotherapy dosimetry of patients was carefully reviewed and reconstructed in the MRC report.[22] The most likely treatment points, x-ray qualities, dose fractions, etc., were used, and depth-dose data for conventional orthovoltage x-ray therapy were used.[35] The patients were classified in two groups, on the assumption that about one-third in the series were treated in the late 1930s and early 1940s, and two-thirds were treated in the late 1940s and early 1950s. For the earlier patients in the series, it was assumed that the equipment used was a 100-kVp radiotherapy x-ray machine with a half-value layer (HVL) of 2-mm aluminum at a 30-cm focus-skin distance (FSD). For the later patients in the series, it was assumed that the equipment used was a 200-kVp radiotherapy x-ray machine with an HVL of 1-mm copper at a 50-cm FSD.

It was assumed that the entire spine (cervical, thoracic, lumbar, and sacral) and sacroiliac joints were treated in the single course of radiotherapy. Therefore, on the basis of the range of rectangular skin-field dimensions for spinal and sacroiliac fields described in the MRC report[22] and the fact that multiple fields were used, it was assumed that in this analysis depth-dose data for a 200-cm^2 field would be appropriate to estimate organ-dose characteristics. The position of the cervical, thoracic, lumbar, and sacral spine were determined from the data of Brinkley and Masters,[15] on the assumption that all ankylosing-spondylitis patients were treated in the prone position. The positions of the various organs and tissues of the thorax, abdomen, and pelvis were determined from contours and relationships from computed-tomography images, cadaver correlative anatomy published by Gambarelli *et al.*,[30] and descriptions in Gray's *Anatomy*;[31] it was recognized that computed-tomography patients and cadaver transverse sections were examined in the supine position.

Patient Selection, Treatment, and Clinical Course

To obtain some understanding of the rationale of patient selection and course of radiotherapy, the histories of the 52 ankylosing-spondylitis patients who developed leukemia outlined in the MRC report[22] were carefully reviewed. No selection process could be associated with the severity of disease at the time of the first course of radiotherapy or with the failure of palliation that warranted a second or additional courses of radiotherapy. Thus, it was assumed that these leukemia patients were no different from all other ankylosing-spondylitis patients when they began radiotherapy. For the purposes of the following analysis, therefore, it was assumed that these patients were not selected on the basis of severity of their disease or of any other predisposing factors and were therefore representative of all 14,558 patients in the study at the start of their radiotherapy for ankylosing spondylitis.

It was further assumed that the large group of patients who required retreatment (7,453, or 51.2%)[26,73] did not enter into their initial course of radiotherapy with a plan for retreatment. In other words, all patients were treated in the hope of palliating their disease in the first course of radiotherapy. There was no way for the radiotherapist to predict that a given patient in the series would require more than a single course of therapy for palliation, and thus each patient who received multiple courses of radiotherapy was initially judged to be a single-course patient and treated accordingly. This would obtain for all 14,558 patients, and therefore for the 52 patients who ultimately developed leukemia. It is of interest that 34 (65.4%) of the initial 52 patients who developed leukemia did receive additional courses of x-ray treatment.

Estimation of Mean Exposure of the Spinal Bone Marrow

On the basis of the clinical and radiotherapeutic histories of the 52 ankylosing-spondylitis patients, it was assumed that the general clinical trend was to begin with a single course of treatment. This resulted in a mean spinal bone-marrow exposure for the initial course of therapy for all 52 patients of 542 ± 355 R (Figure V-1). Thereafter, if a patient returned for additional radiotherapy because of recurring disease, each additional course of therapy up through the fourth course resulted in an increment of 346 ± 319 R in mean spinal bone-marrow exposure. Both in the initial treatment course and in later courses, the standard deviations were extremely large. It was assumed, therefore, that the average spinal-marrow exposure for all 4,420 patients who received a single

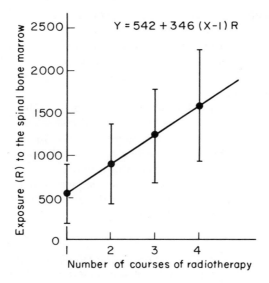

$$Y = 542 + 346\,(X-1)\,R$$

FIGURE V-1 Estimation of mean exposure of the
spinal bone marrow based on initial course of therapy
in 52 cases of ankylosing spondylitis.

course of therapy was 542 ± 355 R. The average exposure of the spinal bone marrow represented a very wide spectrum of exposures selected by the radiotherapist. This suggested extreme variability in treatment techniques and in the clinical response of patients.

The average exposure of the spinal bone marrow in the 18 single-course patients who developed leukemia[22] was 668 ± 325 R. This value is not significantly different from the mean spinal-marrow exposure of all 52 patients after the first course of radiotherapy.

Estimation of Radiation Doses Absorbed in Tissues and Organs in Heavily Irradiated Sites

On the basis of the assumptions outlined above, the average radiation doses to the spinal bone marrow and to the organs and tissues in heavily irradiated sites have been calculated (Table V-7). It was assumed that all radiation-induced leukemias and cancers arose in irradiated tissues and organs.

The estimated absorbed dose for leukemia was based on the assumption that spinal bone marrow constitutes 42.3% of the active bone marrow and the assumption that leukemia arose in irradiated bone marrow in the spine.

The estimated absorbed dose for lymphoma (excluding Hodgkin's disease) was based only on the position of the most prominent lymph nodes in the mediastinum of the thorax. These included the lymph nodes lying in and around the trachea and the bifurcation of the main bronchi. If lymphomas arose in the lymph nodes of the posterior mediastinum, the dose would have been much higher, and the risk per rad would be reduced; if the lymph nodes of the anterior mediastinum were involved, the risk would be increased.

TABLE V-7   Estimated Radiation Doses in Tissues and Organs in Heavily Irradiated Sites in Patients with Ankylosing Spondylitis after a Single Treatment Course with X Rays

Site of Cancer	Dose, rads[a]
Spinal bone marrow (leukemia)	214
Lymphoma, mediastinal, lymph nodes excluding Hodgkin's disease	306
Esophagus	306
Stomach	67[b] / 89[c]
Colon	57
Pancreas	90
Bronchus	197[d]
Vertebral bone	505[e]
Spinal cord and nerves	698
Kidney	46[f]
Bladder	31[g]

[a] Based on average spinal bone-marrow dose of 505 rads.
[b] Assumes 50% of stomach irradiated; hypersthenic configuration.
[c] Assumes 67% of stomach irradiated; asthenic configuration.
[d] Assumes 80% of bronchial epithelium irradiated.
[e] Dose to spinal bone-marrow cells and endosteal lining cells of bone-marrow cavities.
[f] Lightly irradiated site; assumes 10% of organ (both kidneys) in irradiated field.
[g] Lightly irradiated site; assumes 33% of organ in irradiated field.

The position of the esophagus varies considerably in the thorax; and kyphosis in ankylosing-spondylitis patients would affect its position. If neoplasms arose in the upper esophagus, the radiation dose could have been higher, and the risk lower. Because many patients received lumbar-spine irradiation, it is possible that the distal esophagus, although more anterior (and thus receiving a smaller dose), could have been in the irradiated fields more frequently. The cervical portion of the esophagus was not included in this analysis.

The lower value for the absorbed dose in stomach assumed that half of it was irradiated: this may have occurred in hypersthenic patients. The higher value assumed that two-thirds of the stomach was irradiated; this may have occurred in asthenic patients.

The absorbed dose in the colon was based on the assumption that one-third to one-half the colon was in the irradiated field, primarily the transverse colon, the sigmoid, and the rectum. The dose in the pancreas assumed irradiation of the head and the portion of the body of the pancreas anterior to the lumbar spine; this accounts for two-thirds of the organ.

The dose estimated in the bronchus assumed that bronchial cancers arose in the primary and secondary branches. Further branching—say, to the tertiary portions—would increase the amount of bronchial epithelium, but decrease the probability of the epithelium's being situated in the irradiated field. It was assumed that 80% of the bronchial epithelium was irradiated.

The absorbed dose in bone was low, with a large range. Corrections were made for x-ray absorption in bone relative to soft tissue, i.e., for osteocyte lacunae and bone-marrow cell spaces. The dose estimate refers to the bone marrow of the vertebral bodies, transverse and spinous processes, pedicles, etc. It was assumed that only the spine and the sacroiliac joints were irradiated, and no corrections were made for irradiation of other bony structures, such as ribs.

The absorbed dose estimated in the spinal cord and spinal nerves (nerve root and dorsal and ventral branches) originating from the cord assumed that these structures were in the field of irradiation. The dose could have been higher, because the cord and the origins of the spinal nerves are closely related to the surrounding bone of the spinal column.

Conclusions

The estimates of absorbed doses of x rays in bone marrow and heavily irradiated sites in the radiotherapy patients with ankylosing spondylitis in England and Wales after a single treatment course of x rays are

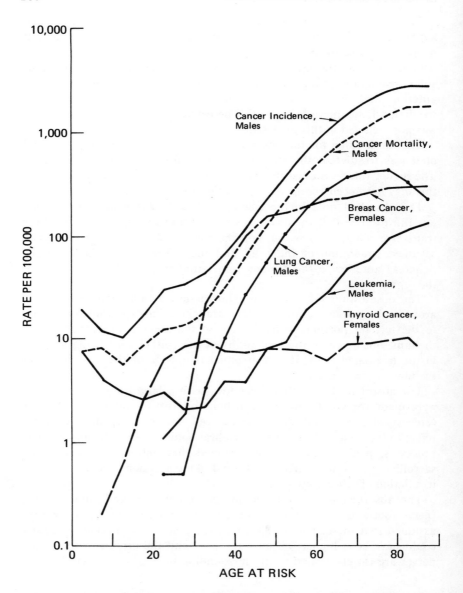

FIGURE V-2 Age-specific mortality and incidence rates for all malignant neoplasms and incidence rates for specific cancers, U.S. whites. Top curve (Cancer Incidence, Males) excludes nonmelanoma skin cancer and carcinoma *in situ*.

extremely crude and are based on very limited data and on a number of assumptions. Some of these assumptions may later prove to be incorrect, but the general principles are valid and are probably reasonably appropriate. It is therefore important to place these estimates of absorbed dose into perspective, recognizing that they may be somewhat inaccurate, but not grossly so. It is probable that they are correct to within a factor of 2. This is particularly important for cancers of heavily irradiated sites with long latent periods. The mean followup period for the single-treatment-course ankylosing-spondylitis patients was 16.2 yr, and an increase in cancers of heavily irradiated sites may appear in these patients after 1969 in tissues and organs with long latent periods for the induction of cancer.

The accuracy of these estimates is severely limited by the inadequacy of information on doses absorbed by the tissues at risk in the irradiated patients. The information on absorbed dose is essential for an accurate assessment of dose-related cancer incidence analysis, which could provide valuable insights into the mechanisms of cancer induction in man. Furthermore, in this unusually valuable human series of irradiated patients, the information on radiation dosimetry entered on the clinical radiotherapy charts is central to any reliable determination of somatic risks of radiation with regard to carcinogenesis in man. The work necessary to obtain these data is under way in England; only when they are available can more precise estimates of risk of human cancer induction by radiation be obtained.

AGE AS A FACTOR IN RADIATION CARCINOGENESIS

Age at exposure to ionizing radiation is a major factor in the carcinogenic response. This is perhaps not surprising, in view of the regular increase in the spontaneous risk of most cancers with age, especially after the first two or three decades of life. Figure V-2 shows age-specific mortality from all malignant neoplasms in U.S. white males in 1970 and age-specific incidence rates from the Third National Cancer Survey[83] for cancers of all sites, for leukemia, and for cancer of lung, trachea, and bronchus in white males and for cancer of the breast and thyroid in white females. These organ sites have been chosen largely on the basis of their sensitivity to the carcinogenic action of ionizing radiation.

The age-specific curves for cancer mortality and incidence provide a point of reference for examining theories of carcinogenesis. Doll[24] has specified four patterns of relationship between age and the incidence of specific cancers: (1) a rise to a peak in childhood, adolescence, or early adult life, followed by a decline, e.g., Wilms's tumor; (2) a rapid, regular increase from adolescence to old age, with practically no cases in childhood, e.g., cancer of the stomach; (3) same as (2), but with a marked turndown at the highest ages, e.g., bronchogenic carcinoma; and (4) a sharp rise until middle life, after which the increase slows down or ceases, e.g., cancers of the breast and cervix uteri in women. Thus, the influence of age is not uniform among cancers, but varies presumably in response to other host factors, such as hormonal and genetic influences, and to environmental conditions. Doll concluded that the influence of age on cancer rates is not direct, but is a measure of previous exposure to carcinogenic agents.

For chemical carcinogens, to which exposure is often prolonged, as in many occupational exposures, the literature gives a mixed picture of the sensitivity of people of different ages.[17,25,29,34,43] In most studies, but not all, the carcinogenic risk of such exposure increases with the age when exposure started. In one study of nickel refiners, for example, occupational nasal sinus cancer was observed to increase sharply with age at first employment, but the incidence of occupational cancer of the lung rose to a peak among men first employed in their early twenties and then fell markedly in men first employed later in life.[25] Fears *et al.*,[28] in analyzing the age-specific incidence of both melanoma and non-melanoma skin cancer, found that a simple power relationship between age and incidence fits the data well for nonmelanoma skin cancer and concluded that age represents the cumulative lifetime exposure to ultraviolet (UV) radiation. For skin melanomas, however, they found that incidence is related to the annual amount of UV exposure—i.e., intensity of exposure, rather than its duration.

Experimental Data on Role of Age in Radiation Carcinogenesis

Although the influence of age has not held the importance for experimental radiation biologists that it has for epidemiologists studying radiation carcinogenesis in man, there is nevertheless a substantial body of animal data on this topic. For the most part, in both mice and rats, investigators have reported a decreasing sensitivity to the carcinogenic influence of ionizing radiation with age. Although in some instances[18] the apparent advantage of the oldest animals may arise from the fact that the latent period for a particular radiogenic cancer exceeds their life expectation at irradiation, this is clearly not the entire explanation for the frequently observed relationship. But the picture is far from uniform, as illustrated by the results of Lindop and Rotblat[45] and of Vesselinovitch *et al.*[86] The former reported a decreasing risk of leukemia with increasing age at exposure in the rat, in contrast with an increasing risk of pulmonary tumors, and a peaking of risk of ovarian tumors at about age 15 wk, with a decline thereafter. Vesselinovitch *et al.,* in experiments on mice (newborn, 15 d old, and 42 d old at exposure), found little or no association of risk with age at irradiation for leukemia and lung adenoma, an increase with age for Harderian gland cystadenoma and ovarian tumors, and a decrease with age for hepatoma in males and for tumors other than those mentioned. Both Upton *et al.*[85] and Vesselino-vitch *et al.*[86] attributed much of the apparent influence of age to specific biologic factors present at the time of irradiation. Upton and Furth[84] found that susceptibility to induction of thymic lymphomas declined with natural or hormone-induced thymic involution and increased with thymic hyperplasia.

Peto *et al.*[61] reported on an experiment in which benzo[*a*]pyrene was applied twice a week to the skin of mice of different ages, and the animals were carefully observed for the occurrence of malignant epithelial tumors over time. The incidence of tumors increased steeply with time in direct association with duration of exposure, but, for fixed duration of exposure, was independent of age at start of exposure. The authors postulated that it is the rate at which somatic mutations are generated that determines both the rate of aging and the age-specific incidence of cancer.

Human Data on Role of Age in Radiation Carcinogenesis

Analytic Considerations If, to spontaneous incidence, irradiation at any age (X_e) adds an increment of cancer incidence, after a minimal latent period (*l*) this increment may be proportional to the spontaneous incidence (relative-risk model) or may be a constant number of cases (absolute-risk model). Figure V-3 shows the time course of such a radiation effect. The minimal latent period may or may not depend on age at irradiation. Figure V-4 shows some possible patterns by which the effect of radiation may be added to spontaneous incidence, depending on age at irradiation. In the upper panels (a and b), irradiation at age X_e occurs before spontaneous incidence normally becomes perceptible. The two upper figures show graphically the difference between the absolute-risk and relative-risk models. According to the former (a), the excess rate is independent of the spontaneous rate; according to the latter (b), the excess is proportional to the spontaneous rate. Figure V-4 might represent any form of cancer except leukemia, for which the natural incidence (Figure V-2) is high even in the first few years of life.

In panels a and b of Figure V-4, the excess age-specific risk is assumed to continue for the lifetime of the irradiated persons. In panels c and d,

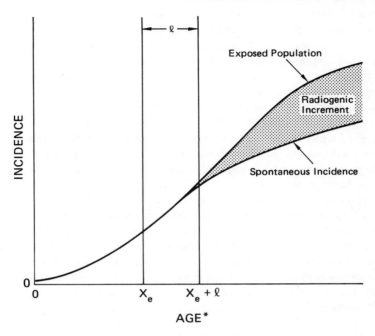

*X_e is age at exposure, ℓ is the minimal latent period.

FIGURE V-3   Superimposition of radiogenic effect on spontaneous incidence.

the excess is assumed to disappear after some period, as has been observed, for example, with radiation-induced leukemia. Panels c and d contrast the effect of different ages of exposure with the total excess risk proportional to the spontaneous rate at the ages when the radiation effect is observed.

   If irradiation occurs at an age when the expectation of life is less than the minimal latent period, l, the risk estimates for that older group will approach zero. If the duration of followup is less than l, no age group will show an appreciable effect. More important is the determination of the basis for comparing different age groups as to their sensitivity to the carcinogenic effect of ionizing radiation. If observations are made on the basis of a completed lifetime experience for each group, the options are clear and easy to understand, provided that there are appropriate controls and the radiation dose is the same for both groups. In that event, comparison with the controls provides an estimate of the excess for each group, and they may then be contrasted in terms of the probability of excess cancers per spontaneously occurring cancer. Only repetition at different doses, however, will determine whether relative sensitivity varies with dose. If doses are not fixed, then the comparison must be adjusted for dose on the basis of a dose-response function. Sensitivity to the effects of ionizing radiation for particular cancers is not the only element of interest in the experience of exposed subjects of different ages. Other characteristics of the radiogenic increase in risk are the minimal latency and the duration of the effect.

For no human series with dosimetry is observation complete for the lifetime of the subjects after exposure, however. When this condition is not met, but observations are complete throughout the period when the radiogenic excess is being expressed, the excess for each group can be estimated, and the age groups compared in terms of excess cancers per person or in terms of excess cancers per spontaneous cancer, if the latter can be estimated from other data, such as life tables and cancer death rates. Of the available human data, it would appear that only those for leukemia in the atomic-bomb survivors meet these conditions, because the excess seems to have disappeared among those under age 10 in 1945, and those 50 or older have been reduced by death to negligibly small numbers.

Leukemia   Figure V-5 shows the age-specific data for leukemia induction for atomic-bomb survivors[9] and for the British spondylitis cases, both expressed in rads to the marrow;[73] the marrow dose for the spondylitics is from Table V-7. The British experience does not cover the first two decades of life, but thereafter there is an upward movement with age that parallels that of the Japanese survivors. The age-specific incidence of the spontaneous disease differs greatly in the two areas[68] after the age of 50, with much higher rates in England and Wales than in Japan.

The fact that the excess radiation-induced rates are so much higher in the Japanese than in the British patients at all ages shows that increased incidence is not simply proportional to the spontaneous rate in a particular population.

Ichimaru *et al.*[36] recently analyzed the leukemia experience of atomic-bomb survivors from the standpoint of age in 1945, latent period, and type of leukemia. Their analysis shows that those under age 15 had the highest incidence of both acute and chronic leukemia early and that older survivors experienced a lower incidence of chronic leukemia only somewhat later than those under age 15. In the older survivors, however, the appearance of an excess of acute leukemia was progressively delayed, depending on age, with the oldest subjects experiencing the greatest delay.

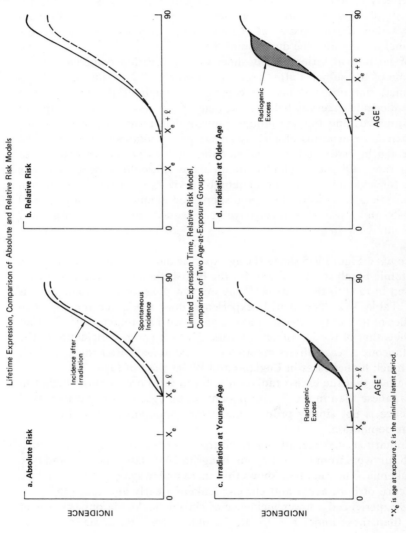

FIGURE V-4 Radiation-induced cancer effect superimposed on spontaneous cancer incidence by age. Illustrations of various possibilities.

*X_e is age at exposure, ℓ is the minimal latent period.

Thyroid Cancer   Although there is some evidence among the Japanese bomb survivors of a higher excess risk of thyroid cancer among those under 20 in 1945 than among those 20 or older,[60] the contrast is based on small numbers and is therefore subject to considerable sampling variability. In their 1975 report on medical findings in the Marshallese exposed to radioactive fallout in 1954, Conard *et al.*[19] reported 22 thyroid lesions, 19 benign and three malignant, among 64 inhabitants of Rongelap, where fallout was heaviest. Those under age 10 at exposure had far more benign lesions than older children or adults; even after adjustment for differences in estimated organ dose, the risk of benign lesions in children under 10 was about 4 times that in older inhabitants. There was one malignant lesion among the children under 10 and two among the 45 older inhabitants. Further observations will be required to determine whether age at irradiation plays an important role in radiation-induced thyroid cancer.

Lung Cancer   Only the reports on Czechoslovakian uranium miners,[70] fluorspar miners, and atomic-bomb survivors[9] contain information on age differentials. Although the reported data are still incomplete for the purpose of comparing age groups as to sensitivity to the carcinogenic effect of ionizing radiation, the observations on atomic-bomb survivors show clearly that the appearance of lung cancer in younger cohorts is not appreciably accelerated in time and that excess deaths begin to appear only at the ages at which mortality from lung cancer of other etiology normally becomes apparent. Thus, the period from irradiation to a perceptible increase in risk depends markedly on age among the Japanese survivors.

Age at exposure influences the reported risk of lung cancer among underground miners. Czechoslovakian miners[70] showed a marked effect of age at initial exposure on lung-cancer excess from radon daughters. DeVilliers and Wigle (unpublished manuscript), reporting on lung-cancer mortality in Canadian fluorspar miners, found an increasing risk of radiogenic cancer with increasing age at entry into the mines. They also found no relationship between age at entry and mean latent period, which is in contrast with the findings on Japanese atomic-bomb survivors. One explanation for this difference is that the miners were exposed to alpha radiation, with a high RBE and thus a high rem dose, compared with the Japanese bomb survivors. Another possibility is that differences in smoking experience can account for this discrepancy. Smoking is known to affect the latent period for lung cancer in the miners.[1,4] The Japanese have been relatively light smokers with a high proportion of nonsmokers,[38] whereas the miners were heavy smokers with relatively few nonsmokers. This marked difference in the effect of age on the minimal latent period for lung cancer may therefore be related to cigarette-smoking. In any

174

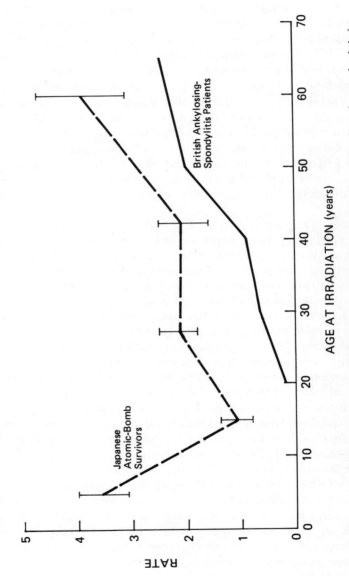

FIGURE V-5 Excess deaths per million persons per year per rad to bone marrow, atomic-bomb survivors and ankylosing-spondylitis patients, by age at irradiation. Vertical bars indicate 90% confidence intervals.

case, in all the groups studied, lung-cancer induction by radiation depends markedly on age at exposure, with no evidence as yet of excess risk before age 35.

Breast Cancer   There are four major series with age-specific risk information, but two are sharply restricted as to age range and amount of information at higher ages. Boice and Monson,[14] reporting on a followup study on the effect of repeated fluoroscopy of tuberculosis patients, observed the highest excess risk in women aged 15–19 at the start of therapy. In the sample of women treated with x rays for mastitis,[72] there were very few women under age 20, and absolute risks were 7.9 excess breast cancers per million women per year per rad for women under 30 versus 9.2 for women 30–44 at irradiation—a small enough difference to be compatible with random variation. The Swedish series[5] represents women treated during 1927–1957 for fibroadenomatosis, acute mastitis, chronic mastitis, and unilateral breast hypertrophy (13 young females) and followed to 1975. Although age and dose are highly correlated, incidence per rad is highest in women aged 20–24 at therapy and declines progressively thereafter when the analysis is confined to those treated for 1 yr or less, with 88 observed cancers of the breast versus 24 expected. In the Tokunaga *et al.* series,[80] observations on atomic-bomb survivors over the complete interval 1950–1974 reveal considerable variation in risk by age. The rate is highest for women 10–19 yr old in 1945 and declines progressively until age 50, after which it is again high, but based on very few cases. Especially important are the absence of breast cancer among those who were under 10 in 1945, regardless of dose, and the absence of any excess breast cancer among those aged 40–49 at that time.

Other Cancers   The radiogenic mortality from gastrointestinal cancer among atomic-bomb survivors[9] is rather small, both absolutely (about 50 deaths) and relatively (50 among 770 cancers in those exposed to 10+ rads), so sampling errors are large; but there is an increasing trend of risk with age. Polednak *et al.* reported much higher excess mortality from bone cancer associated with radium-dial-painting in women first employed under age 20 than among those first employed at age 20 or older.[62]

Summary   This review shows that age at exposure to ionizing radiation has an important influence on its carcinogenic effect, but that the available data are inadequate to resolve all the important issues. The influence of age is not uniform in human carcinogenesis, but more information may make it possible to perceive regularities that occur in the age-specific incidence of spontaneous tumors. The clearest evidence of a very high risk in those irradiated in the first years of life, for example, is found in the leukemia experience of atomic-bomb survivors. Women exposed in the second decade of life, when major hormonal changes are taking place, appear to be at highest risk of breast cancer. The rising excess of risk of leukemia with increasing age of irradiated ankylosing-spondylitis patients in the United Kingdom now seems not too different from the trend for atomic-bomb survivors, despite the difference in spontaneous-leukemia incidence in the two countries.

For the major solid tumors, whatever the age differential in susceptibility, a differential in length of the minimal latent period seems quite well established, younger subjects generally taking longer to begin to show the effect. Although in some animal experiments tumors appear to be accelerated in their appearance, in comparison with the age distribution of spontaneous tumors, little evidence of this is seen in the human data on solid tumors. However, radiation-induced leukemias have shorter appearance times in children than in adults.

It is somewhat difficult to measure age differences in susceptibility to the tumorigenic action of ionizing radiation, because the different age groups do not have equal life expectancies—i.e., equal opportunities for expressing the effect—but useful indexes can be derived, especially when the effect has subsided before observations cease.

Doll[24] has argued that age may serve merely as a measure of the opportunity for inciting events to occur; this seems doubtful, in view of the varied ways in which age appears to influence the carcinogenic response to ionizing radiation. There is now considerable evidence that younger people are at higher risk of some tumors than older people. It is clear that other factors must also be involved in controlling expression of cancer after irradiation.

ESTIMATING THE TOTAL CANCER RISK OF LOW-DOSE, LOW-LET, WHOLE-BODY RADIATION

Those responsible for determining radiation-protection policy and procedures must take some position on the somatic risks of low doses of ionizing radiation, e.g., doses of a few rads. That direct observations on the effects of such small doses in man are lacking does not remove this responsibility if society is to benefit from the applications of nuclear and radiation technology in industry and medicine. The potential risk of cancer induction in man from low-dose radiation exposure and the development of nuclear technology require continuing scrutiny of the scientific basis for maximal permissible dose limits.

After a review of what was known of the effects of ionizing radiation, the 1972 BEIR Committee[56] made the first effort to provide quantitative estimates of the possible effect on cancer mortality of increasing the whole-body exposure of an entire population by a small finite amount, 0.1 rem/yr. The estimating process was stated to be arbitrary and of uncertain validity, especially because values for the low-dose region were estimated by linear extrapolation from data on the high-dose region (100 rads or more) to which most human data pertained. The 1972 BEIR Committee recognized that such linear estimates might be regarded as upper limits of risk for low-LET radiation at low dose rates (see Chapter VII, Section IV.A., 1972 BEIR report). These estimates were disputed by those who argued that scientific estimation should be confined to the dose range for which there were direct observations[55] and that the linear hypothesis was not consistent with much of the experimental evidence on the effects of low-LET radiation.

The 1972 BEIR Committee, in its final estimates of total cancer risk from whole-body exposure, distinguished between low-LET and high-LET radiation only in that it used various quality factors for neutrons and alpha particles. The principal concern of the present Committee is the cancer risk from low-dose, low-LET radiation. The Committee recognizes that the scientific basis for making such estimates is inadequate, but it also recognizes that policy decisions cannot be reached or regulatory authority exercised without someone's taking a position on the probable cancer risk associated with such radiation. Because critical analysis of the different data bases disclosed major inadequacies, however, the Committee decided to emphasize the assumptions, procedures, and uncertainties involved in the estimation process, and not specific numerical estimates. The variety of mathematical functions that could be used to express dose-response relationships reflects additional uncertainty. Therefore, the Committee concluded that the best method of expressing the range of uncertainty associated with these problems would be to present an envelope of risk estimates.

AVAILABLE DATA BASE FOR LOW-LET RADIATION

The evidence of human cancer effects from low-dose, low-LET radiation is incomplete, and human studies provide only an approximate guide for risk estimation. Demographic information on the available population cohorts is especially deficient; age-specific risks are not generally available or do not ordinarily pertain to a wide age span. Whole-body exposure to man-made radiation is rare; the exceptions are the circumstances surrounding nuclear explosions, fetal irradiation, and some occupational exposure. Most human exposure above background is from diagnostic or therapeutic medical radiation and is limited to specific organs or tissues. The atomic-bomb exposures included both high- and low-LET radiation and provide no information on the effect of dose rate or dose fractionation.

Most human data on low-LET radiation result from medical radiation of fairly high dose rate—e.g., radiotherapy of the spine and pelvis in adults with ankylosing spondylitis,[73] of the thymus in infants,[33,65] of the scalp in children with tinea capitis,[52,71] and of the breast in women with postpartum mastitis;[72] fluoroscopy for pneumothorax in women with tuberculosis;[14] and radiographic pelvimetry in obstetrics.[48]

The best available human data on dose fractionation are related to female breast cancer and to repeated fluoroscopic examination of pulmonary-tuberculosis patients treated with pneumothorax,[14] for which total doses cover a wide range. Occupational exposure of radiologists[49] is of special interest because of dose fractionation, but yields no truly quantitative data.

The 1950–1974 Life Span Study (LSS) of Hiroshima and Nagasaki survivors[8] is a major data base, but is not completely satisfactory for estimating the risk of low-dose, low-LET radiation, although it provides in this report the major data base for the mortality estimates of risk. The leukemia observations provide the strongest body of data relating radiation dose to incidence, latent period, expression time, and type of disease induced. The data are statistically robust in the high-dose region, but not at low doses, especially in Nagasaki,[8] and it is the Nagasaki data that determine the estimates for low-LET radiation. The mortality data are less robust for solid cancers than for leukemia, and they produce estimates for low-LET radiation with minimal variance only when a fixed RBE for neutrons is assumed. The advantages of this series are its relatively large size, the full range of dose from 0 to 600 rads (kerma), the almost exclusively gamma-ray exposure in Nagasaki, the complete and unbiased ascertainment of death, and the proven validity of death-certificate information on cancer. The disadvantages include the mixed and highly correlated types of radiation in Hiroshima and the major differences between Japan and the United States with respect to the normal pattern of cancer incidence.

The Committee has considered several sources of data related to cancer risk expressed in terms of incidence. One data base was developed from the mortality estimates modified by sex-specific expansion factors that take into account the normal site-specific ratio of incidence to mortality and the relative size of site-specific estimates of excess cancer risk.

The Nagasaki Tumor Registry data[8] are useful for low-LET incidence estimation because the exposure was almost entirely to gamma rays, and the Registry is said to be among the most reliable in Japan (I. M. Moriyama, personal communication); in 50% of cases there was pathologic confirmation, and in only 13% was the death certificate the only information available (T. Itoga, personal communication). However, there are no reports of case-by-case comparisons in which representative Tumor Registry cases have been examined for validity of diagnosis and dose-related bias of ascertainment. The Tumor Registry data also have an underascertainment bias caused by migration from the registration area. The Committee concluded that the Tumor Registry data could be used only tentatively and not accorded the importance attributed to the mortality data.

The last data source is the site-specific estimates in Appendix A. They have potential value in their provision of the only means of considering dose-incidence data other than those from the atomic bombs. Many of the site-specific estimates are based on data from samples exposed to fairly high doses of partial-body irradiation. Application of these estimates to low-dose exposures requires an assumed dose-response relationship, the usual assumption being that of linearity. Because these data are not generally age- and sex-specific, it is necessary to use data from other series—e.g., the atomic-bomb survivors—to adjust the site-specific coefficients to an age- and sex-specific basis. The data are reasonably firm for only a few organs.

Whether individual-organ risks derived from partial-body data from many diverse sources and types of radiation can, in fact, be summed to predict radiation risk after whole-body irradiation is not resolved; the Committee remained divided on this procedure. Some members of the Committee, recognizing the many uncertainties in the quantitation of the carcinogenic risk of low-dose, low-LET radiation, believe that the partial-body data provide the broadest scientific basis from the vast body of epidemiologic data on human populations; they argue that there is reasonable concordance in the site-specific coefficients for estimation of whole-body risk and thereby lessen the dependence on solely the Japanese atomic-bomb survivor data. Other members of the Committee dispute the validity of the summed-sites method to estimate the cancer risk of low-dose, low-LET, whole-body radiation; they argue that the variation in radiation-induced cancer incidence in different human populations and in different animal strains precludes summing of the partial-body data for whole-body risk estimation. The problem is particularly complex in the case of endocrine-dependent tumors, such as those of the thyroid. If thyroid cancer is to be induced experimentally by iodine-131, high concentrations of thyroid-stimulating hormone (TSH) must also be present.[27] If pituitary function is reduced, as might occur after whole-body irradiation, the effectiveness of concomitant thyroid irradiation would be less than that after direct irradiation of the thyroid (i.e., partial-body irradiation).

Duration of followup is important in relating risk estimates to an appropriate period during which radiation-induced cancer develops. Data on the atomic-bomb survivors are currently available through 1974, 29 yr after the bombings. Risk estimates can be derived for 1955–1974 to approximate the effect of a latent period of 10 yr. The average duration of followup of patients irradiated for enlarged thymus glands in infancy is 24 yr; that of ankylosing-spondylitis patients is 16 yr for those with a single course of treatment. The followup period for most epidemiologic surveys is 20 yr or less.

CANCER MORTALITY AND CANCER INCIDENCE

The risk estimates presented in Appendix A are for human cancers for which there is epidemiologic evidence that exposure to low-LET radiation may increase the risk. Epidemiologic surveys on radiation-induced cancer in man use both mortality and incidence data as end points in the estimation of risk. In the past, estimates of the carcinogenic risk of whole-body exposure to ionizing radiation have been based principally on mortality data; recent reports have provided more information on cancer incidence.

For some sites of cancer—e.g., of the esophagus, pancreas, and lung—incidence is fairly well approximated by mortality, but for other sites, such as the thyroid and breast, this is not the case. Except for leukemia, thyroid cancer, and breast cancer, most site-specific data are available only as mortality information.

Many members of the Committee believe that the incidence of radiation-induced cancer provides a more complete expression of the total social cost than does mortality. Three important sites contribute to differences between cancer incidence and cancer mortality: the thyroid (a major effect with low associated mortality), the female breast (a major effect with moderately high mortality), and the skin (rarely fatal). It is primarily because of breast and thyroid cancer that whole-body risk estimates based on incidence are higher than those based on mortality.

DOSE-RESPONSE RELATIONSHIP

Knowledge of cancer induction is inadequate to derive carcinogenic risks at low doses of low-LET radiation from those at higher doses with confidence. There are a number of possible dose-response functions, but there is no compelling evidence of the validity of any one. Although none can be proved to be inapplicable to carcinogenesis, in its estimates of low-dose risk the Committee chose not to include the class of functions with a threshold, i.e., functions in which the cancer risk is zero up to some positive value of the dose scale.

Dose-response functions for radiation carcinogenesis may be broadly classified as linear, nonlinear with upward curvature, nonlinear with downward curvature, and nonlinear with both upward and downward curvature (cf. Chapter II). Human data on radiation-induced cancer are seldom extensive enough to provide dose-response data extending into the low-dose region and, when they do, they do not permit discrimination among the possible mathematical models. Some studies provide only a single determination in the high-dose region, but there are some important exceptions in which observations range over a wide interval of dose (as in the Japanese atomic-bomb survivors) and can be fitted to a variety of dose-response models. For example, the breast-cancer data (cf. Appendix A) are best described by a linear dose response.[14,80] The leukemia data of the Life Span Study (LSS) sample are consistent with a linear-quadratic response to the gamma-ray component of dose. The mortality data for all forms of cancer except leukemia among the Nagasaki atomic-bomb survivors do not strongly support any particular relationship with radiation dose, so that it is difficult to discriminate among various possible functional forms of the dose-response relationship. In the Nagasaki Tumor Registry data, the relationship between the radiation dose and the total incidence of all major cancers except leukemia is highly significant, and the observed dose-response relationship appears linear, with no suggestion of upward curvature.

It has been proposed that man may be genetically heterogeneous with respect to susceptibility to radiation carcinogenesis; the response of the most susceptible people would be steep and would saturate at fairly low doses, causing a steeper slope in the initial portion of the dose-response curve for the population as a whole.[6] Such dose-response curves would be nonlinear with downward curvature, and linear extrapolations in the region of very low dose would underestimate actual risk. However, little is known about such subpopulations of varied genetic sensitivity; until they can be identified as a significant fraction of the population with a significantly greater risk, their importance as special groups apart from the general population cannot be defined.

It seems unlikely that epidemiologic studies on low-dose exposure will ever be adequate for direct observation of excess cancer risk associated with very low doses of low-LET radiation. The choice of a mathematical dose-response function that uses statistically stronger data at high doses to estimate cancer risk at low doses therefore becomes an important part of the estimation process. The general dose-response relationship currently favored by radiation biologists for radiation-induced cancer (cf. Chapter II, Equation II-6),

$$F(D) = (\alpha_0 + \alpha_1 D + \alpha_2 D^2)\exp(-\beta_1 D - \beta_2 D^2), \qquad \text{(V-1)}$$

requires more data to determine the coefficients than are available from existing epidemiologic studies. For whole-body exposures, the competing effect of cell-killing is represented by the exponential factor in Equation V-1. For the low-dose range, the simpler quadratic function with a linear term (linear-quadratic, LQ),

$$F(D) = \alpha_0 + \alpha_1 D + \alpha_2 D^2, \qquad \text{(V-2)}$$

may adequately represent the dose-response relationship.

Further simplification of the linear-quadratic dose-response relationship involves arbitrary choices, e.g., by fixing the value of the ratio, $r = \alpha_2/\alpha_1$. Experimental studies provide data consistent with a wide range of values for this ratio, and it is difficult on those grounds alone to choose any one value of r, as in the form,

$$F(D) = \alpha_0 + \alpha_1 (D + rD^2). \qquad \text{(V-3)}$$

The linear model (L),

$$F(D) = \alpha_0 + \alpha_1 D, \qquad \text{(V-4)}$$

and the pure quadratic model (Q),

$$F(D) = \alpha_0 + \alpha_2 D^2, \qquad \text{(V-5)}$$

are further simplifications.

In experimental studies with high-LET radiation, the linear term of Equation V-2 generally predominates; but with low-LET radiation delivered in a single dose the linear term usually gives way increasingly to the quadratic term, until, when $D = \alpha_1/\alpha_2$, their contributions are equal. If the true dose-response relationship is one of upward curvature, the fitted linear-model estimate no longer corresponds to the excess risk per rad at low doses, but only to the average increase per rad over the entire dose range. This estimate is therefore biased upward; if the linear model is fitted only to low-dose data—e.g., for D less than α_1/α_2—the upward bias is less, but so is the stability of the estimate, because it is based on fewer data. Many radiation biologists believe that the relationship probably has a linear component that would predominate at low doses. For those who hold this view, the pure quadratic (Q) model (Equation V-5) would therefore underestimate the risk at low doses. There is also support from radiobiologic theory and data for the view that the RBE varies inversely with the square root of neutron dose. A linear dose-effect relationship for high-LET radiation would then lead to the conclusion that the gamma-dose relationship should have a quadratic component. The pure quadratic model implies a much smaller excess risk at low doses than would be predicted on the basis of linearity.

The data on atomic-bomb survivors[8] are the most amenable to analysis by complex dose-response functions, but the presence of a neutron component of dose, which was small in Nagasaki but substantial in Hiroshima, and its high correlation with gamma dose within each city complicate analysis. Unless the relationship between the effects of the two kinds of radiation can be established, risk estimates for gamma radiation below 100 rads must depend almost entirely on the Nagasaki data. The Hiroshima data are much stronger than the Nagasaki data in the sampling sense. Analyses of the data for both cities combined in which the RBE of neutrons is determined from the data are only marginally more useful for estimating the effects of low-LET radiation than are analyses of the Nagasaki data alone. There are other differences between the two cities, including the natural level of cancer risk; in regression analyses, it is therefore desirable to provide for separate intercept values (α_0) for each city.

The following analyses of the Japanese data use city- and dose-specific rates adjusted to the common age and sex distributions of the combined cities, on the assumption that the form, although not the magnitude, of the dose-response relationship is unlikely to depend on sex or on age at exposure. Because neutrons contribute to the total dose in each city, especially in Hiroshima, the Japanese data are analyzed with respect to analogues of the models in Equations V-2, V-4, and V-5, in which the gamma and neutron doses (D_γ and D_n, respectively) are treated as independent quantities. In each case, the effect of the neutron dose is represented by a linear term, because all the evidence suggests that the appropriate dose-response function is linear or nearly so. These models can be denoted as the linear-quadratic gamma, linear neutron model (LQ-L),

$$F(D_\gamma, D_n) = \alpha_0 + \alpha_1 D_\gamma + \alpha_2 D_\gamma^2 + \beta_1 D_n; \qquad \text{(V-6)}$$

the linear gamma, linear neutron model (L-L),

$$F(D_\gamma, D_n) = \alpha_0 + \alpha_1 D_\gamma + \beta_1 D_n; \qquad\qquad \text{(V-7)}$$

and the quadratic gamma, linear neutron model (Q-L),

$$F(D_\gamma, D_n) = \alpha_0 + \alpha_2 D_\gamma{}^2 + \beta_1 D_n. \qquad\qquad \text{(V-8)}$$

The LSS leukemia-incidence data regression analyses (Appendix A) are summarized in Table V-8, with regression coefficients given in terms of excess cases per million persons per year per rad (or per rad[2]) to bone marrow. The data strongly suggest that leukemia risk is increased by exposure to gamma radiation. Although the differences among models with respect to goodness of fit are not large, they suggest dependence on both gamma dose and its square, and they suggest that risk from low-LET radiation may be estimated by using the gamma-dose coefficients in the fitted linear-quadratic gamma, linear neutron (LQ-L) model.

The mortality data for cancers other than leukemia are much less satisfactory for purposes of dose-response analysis. There is an obvious difference between the observed dose-response curves for Hiroshima and Nagasaki (Figure V-6), but these data contain little information for discriminating among the various dose-response models for low-LET radiation. The regression analyses are summarized in Table V-9, with estimated coefficients for risk in terms of excess cancer deaths per million persons per year per rad or rad[2] of average tissue dose. On the basis of χ^2 values for goodness of fit, there is no reason to choose any of these models over the others. In part, this is because these data provide no statistically significant evidence of a gamma-dose effect.

The difficulties of relying on the LSS mortality data for estimates of the low-LET radiation cancer risk (other than leukemia) without a known RBE can be better understood by considering the change in the regression coefficients as the higher-dose data are progressively removed from the analyses. The data do not discriminate among the various dose-response

TABLE V-8   Regression Analyses of Leukemia Incidence, Hiroshima and Nagasaki, 1950–1971

Model (Equation)	Coefficient ± SD			Goodness of Fit	
	α_1	α_2	β_1	χ^2, df	(p)
LQ-L (V-6)	0.99 ± 0.93	0.0085 ± 0.0056	27.5 ± 7.5	10.4, 11	(0.49)
L-L (V-7)	2.24 ± 0.60		25.4 ± 7.5	11.5, 12	(0.49)
Q-L (V-8)		0.014 ± 0.004	31.1 ± 6.9	12.3, 12	(0.42)

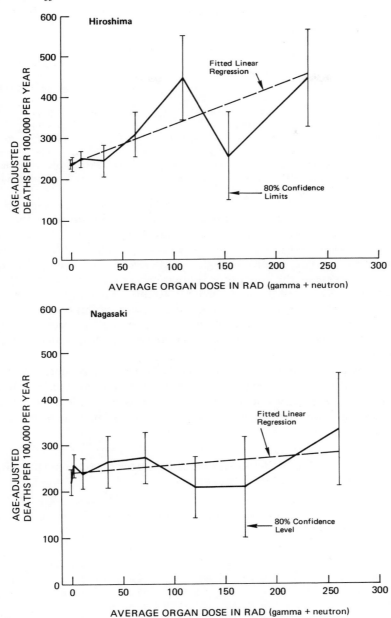

FIGURE V-6   Dose-response plots and fitted linear regressions for deaths from all cancers except leukemia, Hiroshima and Nagasaki, 1955–1974.

TABLE V-9   Regression Analyses for LSS Mortality Data, 1955–1974 (Excluding Leukemia)

Model (Equation)	Coefficient ± SD			Goodness of Fit	
	α_1	α_2	β_1	χ^2, df	(p)
LQ-L (V-6)	1.40 ± 4.56	0 ± [a]	61.9 ± 26.2	14.0, 11	(0.23)
L-L (V-7)	1.40 ± 2.18		61.9 ± 24.6	14.0, 12	(0.30)
Q-L (V-8)		0.0047 ± 0.0104	67.3 ± 21.9	14.3, 12	(0.28)

[a] Boundary-value estimate; α_2 constrained to be nonnegative. The calculated standard deviations of estimates do not allow for the fact that an active constraint is operating in this equation, and they may therefore be misleading.

models for any dose range. To show the effect of dose range on the risk coefficients, it is sufficient to choose the simplest model, i.e., the one that is linear in both gamma and neutron (*L-L*) dose (Table V-10). Radiobiologic theory would predict that the gamma regression coefficient (α_1) would decrease as the dose range is progressively shortened and that the neutron coefficient (β_1) would remain constant. That such is not the case here, and that in fact no consistent pattern emerges from these data, is another indication that they are not strong enough to provide stable estimates of low-dose, low-LET radiation cancer risk when analyzed in this fashion.

No better discrimination among models, in terms of goodness of fit, is seen when the Nagasaki data are analyzed alone with an arbitrary fixed RBE. Because the neutron component of dose is so small, use of an arbitrary fixed RBE for neutrons gives essentially an analysis in terms of gamma dose.

TABLE V-10   *L-L* Model Regression Analyses for LSS Cancer-Mortality Data, 1955–1974 (Excluding Leukemia)

Dose Range, kerma	Coefficient ± SD		Goodness of Fit	
	α_1	β_1	χ^2, df	(p)
0–600	1.40 ± 2.18	61.9 ± 24.6	14.0, 12	(0.30)
0–399	0 ± [a]	79.6 ± 37.4	12.9, 10	(0.23)
0–299	0 ± [a]	122.2 ± 40.2	7.0, 8	(0.54)
0–199	4.76 ± 3.66	45.6 ± 46.6	3.5, 6	(0.74)
0–99	5.58 ± 8.60	0 ± [a]	3.0, 4	(0.56)

[a] See footnote to Table V-9.

Although the *LQ-L* model, when fitted to the Japanese leukemia data, gave a dose-response curve that depended on both gamma dose and its square, that was not the case for solid cancers. In order to obtain a fitted curve intermediate between those corresponding to the *L-L* and *Q-L* models, the ratio $r = \alpha_2/\alpha_1$ was fixed. Although values of this ratio can be derived from experimental data, the Committee preferred to rely on human data, and chose $r = 0.0086$, obtained from the *LQ-L* model fitted to the Japanese leukemia data. This, then, yielded a new function, *LQ*-L*:

$$F(D_\gamma, D_n) = \alpha_0 + \alpha_1(D_\gamma + 0.0086\,D_\gamma^2) + \beta_1 D_n. \qquad \text{(V-9)}$$

When Equation V-9 was fitted to the Hiroshima and Nagasaki data for all forms of cancer except leukemia, however, it was found that the resulting RBE, expressed as a function of dose, became quite high (e.g., 91 at 1 rad of neutrons). These RBE values were much higher than corresponding *LQ-L* values calculated for leukemia (e.g., 23 at 1 rad of neutrons). The coefficients were $\alpha_1 = 0.398 \pm 0.709$ and $\beta_1 = 64.9 \pm 22.7$, and in the test of goodness of fit $\chi^2 = 14.4$ with 12 df, for which $p = 0.25$. Thus, the low-LET cancer-risk estimate obtainable from Equation V-9 is very unstable, as are the coefficients obtained from the *L-L* (Equation V-7) and *Q-L* (Equation V-8) models. In addition, the ratio of excess solid tumors to leukemia, about 2.4 for neutrons, was reversed to about 0.4 for gamma radiation. Finally, all these models appeared out of line with the incidence estimates. Although not all these objections seemed cogent to all members of the Committee, it was agreed that a further modification of the *LQ-L* model would be desirable and that the leukemia experience might provide a reasonable, if arbitrary, guide. This model, denoted $\overline{LQ\text{-}L}$, is

$$F(D_\gamma, D_n) = \alpha_0 + \alpha_1(D_\gamma + 0.0086\,D_\gamma^2 + 27.8\,D_n). \qquad \text{(V-10)}$$

That is, not only is the ratio $r = \alpha_2/\alpha_1$ in Equation V-3 fixed at the leukemia value of 0.0086, but the neutron RBE for leukemia, expressed as a function of dose, is implicit in the model. This further change yields a more stable estimate of the coefficient, $\alpha_1 = 1.40 \pm 0.38$. It also provides about the same ratio of solid-tumor excess to leukemia excess for both neutrons and gamma radiation. Because the *L-L* and *Q-L* functions were also open to some of the same objections as apply to the *LQ*-L* model, it was decided to use in their stead, for purposes of estimation, modified functions that were constrained by the RBE values for leukemia derived from the parallel functional forms. That is, the *L-L* form (Equation V-7), rewritten as $\overline{L\text{-}L}$, became

$$F(D_\gamma, D_n) = \alpha_0 + \alpha_1(D_\gamma + 11.3\,D_n), \qquad (V\text{-}11)$$

and the $Q\text{-}L$ form (Equation V-8), rewritten as $\overline{Q\text{-}L}$, became

$$F(D_\gamma, D_n) = \alpha_0 + \alpha_2(D_\gamma^2 + 2{,}265\,D_n). \qquad (V\text{-}12)$$

The coefficients (excess deaths per million persons per year per rad or per rad^2) obtained in fitting Equations V-10, V-11, and V-12 are shown in Table V-11.

Figure V-7 gives a plot of cancer-incidence data (excluding leukemia and bone cancer) from the Nagasaki Tumor Registry. The $\overline{LQ\text{-}L}$, $\overline{L\text{-}L}$, and $\overline{Q\text{-}L}$ models were all fitted. The neutron component is so small that it contributes little to the result. The regression coefficients, in terms of excess cases per million persons per year per rad or per rad^2, are shown in Table V-12. The best-fitting function is the linear ($\overline{L\text{-}L}$) model, but neither the $\overline{LQ\text{-}L}$ nor the $\overline{Q\text{-}L}$ model can be rejected on the basis of these data.

Although, as noted earlier, the available human data provide no adequate basis for choosing among dose-response models, the foregoing analysis puts into perspective the implications for estimation that derive from any such choice. In addition, because the leukemia data are consistent with a linear-quadratic response to the gamma-ray component of dose, the analysis provides a way of adapting the $LQ\text{-}L$ model to the Japanese data for forms of cancer other than leukemia. Estimates are given for the modified linear-quadratic ($\overline{LQ\text{-}L}$), linear ($\overline{L\text{-}L}$), and pure quadratic ($\overline{Q\text{-}L}$) models. The Committee regards the latter two models as providing an envelope of estimates within which the probable true values fall.

Some members of the Committee hold the opinion that the quadratic component of dose in the true dose-response relationship probably dominates over much if not all of the dose range, not only for leukemia but also for most other forms of cancer induced by low-LET radiation. These members would prefer to regard the linear (L or $L\text{-}L$) model not as

TABLE V-11   Regression Analyses for LSS Mortality Data, 1955–1974 (Excluding Leukemia)

Model (Equation)	Coefficient ± SD		Goodness of Fit	
	α_1	α_2	χ^2, df	(p)
$\overline{LQ\text{-}L}$ (V-10)	1.40 ± 0.38	—	16.3, 13	(0.23)
$\overline{L\text{-}L}$ (V-11)	3.47 ± 0.88	—	15.1, 13	(0.30)
$\overline{Q\text{-}L}$ (V-12)		0.0184 ± 0.0052	17.0, 13	(0.20)

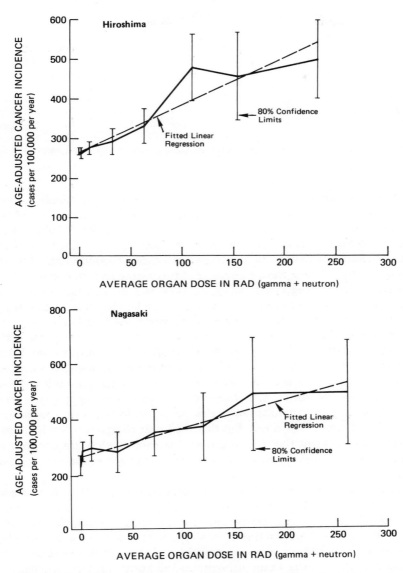

FIGURE V-7 Dose-response plots and fitted linear regressions for incidence of cancer, except leukemia and bone cancer, Hiroshima and Nagasaki Tumor Registry, 1959–1970.

TABLE V-12   Nagasaki Tumor Registry Regression Analyses, 1959–1970 (Major Sites, Excluding Leukemia and Bone Cancer)

Model (Equation)	Coefficient $\pm$ SD		Goodness of Fit	
	α_1	α_2	χ^2, df	(p)
$\overline{LQ\text{-}L}$ (V-10)	3.34 ± 1.00	a	4.9, 6	(0.56)
$\overline{L\text{-}L}$ (V-11)	9.20 ± 2.24	—	3.7, 6	(0.72)
$\overline{Q\text{-}L}$ (V-12)	—	0.042 ± 0.015	6.2, 6	(0.40)

a In the $\overline{LQ\text{-}L}$ model, the coefficient for the square of gamma dose is assumed to be equal to 0.0086 times the value of α_1.

central, but rather as one extreme on which credible upper bounds (in the form of confidence limits) could be based; the other extreme would be provided by the pure quadratic (Q or $Q\text{-}L$) model, on which credible lower bounds could be based. Proponents of this view argue that the linear and pure quadratic relations fit the Nagasaki cancer-mortality data (Figure V-6) equally well over the entire range of tissue dose and that even for the Nagasaki Tumor Registry data (Figure V-7) the pure quadratic cannot be excluded by goodness-of-fit criteria. Their opinion also draws major support from radiobiologic data and theory discussed in Chapter II suggesting that the RBE of high-LET radiation usually varies inversely with dose; for example, in some systems the RBE of fast neutrons has been reported to reach 100 or more.[66] These considerations, coupled with the generally observed linearity of dose response for high-LET radiation for both human and animal radiation-induced cancer (see Appendix A and UNSCEAR[82]), imply that the low-LET response would be a linear-quadratic function of dose at low doses, but that the linear component would dominate. To the extent that the true dose-response relationship contains a linear term, a pure quadratic fit will underestimate excess risk at low doses.

Just as some members of the Committee believe that the linear dose-response function is probably not generally valid for radiation carcinogenesis in man, so others believe that any dose-response function for low-LET radiation that lacks a linear term dominant at low doses may well be unrealistic for radiation carcinogenesis in man. It is for these reasons that most members of the Committee prefer the linear-quadratic model for cancer-risk estimation.

When various dose-response models are used to estimate risk, the possible effect of dose rate on cancer risk could be important. For high-LET radiation, there is some evidence that protraction of the dose—i.e., exposure at low dose rates—increases the cancer risk per rad, compared with exposure at higher dose rates. For low-LET radiation, there are as yet no quantitative data on human populations exposed chronically at low dose rates that permit estimation of the effect of dose rate alone. Experiments measuring cancer induction in animals suggest that a given dose of low-LET radiation would have less effect at a low dose rate than at a high dose rate.[81] Autonomous cell populations show various dose-rate dependences. Dose-rate effects with low-LET radiation are not seen in some recent cell-transformation studies.[51,78] Such effects have long been observed in studies of radiation-induced chromosomal exchanges and have recently been documented for human cells.[46,63] The breast-cancer data obtained from human subjects exposed to fractionated doses of x rays—i.e., where the exposure was to small doses repeated over a period of weeks to years[14]—do not indicate a significant difference in cancer risk per rad, compared with the effects in groups exposed to acute doses of low-LET radiation. Thus, most members of the Committee conclude that it is not now possible to assign a numerical value to any dose-rate factor by which risk estimates obtained in populations exposed to low-LET radiation at relatively high dose rates can be corrected to apply to exposures at low dose rates. In cases where protraction of exposure or low dose rate is eventually found to reduce the cancer risk in man per rad of low-LET radiation, estimates based on the linear model in particular must be modified accordingly. The linear-quadratic model makes some allowance for dose rate, in that, whereas the linear component is assumed to be invariant with dose fractionation, the dose-squared component decreases with increased fractionation until at some point it becomes negligible.

REQUIREMENTS FOR CALCULATING CANCER-RISK ESTIMATES

To calculate cancer-risk estimates, it is necessary to select a population of interest and to specify a variety of parameters of risk. Illustrative calculations are most useful for the working population that is occupationally exposed to radiation and for the general population itself. Here calculations are based on the 1969–1971 life tables for the United States[54] including all ages; risk estimates are calculated separately for the two sexes. The life-table population can be segmented to reflect onset of exposure at any age, as for occupational exposure, and may be used to reflect a single exposure or continuous exposure over extended periods. Risk coefficients are expressed per million men or women of the life-table population.

Three important parameters that influence calculations of cancer risk are the minimal latent period, the magnitude of the effect, and the duration of the effect. Each of these must be age- and sex-specific to be applied to the demographic model. Figure V-8 is a schematic representation of these risks for a particular age-at-exposure group. Exposure occurs at age *a*, and the latent period (no increased risk) ends at age *b*. Thereafter, the excess risk may be represented by a constant *absolute* risk ending at age *c*, or continuing throughout the life of this age cohort (the solid line beyond *c*); in the latter case, the excess is taken to be independent of age. Or the excess risk may increase gradually and continuously to reflect a constant *relative* risk throughout the life of the cohort (dashed line); that is, the relative risk is proportional to the spontaneous risk, which increases with age for nearly all cancers. Thus, the effect of exposure of a population at age *a* eventually appears at age *b* as an increase in cancer risk that lasts for some period. Similarly, an exposure at age *a* + 1 will have an effect beginning at age *b* + 1, etc. In this way, the process of risk estimation can accommodate a single exposure at any age or a continuous exposure beginning at any given age and extending over any given period. But the latent period, magnitude of risk, and duration of risk must all be age- and sex-specific in order to make the calculation. In the 1972 BEIR report[56] calculations, separate risk coefficients were used for three age periods of exposure to radiation: *in utero,* under 10 yr, and 10 yr or more; but both sexes were combined.

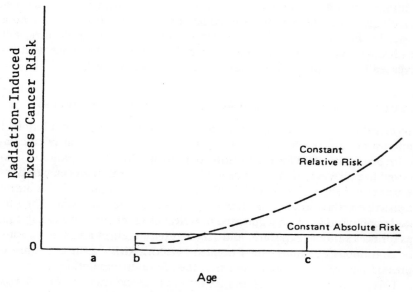

FIGURE V-8   Carcinogenic risk following radiation, absolute and relative risk models. a, age at irradiation. b, age at end of minimal latent period. c, any age after age 6.

The minimal latent period for most radiation-induced cancers is long—10 yr or more after exposure. For some types—cancers arising after *in utero* irradiation, leukemia in children or adults, and bone cancer after exposure to radium-224 alpha radiation—excess cancers have been observed within 2–4 yr after irradiation. Moreover, there is evidence that the increased risk of leukemia and bone cancer does not persist indefinitely, but becomes negligible 25–30 yr after the end of irradiation. For all the other radiation-induced cancers reviewed in Appendix A, the minimal latent period is 10 yr or more, and there is as yet no indication that the increased risk of cancer eventually declines. There are, however, no epidemiologic studies in which followup was carried out to the end of life for the entire population cohort. Hence, any projection of risk over the lifetime of exposed persons involves considerable uncertainty.

COMPUTATION

Selection of dose increments for which cancer-risk estimates are made was guided by existing maximal permissible dose limits, information on occupational exposure recorded in recent surveys (cf. Chapter III), concern for a hypothetical situation in which some part of the general population might be exposed to a single dose of 10 rads, and uncertainty as to whether a total dose of, say, 1 rad would have any effect at all. The dose increments are continuous exposure at 1 rad/yr for selected age intervals and general population exposure to a single dose of 10 rads or continuously at 1 rad/yr for a lifetime.

No allowance was made for "wasted" radiation (i.e., dose increments received after the carcinogenic event has already occurred) in calculations based on continuous exposure, but its potential influence is greatly reduced by the assumption of a 10-yr latent period.

The final estimates were calculated with a modified computer program[20] originally designed for life-table calculations of the effect of any additional risk on survival. Although that program was designed for linear estimation of risk from continuous exposures, it was adapted to permit the use of the other dose-response models and of single, acute exposures as well. The application of nonlinear dose-response models derived from the atomic-bomb survivor data to continuous exposure presented a problem. Although there seems to be general agreement that the linear component of the linear-quadratic model should represent the part of the dose response unaffected by dose fractionation and that the quadratic (dose-squared) component should become smaller as fractionation increases,[16] the precise numerical dependence of this reduction on fractionation is unknown. For exposures of around 1 rad/yr, however, the Committee felt that the quadratic component would be so small that it could be safely ignored. Thus, only the linear portion of the estimated LQ-L model curve was used to estimate risks from continuous exposures to 1 rad/yr; and the Q-L model, which has no linear component, was not applied to such exposures. In these calculations for risk estimation at relatively low doses, the Committee has not had to deal with the difficult problem of applying these dose-response models to continuous exposures amounting to tens of rads per year.

The 1969–1971 U.S. life tables[54] are used, with 1969–1971 U.S. cancer death rates by sex and 5-yr age groups. For any age-sex group in the mortality calculations, an increment of risk representing the radiation hazard is introduced to obtain a number of deaths attributable to radiation, which is then used to decrease the number of survivors entering the next age interval. In this way, the reference life table is modified to show the effect of the additional risk for the period to which it pertains or to any part of it, with no distortion of the normal probabilities of dying.

In a life-table population cohort, all members eventually die. To calculate the effects of incremental risk, therefore, is to estimate the number who would die prematurely as a result of the additional hazard. The final estimates are expressed as the numbers of excess cancers or of excess cancer deaths in an exposed population of 1 million people followed from the onset of exposure until the end of life. These numbers may also be expressed as percentages of the numbers of cancers normally expected for a population cohort of that size over the period under consideration and in the absence of the additional radiation exposure. Their expression per rad is generally avoided in the final tables, because it would suggest a commitment to the linear hypothesis that some members of the Committee wished to avoid, believing that the effect per rad is most probably variable, an increasing function of dose in the region from zero rads up to the point where cell-killing becomes important.

Average cumulative doses from continuous exposure to 1 rad/yr range, for example, from 67 rads for males exposed continuously from birth and 75 rads for females, to 13 rads for males and 14 for females exposed from ages 50 to 65. These dose ranges reflect the life expectancies of cohorts defined at birth and at age 50, respectively.

The information on the atomic-bomb survivors now extends to 29 yr after exposure, and followup is less prolonged for most other surveys. Within the first 30 yr after irradiation, however, excess incidence for radiation-induced cancers of the lung and female breast appears to follow the same temporal patterns as the natural age-specific incidence or mortality; this suggests that the pattern may continue beyond the period of information, perhaps even throughout life. This may not apply to all radiation-induced cancers, or it may apply only to individual cancers and not to groups of cancers.

The method used to project risk forward in time after the 30 yr of followup on which the risk coefficients for postnatal exposure were based has an important influence on the estimates. This presents no problem for leukemia, for which the excess risks are believed to end within 30 yr after exposure; but it is important for cancers for which the excess risk is assumed to extend throughout life. The 1972 BEIR report[56] calculations were premised on two alternative assumptions as to the expression time for radiation-induced cancers other than leukemia, 30-yr and lifetime risks. In the present calculations, the shorter, 30-yr period was not used.

Two methods have been used to project risk forward in time beyond the period represented by followup data. The first is the absolute-risk model, which assumes that the dose-related excess risk per year observed during years 11–30 of followup continues until the end of life (Figure V-8). The second is the relative-risk model, which assumes that this risk increases or decreases as the normal age-specific risk varies relative to that corresponding to 11–30 yr after exposure. In general, population rates for cancer mortality and incidence increase with increasing age. Projections using the absolute-risk model may lead to underestimates if, as seems to be the case for radiation-induced breast cancer and lung cancer, the temporal pattern of excess risk after exposure parallels age-specific population rates. On the other hand, few exposed populations, and especially few populations exposed at young ages, have been followed until the end of life. Thus, the assumption of a constant relative risk throughout life also is arbitrary and could introduce bias. For ages under 10 yr at exposure, the relative-risk ratios thus obtained appeared unreliable, and the ratios for ages 10–19 at exposure were substituted for them. This is numerically the most important departure from the method of the 1972 BEIR report,[56] in which it was assumed that the large, but statistically unstable, relative-risk estimates for cancers other than leukemia, calculated from observations 15–25 yr after exposure for the cohort exposed at less than 10 yr of age, applied to the remainder of life. About half the total lifetime excess deaths from solid tumors estimated according to the relative-risk model for an exposed population of all ages corresponded to exposures received before the age of 10. No population exposed at such early ages has been followed long enough to provide a firm basis for evaluating the assumption of a constant relative risk holding throughout life. However, the most recent followup data on the LSS sample fail to suggest an increase in absolute risk over time that is commensurate with a constant relative risk for the cohort aged 0–9 yr at the time of bombing.

Cancer of the prostate and melanoma and other skin cancer have been subtracted from the age-specific rates (provided by T. Mason, National Cancer Institute, personal communication) used for the relative-risk approach. There is little evidence that these cancers are an important part of the carcinogenic response to radiation, and therefore their subtraction removed some age variation, presumably irrelevant to this response. Beyond this, no attempt was made to weight the population rates on a site-specific basis to reflect the relative importance of different cancers to the aggregate radiation effect, because of a lack of data on how the spectrum of radiation-induced cancers might change with followup past 30 yr. It seemed possible, for example, that the marked increase in population rates for digestive cancer with increasing age might exaggerate the relative-risk projection of excess cancer incidence among women exposed at ages 10–19, for whom 80% of the estimated excess during the first 30 yr of followup is due to thyroid and breast cancer. For leukemia and bone cancer, both absolute- and relative-risk projection models would have produced exaggerated estimates if calculations were made on the basis of knowledge at earlier times before the decrease in incidence was determined. To the extent that other cancers may be shown to have expression times shorter than the life span of irradiated subjects, the absolute- and relative-risk models would result in overestimates. Thus, both the absolute- and relative-risk projection models may provide high risk estimates.

The age- and sex-specific risk coefficients used in the various calculations of excess cancer-mortality risk appear in later tables. The regression coefficients were obtained from the atomic-bomb survivor data and were fitted to age-adjusted rates on the assumption that the shape, although not necessarily the magnitude, of the dose response should be independent of both sex and age at exposure. For purposes of calculation, the Committee used the following models discussed above:

leukemia: $LQ\text{-}L$, $L\text{-}L$, and $Q\text{-}L$ (Equations V-6, V-7, and V-8);

other cancers: $\overline{LQ\text{-}L}$, $\overline{L\text{-}L}$, and $\overline{Q\text{-}L}$ (Equations V-10, V-11, and V-12).

Regression analyses were made in sex-specific fashion for all ages, and in age-specific fashion for both sexes combined, and the resulting coefficients were converted to age- and sex-specific arrays on a proportional basis. Finally, the mortality coefficients were expanded by a factor of 1.23 (derived from comparison of autopsy and death-certificate diagnoses[75]) to adjust for incomplete death-certificate ascertainment of cancer.

In its temporal pattern of expression, radiation-induced bone cancer is different from most other solid cancers, but similar to leukemia. For bone cancer, there are no low-LET dose-response data suitable for fitting models. As for all solid cancers as a group, the leukemia analyses were used to provide the needed structure for linear-quadratic, linear, and quadratic model risk estimates for bone cancer. The corresponding leukemia risk coefficients were multiplied by the ratio 0.05/2.24, where 0.05 is the linear estimate for bone cancer with regard to endosteal dose from low-LET radiation given in Appendix A, and 2.24 the *L-L* model estimate for leukemia.

The regression estimates for the Japanese data correspond to the periods 1950-1971 for leukemia, 1959-1970 for the Nagasaki Tumor Registry, and 1955-1974 for nonleukemia cancer mortality. In the life-table calculations, these estimates were then applied for the period 3-27 yr after exposure for leukemia and bone cancer, and for the period 11-30 yr after exposure for other cancers. Projection beyond 30 yr for cancer other than leukemia and bone cancer was made with the relative- and absolute-risk models discussed above.

The estimated coefficient for any age interval was assigned to each single year of age within that interval.

The kerma-to-tissue dose conversions used to generate the data from which the regression estimates were obtained are as shown in Table V-13.

Excess cancer risk in terms of incidence was approached in three ways: (1) mortality-risk estimates multiplied by a factor depending on the normal site-specific ratio of cancer incidence to cancer mortality and on the site-specific risk coefficients derived from Appendix A (Table V-14); (2) the Nagasaki Tumor Registry data for 1959-1970 referred to the LSS sample of atomic-bomb survivors;[8] and (3) the site-specific risk estimates of Appendix A summed to approximate the effect of whole-body irradiation (Table V-14).

TABLE V-13 Factors for Converting from Kerma to Tissue Dose[a]

Radiation	Tissue Dose/Kerma Dose	
	Leukemia	Other Cancers
Gamma	0.56	0.50
Neutron	0.28	0.22
Neutron-capture gamma	0.07	0.07

[a] Based on Kerr.[41]

TABLE V-14   Estimated Excess Cancer Incidence (Excluding Leukemia and Bone Cancer) per Million Persons per Year per Rad, 11–30 Yr after Exposure, by Site, Sex, and Age at Exposure

Site	Age at Exposure, yr					Age-Weighted Average[a]
	0–9	10–19	20–34	35–49	50+	
Males						
Thyroid[b,c]	2.20	2.20	2.20	2.20	2.20	2.20
Lung[d,e]	0.00	0.54	2.45	5.10	6.79	3.64
Esophagus[e,f]	0.07	0.07	0.13	0.21	0.56	0.26
Stomach[e,f]	0.40	0.40	0.77	1.27	3.35	1.53
Intestine[e,f]	0.26	0.26	0.52	0.84	2.23	1.02
Liver[e,f]	0.70	0.70	0.70	0.70	0.70	0.70
Pancreas[e,f]	0.24	0.24	0.45	0.75	1.97	0.90
Urinary[e,g]	0.04	0.23	0.50	0.92	1.62	0.81
Lymphoma[c,e]	0.27	0.27	0.27	0.27	0.27	0.27
Other[h]	0.62	0.38	1.12	1.40	2.90	1.52
All sites[i]	4.80	5.29	9.11	13.66	22.59	12.85
Females						
Thyroid[b,c]	5.80	5.80	5.80	5.80	5.80	5.80
Breast[j]	0.00	7.30	6.60	6.60	6.60	5.82
Lung[d,e]	0.00	0.54	2.45	5.10	6.79	3.94
Esophagus[e,f]	0.07	0.07	0.13	0.21	0.56	0.28
Stomach[e,f]	0.40	0.40	0.77	1.27	3.35	1.68
Intestine[e,f]	0.26	0.26	0.52	0.84	2.23	1.12
Liver[e,f]	0.70	0.70	0.70	0.70	0.70	0.70
Pancreas[e,f]	0.24	0.24	0.45	0.75	1.97	0.99
Urinary[e,g]	0.04	0.23	0.50	0.92	1.62	0.88
Lymphoma[c,e]	0.27	0.27	0.27	0.27	0.27	0.27
Other[h]	0.62	0.38	1.12	1.40	2.90	1.64
All sites[i]	8.40	16.19	19.31	23.86	32.79	23.10

[a] Average of the age-specific coefficients, weighted according to the age distribution, by sex, of the 1969–1971 U.S. life-table population.

[b] Estimate of 4 excess cases per million persons per year per rad adjusted by the observed male:female relative-risk ratio of 0.38 for atomic-bomb survivors.

[c] Risk assumed not to depend on age at exposure.

[d] Estimates are based on the expression, (attained age −35) × 0.2, with a risk of 0 to attained age 35, latent periods of 15 yr for ages 20–34 at irradiation, and 10 thereafter, except that a risk of 7.0 is used for those irradiated at age 65 or older.

[e] Risk assumed not to depend on sex.

[f] Age variation assumed proportional to linear estimates of atomic-bomb survivors for all gastrointestinal cancers.

[g] Age variation assumed proportional to smoothed risk estimates for cancers of urinary organs among atomic-bomb survivors.

TABLE V-14
Continued

---

h Although cancers of other sites—especially pharynx, larynx, salivary glands, and brain—are thought to be produced by low-dose, low-LET radiation, good estimates of absolute risk are not available. An arbitrary average of 1.0 excess cancer per million persons (of the age and sex distribution of atomic-bomb survivors) per year per rad is assumed. Age-specific coefficients are proportional to those for deaths from all malignant neoplasms, except leukemia, in the atomic-bomb survivors of both sexes combined.

i The total, for "all sites," is one possible measure of the effect (excluding leukemia and bone cancer) of whole-body radiation with all tissues receiving 1 rad.

j The value for ages 10–19 has been reduced to allow for dilution of effect arising from inclusion of pre-age-30 yr of exposure when latent period is set at 10 yr for all ages.

In deriving a suitable ratio of incidence to mortality for radiation-induced cancers, it was necessary to use some incidence measure of site-specific radiation risk (α_i) and to relate this to average U.S. rates for both site-specific incidence (I_i) and site-specific mortality (M_i) to yield the weighted ratio:

$$\Sigma\alpha_i / \Sigma\alpha_i R_i, \text{ where } R_i = M_i / I_i.$$

Although M_i / I_i varies somewhat with age, for the present purpose it was thought adequate to use I_i and M_i calculated as lifetime expectations at birth.[69] The estimated value of the weighted ratio was assumed to be invariant with respect to age at exposure, but was calculated separately for the two sexes: 1.54 for males, 2.00 for females (Table V-15). The life-table estimates of excess mortality were then multiplied by these weighted ratios to yield the incidence estimates.

The Nagasaki Tumor Registry data for major sites were fitted to the $\overline{LQ\text{-}L}$, $\overline{L\text{-}L}$, and $\overline{Q\text{-}L}$ models, as was done for the mortality data. Ratios derived from the age-adjusted and sex- and age-specific coefficients for cancer mortality and the sex-specific mortality-to-incidence expansion factors discussed above were used to calculate sex- and age-specific coefficients for cancer incidence.

In the third approach, each site-specific estimate of Appendix A was expressed in age- and sex-specific form on the basis of information given there, or by assuming that the array of risk coefficients with respect to age and sex was proportional to the parallel array of mortality-risk coefficients in the most recent atomic-bomb mortality analysis.[8] In two cancer sites (liver and lymphoma), the data for estimation of risk provided no basis for estimating variation by age and sex, and the atomic-bomb data provided no guide; in these instances, a constant value was used throughout. For

TABLE V-15 Derivation of Ratios for Transforming Mortality-Risk Estimates to Incidence-Risk Estimates, by Sex

Site (i)	Males				Females			
	Weight $(\alpha_i)^a$	Percentage Expectation at Birth of Eventually Developing or Dying from Cancer		Ratio $(R_i = M_i/I_i)^c$	Weight $(\alpha_i)^a$	Percentage Expectation at Birth of Eventually Developing or Dying from Cancer		Ratio $(R_i = M_i/I_i)^c$
		Mortality $(M_i)^b$	Incidence $(I_i)^b$			Mortality $(M_i)^b$	Incidence $(I_i)^b$	
Esophagus	0.26	0.4	0.4	1.00	0.28	0.2	0.2	1.00
Stomach	1.53	0.9	1.2	0.75	1.68	0.7	0.9	0.78
Intestine	1.02	2.3	4.4	0.52	1.12	2.8	5.1	0.55
Pancreas	0.90	1.0	1.1	0.91	0.99	0.9	1.0	0.90
Lung	3.64	4.9	5.9	0.83	3.94	1.2	1.6	0.75
Urinary	0.81	1.0	2.7	0.37	0.88	0.6	1.3	0.46
Lymphoma	0.27	1.1	1.5	0.73	0.27	0.9	1.2	0.75
Breast	0	0	0	—	5.82	3.0	7.7	0.39
Thyroid	2.20	0.03	0.17	0.18	5.80	0.09	0.46	0.20
Liver	0.70	0.20	0.20	1.00	0.70	0.18	0.18	1.00
Sum^d	11.33	—	—	—	21.48	—	—	—
Weighted sum^d	—	—	—	7.37	—	—	—	10.76
Expansion factor^d	—	1.54		—	—	2.00		—

^{a} Age-adjusted risk estimate, Table V-14.

^{b} Data from Seidman et al.[69] or calculated directly (thyroid, liver).

^{c} The ratios of mortality to incidence for specific types of cancer derived from vital statistics are not generally in close agreement with survival probabilities based on long-term clinical followup studies of cancer patients, nor is there any reason that they should be.

^{d} Sum $= \Sigma \alpha_i$, weighted sum $= \Sigma \alpha_i R_i$, expansion factor $= \Sigma \alpha_i / \Sigma \alpha_i R_i$.

the thyroid, the site-specific analysis of Appendix A provided the basis for differentiation by sex, but not by age. For the lung, esophagus, intestine, urinary organs, and a small residual group, it was possible to derive numerical risk coefficients specific for age, but not for sex. An example of the use of atomic-bomb data to produce the array is provided by the Appendix A estimate of 0.65 for the pancreas, derived from the experience of the ankylosing-spondylitis patients of average age 36 at radiation. The array, with respect to age, of risk coefficients for cancers of the digestive tract and peritoneum among the atomic-bomb survivors is as follows:

age, yr	0–9	10–19	20–34	35–49	50+
coefficient	0.49	0.49	0.94	1.55	4.10

For age 36, these coefficients suggested an approximate value of 1.35, and the ratio 0.65/1.35 was multiplied by the foregoing array to produce the age-specific risk coefficients of Table V-14. Clearly, the expression of the site-specific estimates of Appendix A in age- and sex-specific format is somewhat arbitrary and yields results of uncertain validity, especially at the youngest ages, which are seldom represented in the partial-body irradiation groups. Other limitations on these data are discussed earlier in this section. The age- and sex-specific sums of the site-specific estimates were then used as though they were the parallel whole-body risk estimates.

In other respects, calculation procedures for estimating the risk of excess cancer in terms of incidence were the same as those used for mortality. As in the mortality calculations, incident cases were withdrawn from the exposed life-table population subject to the risk coefficient for that year, modified by reduction factors to allow for the fact that not all incident cases lead to death from the cancers. The reduction factors were the sex-specific, age-adjusted ratios of the risk coefficients for incidence and mortality from comparable tables. For example, these ratios were set at 1.54 for males and 2.00 for females (Table V-15), between the various incidence coefficients used for Table V-26 and the corresponding mortality coefficients in Tables V-19, V-20, and V-21. Where the age-specific incidence coefficients are constant multiples of the corresponding mortality coefficients, the absolute-risk projections are in the same proportion between mortality and incidence. No such simple relationship holds between mortality and incidence for the relative-risk projections, however, because the underlying population rates for mortality and incidence vary differently with respect to age. In general, although age-specific incidence is greater than mortality, it increases less steeply with age. A somewhat paradoxical consequence of this, and of our incomplete knowledge of lifetime cancer risk from exposures at young ages, is that, for some hypothetical exposures of young populations, the projected excess lifetime cancer incidence may be less than the projected excess mortality, even though the age-specific estimates of excess risk corresponding to the first 30 yr of followup are higher for incidence than for mortality.

RESULTS

Because the risk estimates in this report are expressed in terms of whole-body or specific-organ absorbed dose, it is essential that they be applied in terms of absorbed dose received in any given situation; e.g., if a dose in a given exposure is expressed in kerma air dose or surface dose, a correction factor should be applied to estimate mean whole-body or specific-organ dose.

Mortality from Leukemia and Bone Cancer

Tables V-16, V-17, and V-18 present the life-table estimates for leukemia and bone cancer (incidence assumed to equal mortality) according to three dose-response models, LQ-L, L-L, and Q-L. In each instance, the tables give (1) the dose-response relationships for leukemia and for bone cancer; (2) the age- and sex-specific coefficients derived from the parameter estimates; and (3) the final life-table estimates corresponding to the specified exposure conditions. For example, under the LQ-L model, which fits the leukemia and bone-cancer data best (see above), exposure of a life-table population to a single dose of 10 rads generates lifetime excess deaths (or cases) of 274 per million males and 186 per million females. For leukemia and bone cancer, the projection models are not used, because the effect is assumed to cease within 30 yr. For each exposure condition, three lines are given: the expected normal incidence of leukemia and bone cancer (apart from the excess risk from radiation), the estimated excess number of radiation-induced cases, and the excess number expressed as a percentage of normal expectation of leukemia and bone cancer. For example, in Table V-16, 274 excess cancers are estimated to result from a single exposure of a million men of all ages to 10 rads; this is to be compared with the 9,860 cases of leukemia and bone cancer expected after exposure independently of the radiation effect; in other words, the estimated excess is $100 \times 274/9,860 = 2.8\%$ of the normal expectation for these cancers. For continuous irradiation at 1 rad/yr, the risk estimates depend on the duration of exposure. For example, in Table V-16, under the LQ-L model, exposure to 1 rad/yr throughout life results in a 15% or 13% excess, depending on sex; exposure at ages 20–65, in 9% or 8%; exposure at ages 35–65, in 6%; and exposure at

TABLE V-16 Estimated Excess Incidence of (and Mortality from) Leukemia and Bone Cancer from Low-LET Radiation Dose (D): *LQ-L* Model[a]

Estimated dose-response relationship:
  Leukemia:[b] Excess risk $= 0.9892D + 0.008508D^2$
  Bone cancer:[c] Excess risk $= 0.02209D + 0.000190D^2$

Estimated age- and sex-specific regression coefficients for D and D^2:

Sex		Age at Exposure, yr					
		0–9	10–19	20–34	35–49	50+	All[d]
M	D	1.829	0.7855	1.138	0.8511	1.937	1.367
	D^2	0.01575	0.006766	0.009798	0.007331	0.01669	0.0117
F	D	1.169	0.5067	0.7301	0.5483	1.238	0.9039
	D^2	0.01007	0.004364	0.006289	0.004723	0.01047	0.007717

Life-table estimates of excess cases per million persons:

	Male	Female
Single exposure to 10 rads:		
Normal expectation	9,860	8,018
Excess cases: number	274	186
% of normal	2.8	2.3
Continuous exposure to 1 rad/yr, lifetime:		
Normal expectation	10,600	9,050
Excess cases: number	1,592	1,209
% of normal	15.0	13.4
Continuous exposure to 1 rad/yr, ages 20–65:		
Normal expectation	10,020	8,545
Excess cases: number	940	705
% of normal	9.4	8.3
Continuous exposure to 1 rad/yr, ages 35–65:		
Normal expectation	9,828	8,372
Excess cases: number	587	465
% of normal	6.0	5.6
Continuous exposure to 1 rad/yr, ages 50–65:		
Normal expectation	9,667	8,124
Excess cases: number	370	306
% of normal	3.8	3.8

[a] D is in rads; coefficients for D and D^2 are per million persons per year.
[b] Based on Leukemia Registry cases, 1950–1971, in LSS sample.
[c] With regard to endosteal dose from low-LET radiation.
[d] Weighted average for U.S. life-table population, 1969–1971.

TABLE V-17   Estimated Excess Incidence of (and Mortality from) Leukemia and Bone Cancer from Low-LET Radiation Dose (D): *L-L* Model[a]

Estimated dose-response relationship:
   Leukemia:[b]   Excess risk = 2.239D
   Bone cancer:[c] Excess risk = 0.05D

Estimated age- and sex-specific regression coefficients for D:

| Sex | Age at Exposure, yr | | | | | |
	0-9	10-19	20-34	35-49	50+	All[d]
M	3.977	1.849	2.596	1.921	4.319	3.051
F	2.542	1.192	1.666	1.237	2.760	2.025

Life-table estimates of excess cases per million persons:

	Male	Female
Single exposure to 10 rads:		
Normal expectation	9,860	8,018
Excess cases: number	566	384
% of normal	5.7	4.8
Continuous exposure to 1 rad/yr, lifetime:		
Normal expectation	10,600	9,050
Excess cases: number	3,568	2,709
% of normal	33.7	29.9
Continuous exposure to 1 rad/yr, ages 20–65:		
Normal expectation	10,020	8,545
Excess cases: number	2,119	1,589
% of normal	21.1	18.6
Continuous exposure to 1 rad/yr, ages 35–65:		
Normal expectation	9,828	8,372
Excess cases: number	1,315	1,041
% of normal	13.4	12.4
Continuous exposure to 1 rad/yr, ages 50–65:		
Normal expectation	9,667	8,124
Excess cases: number	826	682
% of normal	8.5	8.4

[a] D is in rads; coefficients for D are per million persons per year.
[b] Based on Leukemia Registry cases, 1950–1971, in LSS sample.
[c] With regard to endosteal dose from low-LET radiation.
[d] Weighted average for U.S. life-table population, 1969–1971.

TABLE V-18 Estimated Excess Incidence of (and Mortality from) Leukemia and Bone Cancer from Low-LET Radiation Dose (*D*): *Q-L* Model[a]

Estimated dose-response relationship:
  Leukemia:[b] Excess risk = $0.01372D^2$
  Bone cancer:[c] Excess risk = $0.000306D^2$

Estimated age- and sex-specific regression coefficients for D^2:

| Sex | Age at Exposure, yr | | | | | |
	0–9	10–19	20–34	35–49	50+	All[d]
M	0.02639	0.01068	0.01578	0.01182	0.02706	0.01906
F	0.01686	0.006893	0.01013	0.007621	0.01729	0.01265

Life-table estimates of excess cases per million persons:

	Male	Female
Single exposure to 10 rads:		
Normal expectation	9,860	8,018
Excess cases: number	35	24
% of normal	0.35	0.30

---

[a] *D* is in rads; coefficients for D^2 are per million persons per year.
[b] Based on Leukemia Registry cases, 1950–1971, in LSS sample.
[c] With regard to endosteal dose from low-LET radiation.
[d] Weighted average for U.S. life-table population, 1969–1971.

ages 50–65, 4%. Under the *L-L* model (Table V-17), these values are about doubled; and under the *Q-L* model (Table V-18), which was applied only to the single 10-rad exposure, they are reduced to about one-eighth.

Mortality from Cancer Other than Leukemia and Bone Cancer

Tables V-19, V-20, and V-21 pertain to excess mortality from cancer other than leukemia and bone cancer and have the same format as Tables V-16, V-17, and V-18. For the $\overline{LQ\text{-}L}$ model (Table V-19) and a single exposure to 10 rads, the excess cancers are estimated as 0.25% of the normal expectation of cancer mortality for males and 0.47% for females by the absolute-risk model, and both are increased by factors of 3–4 in the relative-risk model. For continuous lifetime exposure to 1 rad/yr, the values range from 1.5% to 7.9%; and for shorter durations of exposure, the values are correspondingly less. The $\overline{L\text{-}L}$ model yields values for continuous lifetime exposure that are about 2–3 times those of the $\overline{LQ\text{-}L}$

TABLE V-19   Estimated Excess Fatal Cancers Other than Leukemia and Bone Cancer from Low-LET Radiation Dose (D): $\overline{LQ\text{-}L}$ Model[a]

Estimated dose-response relationship:[b] Excess risk $= 1.397(D + 0.008614D^2)$

Estimated age- and sex-specific regression coefficients for D and D^2:

Sex		Age at Exposure, yr					
		0–9	10–19	20–34	35–49	50+	All[c]
M	D	0.8972	0.6095	1.774	2.278	3.446	2.076
	D^2	0.007728	0.005250	0.01528	0.01962	0.02968	0.01788
F	D	1.169	0.7940	2.311	2.968	4.489	2.858
	D^2	0.01007	0.006839	0.01990	0.02556	0.03867	0.02462

Life-table estimates of excess cases per million persons:

	Absolute-Risk Projection Model		Relative-Risk Projection Model	
	M	F	M	F
Single exposure to 10 rads:				
Normal expectation	170,400	139,400	170,400	139,400
Excess deaths: number	421	652	1,917	2,133
% of normal	0.25	0.47	1.1	1.5
Continuous exposure to 1 rad/yr, lifetime:				
Normal expectation	165,700	149,200	165,700	149,200
Excess deaths: number	2,459	4,243	9,287	11,850
% of normal	1.5	2.8	5.6	7.9
Continuous exposure to 1 rad/yr, ages 20–65:				
Normal expectation	171,600	152,800	171,600	152,800
Excess deaths: number	1,788	3,104	3,694	5,677
% of normal	1.0	2.0	2.2	3.7
Continuous exposure to 1 rad/yr, ages 35–65:				
Normal expectation	175,700	153,300	175,700	153,300
Excess deaths: number	1,005	1,848	1,214	2,301
% of normal	0.57	1.2	0.69	1.5
Continuous exposure to 1 rad/yr, ages 50–65:				
Normal expectation	178,000	147,300	178,000	147,300
Excess deaths: number	410	818	419	862
% of normal	0.23	0.56	0.24	0.59

[a] D is in rads; coefficients for D and D^2 are per million persons per year.
[b] Based on Hiroshima and Nagasaki LSS data, 1955–1974.
[c] Weighted average for U.S. life-table population, 1969–1971.

TABLE V-20   Estimated Excess Fatal Cancers Other than Leukemia and Bone Cancer from Low-LET Radiation Dose (D): $\overline{L\text{-}L}$ Model[a]

Estimated dose-response relationship:[b] Excess risk $= 3.470D$

Estimated age- and sex-specific regression coefficients for D:

Sex	Age at Exposure, yr					
	0–9	10–19	20–34	35–49	50+	All[c]
M	1.920	1.457	4.327	5.291	8.808	5.087
F	2.576	1.955	5.807	7.102	11.823	7.254

Life-table estimates of excess cases per million persons:

	Absolute-Risk Projection Model		Relative-Risk Projection Model	
	M	F	M	F
Single exposure to 10 rads:				
Normal expectation	170,400	139,400	170,400	139,400
Excess deaths: number	919	1,473	4,226	4,852
% of normal	0.54	1.1	2.5	3.5
Continuous exposure to 1 rad/yr, lifetime:				
Normal expectation	165,700	149,200	165,700	149,200
Excess deaths: number	5,827	10,400	22,080	29,030
% of normal	3.5	7.0	13.3	19.5
Continuous exposure to 1 rad/yr, ages 20–65:				
Normal expectation	171,600	152,800	171,600	152,800
Excess deaths: number	4,324	7,745	8,916	14,100
% of normal	2.5	5.1	5.2	9.2
Continuous exposure to 1 rad/yr, ages 35–65:				
Normal expectation	175,700	153,300	175,700	153,300
Excess deaths: number	2,420	4,603	2,905	5,685
% of normal	1.4	3.0	1.7	3.7
Continuous exposure to 1 rad/yr, ages 50–65:				
Normal expectation	178,000	147,300	178,000	147,300
Excess deaths: number	1,046	2,153	1,069	2,265
% of normal	0.59	1.5	0.60	1.5

[a] D is in rads; coefficients for D are per million persons per year.
[b] Based on Hiroshima and Nagasaki LSS data, 1955–1974.
[c] Weighted average for U.S. life-table population, 1969–1971.

TABLE V-21   Estimated Excess Fatal Cancers Other than Leukemia and Bone Cancer from Low-LET Radiation Dose (D): $\overline{Q\text{-}L}$ Model[a]

Estimated dose-response relationship:[b] Excess risk $= 0.01825D^2$

Estimated age- and sex-specific regression coefficients for D^2:

Sex	Age at Exposure, yr					
	0–9	10–19	20–34	35–49	50+	All[c]
M	0.01294	0.008179	0.02332	0.03091	0.04357	0.02717
F	0.01653	0.01045	0.02980	0.03950	0.05567	0.03652

Life-table estimates of excess cases per million persons:

	Absolute-Risk Projection Model		Relative-Risk Projection Model	
	M	F	M	F
Single exposure to 10 rads:				
Normal expectation	170,400	139,400	170,400	139,400
Excess deaths: number	52	79	236	257
% of normal	0.031	0.057	0.14	0.18

[a] D is in rads; coefficients for D^2 are per million persons per year.
[b] Based on Hiroshima and Nagasaki LSS data, 1955–1974.
[c] Weighted average for U.S. life-table population, 1969–1971.

model; the $\overline{Q\text{-}L}$ model, for a single exposure to 10 rads, gives values about one-eighth as large as those from the $\overline{LQ\text{-}L}$ model.

Mortality from All Forms of Cancer

There are various ways of combining estimates of leukemia and bone cancer and other solid-cancer mortality projections. For example, in Table V-22 the $\underline{LQ\text{-}L}$ estimates for leukemia and bone cancer are combined with the $\overline{LQ\text{-}L}$ estimates for other cancers. Much of the variation is due to the total dose received—10 rads for the single dose, 13–75 rads for the continuous exposures. The next most important source of variation is the projection model. For continuous lifetime exposure, the relative-risk model yields estimates that are 2–3 times those of the absolute-risk model; for exposure at ages 35–65 and 50–65, the two models are in closer agreement.

An alternative way of combining estimates of leukemia and solid-cancer mortality projections is given in Table V-23, where the dose-response models are compared for a single exposure to 10 rads. The estimates range from 0.06 to 3.1 when expressed as percentages of normal expectation, depending on dose-response model and on projection model.

Of particular interest is a comparison of the dose-response models from the standpoint of the ratio of excess fatal cancers other than leukemia and bone cancer to deaths from leukemia and bone cancer. In the 1977 UNSCEAR report,[82] this ratio plays a prominent role in the estimation of mortality from radiation-induced solid tumors. In the ankylosing-spondylitis series, the ratio is 4.7; in the study of U.S. radiologists, 4.3; and in the atomic-bomb survivors, 1.15. The ratio is very sensitive to the age distribution of the subjects under study and to the duration of followup. In general, the younger the subjects, and the shorter the period of followup, the smaller the ratio will be. Ankylosing-spondylitis patients and radiologists have an older age distribution at exposure than the atomic-bomb survivors with their heavy representation of ages under 20 at exposure, and for none of these did followup cover an entire lifetime. In the 1977 UNSCEAR report, it was estimated that lifetime excess mortality

TABLE V-22 Estimated Excess Mortality per Million Persons from All Forms of Cancer, Linear-Quadratic Dose-Response Model[a] for Low-LET Radiation

	Absolute-Risk Projection Model	Relative-Risk Projection Model
Single exposure to 10 rads:		
Normal expectation	163,800	163,800
Excess cases: number	766	2,255
% of normal	0.47	1.4
Continuous exposure to 1 rad/yr, lifetime:		
Normal expectation	167,300	167,300
Excess cases: number	4,751	11,970
% of normal	2.8	7.2
Continuous exposure to 1 rad/yr, ages 20–65:		
Normal expectation	171,500	171,500
Excess cases: number	3,268	5,508
% of normal	1.9	3.2
Continuous exposure to 1 rad/yr, ages 35–65:		
Normal expectation	173,600	173,600
Excess cases: number	1,952	2,283
% of normal	1.1	1.3
Continuous exposure to 1 rad/yr, ages 50–65:		
Normal expectation	171,600	171,600
Excess cases: number	952	978
% of normal	0.55	0.57

[a] $LQ\text{-}L$ for leukemia, $\overline{LQ\text{-}L}$ for other.

TABLE V-23 Estimated Excess Mortality per Million Persons from All Forms of Cancer, Single Exposure to 10 Rads of Low-LET Radiation, by Dose-Response Model

Dose-Response Model				
Leukemia and Bone	Other Cancer	Projection Model	Absolute-Risk Projection Model	Relative-Risk Projection Model
		Normal expectation of cancer deaths	163,800	163,800
LQ-L	$\overline{LQ\text{-}L}$	Excess deaths: number	766	2,255
		% of normal	0.47	1.4
L-L	$\overline{L\text{-}L}$	Excess deaths: number	1,671	5,014
		% of normal	1.0	3.1
Q-L	$\overline{Q\text{-}L}$	Excess deaths: number	95	276
		% of normal	0.058	0.17

from solid tumors would be approximately 3–5 times that for leukemia. Estimates made in this report for lifetime risk are slightly lower for the absolute-risk projection model, but considerably higher for the relative-risk model, reflecting the increasing importance of solid tumors in the latter decades of life. With exposure beginning at ages 20, 35, and 50, the ratio declines sharply, as the expression time for solid tumors is shortened relative to that for leukemia. There is little difference among the models used here. The ratios for continuous exposure over a full lifetime and from ages 20–65, for the *LQ-L* and *L-L* models, are shown in Table V-24.

Comparison of Mortality-Risk Estimates with Those in the 1972 BEIR Report[56] and the 1977 UNSCEAR Report[82]

In the 1972 BEIR report,[56] annual estimates of possible carcinogenic effect were made on the basis of the linear hypothesis and exposure of the 1967 U.S. population of 197.9 million to 0.1 rem/yr, with absolute- and relative-risk projection models for a 30-yr period and a lifetime. To represent the information available at the time of the 1972 report and the assumptions on which it was based, the 1972 age-specific risk coefficients were used in the life-table calculation procedures of the present report (Table V-25). The current estimates are exemplified by the *LQ-L* model for leukemia and bone cancer and the $\overline{LQ-L}$ model for other forms of cancer. The estimates obtained for other models used in the present calculations are shown in the first footnote of Table V-25. The present estimates for the linear-quadratic models are below the 1972 estimates, and especially so when the comparison is based on the relative-risk projection model. Much of the difference between the 1972 estimates and the present

TABLE V-24 Ratios of Excess Deaths from Radiation-Induced Cancers Other than Leukemia and Bone Cancer to Excess Deaths from Radiation-Induced Leukemia and Bone Cancer

Dose-Response Model		Projection Model	
Leukemia and Bone Cancer	Other Cancer	Relative-Risk	Absolute-Risk
Lifetime exposure: 1 rad/yr			
LQ-L	$\overline{LQ-L}$	7.5	2.4
L-L	$\overline{L-L}$	8.1	2.6
Exposure at ages 20–65: 1 rad/yr			
LQ-L	$\overline{LQ-L}$	5.7	3.0
L-L	$\overline{L-L}$	6.2	3.3

TABLE V-25   Comparative Estimates of the Lifetime Risk of Cancer
Mortality Induced by Low-LET Radiation—Excess Deaths per Million,
Average Value per Rad by Projection Model, Dose-Response Model,
and Type of Exposure

| | | Projection Model | | | |
| | | Single Exposure to 10 Rads | | Continuous Lifetime Exposure to 1 Rad/Yr | |
Source of Estimate	Dose-Response Models[a]	Absolute	Relative	Absolute	Relative
BEIR, 1980[b]	LQ-L, $\overline{LQ\text{-}L}$	77	226	67	169
1972 BEIR report factors[c]	Linear	117	621	115	568
UNSCEAR 1977[d]	Linear		75–175		

[a] For BEIR 1980, the first model is used for leukemia, the second for other forms of cancer.
The corresponding estimates when the other models are used (thereby providing an envelope
of risk estimates) are:

	L-L, $\overline{L\text{-}L}$	167	501	158	403
	Q-L, $\overline{Q\text{-}L}$	10	28		

[b] The values are average values per rad, and are not to be taken as estimates at only 1 rad
of dose.

[c] 1972 BEIR[56] postnatal, age-specific risk factors used with 1969–1971 life tables, with
plateau extending throughout the lifetime remaining after irradiation, estimate (b) in the
1972 report. The average age of the 1969–1971 life-table population exceeds that of the 1967
U.S. population used for the 1972 BEIR report. For this reason, the numbers shown here for
continuous exposure are larger, on a per-rad basis, than those obtainable from Tables 3-3
and 3-4 of the 1972 BEIR report.

[d] UNSCEAR range of estimates for low-dose, low-LET radiation (p. 414, para. 318).[82]

estimates results from the more precise handling of age in the present analysis and the substitution of relative-risk ratios derived from exposure in the second decade of life for those derived from exposure in the first decade.

The effect of *in utero* radiation is uncertain (see Appendix A) and has not been included in the foregoing summary of results. Evidence with respect to radiation risk to the human fetus comes from atomic-bomb survivors and from the children of patients receiving x-ray pelvimetry. Risk estimates vary from zero added cancers to about 50 cancers per million children exposed *in utero* per rad per year during the first decade of life. When the Oxford Childhood Cancer Survey risk estimates were used in a fashion parallel to that used here for postnatal exposure, the exposure of 1 million women of life-table age composition to a single dose of 10 rads was estimated to yield about 110 cancer deaths among their progeny. With continuous population exposure to 1 rad/yr, all persons are exposed to about 0.75 rad *in utero*. This exposure might result in about 425 excess cancer deaths per million population in addition to deaths induced by postnatal exposure.

Incidence of Cancer Other than Leukemia and Bone Cancer

Table V-26 contains the incidence estimates obtained from the mortality estimates in the fashion described above. The mortality estimates in Tables V-19, V-20, and V-21 corresponding to all three dose-response models were multiplied by 1.54 for males and 2.0 for females (cf. Table V-15) to provide the estimates in Table V-26. For a single dose of 10 rads, the estimates range from 0.03% of normal expectation to 2.4%, depending primarily on dose-response model and to a lesser extent on projection model and sex. For a lifetime exposure to 1 rad/yr, the estimated excess ranges from 1.4% of normal expectation to 9% for the $\overline{LQ\text{-}L}$ model and from 3.3% to 22% for the *L-L* model.

The Nagasaki Tumor Registry data for 1959–1970 yield the estimates of Tables V-27, V-28, and V-29, each for a particular dose-response model. The incidence estimates for the $\overline{LQ\text{-}L}$ and $\overline{L\text{-}L}$ dose-response models differ by a factor of about 2.5; for a single exposure to 10 rads, the $\overline{Q\text{-}L}$ model estimates are about one-eighth the corresponding $\overline{LQ\text{-}L}$ model estimates. For a particular dose-response model, the relative-risk projection model gives values that are 2–3 times those obtained from the absolute-risk projection model for the single exposure to 10 rads and for continuous lifetime exposure to 1 rad/yr; for exposure progressively later in life, the two projection models give increasingly similar values.

In Table V-30, the summed-sites approach yields estimates of excess cancer that range from 0.8% to 4.3% of normal cancer incidence when based on a single dose of 10 rads, 5.4% to 64% when based on a lifetime exposure of 1 rad/yr, and correspondingly lower percentages when based on exposures of shorter duration. When the leukemia estimates of Table V-16 are added, the percentages in Table V-30 change little.

For comparison, Table V-31 brings together the results of the various approaches to the estimation of the incidence of all cancers other than leukemia and bone cancer for the exposure situation of 1 rad/yr throughout life. Although excess incidence is considered to be a more complete index of radiation-induced cancer than mortality, the uncertainties surrounding the data bases for incidence are greater than those for mortality. The estimates based on the summed-sites data of Appendix A are the highest; those derived from the other two data sources are similar to each other. The strengths and weaknesses of these several approaches have already been reviewed. Overall, however, the summed-sites approach seems to have considerable upward bias, the Nagasaki Tumor Registry data do not seem to have been sufficiently well evaluated to be relied on strongly, and the most reliable approach may be the indirect conversion of mortality estimates to incidence estimates.

TABLE V-26 Estimated Excess Cases of Cancer Other than Leukemia and Bone Cancer, Based on Expansion of Mortality Estimates in Tables V-19, V-20, and V-21—Excess Cases per Million Exposed to Low-LET Radiation, by Exposure, Risk Model, Projection Model, and Sex

	Absolute-Risk Projection Model		Relative-Risk Projection Model	
	M	F	M	F
Single exposure to 10 rads:				
Normal expectation	274,900	252,400	274,900	252,400
$\overline{LQ\text{-}L}$ model estimate: number	648	1,304	2,105	2,627
% of normal	0.24	0.52	0.77	1.0
$\overline{L\text{-}L}$ model estimate: number	1,415	2,946	4,632	5,998
% of normal	0.51	1.2	1.7	2.4
$\overline{Q\text{-}L}$ model estimate: number	81	158	245	315
% of normal	0.029	0.063	0.089	0.12
Continuous exposure to 1 rad/yr, lifetime:				
Normal expectation	272,800	285,600	272,800	285,600
$\overline{LQ\text{-}L}$ model estimate: number	3,787	8,486	6,624	25,950
% of normal	1.4	3.0	2.4	9.1
$\overline{L\text{-}L}$ model estimate: number	8,974	20,800	15,900	63,530
% of normal	3.3	7.3	5.8	22.2
Continuous exposure to 1 rad/yr, ages 20–65:				
Normal expectation	281,600	291,500	281,600	291,500
$\overline{LQ\text{-}L}$ model estimate: number	2,751	6,208	3,874	12,070
% of normal	0.98	2.1	1.4	4.1
$\overline{L\text{-}L}$ model estimate: number	6,659	15,490	9,358	29,970
% of normal	2.4	5.3	3.3	10.3
Continuous exposure to 1 rad/yr, ages 35–65:				
Normal expectation	286,000	287,200	286,000	287,200
$\overline{LQ\text{-}L}$ model estimate: number	1,547	3,696	1,736	4,402
% of normal	0.54	1.3	0.61	1.5
$\overline{L\text{-}L}$ model estimate: number	3,727	9,206	4,163	10,880
% of normal	1.3	3.2	1.5	3.8
Continuous exposure to 1 rad/yr, ages 50–65:				
Normal expectation	285,800	260,400	285,800	260,400
$\overline{LQ\text{-}L}$ model estimate: number	631	1,636	643	1,655
% of normal	0.22	0.63	0.23	0.64
$\overline{L\text{-}L}$ model estimate: number	1,611	4,306	1,642	4,350
% of normal	0.56	1.7	0.57	1.7

TABLE V-27   Estimated Excess Cases of Cancer Other than Leukemia and Bone Cancer from Low-LET Radiation Dose (D): $\overline{LQ\text{-}L}$ Model[a]

Estimated dose-response relationship:[b]
   Excess risk $= 3.335(D + 0.008614D^2)$

Estimated age- and sex-specific regression coefficients for D and D^2:

Sex		Age at Exposure, yr					
		0–9	10–19	20–34	35–49	50+	All[c]
M	D	1.478	1.004	2.922	3.753	5.677	3.420
	D^2	0.01273	0.008635	0.02517	0.03227	0.04882	0.02942
F	D	2.501	1.699	4.944	6.349	9.606	6.115
	D^2	0.02151	0.01461	0.04252	0.05461	0.08261	0.05261

Life-table estimates of excess cases of cancer per million persons:

	Absolute-Risk Projection Model		Relative-Risk Projection Model	
	M	F	M	F
Single exposure to 10 rads:				
Normal expectation	274,900	252,400	274,900	252,400
Excess cases: number	694	1,394	2,251	2,811
% of normal	0.25	0.55	0.82	1.1
Continuous exposure to 1 rad/yr, lifetime:				
Normal expectation	272,800	285,600	272,800	285,600
Excess cases: number	4,051	9,078	7,086	27,760
% of normal	1.5	3.2	2.6	9.7
Continuous exposure to 1 rad/yr, ages 20–65:				
Normal expectation	281,600	291,500	281,600	291,500
Excess cases: number	2,944	6,641	4,144	12,910
% of normal	1.0	2.3	1.5	4.4
Continuous exposure to 1 rad/yr, ages 35–65:				
Normal expectation	286,000	287,200	286,000	287,200
Excess cases: number	1,657	3,953	1,857	4,709
% of normal	0.58	1.4	0.65	1.6
Continuous exposure to 1 rad/yr, ages 50–65:				
Normal expectation	285,800	260,400	285,800	260,400
Excess cases: number	675	1,750	688	1,770
% of normal	0.24	0.67	0.24	0.68

[a] D is in rads; coefficients for D and D^2 are per million persons per year.
[b] Based on Nagasaki Tumor Registry, 1959–1970, and LSS sample.
[c] Weighted average for U.S. life-table population, 1969–1971.

TABLE V-28 Estimated Excess Cases of Cancer Other than Leukemia and Bone Cancer from Low-LET Radiation Dose (D): $\overline{L\text{-}L}$ Model[a]

Estimated dose-response relationship:[b]
  Excess risk $= 9.202D$

Estimated age- and sex-specific regression coefficients for D:

Sex	Age at Exposure, yr					
	0–9	10–19	20–34	35–49	50+	All[c]
M	3.509	2.663	7.909	9.673	16.102	9.299
F	6.117	4.642	13.79	16.86	28.07	17.22

Life-table estimates of excess cases of cancer per million persons:

	Absolute-Risk Projection Model		Relative-Risk Projection Model	
	M	F	M	F
Single exposure to 10 rads:				
Normal expectation	274,900	252,400	274,900	252,400
Excess cases: number	1,680	3,498	5,499	7,124
% of normal	0.61	1.4	2.0	2.8
Continuous exposure to 1 rad/yr, lifetime:				
Normal expectation	272,800	285,600	272,800	285,600
Excess cases: number	10,650	24,690	18,870	75,410
% of normal	3.9	8.6	6.9	26.4
Continuous exposure to 1 rad/yr, ages 20–65:				
Normal expectation	281,600	291,500	281,600	291,500
Excess cases: number	7,904	18,390	11,110	35,570
% of normal	2.8	6.3	3.9	12.2
Continuous exposure to 1 rad/yr, ages 35–65:				
Normal expectation	286,000	287,200	286,000	287,200
Excess cases: number	4,423	10,930	4,942	12,920
% of normal	1.5	3.8	1.7	4.5
Continuous exposure to 1 rad/yr, ages 50–65:				
Normal expectation	285,800	260,400	285,800	260,400
Excess cases: number	1,912	5,111	1,949	5,163
% of normal	0.67	2.0	0.68	2.0

[a] D is in rads; coefficients for D are per million persons per year.
[b] Based on Nagasaki Tumor Registry, 1959–1970, and LSS sample.
[c] Weighted average for U.S. life-table population, 1969–1971.

TABLE V-29   Estimated Excess Cases of Cancer Other than Leukemia and Bone Cancer from Low-LET Radiation Dose (D): $\overline{Q\text{-}L}$ Model[a]

Estimated dose-response relationship:[b]
  Excess risk $= 0.04191D^2$

Estimated age- and sex-specific regression coefficients for D^2:

Sex	0–9	10–19	20–34	35–49	50+	All[c]
			Age at Exposure, yr			
M	0.02048	0.01295	0.03692	0.04894	0.06898	0.04301
F	0.03460	0.02187	0.06237	0.08267	0.1165	0.07645

Life-table estimates of excess cases of cancer per million persons:

	Absolute-Risk Projection Model		Relative-Risk Projection Model	
	M	F	M	F
Single exposure to 10 rads:				
Normal expectation	274,900	252,400	274,900	252,400
Excess cases: number	82	165	252	330
% of normal	0.030	0.065	0.092	0.13

[a] D is in rads; coefficients for D^2 are per million persons per year.
[b] Based on Nagasaki Tumor Registry, 1959–1970, and LSS sample.
[c] Weighted average for U.S. life-table population, 1969–1971.

TABLE V-30   Estimated Excess Cases of Cancer Other than Leukemia and Bone Cancer from Low-LET Radiation Dose (*D*), Based on Sum of Site-Specific Linear Coefficients (Linear Age- and Sex-Specific Risk Coefficients Obtained from Table V-14)[a]

Estimated age- and sex-specific risk coefficients for low-LET dose:

| | Age at Exposure, yr | | | | | |
Sex	0-9	10-19	20-34	35-49	50+	All[b]
M	4.80	5.29	9.11	13.66	22.59	12.85
F	8.40	16.19	19.31	23.86	32.79	23.10

Life-table estimates of excess cases of cancer per million persons:

	Absolute-Risk Projection Model		Relative-Risk Projection Model	
	M	F	M	F
Single exposure to 10 rads:				
Normal expectation	274,900	252,400	274,900	252,400
Excess cases: number	2,312	5,356	8,527	15,970
% of normal	0.84	2.1	3.1	6.3
Continuous exposure to 1 rad/yr, lifetime:				
Normal expectation	272,800	285,600	272,800	285,600
Excess cases: number	14,640	37,540	29,530	184,000
% of normal	5.4	13.1	10.8	64.4
Continuous exposure to 1 rad/yr, ages 20-65:				
Normal expectation	281,600	291,500	281,600	291,500
Excess cases: number	10,200	24,720	14,030	48,810
% of normal	3.6	8.5	5.0	16.7
Continuous exposure to 1 rad/yr, ages 35-65:				
Normal expectation	286,000	287,200	286,000	287,200
Excess cases: number	6,232	14,260	6,963	17,064
% of normal	2.2	5.0	2.4	5.9
Continuous exposure to 1 rad/yr, ages 50-65:				
Normal expectation	285,800	260,400	285,800	260,400
Excess cases: number	2,683	5,973	2,734	6,034
% of normal	0.94	2.3	0.96	2.3

[a] *D* is in rads; coefficients are per million persons per year.
[b] Weighted average for U.S. life-table population, 1969-1971.

TABLE V-31 Comparison of Three Approaches to Estimating Excess Incidence of Cancer Other than Leukemia and Bone Cancer—Estimated Numbers of Excess Cancer Cases per Million Persons Exposed to Low-LET Radiation, by Source of Data, Dose-Response Model, Projection Model, and Sex

Source of Data	Dose-Response Model	Absolute-Risk Projection Model		Relative-Risk Projection Model	
		Male	Female	Male	Female
Single exposure to 10 rads of low-LET radiation, entire population:					
Normal expectation	—	274,900	252,400	274,900	252,400
LSS mortality, 1955–1974, expanded (Tables V-22 and V-23)	$\overline{LQ\text{-}L}$ estimate: no.	648	1,304	2,105	2,627
	%	0.24	0.52	0.77	1.0
	$\overline{L\text{-}L}$ estimate: no.	1,415	2,946	4,632	5,998
	%	0.51	1.2	1.7	2.4
	$\overline{Q\text{-}L}$ estimate: no.	81	158	245	315
	%	0.029	0.063	0.089	0.12

Nagasaki Tumor Registry, 1959–1970 (Table V-26)				
$\overline{LQ\text{-}L}$ estimate: no.	694	1,394	2,251	2,811
%	0.25	0.55	0.82	1.1
$\overline{L\text{-}L}$ estimate: no.	1,680	3,498	5,499	7,124
%	0.61	1.4	2.0	2.8
$\overline{Q\text{-}L}$ estimate: no.	82	165	252	330
%	0.030	0.065	0.092	0.13
Summed sites (Appendix A)				
$\overline{L\text{-}L}$ estimate: no.	2,312	5,356	8,527	15,970
%	0.84	2.1	3.1	6.3
Continuous lifetime exposure to 1 rad/yr:				
Normal expectation —	272,800	285,600	272,800	285,600
Lss mortality, 1955–1974, expanded (Tables V-22 and V-23)				
$\overline{LQ\text{-}L}$ estimate: no.	3,787	8,486	6,624	25,950
%	1.4	3.0	2.4	9.1
$\overline{L\text{-}L}$ estimate: no.	8,974	20,800	15,900	63,530
%	3.3	7.3	5.8	22.2
Nagasaki Tumor Registry, 1959–1970 (Table V-26)				
$\overline{LQ\text{-}L}$ estimate: no.	4,051	9,078	7,086	27,760
%	1.5	3.2	2.6	9.7
$\overline{L\text{-}L}$ estimate: no.	10,650	24,690	18,870	75,410
%	3.9	8.6	6.9	26.4
Summed sites (Appendix A)				
$\overline{L\text{-}L}$ estimate: no.	14,640	37,540	29,530	184,000
%	5.4	13.1	10.8	64.4

REFERENCES

1. Archer, V. E., J. K. Wagoner, and F. E. Lundin, Jr. Uranium mining and cigarette smoking effects on man. J. Occup. Med. 15:204-211, 1973.
2. Auxier, J. A. Physical dose estimates for A-bomb survivors. Studies at Oak Ridge, U.S.A. J. Radiat. Res. (Tokyo) 16(Suppl.):1-11, 1975.
3. Awa, A. A. Chromosome aberrations in somatic cells (in atomic bomb survivors). J. Radiat. Res. (Tokyo) 16(Suppl.):122-131, 1975.
4. Axelson, O., and L. Sundell. Mining, lung cancer and smoking. Scand. J. Work Environ. Health 4:46-52, 1978.
5. Baral, E., L.-E. Larsson, and B. Mattsson. Breast cancer following irradiation of the breast. Cancer 40:2905-2910, 1977.
6. Baum, J. W. Population heterogeneity hypothesis on radiation induced cancer. Health Phys. 25:97-104, 1973.
7. Beebe, G. W., H. Kato, and C. E. Land. Life Span Study, Hiroshima-Nagasaki Report 5. Mortality and Radiation Dose, October 1950-September 1966. Atomic Bomb Casualty Commission Technical Report TR 11-70. Hiroshima: Atomic Bomb Casualty Commission, 1970.
8. Beebe, G. W., H. Kato, and C. E. Land. Life Span Study Report 8. Mortality Experience of Atomic Bomb Survivors, 1950-74. Radiation Effects Research Foundation Technical Report TR 1-77. Hiroshima: Radiation Effects Research Foundation, 1978.
9. Beebe, G. W., H. Kato, and C. E. Land. Studies of the mortality of A-bomb survivors. 6. Mortality and radiation dose, 1950-1974. Radiat. Res. 75:138-201, 1978.
10. Beebe, G. W., M. Ishida, and S. Jablon. Studies of the mortality of A-bomb survivors. I. Plan of study and mortality in the medical subsample (selection I), 1950-58. Radiat. Res. 16:253-280, 1962.
11. Beebe, G. W., C. E. Land, and H. Kato. The hypothesis of radiation-accelerated aging and the mortality of Japanese A-bomb victims, pp. 3-27. In Late Biological Effects of Ionizing Radiation. Vol. 1. Vienna: International Atomic Energy Agency, 1978.
12. Beebe, G. W., and M. Usagawa. The Major ABCC Samples. Atomic Bomb Casualty Commission Technical Report TR 12-68. Hiroshima: Atomic Bomb Casualty Commission, 1968.
13. Belsky, J. L., K. Tachikawa, and S. Jablon. The health of atomic-bomb survivors. Yale J. Biol. Med. 46:284-296, 1973.
14. Boice, J. D., Jr., and R. R. Monson. Breast cancer in women after repeated fluoroscopic examinations of the chest. J. Natl. Cancer Inst. 59:823-832, 1977.
15. Brinkley, D., and H. E. Masters. The depth of the spinal cord below the skin. Brit. J. Radiol. 40:66-67, 1967.
16. Brown, J. M. Linearity versus non-linearity of dose response for radiation carcinogenesis. Health Phys. 31:231-245, 1976.
17. Case, R. A. M., M. E. Hosker, D. B. McDonald, and J. T. Pearson. Tumours of the urinary bladder in workmen engaged in the manufacture and use of certain dyestuff intermediates in the British chemical industry. Brit. J. Ind. Med. 11:75-104, 1954.
18. Castanera, T. J., D. C. Jones, D. J. Kimeldorf, and V. J. Rosen. The effect of age at exposure to a sublethal dose of fast neutrons on tumorigenesis in the male rat. Cancer Res. 31:1543-1549, 1971.
19. Conard, R. A., et al. A Twenty-Year Review of Medical Findings in a Marshallese Population Accidentally Exposed to Radioactive Fallout. Report No. 50424. N.Y.: Brookhaven National Laboratory, 1975.
20. Cook, J. R., B. M. Bunger, and M. K. Barrisk. A Computer Code for Cohort Analysis

of Increased Risks of Death (CAIRD). ORP Technical Report 520/4-78-012. Washington, D.C.: Environmental Protection Agency, 1978.

21. Court Brown, W. M., and J. D. Abbatt. The incidence of leukemia in ankylosing spondylitis treated with x-rays. A preliminary report. Lancet 1:283–285, 1955.

22. Court Brown, W. M., and R. Doll. Leukaemia and Aplastic Anaemia in Patients Irradiated for Ankylosing Spondylitis. Medical Research Council Special Report Series No. 295. London: H. M. Stationery Office, 1957.

23. Court Brown, W. M., and R. Doll. Mortality from cancer and other causes after radiotherapy for ankylosing spondylitis. Brit. Med. J. 2:1327–1332, 1965.

24. Doll, R. Cancer and aging: The epidemiologic evidence, pp. 1–28. In R. L. Clarke, R. W. Cumley, J. E. McCay, and M. M. Copeland, Eds. Oncology 1970. Chicago: Year Book Medical Publishers, 1971.

25. Doll, R., L. G. Morgan, and F. E. Speizer. Cancers of the lung and nasal sinuses in nickel workers. Brit. J. Cancer 24:623–630, 1970.

26. Doll, R., and P. G. Smith. Mortality from Cancer and Other Causes after Radiotherapy for Ankylosing Spondylitis. Further Observations. (Cited in Sources and Effects of Ionizing Radiation.) New York: United Nations, 1977.

27. Doniach, I. The effect of radioactive iodine alone and in combination with methylthiouracile and acetylaminofluorene upon tumor production in the rat's thyroid gland. Brit. J. Cancer 4:223–234, 1950.

28. Fears, T. R., J. Scotto, and M. A. Schneiderman. Mathematical models of age and ultraviolet effects on the incidence of skin cancer among whites in the United States. Amer. J. Epidemiol. 105:420–427, 1977.

29. Fisher, R. E. W. The effects of age upon the incidence of tar warts, p. 402. In The Twelfth International Congress on Occupational Health. Vol. 3. Helsinki: Valtioneuvston Kirjapaino, 1958.

30. Gambarelli, J., G. Guerinel, L. Chevrot, and M. Mattei. Computerized Axial Tomography. New York: Springer-Verlag, 1977.

31. Gray, H. Anatomy of the Human Body. 29th ed. Charles M. Goss, Ed. Philadelphia: Lea & Febiger, 1973.

32. Hashizume, T., and T. Maruyama. Physical dose estimates for atomic-bomb survivors. Studies at Chiba, Japan. J. Radiat. Res. (Tokyo) 16(Suppl.):12–23, 1975.

33. Hempelmann, L. H., W. J. Hall, M. Phillips, R. A. Cooper, and W. R. Ames. Neoplasms in persons treated with x-rays in infancy. Fourth survey in 20 years. J. Natl. Cancer Inst. 55:519–530, 1975.

34. Hoover, R., and P. Cole. Temporal aspects of occupational bladder carcinogenesis. N. Engl. J. Med. 288:1040–1043, 1973.

35. Hospital Physicists' Association. Central Axis Depth Dose Data for X Radiations of Half Value Layers from 0.01 mm Al to 15.0 mm Cu, Cobalt 60 Radiation, H.V.L. 11 mm Pb, and Betatron Radiation, 22 MeV. A Survey Made by the Scientific Subcommittee. C. B. Allsopp, Ed. London: British Institute of Radiology, 1953. 41 pp. (Brit. J. Radiol. Suppl. 5.)

36. Ichimaru, M., T. Ishimaru, and J. L. Belsky. Incidence of leukemia in atomic bomb survivors belonging to a fixed cohort in Hiroshima and Nagasaki, 1950–71. Radiation dose, years after exposure, age at exposure, and type of leukemia. J. Radiat. Res. (Tokyo) 19:262–282, 1978.

37. Ishida, M., and G. W. Beebe. Joint JNIH-ABCC Study of Life-Span in Atomic Bomb Survivors. Research Plan. Atomic Bomb Casualty Commission Technical Report TR 04-59. Hiroshima: Atomic Bomb Casualty Commission, 1959.

38. Ishimaru, T., R. W. Cihak, C. E. Land, A. Steer, and A. Yamada. Lung cancer at

autopsy in A-bomb survivors and controls, Hiroshima and Nagasaki, 1961-1970. II. Smoking, occupation, and A-bomb exposure. Cancer 36:1723-1728, 1975.

39. Jablon, S., and H. Kato. Childhood cancer in relation to prenatal exposure to atomic bomb radiation. Lancet 2:1000-1003, 1970.

40. Jablon, S., and H. Kato. Studies of the mortality of A-bomb survivors. 5. Radiation dose and mortality, 1950-70. Radiat. Res. 50:649-698, 1972.

41. Kerr, G. D. Organ dose estimates for the Japanese atomic-bomb survivors. Health Phys. 37:487-508, 1979.

42. Kneale, G. W., and A. M. Stewart. Pre-cancers and liability to other diseases. Brit. J. Cancer 37:448-457, 1978.

43. Knox, J. F., S. Holmes, R. Doll, and I. D. Hill. Mortality from lung cancer and other causes among workers in an asbestos textile factory. Brit. J. Ind. Med. 25:293-303, 1968.

44. Lange, R. D., W. C. Moloney, and T. Yamawaki. Leukemia in atomic bomb survivors. I. General observations. Blood 9:574-585, 1954.

45. Lindop, P. J., and J. Rotblat. The age factor in the susceptibility of man and animals to radiation. 1. The age factor in radiation sensitivity in mice. Brit. J. Radiol. 35:23-42, 1962.

46. Lloyd, D. C., R. J. Purrott, G. W. Dawson, D. Bolton, and A. A. Edwards. The relationship between chromosome aberrations and low LET radiation dose to human lymphocytes. Int. J. Radiat. Biol. 28:75-90, 1975.

47. MacMahon, B. Prenatal x-ray exposure and childhood cancer. J. Natl. Cancer Inst. 28:1173-1191, 1962.

48. MacMahon, B., and G. B. Hutchison. Prenatal x-ray and childhood cancer. A review. Acta Univ. Int. Contra Cancrum 20:1172-1174, 1964.

49. Matanoski, G. M., R. Seltser, P. E. Sartwell, E. L. Diamond, and E. A. Elliott. The current mortality rates of radiologists and other physician specialists. Specific causes of death. Amer. J. Epidemiol. 101:199-210, 1975.

50. McGregor, D. H., C. E. Land, K. Choi, S. Tokuoka, P. I. Liu, T. Wakabayashi, and G. W. Beebe. Breast cancer incidence among atomic bomb survivors, Hiroshima and Nagasaki, 1950-69. J. Natl. Cancer Inst. 59:799-811, 1977.

51. Miller, R., and E. J. Hall. X-ray dose fractionation and oncogenic transformations in cultured mouse embryo cells. Nature 272:58-60, 1978.

52. Modan, B., D. Baidatz, H. Mart, R. Steinitz, and S. G. Levin. Radiation-induced head and neck tumours. Lancet 1:277-279, 1974.

53. Moloney, W. C., and R. D. Lange. Leukemia in atomic bomb survivors. II. Observation on early phases of leukemia. Blood 9:663-685, 1954.

54. National Center for Health Statistics. United States Life Tables: 1969-71. DHEW Publ. No. (HRA) 75-1150. Washington, D.C.: U.S. Government Printing Office, 1975.

55. National Council on Radiation Protection and Measurements. Review of the Current State of Radiation Protection Philosophy. Report No. 43. Washington, D.C., 1975.

56. National Research Council, Advisory Committee on the Biological Effects of Ionizing Radiations. The Effects on Populations of Exposure to Low Levels of Ionizing Radiation. Washington, D.C.: National Academy of Sciences, 1972.

57. Noble, K. B., Ed. Shielding. Survey and Radiation Dosimetry Study Plan, Hiroshima-Nagasaki. Atomic Bomb Casualty Commission Technical Report TR 07-67. Hiroshima: Atomic Bomb Casualty Commission, 1967.

58. Okada, S., H. B. Hamilton, N. Egami, S. Okajima, W. J. Russell, and K. Takeshita, Eds. A review of thirty years study of Hiroshima and Nagasaki atomic bomb survivors. J. Radiat. Res. (Tokyo) 16(Suppl.):1-164, 1975.

59. Oughterson, A. W., and S. Warren. Medical Effects of the Atomic Bomb in Japan. New York: McGraw-Hill, 1956.

60. Parker, L. N., J. L. Belsky, T. Yamamoto, S. Kawamoto, and R. J. Keehn. Thyroid carcinoma after exposure to atomic radiation. A continuing survey of a fixed population, Hiroshima and Nagasaki, 1958-1971. Ann. Int. Med. 80:600-604, 1974; and Atomic Bomb Casualty Commission Technical Report TR 5-73. Hiroshima: Atomic Bomb Casualty Commission, 1974.

61. Peto, R., F. J. C. Roe, P. N. Lee, L. Levy, and J. Clark. Cancer and aging in mice and man. Brit. J. Cancer 32:411-426, 1975.

62. Polednak, A. P., A. F. Stehney, and R. E. Rowland. Mortality among women first employed before 1930 in the U.S. radium dial-painting industry. A group ascertained from employment lists. Amer. J. Epidemiol. 107:179-195, 1978.

63. Purrott, R. J., and E. Reeder. Chromosome aberration yields in human lymphocytes induced by fractionated doses of alpha-radiation. Mutat. Res. 34:437-445, 1976.

64. Radford, E. P., R. Doll, and P. G. Smith. Mortality among patients with ankylosing spondylitis not given x-ray therapy. N. Engl. J. Med. 297:572-576, 1977.

65. Refetoff, S., J. Harrison, B. T. Karanfilski, E. L. Kaplan, L. J. DeGroot, and C. Bekerman. Continuing occurrence of thyroid carcinoma after irradiation to the neck in infancy and childhood. N. Engl. J. Med. 292:171-175, 1975.

66. Rossi, H. H. The effects of small doses of ionizing radiation. Fundamental biophysical characteristics. Radiat. Res. 71:1-8, 1977.

67. Rotblat, J. The puzzle of absent effects. New Sci. 75:475-576, 1977.

68. Segi, M., and M. Kurihara. Cancer Mortality for Selected Sites in 24 Countries. No. 3. 1960-1961. Sendai, Japan: Tohoku Univ. School of Medicine, 1964.

69. Seidman, H., E. Silverberg, and A. Bodden. Probabilities of eventually developing and of dying of cancer (risk among persons previously undiagnosed with the cancer). CA 28:33-46, 1978.

70. Ševc, J., E. Kunz, and V. Plaček. Lung cancer in uranium miners and long-term exposure to radon daughter products. Health Phys. 30:433-437, 1977.

71. Shore, R. E., R. E. Albert, and B. S. Pasternack. Follow-up study of patients treated by x-ray epilation for tinea capitis. Resurvey of post-treatment illness and mortality experience. Arch. Environ. Health 31:17-28, 1976.

72. Shore, R. E., L. H. Hempelmann, E. Kowaluk, P. G. Mansur, B. S. Pasternack, R. E. Albert, and G. E. Haughie. Breast neoplasms in women treated with x-rays for acute postpartum mastitis. J. Natl. Cancer Inst. 59:813-822, 1977.

73. Smith, P. G., and R. Doll. Age- and time-dependent changes in the rates of radiation-induced cancers in patients with ankylosing spondylitis following a single course of x-ray treatment, pp. 205-214. In Late Biological Effects of Ionizing Radiation. Vol. 1. Vienna: International Atomic Energy Agency, 1978.

74. Smith, P. G., R. Doll, and E. P. Radford. Cancer mortality among patients with ankylosing spondylitis not given x-ray therapy. Brit. J. Radiol. 50:728-734, 1977.

75. Steer, A., I. M. Moriyama, and K. Shimizu. ABCC-JNIH Pathology Studies, Hiroshima and Nagasaki. Report 3. The Autopsy Program and the Life Span Study: January 1951-December 1970. Atomic Bomb Casualty Commission Technical Report TR 16-73. Hiroshima: Atomic Bomb Casualty Commission, 1973.

76. Stewart, A., and G. W. Kneale. Radiation dose effects in relation to obstetric x-rays and childhood cancers. Lancet 1:1185-1188, 1970.

77. Tachikawa, K., and H. Kato. Mortality Among Atomic Bomb Survivors: October 1945-September 1964. Based on 1946 Hiroshima City Casualty Survey. Atomic Bomb Casualty Commission Technical Report TR 6-69. Hiroshima: Atomic Bomb Casualty Commission, 1969.

78. Terzaghi, M., and J. B. Little. Repair of potentially lethal radiation damage in mammalian cells is associated with enhancement of malignant transformation. Nature 253:548–549, 1975.
79. The Hazards to Man of Nuclear and Allied Radiations. Cmnd. 9780. London: H. M. Stationery Office, 1956.
80. Tokunaga, M., J. E. Norman, Jr., M. Asano, S. Tokuoka, H. Ezaki, I. Nishimori, and Y. Tsuji. Malignant breast tumors among atomic bomb survivors, Hiroshima and Nagasaki, 1950–74. J. Natl. Cancer Inst. 62:1347–1359, 1979.
81. Ullrich, R. L., and J. B. Storer. Influence of gamma irradiation on the development of neoplastic disease in mice. III. Dose-rate effects. Radiat. Res. 80:303–316, 1979.
82. United Nations Scientific Committee on the Effects of Atomic Radiation. Sources and Effects of Ionizing Radiation. 1977 Report to the General Assembly. New York: United Nations, 1977.
83. U.S. Department of Health, Education, and Welfare. Third National Cancer Survey. Incidence Data. Edited by S. J. Cutler and J. L. Young, Jr. Natl. Cancer Inst. Monogr. 41. DHEW Publ. No. (NIH) 75-787. Washington, D.C.: U.S. Government Printing Office, 1975.
84. Upton, A. C., and J. Furth. Host factors in the pathogenesis of leukemia in animals and in man, pp. 312–324. In Proceedings of Third National Cancer Conference. Philadelphia: J. B. Lippincott, 1957.
85. Upton, A. C., T. T. Odell, Jr., and E. P. Sniffen. Influence of age at time of irradiation on induction of leukemia and ovarian tumors in RF mice. Proc. Soc. Exp. Biol. Med. 104:769–772, 1960.
86. Vesselinovitch, S. D., E. L. Simmons, N. Mihailovich, K. V. Rao, and L. S. Lombard. The effect of age, fractionation, and dose on radiation carcinogenesis in various tissue of mice. Cancer Res. 31:2133–2142, 1971.
87. Walburg, H. E., Jr. Radiation-induced life shortening and premature aging, pp. 145–179. In J. T. Lett and H. Adler, Eds. Advances in Radiation Biology. Vol. 5. New York: Academic Press, 1975.

Statement Concerning the Current Version of Cancer Risk Assessment in the Report of the Advisory Committee on the Biological Effects of Ionizing Radiations (BEIR III Committee)

EDWARD P. RADFORD, M.D.
Professor of Environmental Epidemiology
Graduate School of Public Health, University of Pittsburgh

Chairman, BEIR III Committee *and*
Chairman, Subcommittee on Somatic Effects

The present version of the report of the Advisory Committee on the Biological Effects of Ionizing Radiations (the BEIR III Report) is a modification of the draft report approved by the Academy in April 1979 and released at a press conference at the Academy on May 2, 1979. Subsequent modifications of this approved draft have been prepared by a group appointed by Dr. Philip Handler, President of the Academy, consisting of six members of the somatic effects subcommittee and one member of the genetic effects subcommittee. The modifications involve principally the section of the report summarizing cancer risk estimates (the third and final section of Chapter V) and some of the conclusions that flow from this section. Cancer is a somatic effect of radiation, that is an effect on the body cells of individuals exposed, as distinct from effects on the germ cells or genetic effects. Thus, the sections at issue have been the responsibility of the subcommittee on somatic effects of the full BEIR III Committee. This subcommittee originally consisted of seventeen members whose names are given in the front of the report. This number has been reduced to fifteen by the deaths of Dr. Benjamin Trimble in November 1977 and Dr. Cyril Comar in June 1979.

The material prepared by the subcommittee on somatic effects was written largely during 1977–1978, with occasional one or two-day meetings of the subcommittee to review draft material as it was prepared. It is important to note that the last meeting of the full subcommittee was held on December 15, 1978, a one-day meeting. The new material incorporated in the report since May 1979 has, therefore, not been approved by the subcommittee as a whole except by the process of asking for comments by mail. Perhaps because completion of the BEIR III report has been delayed for such a long time, few members of the subcommittee have responded. Nevertheless, the present version of the report includes very major change from the earlier draft and from the BEIR I report of 1972. That is the decision to adopt the so-called linear-quadratic model (excess cancer risk = $aD + bD^2$, where a and b are constants and D is radiation dose) as the basis for calculating risk at low doses of low LET radiation for all cancers, and not just leukemia as in the previous draft. In addition, risk estimates calculated from a model in which the excess cancer was assumed to be proportional to the dose squared (the so-called pure quadratic model, excess risk = bD^2) were also included. The effect of adopting the linear-quadratic model is to reduce the risk estimates at low doses somewhat. The pure quadratic model implies a very low risk at low doses.

The decision to use the *linear* (straight line) no-threshold model (excess cancer risk = aD), which implies a risk directly proportional to dose at all levels, for all radiation types and for all cancers except leukemia was the result of a vote taken in a meeting of the subcommittee in October 1977. This vote has never been rescinded by action of the whole subcommittee, and thus as chairman of the subcommittee, I cannot consider that the present version is in accord with the perceptions of at least several of its members.

The most serious consequence of this alteration in the conclusions of the earlier draft, however, is that all of the discussions and evaluations of the data on cancer risks that took place among subcommittee members as the draft material for the report accumulated during 1978, did so on the basis that the linear model would be applied. In this regard the subcommittee was adhering to a principle adopted by the BEIR I Committee, and as an expedient measure, in view of the limited amount of time available, I had felt that we would not spend our time reviewing in detail the scientific basis for those conclusions which agreed with the BEIR I report. In short, the requirement to complete the report in 1978 imposed by the Academy staff meant that the extent of discussions of fundamental issues had to be limited, particularly for matters that had been thoroughly presented in BEIR I. Thus, a detailed and critical discussion by the subcommittee of the scientific basis of deciding whether one or another dose-response model was applicable to cancer risks was not undertaken.

One exception to the above statement was the data from the Japanese A-bomb survivors. The results of the follow-up of cancer experience through 1974 in this important study population had been made available to subcommittee members in page proof by Dr. Gilbert Beebe in 1977, but in this form it was used primarily to provide an important source of data for the individual cancer risk sections being prepared by several members of the subcommittee and now found in Appendix A of Chapter V. Bound copies of this report (Life Span Study Report 8, Technical Report RERF TR 1-77) were distributed by the Academy staff in mid-1978. The significance of this distribution was that for the first time all the members of the subcommittee had, in an easily readable form, the latest information concerning cancer risk in this population. At about the same time we obtained the Oak Ridge calculations of factors by which kerma doses could be converted to specific tissue doses for both gamma ray and neutron exposures in the two cities. Subsequently, a large amount of time during the remaining few meetings of the subcommittee was spent in discussion of cancer data from this report in terms of the tissue dose-response relationships that could be inferred from the data as presented. Since such a process amounts at best to fitting theoretical lines to data points, in these discussions the subcommittee did not address the fundamental scientific basis of any of the models proposed to fit the Japanese data.

In my view, new data, obtained since the BEIR I report in 1972, strongly supported the decision of the BEIR I committee to adopt the linear no-threshold model for cancer induction by radiation. 1) New human studies were available giving stronger evidence of effects in the 10 to 50 rad range, and these studies generally gave about the same risk of excess cancer per unit dose as the higher dose data had. 2) The range of exposure patterns to low LET radiation included more studies of multiple small doses which could be compared to effects of single doses. 3) Studies of individuals especially susceptible to cancer induction by radiation and other carcinogens were being expanded (e.g., see Chapter II, the section entitled "Cell Mutation or Transformation"), and there was a possibility that these susceptible populations might be fairly large and not identifiable in advance. This possibility suggested at least that cancer risk estimates at low doses for this population subset could be somewhat higher than would be inferred from studies of unselected populations. 4) Studies of oncogenic transformations of human and animal cells in culture had been greatly expanded, with startling new results that challenged many of the traditional radiobiologic concepts that had formed a scientific basis for extrapolation of effects of higher doses of low LET radiation into the low dose range. These results suggested, for example, that DNA repair did not necessarily imply that low doses of low LET radiation would be less carcinogenic per unit dose than high doses. 5) Finally, new evidence of cytogenetic changes observed in populations living in areas of high background radiation exposure had been obtained. At my suggestion this last evidence was not considered extensively by the subcommittee, primarily for the same reason they were not by the BEIR I Committee; that is, the significance of these changes observed in circulating lymphocytes in terms of human disease had not yet been defined. But these observations indicated that effects of radiation exposure at doses and dose rates moderately above background could be detected.

All of the above considerations indicated not only that the decision of the present subcommittee to reaffirm the applicability of the linear no-threshold dose response relationship was the correct one, but also that such a decision was not so conservative as had been thought at the time of the BEIR I report. That is, the cancer risk estimates for exposure to low doses based on the straight line extrapolation could be somewhat lower than might be found eventually to apply, especially to susceptible subsets of the population. Such an underestimation of risk, the subcommittee agreed, would be unlikely for low LET radiation, but the view that the linear extrapolation greatly overestimated the risk of low LET radiation at low doses appeared to me to be equally unwarranted. For high LET radiation, such as alpha radiation, the straight line extrapolation could underestimate the risk at low doses, but the evidence was not strong that such underestimation was very significant except in its theoretical inferences.

I now proceed to consider in some detail the scientific evidence pertinent to estimates of cancer risk in human populations from low doses of radiation. Of special importance are two questions that have divided the subcommittee. First, what is the experimental evidence to support the linear no-threshold dose-response relationship of cancer induction? Second, to what extent are the data from the Japanese A-bomb survivors concordant with all other human studies, and also consistent with linear or other dose-response models? A problem related to this last question is the degree of concordance of results from the two cities, Hiroshima and Nagasaki, and from comparison between the two cities the appropriate inferences to draw about the relative effectiveness of the neutron component of exposure in Hiroshima. (The type of bomb exploded in the two cities differed: both resulted in exposure to gamma radiation, but the Hiroshima bomb had a significant fraction of the radiation exposure from neutrons.)

Some general comments are in order at this point. First, there was *no* disagreement among the members of the somatic effects subcommittee to accept the linear no-threshold dose-response relationship to define *genetic* effects of radiation at low doses, a position firmly taken by the BEIR III subcommittee on genetic effects (Chapter IV) in agreement with the BEIR I report. Based especially on the mouse studies of William L. Russell,[1,2] a member of the subcommittee on genetic effects for both BEIR I and BEIR III, the subcommittee did recommend that for low LET radiation exposure at low dose rates, the mutational risk per unit dose for radiation of the male testis is probably less by a factor of three at low dose rates than for equivalent doses given at a higher rate.

In the present version of the report, there is an inconsistency between the conclusions of the two subcommittees with regard to the appropriate dose-response relationship to be applied for genetic and carcinogenic effects of radiation. Consistency in evaluating these two effects of radiation is reasonable because there is now wide agreement among the scientific community studying cancer (for a summary of the evidence see *Origins of Human Cancer*[3]) that a necessary condition for induction of cancer is production of one or more mutations in the DNA of one or more cells in a tissue. This mutational change in somatic cells as a condition for carcinogenesis is the foundation of the use of short-term testing of mutations produced by environmental agents as a screening test for carcinogenic potency.[4]

The entire process of carcinogenesis is a complex one, however, and an initiating event, such as a somatic cell mutation, is not the only condition required for cancer to arise, whereas a mutation in a germ cell that retains its viability is the sole condition of a transmitted hereditary defect. For this reason one might anticipate that the dose-response relationship for cancer induction could differ in certain ways from that of genetic mutation. But it is important to note that the differences in the two processes arise because of host factors or other biological factors in cancer expression that are essentially *independent of the initiating event or events,* thus not necessarily related either in space or time to the dose of radiation. If, therefore, one argues from the above-mentioned difference that the dose-response curve for cancer induction should differ from that for genetic effects, such argument cannot be based on biophysical principles that relate to the initiating mutational event. Indeed, because we suspect that many unrelated biological factors influence the probability of subsequent development of human cancer after exposure to radiation (see Chapter II, the section entitled "Host Factors in Radiation Carcinogenesis"), it is far from obvious in which way one would postulate that the dose-response curve should be modified at low doses. If evidence existed that a significantly large group were especially susceptible because of differences in some of the host factors related to carcinogenesis, we would expect that any cancer initiator such as radiation could be more effective per unit dose at low doses than at high doses, where all or most of the susceptible group could already have cancer induced.

The fact that we do not yet understand all the factors governing cancer development in man was an important reason why the subcommittee unanimously agreed to depend primarily on studies of human populations to define cancer risk from radiation exposure. The number of studies available is impressive, about 50 investigating cancer at various sites from irradiation for various reasons. In a few instances the results are negative, as one might expect on statistical grounds, or because epidemiologic criteria such as a suitable control population were difficult to meet. Yet, the remarkable fact is that the cancer risk estimates derived from a majority of the studies, involving widely different ethnic groups irradiated in different ways for different reasons, show a considerable agreement (see Chapter V, Appendix A), at least in the higher range of radiation doses where it has been possible to detect clear effects. The cancer mortality data from the Nagasaki A-bomb survivors are perceived by some members of the subcommittee as an exception, and this point will be discussed in detail below.

EXPERIMENTAL BASIS FOR DOSE-RESPONSE MODELS

The present version of the report has departed to some extent from the subcommittee decision to depend primarily on human studies for cancer risk estimates, in that adoption of the linear-quadratic dose-response model as the primary model to use for extrapolation of low dose effects of low LET radiation has been strongly influenced by data obtained on laboratory animals, which usually show cancer dose-response relationships curvilinear upward within, say, 200 rad. This influence is understandable if one considers that the human evidence of cancer risk is sparse for low radiation doses, but there are many reasons why animal studies are of limited value, and indeed may be misleading, with regard to dose-response information for human cancers.

These reasons include: 1) Animal cancers at particular sites may differ morphologically and in growth characteristics from human tumors at the same site, and for this reason initiating and promoting processes could be quantitatively different. 2) The strains of experimental animals used for nearly all research are highly inbred, and for each strain susceptibility to cancer induction is likely to be more homogeneous than in man. Human populations have variable genetic makeup and it is known that genetic factors influence cancer susceptibility.[5] This variability would have the effect of making the response at low doses greater per unit dose than at higher doses where the proportion of cancer-sensitive groups affected would be less. 3) The life span of most species such as rodents widely used for experimental studies of cancer is short, generally two to three years, and the latent period between exposure to radiation and onset of increased cancer incidence is proportionately a larger fraction of the life span in these species than in man. 4) Because animals used for lifetime studies of cancer development are kept in artificial surroundings, on a fixed nutritional regimen, and protected from intercurrent infections such as from viruses, exposure to a wide range of cancer-promoting or other factors which could modify cancer expression is thereby kept to a minimum. Such exposure is considered to be the almost daily lot of human existence, and may be an important contributor to the very marked influence of age on incidence of most cancers in man.[6] One consequence of this artificial environment of experimental animals is that for any single chemical or physical agent under study to lead to frank cancer, both initiating and promoting factors must be provided by the carcinogen; in the parlance of cancer research, the agent tested must be a complete carcinogen. There are two important consequences of this condition: first, the latent period may be inversely related to dose,[7] and second, one would expect that the cancer rate would more likely be proportional to the square of the dose, rather than to the first power of dose anticipated if only random initiating events were required for cancers to appear. Both these reasons, as well as the longer latent period in proportion to the short life span of these animals, lead to the dose-response curve at any time after the onset of excess cancer being likely to be strongly curvilinear upward. That is, low doses will appear to be less effective per unit dose than higher doses, even if the probability of cancer initiation were random and followed a linear, no-threshold relationship. It is significant that in human studies of radiogenic cancer where an effect of dose on latent period was looked for (Appendix A), the inverse dependence of latent period on radiation dose appears to be slight at most, consistent with the idea that the promoting step of radiation carcinogenesis in man is independent of the initiating event.

For the above reasons, therefore, I believe it is unwise to rely on dose-response data for cancer induction in experimental animals to support use of any particular dose-response model for human risk estimates from radiation exposure at low doses.

In the above discussion it is evident that the step of cancer initiation by radiation is an important element in quantitative understanding of risks of radiation exposure. Because this process is believed to be a cellular phenomenon, albeit influenced by tissue and host factors, quantitative assessment of dose-response relationships for the process of oncogenic transformation of cells has been actively pursued both in theoretical terms and experimentally, especially since the BEIR I report. One of the most widely discussed theoretical concepts in recent years has been the Kellerer-Rossi theory of dual radiation action.[8] The essence of this theory is found in Chapter II in the section entitled "Physical Aspects of the Biologic Effects of Ionizing Radiation."

It is important to note at the outset the fundamental assumption underlying the theory, which is that *pairs* of sublesions, produced by radiation in critical sites in the cell, combine to form lesions which are eventually expressed as a permanent change in the cell, such as a mutation or oncogenic transformation. This assumption is an extension of the theory of Lea,[9] developed to account for effects of gamma and neutron radiation in producing gross chromosomal aberrations. In this particular case the assumption that two breaks (or sublesions) are required to produce the effect is very plausible. For chromosomal aberrations in human lymphocytes a dose-squared dependence of effects has been observed for low LET radiation,[10,11] consistent with Lea's theory. To extend the assumption of two sublesions being required for other effects of radiation than gross chromosomal aberrations requires that experimental evidence of an effect proportional to the dose-squared be observed for such effects. This experimental evidence, as referenced, is derived from studies of chromatid aberrations in *Tradescantia,* the spiderwort plant,[12] effects on bacterial spores,[13] and radiation induced life-shortening in animals.[14] (This last effect of radiation would be expected to involve non-stochastic processes, in sharp contrast to cancer induction; moreover, the subcommittee has concluded on the basis of available human data, that no nonspecific life-shortening effect of radiation has been observed in man.) This array of evidence is far from convincing justification of the assumption that two sublesions are required to produce lesions in the DNA of mammalian cells that may lead, for example, to oncogenic transformation, unless such transformation is consistently associated with gross chromosomal aberrations.

If we follow the Kellerer-Rossi formalism, nevertheless, on the further assumption that the sublesions interact to produce a lesion over a range of about 1 mμ in the cell, then the frequency of effects, $E = K(\zeta D + D^2)$, where K is an arbitrary constant and zeta is a variable dependent on the frequency distribution of specific energies produced by single events. The Kellerer-Rossi theory, therefore, leads to a linear-quadratic dependence of effect on dose, a conclusion that is obvious from the fundamental assumption that pairs of sublesions are a necessary condition of ultimate effects. The theory has been applied to the problem of the relative biological effectiveness of different types of radiation at low doses, in which case both K and ζ are variables which are used to fit the experimental data. Experiments of Cox et al.,[15] in which mutation of HF19 human fibroblasts and V79 Chinese hamster cells by various radiations encompassing a wide range of LET was examined, were analyzed by Goodhead in terms of the Kellerer-Rossi theory.[16] Goodhead showed that the RBE values predicted on the Kellerer-Rossi theory were at considerable variance from those observed, and it was apparent that no consistent set of values for K and ζ in relation to LET could be derived from the data, nor were the derived "constants" consistent for similar effects in the two species. Goodhead also pointed out that ζ, which is equivalent to the dose at which the linear and quadratic terms are equal and which thus defines the dose range over which a simple linear fit to data is generally adequate, is very markedly affected by the diameter of the "interaction site," the locus within which the pairs of sublesions are presumed to produce the lesion. For an interaction diameter of 1 mμ, Goodhead's calculations indicate a value of ζ of about 30 rad for Co-60 gamma rays, and about 100 rad for 250 kVp x-rays. For a more likely interaction diameter of 0.4 mμ for cell transformation effects, the corresponding values are about 400 rad for both types of radiation. These latter values are so high that one would conclude that over the range of doses up to 200 rad, the Kellerer-Rossi theory actually *supports* application of the linear no-threshold dose-response relationship for oncogenic transformation.

But even more significant than these theoretical considerations are the results of recent studies of oncogenic transformation in mammalian cells by low doses of x-rays. Borek and Hall first showed[17] in hamster embryo cells that split doses of 210 kVp x-rays were more effective in producing transformations, and this result has been confirmed for doses below 100 rad in mouse 10T1/2 cells[18] and in A31-11 mouse BALB/3T3 fibroblasts.[19] Little and his colleagues[20] have pointed out the complexity of the role of DNA repair in these results, and have concluded from studies in which a phorbol promoter or a protease inhibitor has also been added to mouse 10T1/2 fibroblast cultures that the DNA lesions and repair process associated with cell killing and cell transformation are different. This observation is especially important because the Kellerer-Rossi theory has been mainly applied to studies of cell killing. Little[21] also has postulated that rapid DNA repair mechanisms are error-prone, and result in transformations. A slower, at least partially error-correcting repair process is also present, but if the cell undergoes DNA replication before this latter repair can occur, then the DNA alteration becomes "fixed" or "stabilized" in a heritable form after one cell division. This change becomes expressed as a transformation after a number of subsequent cell divisions, the number influenced by whether the cells are exposed to other non-transforming chemicals or agents during this stage. These results emphasize the importance of exposure to other agents affecting cell proliferation in fixation and expression of transformational damage, a concept in accord with much evidence concerning non-specific factors in promotion of human cancer.

Work on this aspect of oncogenic transformation of cells is progressing rapidly and can be expected to yield important new insights into the relationship between transformations produced by low doses of all types of radiation and the process of carcinogenesis in animals and man. But the important point here is that the data in hand show clearly that biological factors such as DNA repair mechanisms and exposure to other non-transforming agents markedly modify the probability of an oncogenic transformation, and the simple view that repair of initial damage produced by low LET radiation at low dose rates will inevitably reduce the subsequent probability of cancer induction when compared to the same dose given at high dose rates, is clearly untenable.

For both these biological reasons as well as the theoretical points made, for example, by Goodhead, I believe the Kellerer-Rossi theory is quite unacceptable in having any relevance to dose-response relationships for human cancer. Indeed, the cell transformation data suggest that the linear no-threshold dose-response curve as a basis for extrapolating carcinogenic effects from high to low doses of low LET radiation could even somewhat *underestimate* the low-dose risk, as Miller and Hall[18] and Borek[22] have emphasized.

DOSE-RESPONSE DATA FROM EPIDEMIOLOGIC STUDIES OF HUMAN POPULATIONS

The above practical and theoretical problems thus refute the idea that experimental evidence provides any basis for deciding on the particular forms of the dose-response relationship in human radiation carcinogenesis. This situation means that we must rely on epidemiologic evidence to estimate risks at low doses of low LET radiation, as the subcommittee had concluded early in its deliberations. Unfortunately, as the third section of Chapter V points out, good dose-response data in human populations of large enough size to provide statistically reliable risk estimates in the range of doses less than 50 rad are very limited. Such data are needed if extrapolation to lower doses is to have any precision, or even to determine whether the simplest extrapolation curve, the linear no-threshold model adopted by the subcommittee to estimate cancer risks from low LET radiation, is reasonable or not. As the above comments indicate, use of the linear extrapolation can hardly be considered to provide an "extreme" estimate of low-dose risk.

The only population study that does provide dose-response data of this type is that of the Japanese A-bomb survivors. It is not generally recognized that the strength of the Japanese data in epidemiologic terms lies in data obtained for low doses, less than 100 rad kerma. The major part of the number of survivors with significant exposures are in the two dose groups, 10–49 rad kerma, or a mean *tissue* dose of about 11 rad, and 50–99 rad kerma, or a mean *tissue* dose of about 35 rad. For doses greater than 200 rad kerma, about 120 rad mean tissue dose, the numbers of survivors included in the Life Span Study October 1, 1950, and who were over age 20 at the time of the bombing (the group in which nearly all cancer deaths had occurred between 1950 and 1974) were only 942 in Hiroshima and 684 in Nagasaki, numbers that are small enough that if the dose is fractionated further into three dose categories, as has been done in RERF Report 8, the results are likely to lead to statistically unstable estimates of excess cancer risk, especially in Nagasaki. Thus, it is fair to say that in the long run, a principal value of data obtained from this study population will be to permit estimation of cancer risk from acute exposures in a range of 10–35 rad mean tissue dose.

The fact that the A-bomb survivors are the only large group with a wide range of whole body radiation exposure makes them singularly important in dose-response evaluation of the carcinogenic effect of radiation in man. There was general agreement for this position among the subcommittee members, and it was the reason that extensive debate concerning interpretation of the follow-up data through 1974 from RERF Report 8, took place up to the final meeting of the subcommittee.

The areas of discussion revolved especially around interpretation of the Nagasaki data to evaluate effects of low LET radiation. Because the Hiroshima bomb led to a significant neutron exposure whose effect was difficult to assess independently, the Nagasaki data thus became the basis for defining low LET radiation effects. Unfortunately, the Nagasaki study population is much smaller than the Hiroshima group, and is especially small in the zero dose category, the accepted control population for the exposed populations. A better control population can be developed by combining the zero dose group and those exposed to 1–9 rad kerma (mean tissue dose about 1.8 rad), an approach which has been widely used to improve the analysis by investigators reporting results from these studies. Regardless of the control base selected, however, the data from Nagasaki inevitably show quite large statistical error ranges, especially at the higher doses.

Another important issue has been the relative importance of cancer mortality data from the death certificate study compared to the results obtained from the Tumor Registries in the two cities. The results of the dose-response analysis for both cities and for these two data sources are shown for all cancers except leukemia and bone cancer in Figures V-6 and V-7 of Chapter V. The mortality data in Figure V-6 are for the period 1955–1974, while the incidence data are for 1959–1970. The total number of cancer cases in the two instances is about the same, thus the statistical power of analysis of results from the incidence and mortality studies is also about the same.

The mortality data shown in Figure V-6 suggest from the fitted regression lines that the radiation effect in Nagasaki was much less than in Hiroshima, thus implying that the neutron component in Hiroshima may have been of major importance. But it is clear from analysis of the individual data points that a major difference accounting for the low slope of the Nagasaki dose-response is the single point at about 120 rad (200–299 rad kerma). This point shows a quite high cancer rate in Hiroshima and low in Nagasaki. The data points for *both* cities are low for the point at about 160 rad. At the request of the subcommittee Dr. Charles Land ran the correlation for the data below 100 rad (5 data points) and found that the results gave a reasonable linear fit with a difference in slope between the two cities consistent with a constant RBE of about 5 for the neutron component.

While I do not suggest that this type of mathematical manipulation provides a great deal of help in establishing firm conclusions, I do believe that it is important to understand that the apparent difference in response for the two cities indicated by the regression slope in Figure V-6 arises because of differences observed at *high* doses, where the Nagasaki data especially are less reliable on statistical grounds, rather than because of differences at low doses, where the data are somewhat more robust. Moreover, to attribute the difference entirely to a high neutron effectiveness in cancer induction implies that an especially high RBE applies to high doses only, a conclusion entirely at variance with current views of the effect of dose on the RBE of neutrons.

The results of the data from the Tumor Registries, Figure V-7, show a marked difference for the Nagasaki dose-response compared to Figure V-6, and a concordance between the two cities that suggests a constant RBE for neutrons of about 5. It should be noted that the tumor incidence dose-response data depend on the same denominator base of the Life Span Study population as do the mortality data. One problem with the

Tumor Registry data, however, is the fact that they have not yet been "evaluated," that is, it has not been determined whether out-migration from the cities, which would lose cases and therefore provide a lower estimate of risk, is randomly distributed by dose categories, and thus would not affect the slope of the dose-response curve. A random distribution by dose category of loss to follow-up from out-migration occurred in the women studied for breast cancer incidence in the two cities.[23] The loss by out-migration was only 16%, despite the fact that the study population included in 1950 a large number of young women who might be expected to move because of marriage.[24] The Tumor Registry data have the advantage, however, that a high percentage of the cases have either histologic or autopsy confirmation of the cancer diagnosis, and the Nagasaki Registry particularly is believed to be quite complete for the area around the city (Moriyama, I., personal communication to the subcommittee, 1978).

On the other hand, the death certificate data have an important deficiency in that major radiogenic cancers are significantly under-reported. Breast cancer in women is markedly under-reported because breast cancer has a relatively long survival time and thus death is often recorded as from another cause, and thyroid cancer is usually not fatal. Thus in both cases these highly important radiogenic cancers are not well reported in death certificates. Autopsy studies have also confirmed that in the study population lung cancer is misdiagnosed on death certificates in over half the cases, with over 1/3 of cases not even coded as cancer.[25] Thus, three of the major cancers induced by radiation are not accurately represented in the mortality data from death certificates, and for this reason, the advantage of complete ascertainment of death records for the study group is largely lost. While it is unlikely that such under-reporting of cases could by itself alter the dose-response curve, it could have the effect of making the range of uncertainty at any dose greater.

In the final analysis, there are inadequacies for both the death certificate and Tumor Registry data, but when they are all taken together a reasonable concordance appears. For all cases except the Nagasaki mortality data, the linear no-threshold dose-response curve appears to be an adequate description of the results, although as the voluminous discussion and tortured mathematics of the third section of Chapter V attest, it is possible to fit a number of other curves to the data about as well as the linear fit. The Nagasaki mortality data *are* consistent with the rest of the results except for the two data points at high doses in Figure V-6. But the chief point to be made at this stage is that mathematical constructs based on the Japanese data do not really contribute to decisions about the appropriateness of any particular dose-response relationship. The data for all cancers are as yet too imprecise, and thus adoption of a particular dose-response relationship remains an arbitrary choice.

The dose-response data for leukemia mortality from 1950–1974 in Nagasaki are based on only 22 deaths for those exposed above 10 rad kerma, and as anyone familiar with analysis of dose-response is aware, it is impossible to do much more than say that a significant effect of exposure exists with such a limited number of cases. Certainly these data are totally inadequate to define the dose-response curve. Cases from the leukemia registry results presented in RERF Report 8 are more numerous, and suggest a curvilinear dose-response relationship for both cities consistent with a constant RBE for neutrons of about 10.

In the present version of the report, the Leukemia Registry data have been used as a "guide" to define the linear and quadratic coefficients (a and b above) to be used in the linear-quadratic model applied to *all* cancers. In other words, mathematical adjustments to the coefficients, necessary because the results of fitting the theoretical curves to the Japanese mortality data led to unreasonable figures (all the coefficients derived from mortality "appeared out of line with the incidence estimates"), were based on the leukemia "guide." On biological grounds the idea that dose-response relationships for solid tumors must be similar to leukemia is far from reasonable. First, of course, is the markedly different time course for induction of radiation-induced leukemias compared to the much more quantitatively important solid tumors. This fact suggests a major difference in the factors involved in carcinogenesis, which by inference could affect the dose-response relationship. Second is the observation, thoroughly discussed within the full subcommittee, that leukemias are the only human cancers in which distinct chromosomal abnormalities are consistently associated with the disease. In the case of chronic granulocytic leukemia, quantitatively a very important type of leukemia induced by radiation, the great majority of cases ($\sim 85\%$) have the Philadelphia chromosome abnormality present in the leukemic cells, and there is agreement among cytologists and hematologists that the abnormality is causally related to the disease.[26]

In contrast, consistent visible chromosomal abnormalities in the early stages of solid tumors have not been found. The implication is that the somatic mutations in these tumors either involve point mutations or chromosomal changes small enough not to appear as readily visible translocations, deletions, or other abnormalities, or they are not associated with any particular chromosome site. The importance in radiobiological terms of the association of specific chromosomal abnormalities with leukemia is that such abnormalities are well-known to be two-break events, and thus a dose-squared dependence for at least part of the induced leukemias has a biological rationale. This is the main reason I accepted the linear-quadratic model for leukemia in the April 1979 draft. Cytologic differences between leukemia and solid tumors such as those mentioned above, support the view that the dose-response curves may not be the same for all cancer types. This is an idea that Harald Rossi and I both felt was an important contribution of the BEIR III Report; now of course in the present version it has been eliminated. In sum, the approach taken to "adjust" constants to provide risk estimates for solid tumors based on the leukemia "guide" is arbitrary and in my view not scientifically justified. The leukemia tail is still wagging the radiogenic cancer dog.

An important question is the extent to which the Japanese data are consistent with the data from all the other studies described in Chapter V, Appendix A, when expressed as an excess risk of cancer incidence per rad per million person years, and roughly age-adjusted. In general, the concordance is excellent for the major cancers where several data sets exist such as breast, thyroid and lung cancer. Other sites show various degrees of agreement. But the most important comparison is for *total cancer incidence* coefficients derived for each sex from the Nagasaki Tumor Registry data. From data presented in the April 1979 draft, these are found to be about 2/3 as great as the sum-of-sites coefficients summarized in Table V-14. This degree of concordance of results from human studies of a great range of exposure conditions, ethnic makeup and basis for radiation exposure is truly remarkable. The relatively small difference could be accounted for in part by underascertainment of cases in the Nagasaki data, and by a somewhat lesser susceptibility to cancer induction by radiation in Japanese as compared with occidental populations, a reasonable conclusion because of the somewhat lower total cancer rates in Japan compared with the U.S. The fact that the total excess cancer incidence rate per unit dose in the Nagasaki A-bomb survivors is quantitatively similar to the total excess incidence derived on the linear hypothesis from the aggregation of all the other available human studies lends strong support to application of the risk coefficients from the data in Table V-14 for deriving cancer risk estimates from whole-body exposure to low LET radiation.

With regard to concordance of dose-response relationships between the Japanese data and other sources, most of the other studies do not have a sufficient range of doses or sufficient numbers to permit comparison with the Japanese data. For female breast cancer incidence vs dose, Figure A-1 (Appendix A) shows good agreement of the three western studies cited compared with the data from the A-bomb survivors. (In this case the data for both Hiroshima and Nagasaki give a good fit to the linear no-threshold relationship, with no evidence of an RBE for neutrons greater than 1.) For thyroid cancer Hempelmann's data in children[27] do not agree closely with those of Colman[28] but taken together they are consistent with the linear model over a reasonably wide dose range. The lowest dose point at about 7 rad provided by Modan's results from examination of thyroid cancer in 10,900 Israeli children given scalp irradiation for tinea capitis[29] fits reasonably well with the linear extrapolation for the other two studies (see Appendix A). The lung cancer data for underground miners suggest that the dose-response curve from exposure to alpha radiation could be curvilinear downward, that is, low doses may be somewhat more effective in cancer induction per unit dose than high doses, a concept in accord with the idea that high LET radiation may show cell-killing effects at relatively low doses that would progressively reduce the cancer risk/rem as dose increased.

Some members of the subcommittee believe that the "sum-of-sites" method, used by the BEIR I Committee to estimate total cancer risks, overestimates risks somewhat, because out of the numerous epidemiologic studies of radiation-induced cancer at individual sites presented in Appendix A, some would be expected by chance to yield higher than the true estimates, since in any study observed and expected cases have an inherent statistical variability. For this reason selection of only positive results would bias the risk estimates upward. To some extent this problem has been dealt with for several minor cancers by pooling risk estimates for them and striking a balance between high and low estimates, these sites being particularly susceptible to the above problem because risk estimates for them often were derived from a single study. But for two of the most important contributors to the total cancer incidence risk, thyroid and female breast cancers, there are several studies available for each that show excellent agreement, and thus the uncertainty of the risk coefficients is small, and no selection of high values has occurred. For lung cancer there are also several studies, but only two involving low LET radiation, the Nagasaki Tumor Registry and British ankylosing spondylitis studies. These two studies show reasonable concordance, and are also concordant with the studies of the underground miners on the basis, derived independently from dosimetric and radiobiologic principles, that exposure to one Working Level Month is equivalent to a dose of 6 rem to the basal cell layers of the proximal bronchial epithelium.

The only sites contributing significantly to the total in Table V-14 where the above argument could have some merit are those for the digestive tract: esophagus, stomach and intestines, primarily large bowel. Even in these cases there is reasonable concordance among the studies available, and the likelihood that selection of data has biased the risk estimates upward is not great. But this reason for rejecting use of the "summed sites" approach to defining cancer incidence risks from whole-body exposure obscures two important points. First is that ionizing radiation is the only known human cancer-producing agent that has been found to increase the risk of cancer in nearly all the parenchymatous or epithelial tissues of the body (see Appendix A). Indeed it is a reasonable conclusion that at high enough doses, it should be possible to demonstrate a carcinogenic effect of radiation on *any* human tissue. Therefore one may conclude that in human studies where a small excess of cancer is found at a particular dose of radiation but is borderline in statistical significance, it is prudent to consider the effect may be real rather than to dismiss the study as negative.

Second, as the follow-up time of the human study populations in which many organs were irradiated is extended, evidence of excess cancers at many of the minor sites has emerged slowly over time because the excess is set against the usual variability of cancer arising from other causes. Thus "statistically significant" excess cancer in the irradiated population may not occur for those sites where a lesser radiation effect is present until many total cases at that site have accumulated. This phenomenon has been obvious from the successive follow-up reports of the Japanese A-bomb survivors, where the bulk of the cases are observed at relatively low doses. For this reason we must consider *any* quantitative risk estimates, positive or negative, as tentative and could underestimate the risk until a lifetime follow-up is completed. For the above two reasons the idea that Table V-14 risk coefficients are biased upward by an effect of selection of positive results totally ignores the combined strength of the evidence presented in Appendix A.

Another point raised by use of Table V-14 for estimating cancer risks is that it gives cancer incidence rates rather than cancer mortality. The decision to define cancer risks in terms of incidence rather than mortality was adopted early by the subcommittee, and constituted a significant change from the BEIR I report. This decision was based in part on the awareness that cancers of the thyroid and female breast are now major radiation-induced cancers, and for these two sites mortality data give an inadequate indication of risk. This change from BEIR I was also based on the consideration by the subcommittee that any radiation-induced cancer produces a major psychological, social and economic cost to the individual affected, whether or not the cancer is ultimately the cause of death. Thus the idea that cancer deaths alone are the proper measure of radiation impact was rejected. Since the BEIR I report, new information was available which permitted better estimation of excess cancer incidence from radiation exposure to the thyroid and female breast; for other cancers there is little incidence data except from the Japanese Tumor Registries, but because most of the other important radiogenic cancers including leukemia are eventually fatal, mortality gives a reasonable estimate of incidence. For this reason, the other coefficients in Table V-14 have been derived from mortality data.

Because cancer incidence risk estimates are those intended by the subcommittee, the amount of emphasis in the current version of the third section of Chapter V on discussion of cancer mortality data is unwarranted, and indeed the procedure of "indirect conversion of mortality estimates to incidence estimates" is clearly inappropriate for cancer of the thyroid and female breast. In my view the best basis for cancer incidence risk estimates from radiation exposure is Table V-14, because it draws on all the evidence available from Appendix A, much of it obtained in American or British study populations and on this basis more immediately applicable to risk estimates intended to be applied to the U.S. population. As pointed out above, it is supported well by the Nagasaki total cancer incidence data. These risk estimates applied to the 1969–1971 U.S. life table population are presented in Table V-30 of Chapter V, but it should be noted that this table does not include the risk for leukemia and bone cancer incidence. To determine total cancer risk the data from Table V-30 must have added the data from Table V-16, where leukemia and bone cancer incidence are derived using the linear-quadratic model agreed by the subcommittee as appropriate for leukemia only (bone cancer is such a minor cancer that it contributes trivially to total cancer risk regardless of the model used). Failure to provide a single estimate of risk of *total* cancer incidence is another deficiency of the present version of the third section of Chapter V. Table V-30 gives a range of risk calculations for each sex according to the various exposure regimens. This range reflects our uncertainty about the appropriate model by which current estimates of risk are projected forward to a lifetime cumulative risk. The two projection models used are the so-called absolute and relative risk models (see Chapter II, the section entitled "Epidemiologic Studies as the Basis of Risk Estimates for Effects of Ionizing Radiation"). It is evident that these two projection methods give total risk estimates that differ by a factor of about 3 for the projections of total population exposures. There was general agreement among the subcommittee members that at least this degree of uncertainty applied to the estimates of lifetime risk in these instances. For the occupationally exposed groups the two projections agree reasonably well.

In Table 1, I have combined Table V-30 with Table V-16 to give the best estimate of total excess cancer incidence derived for the exposure conditions used in the third section of Chapter V.

The exposure conditions adopted for illustration are unrealistic, in that it is extremely unlikely that 1,000,000 persons in the general population or

TABLE 1 Estimates of Total Lifetime Excess Cancer Incidence from Exposure to Low LET Radiation—Projections Based on 1969-1971 U.S. Life-Table Population of One Million Persons at Start of Exposure, According to Absolute-Risk and Relative-Risk Projection Models. Data Taken from Tables V-30 and V-16 of Chapter V

	Absolute-Risk Projection		Relative-Risk Projection	
	Male	Female	Male	Female
1. *Single exposure to 10 rad* to 1,000,000 persons of all ages				
Expected lifetime cancers without radiation	285,000	260,000	285,000	260,000
Excess cancers induced by radiation	2,600	5,500	8,800	16,200
2. *Continuous exposure to 1 rad/yr* to 1,000,000 persons at outset				
a. *Lifetime exposure from birth*				
Expected lifetime cancers without radiation	283,000	285,000	283,000	285,000
Excess cancers induced by radiation	16,200	37,600	31,100	185,200
b. *Exposure ages 20–65*				
Expected lifetime cancers without radiation	292,000	300,000	292,000	300,000
Excess cancers induced by radiation	11,100	25,400	15,000	49,500
c. *Exposure ages 35–65*				
Expected lifetime cancers without radiation	296,000	296,000	296,000	296,000
Excess cancers induced by radiation	6,800	14,700	7,600	17,500
d. *Exposure ages 50–65*				
Expected lifetime cancers without radiation	295,000	269,000	295,000	269,000
Excess cancers induced by radiation	3,100	6,300	3,100	6,300

among radiation workers would ever be exposed either to a single dose of 10 rad or to continuous doses of 1 rad per year. The numbers of radiation-induced cancers appear to be large in most instances, but it is important to note that except possibly for the case of lifetime exposure to one rad/year, even with these unrealistically high exposures it would be very difficult to detect by epidemiologic methods that the excess cancers had occurred except for those particular sites which are especially sensitive to cancer induction by radiation.

On the linear hypothesis, the data for the single exposure to 10 rad can be converted to conventional "risk per rad" estimates by dividing by ten. This yields a range of 260 to 880 cases per rad per million exposed for males, and 550 to 1620 cases per rad per million exposed for females. If we adopt an intermediate value as more likely to obtain (that is, the relative risk model will only partially be found to be correct), the risk per rad for cancer induction is about 500 cases per million for males and 1000 cases per million for females. These values are higher than the risk estimates from BEIR I, in part because incidence is considered instead of mortality, and in part because the new data indicate somewhat higher lifetime risk than was evident in 1972.

If one applies total cancer risk estimates obtained from the life table projections for a single exposure in Table 1 to the Japanese A-bomb Life Span Study population by use of the linear hypothesis and the same method as was done to produce the estimates in Table 1, some important limitations of the Japanese A-bomb follow-up study become clearer. The Life Span Study population has a greater proportion of younger people than the 1969–1971 U.S. life table population, a circumstance that means the total radiation-induced cancers anticipated per number exposed will be somewhat greater than predicted from the model applied to single dose exposure in Table 1. Nevertheless some approximate conclusions are justified. First is that the number of excess cancers observed to the present follow-up in 1974 constitutes only about one-third of those that eventually will be expected if the time for expression of excess cancer risk is the lifetime of those exposed over the age of ten. In other words, the follow-up period for the Life Span Study group is still too short to define total cancer risks adequately. Second, even on the upper limit assumption that the lifetime *relative* risk model applies, no statistically significant excess of all cancers will ever be observed in the two lowest dose categories of the study population in Nagasaki, that is at mean tissue doses of 2 rad and 10.8 rad. For Hiroshima the same statement can be made for the lowest dose category (mean tissue dose 1.7 rad) regardless of the RBE assumed for neutrons within any reasonable range. For the next dose category, 10–49 rad kerma or a mean tissue dose of 10 rad, if a statistically significant excess of total cancers is observed in Hiroshima compared to the zero dose group, such an observation will be consistent with an RBE for neutrons greater than one, but because the mean tissue dose from neutrons is only one rad in this group, the reliability of any numerical estimate of RBE derived from this excess will always be weak indeed. For the next dose category, 50–99 rad kerma or a mean tissue dose of about 34 rad in each city, a significant lifetime excess of total cancers will be easy to demonstrate in Hiroshima, but for Nagasaki the smaller sample size will probably mean that the statistical significance of the excess will be marginal if the lifetime relative-risk model is found eventually not to hold.

This application of the current total cancer risk estimates to the A-bomb survivor populations again emphasizes the caution that must be applied in interpreting the data for excess cancer risk in this study group, especially at low doses. Another implication of the above analysis is that an excess risk of cancers at particular sites which are sensitive to radiation and have a high natural rate will always be easier to demonstrate, especially in Nagasaki, than will an excess for all cancers, because the inclusion of a large number of cancer types with low or zero radiation sensitivity increases the random "noise" in the data. The above phenomenon is already obvious in the analysis of breast cancer incidence up to the present. In sum, the fact that the Japanese data at any follow-up state may not be strong enough in statistical terms to show a significant effect of low doses on total cancer risk does not prove that effects are not present; the excess cancer risk may be better evaluated by looking at particular cancer sites.

With regard to the appropriate RBE for high LET radiation and its dependence on dose, the data for alpha radiation compared with x-rays or gamma rays give reasonable RBE values of about 10 to 20 for lung and liver cancer (Appendix A). Comparisons of the Hiroshima-Nagasaki results do not allow any definitive statement with regard to the RBE of neutrons for the following reasons: First, the rates for total cancer incidence from the zero dose (control) populations are substantially higher in Hiroshima than in Nagasaki, and thus the assumption that the neutron component is the sole factor accounting for differences in cancer dose-response is untenable. Second, at low doses, where the results are most important, excess cancer rates are not yet statistically strong enough to provide an appropriate estimate of the contribution of neutrons and in some instances are likely never to be strong enough (see above). Third, neutron and gamma ray exposures were highly correlated for Hiroshima, and in the low dose range *tissue* doses for neutrons were only about 1/10 those for gamma radiation, thus random differences in results greatly magnify the imputed neutron effects at low doses. Fourth, the dosimetry for gamma rays and neutrons is estimated to be good only to ±30%, thus any consistent dosimetry errors could also greatly affect the analysis of neutron effects in the comparison.

It should be pointed out that the assumption that the RBE for high LET compared to low LET radiation increases as the dose decreases does not necessarily imply that the dose-response curve for low LET radiation must be curvilinear upward at low doses. It is equally possible that the dose-response curve for high LET radiation is curvilinear downward. The point is that if we assume a fixed RBE independent of dose, we may underestimate somewhat the risk of low doses of high LET radiation and overestimate somewhat the risk of low doses of low LET radiation. But the available human epidemiologic data do not indicate to me that this degree of over- or under-estimation is very great, that is, more than a factor of 2. When we consider that cancer risk estimates may eventually have to take account of a significant subfraction of the population whose radiogenic cancer risk can be expected to be higher than the population at large, any conservatism arising from assumptions that may overestimate the risk by a small amount is justified at this time.

Pertinent to this question of the relative effectiveness of high LET radiation at low doses are the results of chromosome aberration studies in populations living in or otherwise exposed to high background radiation. In those situations where exposure has been especially to radon-222 the alpha radiation can account for these essentially two-break effects on the chromosomes.[30,31] In the Brazilian population living in a village on monazite sands, chromosome abnormalities were found elevated compared to a control group not so exposed.[32] In this case, it was postulated that alpha radiation from the Pb-212 daughter of Rn-220 reached the lungs or blood, and this exposure rather than the high background of gamma rays accounted for this effect. On the other hand, the dose-related chromosomal aberrations observed by Evans et al.[33] in nuclear shipyard workers exposed to relatively low cumulative doses were from exposures to "almost exclusively gamma radiation." It is of interest that 5 rad of acute x or gamma radiation has produced in human lymphocytes significant chromosomal aberrations.[34,35] Luchnik and Sevankaev[35] also observed an anomalous "plateau" of effect at intermediate gamma doses, very similar to that observed for cell transformations by Miller and Hall,[18] an effect which meant that extrapolations from doses of 50 to 400 rad would underestimate the effect at the lowest doses. The production of chromosomal aberrations at low doses cannot be considered pathogenic for any disease as yet, as mentioned above, but these observations indicate that caution is warranted in any assumptions about the relative effectiveness of high and low LET radiation at cumulative doses of 10 rad or less.

SUMMARY AND CONCLUSIONS

It is evident that adoption by the somatic effects subcommittee of the linear no-threshold dose-response model for defining radiation-induced cancer risks remains empirical at this time. There is no adequate theoretical model of human carcinogenesis that permits derivation of a dose-response relationship from first principles. The fact that radiation-induced cancer risk estimates from a large number of human studies with great variability of ethnic, cultural, and other environmental factors capable of influencing the results are as consistent as they are when compared on the basis of the linear extrapolation, suggests that radiation acts by increasing the probability of an initiating event, a somatic mutation. Other environmental factors which can modify the subsequent chance of neoplasia are sufficiently widely and randomly distributed in all human populations that the excess cancer risk is defined primarily by the probability of oncogenic cell transformation by radiation exposure. If such a transformation involves a radiation-induced point mutation or other small modification in the cell genome, then on classic target theory the linear no-threshold dose-response curve is entirely appropriate. Until we know more about the process of cancer development in man, we cannot go further with this problem.

The new evidence concerning cellular mechanisms of radiation carcinogenesis available since the BEIR I report represents in my view a major change in emphasis from the past. Whereas biophysical considerations, of which the Kellerer-Rossi theory is an example, have previously dominated the field and played an important role in concepts of effects of low doses of the different types of radiation, it is apparent that much more prominent now are biological variables that can involve the conversion of an initiating event induced by radiation into a fixed or heritable cell transformation, and the subsequent host factors that determine the probability of developing cancer. These biological factors include DNA repair processes and cellular mechanisms that modify them, the action of promoting agents and conditions that affect cell proliferation, the influence of viral infection on transformed cell DNA, immune processes affecting survival of transformed cells, and the effect of age on replication characteristics of the transformed cell or cells.

The above comments appear to be quite straightforward, and I believe were the consensus of the somatic effects subcommittee during the period when the subcommittee was continuing to meet. Contrast this position with that adopted in the third section of Chapter V of the present version. The basis of the ratio of the linear and quadratic coefficients (a/b, in the equation $E = aD + bD^2$) is the leukemia registry data from the Japanese A-bomb survivors, data which do show a definite curvilinearity of dose-response. In addition the RBE assumed for neutrons is taken from the fit of the data to the leukemia results. Thus leukemia, a human cancer with cellular characteristics and time course after irradiation differing markedly from other radiogenic cancer types, is taken as the paradigm governing a number of important inferences for *all* radiation-induced cancers. These factors derived from leukemia are then used to fit the observed data for cancer *mortality*, which as has been discussed above are deficient in important ways for major radiation-sensitive solid tumors. Mortality data are then converted to incidence data by applying factors of cancer mortality by site shown in Table V-15. This approach studiously avoids using the Japanese Tumor Registry data for total cancer incidence which for both Nagasaki and Hiroshima (with adjustment of an RBE for neutrons of about 5) are in excellent agreement with the incidence data derived for individual sites from the extensive international studies described in Appendix A, and summarized in Table V-14.

The roundabout approach taken above in the present version in effect discards all the human studies of radiation-induced cancer except the Japanese data in defining cancer risk from low LET radiation. It also has the effect of reducing the cancer risk estimates sufficiently that it is possible for the conclusion to be drawn that the BEIR III cancer mortality risk estimates are about the same as were derived in BEIR I. This conclusion ignores, of course, the important step of changing to cancer incidence as a basis of defining risk, and also ignores the considerable body of supportive data, especially for cancers of the thyroid and female breast, which indicate that as the follow-up of human study populations has been extended, evidence of cancer risk is increasing, the doses at which effects have been observed have progressively decreased, and the number of different human cancers in which radiation exposure has shown an effect has been extended. The present version of the third section of Chapter V has failed to make these important points, and thus has not provided, in my view, an adequate up-to-date scientific assessment of risk which was the purpose for which the BEIR III Committee was established.

The fact that the human epidemiologic data which are relevant to the dose-response issue are generally consistent with the linear no-threshold dose-response model remains the principal basis for use of this model. It should be emphasized that every effort in presenting epidemiologic evidence of cancer induction by radiation should be as carefully and rigorously done as possible to take account of the dilution effect of non-radiosensitive cancers, age-specific adjustments, effects of confounding variables and the influence of latent period. In Appendix A, and for the Japanese data in Figs. V-6 and V-7, efforts have been made to achieve this aim. The graph of Japanese data presented in Dr. Harald Rossi's dissenting report has not been corrected for age, which is a major correction for cancer evaluation because of the sharp effects of age on cancer rates; the Nagasaki Life Span Study population is younger than the Hiroshima population and the age distribution varies by dose category. In addition the period 1950–1954 has been included by Dr. Rossi for all cancers, when we know that for all cancers except leukemia no excess risk is likely to have occurred during this period. It is time to recognize that epidemiology is a rigorous discipline requiring special attention to detail that characterizes any science.

Finally, I would like to take this opportunity to thank those members of the full committee who have worked hard to produce those parts of the current version that provide a scientific basis for assessing somatic and genetic risks. It is regrettable that the results of their hard work have been so long delayed in being released for general use.

REFERENCES

1. Russell, W. L., L. B. Russell and E. M. Kelly. Radiation dose rate and mutation frequency. *Science* 128:1546–1550, 1958.
2. Russell, W. L. Mutagenesis in the mouse and its applicability to the estimation of the genetic hazards of radiation. In Proceedings of IV International Congress of Radiation Research, Evian, France, June 29–July 4, 1970.
3. *Origins of Human Cancer.* H. H. Hiatt, J. D. Watson and J. A. Winsten, Eds. Cold Spring Harbor Laboratory, 1977. See e.g. R. B. Setlow, F. E. Ahmed and E. Grist, Xeroderma pigmentosum: damage to DNA is involved in carcinogenesis, pp. 889–901. C. Heidelberger, Oncogenic transformation of rodent cell lines by chemical carcinogens, pp. 1513–1519. J. McCann and B. N. Ames, The Salmonella/microsome mutagenicity test: predictive value for animal carcinogenicity, pp. 1431–1449. M. Meselson and K. Russell, Comparisons of carcinogenic and mutagenic potency, pp. 1473–1481.
4. McCann, J., E. Choi, E. Yamasaki and B. N. Ames. Detection of carcinogens as mutagens in the Salmonella/microsome test: Assay of 300 chemicals. Proc. Natl. Acad. Sci. 72:5135–5139, 1975.
5. Knudson, A. G., Jr. Genetic predisposition to cancer. In *Origins of Human Cancer.* H. H. Hiatt, J. D. Watson, J. A. Winsten, Eds. Cold Spring Harbor Laboratory, 1977. pp. 45–52.
6. Peto, R. Epidemiology, multistage models and short-term mutagenicity tests. In *Origins of Human Cancer.* H. H. Hiatt, J. D. Watson, J. A. Winsten, Eds. Cold Spring Harbor Laboratory, 1977. pp. 1403–1428.
7. Druckrey, H. Quantitative aspects in chemical carcinogenesis. In *Potential Car-*

cinogenic Hazards from Drugs. U.I.C.C. Monograph Series, Vol. 7. Springer Verlag, Berlin, 1967. pp. 60–78.

8. Kellerer, A. M., and H. H. Rossi. The theory of dual radiation action. Current Topics in Radiat. Res. Quarterly 8:85–158, 1972.

9. Lea, D. E. *Action of Radiation on Living Cells, 2nd Ed.* University Press, Cambridge, 1956.

10. Schmid, E., G. Rimpl and M. Bauchinger. Dose-response relation of chromosome aberrations in human lymphocytes after in vitro irradiation with 3 MeV electrons. Radiat. Res. 57:228–238, 1974.

11. Kucerova, M., A. J. B. Anderson, K. E. Buckton and H. J. Evans. X-ray-induced chromosome aberrations in human peripheral blood leucocytes: the response to low levels of exposure in vitro. Int. J. Radiat. Biol. 21:389–396, 1972.

12. Neary, G. J., J. R. K. Savage and H. J. Evans. Chromatid aberrations in Tradescantia pollen tubes induced by monochromatic x rays of quantum energy 3 and 1.5 keV. Int. J. Radiat. Biol. 8:1–19, 1964.

13. Powers, E. L., J. T. Lyman and C. A. Tobias. Some effects of accelerated charged particles on bacterial spores. Int. J. Radiat. Biol. 14:313–330, 1968.

14. Sacher, G. A. Dose, dose-rate, radiation quality, and host factors for radiation-induced life shortening. In *Aging, Carcinogenesis, and Radiation Biology.* K. C. Smith, Ed. Plenum Publishing Corp. N.Y. 1975. pp. 493–517.

15. Cox, R., J. Thacker and D. T. Goodhead. Inactivation and mutation of cultured mammalian cells by aluminium characteristic ultrasoft x-rays. II. Dose-responses of Chinese hamster and human diploid cells to aluminium x-rays and radiations of different LET. Int. J. Radiat. Biol. 31:561–576, 1977.

16. Goodhead, D. T. Inactivation and mutation of cultured mammalian cells by aluminium characteristic ultrasoft x-rays. III. Implications for theory of dual radiation action. Int. J. Radiat. Biol. 32:43–70, 1977.

17. Borek, C., and E. J. Hall. Effect of split doses of x-rays on neoplastic transformation of single cells. Nature 252:499–501, 1974.

18. Miller, R., and E. J. Hall. Effect of x-ray dose fractionation on the induction of oncogenic transformations in vitro C3H 10T mouse embryo cells. Nature 272:59–60, 1978.

19. Little, J. B. Quantitative studies of radiation transformation with the A31-11 mouse BALB/3T3 cell line. Cancer Research 39:1474–1480, 1979.

20. Little, J. B., H. Nagasawa and A. R. Kennedy. DNA repair and malignant transformation: effect of x-irradiation, 12-0-tetradecanoyl-phorbol-13-acetate, and protease inhibitors on transformation and sister-chromatid exchanges in mouse 10T cells. Radiat. Res. 79:241–255, 1979.

21. Little, J. B. Radiation carcinogenesis in vitro: implications for mechanisms. In *Origins of Human Cancer.* H. H. Hiatt, J. D. Watson, J. A. Winsten, Eds. Cold Spring Harbor Laboratory, 1977. pp. 923–939.

22. Borek, C. Neoplastic transformation following split doses of x-rays. Brit. J. Radiol. 50:845–846, 1977.

23. Tokunaga, M., J. G. Norman, M. Asano, S. Tokuoka, H. Ezaki, T. Nishimori and Y. Tsuji. Malignant breast tumors among atomic bomb survivors, Hiroshima and Nagasaki, 1950-1974. J. Natl. Cancer Inst. 62:1347–59, 1979.

24. Belsky, J. L., K. Tachikawa and S. Jablon. The health of atomic bomb survivors. A decade of examination in a fixed population. Yale J. Biol. Med. 46:284–296, 1973.

25. Steer, A., C. E. Land, I. M. Moriyama, T. Yamamoto, M. Asano and H. Sanefuji. Accuracy of diagnosis of cancer in the JNIH-ABCC life-span study sample. Radiation Effects Research Foundation Technical Report TR 1-75. Hiroshima: Radiation Effects Research Foundation, 1975.

26. Fialkow, P. J. The origin and development of human tumors studied with cell markers. New Eng. J. Med. 291:26-35, 1974.

27. Hempelmann, L. H., W. J. Hall, M. Phillips, R. A. Cooper and W. R. Ames. Neoplasms in persons treated with x-rays in infancy. Fourth survey in 20 years. J. Nat. Cancer Inst. 55:519-530, 1975.

28. Colman, M., L. R. Simpson, L. K. Patterson and L. Cohen. Thyroid cancer associated with radiation exposure: dose-effect relationship. Proc. Symp. on Biol. Effects of Low Level Radiation Pertinent to Protection of Man and His Environment. (IAEA SM202) Vol. 2. Int. Atomic Energy Agency, Vienna, 1976. pp. 285-289.

29. Modan, B., E. Ron and A. Werner. Thyroid cancer following scalp irradiation. Radiology 123:741-744, 1977.

30. Pohl-Rüling, J. and P. Fischer. The dose-effect relationship of chromosome aberrations to α and γ irradiation in a population subjected to an increased burden of natural radioactivity. Radiat. Res. 80:61-80, 1979.

31. Stenstrand, K., M. Annanmaki and T. Rytömaa. Cytogenetic investigation of people in Finland using household water with high natural radioactivity. Health Phys. 36:441-444, 1979.

32. Barcinski, M. A., M. Do Céu Abreu, J. C. de Almeida, J. M. Naya, L. G. Fonseca and L. E. Castro. Cytogenetic investigation in a Brazilian population living in an area of high natural radioactivity. Am. J. Hum. Genet. 27:802-806, 1975.

33. Evans, H. J., K. E. Buckton, G. E. Hamilton and A. Carothers. Radiation-induced chromosome aberrations in nuclear-dockyard workers. Nature 277:531-534, 1979.

34. Schmickel, R. Chromosome aberrations in leukocytes exposed in vitro to diagnostic levels of x-rays. Am. J. Hum. Genet. 19:1-11, 1967.

35. Luchnik, N. V. and A. V. Sevankaev. Radiation-induced chromosomal aberrations in human lymphocytes. I. Dependence on the dose of gamma-rays and an anomaly at low doses. Mutation Res. 36:363-378, 1976.

Separate Statement
Critique of BEIR III

HARALD H. ROSSI

SUMMARY

The first report of the Committee on the Biological Effects of Ionizing Radiations (BEIR I) has profoundly influenced governmental regulations and the public attitude towards radiation. It is to be expected that the impact of the current report (BEIR III) will be equally significant. The Committee drafting that report has thus been faced with a heavy responsibility because its findings are likely to affect national energy policy and the practice of medicine. In both of these areas overestimates as well as underestimates of the radiation hazard could result in serious detriment.

This is especially important with regard to the risk of radiogenic cancer which is frequently considered to be the major hazard of ionizing radiation. This critique deals with this subject only.

BEIR III represents an advance over BEIR I in a number of respects:

I. The uncertainties of risk estimates are stated more explicitly and it is stressed that the so-called "linear hypothesis" is likely to result in overestimates of the hazard from low-LET radiation.

This has led directly or indirectly to further improvements.

II. It is acknowledged that it is probable that the cancer risk rises with absorbed dose at a rate that is higher than linear and the preferred mathematical model conforms with this postulate.

III. Extrapolations to single whole body doses of less than 10 rads are eschewed.

IV. It is stated that the effects of annual radiation doses of the order of 100 mrads (low LET) are unknown and that it is unlikely that they can be demonstrated.

V. It is recognized that RBE is an important factor and it is frequently assumed that it increases with decreasing level of effect. In most instances data from Hiroshima and Nagasaki are not pooled on the assumptions of equal effectiveness.

BEIR III is however deficient in two major respects:

I. Many of the risk estimates provided are still based on the "linear hypothesis" despite continuing and mounting contrary evidence from radiobiology and epidemiology (much of it quoted in BEIR III). Even though these figures are given somewhat less prominence, they are likely to assume primary importance for standard-setting bodies which, for the sake of prudence, are likely to adopt the highest estimates.

II. BEIR III fails to present explicitly data that indicate risk factors that are less than the lowest given in its report. This does not only again tend to support excessive risk estimates for low-LET radiation, but may also lead to, perhaps even more important, underestimates of neutron hazards.

DETAILED COMMENTS

The inadequacies of the epidemiological information on radiogenic cancer in man permit a wide variety of interpolations and extrapolations of data that are often uncertain, if only in the statistical sense. The deduction of the most likely risk estimates can, however, be facilitated by considerations of theoretical or experimental findings of radiobiology which make certain models more—and sometimes much more—plausible.

Theoretical considerations permit definitive conclusions on the dose-effect relation for individual (autonomous) cells, but at this time they cannot be employed with any assurance to determine this relation for the complicated process of radiation carcinogenesis. They do, however, lead to the conclusion that the RBE of high- relative to low-LET radiations should increase with decreasing level of effect to values which are very substantial and that this should be so not only for autonomous cells, but also for interacting cell systems.

Experimental observations on higher organisms have confirmed this expectation. In line with theoretical predictions, the RBE generally increases with decreasing neutron dose, D_N, according to

$$RBE = K(D_N)^{-1/2}.$$

In a number of systems K has been found to be about 45 if D_N is expressed in rads and the neutrons have energies comparable to the mean energy of the fission spectrum (~ 0.5 MeV). RBE values in excess of 100 have been observed at neutron doses of the order of 100 mrads which are thus equivalent to gamma ray doses of the order of 10 rads.

While experimental radiobiology is in accord with theoretical predictions regarding the dose-RBE relation, it also discloses a wide variety of dose-effect relations for carcinogenesis. Some of these even show a reduction of the natural incidence at moderate doses of low-LET radiation (and even for high-LET radiations). This is only observed when the natural incidence is high; however, statistical limitations would not permit a clear indication of this effect when the natural incidence is low. In most (but not all) instances, the curvature of the relation for low-LET radiation is positive indicating that in addition to any linear dependence on low doses (regardless of sign), there are positive quadratic and perhaps higher order terms in dose at intermediate doses. At high doses, a reduction or even a reversal of slope is often observed.

In summary, radiobiological considerations lead to the expectation that if cancer incidence is related only to terms that are linear and/or quadratic in dose, only a rough approximation may be attainable in many instances. In such approximations the relative magnitude of linear and quadratic terms is likely to differ depending on the type of neoplasm involved and a summation for all neoplasms could have a particularly complicated shape. It would, however, also be expected that, in general, linear extrapolations from doses of several hundred rads lead to an overestimate of the effects of doses of the order of 10 rads. It would furthermore be expected that because of the dose dependent RBE, the shape of any dose-effect relations is not the same for gamma and neutron radiations and in particular that they not *both* be linear above gamma ray doses in excess of about 10 rads or neutron doses that are 100 times less.

BEIR III employs three approaches to the analysis of epidemiological data on radiation carcinogenesis: They involve the "summed sites" method, the mortality (LSS) data for Japanese atomic-bomb survivors and the Nagasaki tumor registry data.

The "summed sites" treatment is based on estimates of the incidence of cancers in individual organs as given in Appendix A. The input data are derived from many sources, most of which involve irradiations with doses in excess of 100 rads. The Japanese data employed are essentially all from Hiroshima with assigned RBE values that vary between sites but are independent of dose for any of them. With the exception of leukemia, the "linear hypothesis" is employed throughout. This treatment evidently conflicts with radiobiological knowledge on several counts. There are further objections to these data largely obtained from diseased individuals of different ethnic backgrounds. For example, the spondylitic population was exposed to very high doses and these were applied only to tissues in or near the spine. This poses problems in the assessment of the "average" dose. If the leukemogenic effect of large doses depends on the square of the x-ray dose (as in fact assumed in BEIR III) and if 40% of the bone marrow (that located in or near the spine) is irradiated with a dose, D_x, with the remainder receiving essentially zero dose, the *effective* dose is not the mean 0.4 D_x, but instead 0.63 D_x. Such discrepancies become even more pronounced if the irradiated fraction of a tissue or organ becomes smaller.

In the absence of other information, these estimates might be considered as crude upper limits of the true risk for individual organs. However, the utilization of their sum in the methods employed to determine the overall cancer risk is one of the principal deficiencies of the BEIR III report. As was to be expected, it results in a substantially larger risk coefficient than those obtained by other methods and this inflated estimate may well be adopted by standard setting bodies who, in the interest of caution, may select the highest estimate provided.

The LSS data are generally considered to be the most reliable source of information on radiogenic cancer in the Japanese cities. They also permit a straightforward assessment of the cancer risk for a period of almost 30 yr following irradiation of a normal (albeit ethnically distinct) population. BEIR III provides this information for leukemia and all other cancers separately. Although this division may be necessary for the risk calculations, it masks the true dose effect relation of the over-all cancer impact for which the statistical fluctuations are substantially less. Figure 1 is a plot of cancer mortality per person year as a fraction of total kerma at Hiroshima and Nagasaki. These curves are not corrected for sex or age, but it may be assumed that such corrections could introduce only minor changes.

It appears that at Nagasaki it is impossible to detect an excess cancer

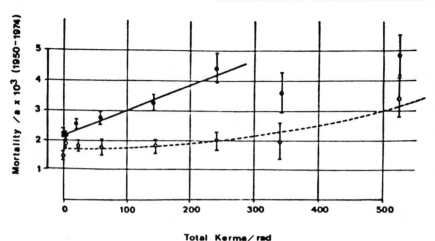

FIGURE 1   Average for the period 1950-1974 of the mortality from all malignant neoplasms per person year versus total free-in-air tissue kerma at Hiroshima (closed circles) and Nagasaki (open circles). The bars represent ± 1 standard deviation.

incidence at kerma values of less than about 300 rads although the populations exposed in each of the low dose intervals were about 1,000 or more. It is also evident that in line with other radiobiological information, the RBE of neutrons was very high. It should be borne in mind that at a given kerma at Hiroshima, only roughly 10% of the total absorbed dose to deep lying organs was due to neutrons. At high kerma, the Hiroshima data exhibit fluctuations which may be due to a variety of reasons, but the low kerma data can be approximated by

$$M_{Hi} = 2.2 \times 10^{-3} + 8 \times 10^{-6} (K_{Hi}/rad),$$

while the Nagasaki data conform to

$$M_{Na} = 1.8 \times 10^{-3} + 5 \times 10^{-9} (K_{Na}/rad)^2.$$

Here M is the mortality due to all malignant neoplasms per person year, K the total free-in-air tissue kerma and the subscripts stand for the two cities.

Because of the high RBE of neutrons and their virtual absence at Nagasaki, it may be assumed that at low doses, all cancers were induced by neutrons at Hiroshima and by gamma radiation at Nagasaki. Employing the dose versus kerma relations given in BEIR III one obtains approximately

$$M_N = 1.3 \times 10^{-4} \, (D_N/rad)$$

and

$$M_\gamma = 1.7 \times 10^{-8} \, (D_\gamma/rad)^2.$$

Neither of these relations (and especially not the second) should be applied to absorbed doses that are less than about 10 rads. The estimate for gamma radiations is lower than any given in BEIR III. The neutron estimate is higher than any values that might be inferred from this report.

The failure to explicitly provide the information in Figure 1 and to derive the above estimates is another major deficiency of BEIR III. Discussions within Committee did not produce substantive reasons for rejecting the validity of this analysis and while there may well be reasons for considering other approaches, it is apparent that this analysis is of considerable significance.

The so-called L-L estimate for these data is not only scientifically contraindicated, but also lacks any foundation in the absence of a sensible linear component for Nagasaki. Efforts to approximate this curve by a linear and a quadratic dose term yield a negative sign for the former. In principle, there is no reason why this should not be so and mortality data for lung cancer at Nagasaki point in this direction. The statistical evidence for this possibility is nevertheless much too weak to provide significant support for the assertion that the natural cancer mortality was reduced by small doses. On the other hand, the LQ-L analysis is based on the relative magnitude of linear and quadratic terms as derived from *leukemia* incidence data from the *tumor registry* and this is being justified by the objective of introducing a linear term into data in which linearity cannot be found.

The mixing of data from the LSS series and the tumor registries is also inappropriate because they appear to be discordant. The reasons for this are not clear at this time. Although the LSS data are generally considered to be more reliable and cancer mortality may be deemed to be more relevant than cancer incidence, BEIR III quite properly decided not to ignore the registry data especially since they indicate higher risk factors. Analysis in terms of all three models can be justified including that by the L-L model since the Nagasaki data are best fitted by a linear relation. However the implied dose independence of RBE casts further doubt on the validity of the registry data.

The arguments presented here lead to the conclusion that the most plausible estimate of the cancer risk from low-LET radiation is lower than any of the ones given in BEIR III. As a corollary BEIR III may motivate potentially dangerous underestimates of the hazards of high-LET radiation.

Comments on Certain Divisive Issues Noted in BEIR III

EDWARD W. WEBSTER*

This commentary is not intended as a dissent from the principal findings of the Committee, but rather to illuminate some of the issues on which the Report notes divisions of opinion within the Committee.

Probably the most important charge to the Committee was to estimate the increased risk of cancer likely to be incurred as a result of low doses of low-LET radiation delivered to the whole body. A linear-quadratic dose/effect relationship, defensible in the light of current radiobiologic findings, has been adopted by most of the Committee members as a reasonable basis for prediction of the risks of radiation-induced cancer. While subscribing to this important change in scientific viewpoint of the BEIR III Committee compared to that of BEIR I (1972), I must express a number of caveats regarding the actual forms of the dose/effect relationships utilized in the BEIR III risk estimates. I recognize, however, that the three estimates of mortality from solid cancer are not inconsistent with the Nagasaki mortality data.

---

* Dr. Ingram subscribes to this statement. Dr. Mays also subscribes to this statement, with the addition of the paragraph that appears at the end of the statement.

1. On page 187 it is noted that in the linear-quadratic relation fitted to the Nagasaki solid cancer mortality data (Figure V-6), the slope of the linear component is about 0.4 excess cancer per million per year per rad. This slope depends on the assumption that the linear term and the square-law term are equal for a gamma dose of 1/0.0086 or 116 rads. This particular linear-quadratic relation was rejected by some Committee members on two main bases: a) the RBE is about 91 for a neutron dose of 1 rad; b) the ratio of solid cancer to leukemia for gamma rays is 0.4 whereas the British ankylosing spondylitis study for high doses of x-rays suggests a ratio of about 5. The relationship was thereupon adjusted to include the RBE for the leukemia LQ model: viz. 23. This arbitrary change caused the slope of the linear component of the LQ relationship to be increased from 0.4 to 1.4; that is, by a factor of 3.5. The solid cancer risk estimates finally propounded in Table V-19 and which are the "preferred" estimates, are based on this larger slope. It is important to note that a) the recent study of leukemia in the A-bomb survivors by Ishimaru et al.[1] estimated the RBE for 1 rad of fission neutrons at 48, based on a quadratic model for gamma response; this is similar to the value of 45 proposed by Rossi on more general grounds;[2] and b) there is no obvious reason why the ratio of solid cancer to leukemia should be 5:1, particularly in the low dose range. The ratio will depend on the specific shapes of the leukemia and solid cancer dose/response curves. Thus in the animal studies by Ullrich et al.[3] the ratio of the incidence of 3 solid tumors (ovarian, pituitary, and Harderian) to the incidence of thymic leukemia varied from 2.4 at 100 rads to 0.8 at 25 rads and 0.1 at 10 rads. The ratio was more nearly constant for neutrons. Moreover it is noted on p. 209 that the solid tumor/leukemia ratio is "very sensitive to the age distribution of the subjects under study and to the duration of followup." For example, the work of Stewart and Kneale[4] on *in utero* exposure indicates a ratio of 28/25 or 1.1. Thus, if the Hiroshima/Nagasaki mortality data is not adjusted for RBE in this arbitrary fashion, the "preferred" risk estimates presented in Table V-19 would fall by a factor of about 3.

2. In the Report, the arguments on p. 187 summarized above were also employed to change the slope of the *linear* dose/effect relation employed for risk estimation. Whereas the slope of the best-fitting line for gamma radiation data shown in Figure V-6 and Table V-9 was 1.40, the actual slope employed in Tables V-11 and V-20 was 3.47, an increase by a factor of about 2.5. Thus the linear model estimates of cancer mortality presented are higher than those suggested by the Hiroshima-Nagasaki study by this factor.

3. In the Report, the arguments on p. 187 were also used to change

the coefficient of the *quadratic* relationship from 0.0047 (Table V-9) to 0.0184 (Table V-11 and V-21), an increase by a factor of 3.9. Again therefore the estimates of excess solid cancer mortality presented for the quadratic (square-law) model are higher by this factor than would be deduced *a priori* from the Hiroshima-Nagasaki data above.

4. The Report fails to state explicitly that the linear risk estimate for excess cancer *incidence* derived from Table V-10 (sum of the individual site risks) is grossly incompatible with the *linear* estimate for excess cancer *mortality* derived from the Hiroshima-Nagasaki study (Figure V-6). The average incidence risk from Table V-14 is 18 cases per million per year per rad, which is about *13 times greater* than the 1.40 fatal cancer cases deduced from the Japanese study, or about 7 times greater than the incidence risk derived from the Japanese study using the expansion factors in Table V-15. This great difference seriously challenges the credibility of the linear risk estimates based on the "summed sites" approach of Table V-30. This writer believes that these values not only have "considerable upward bias" as stated in the Report, but cannot be seriously considered in the light of the Japanese experience.

5. It is stated on page 179 that "the data [for the site-specific estimates in Appendix A] are reasonably firm for only a few organs." One of the important organs to which this applies is the lung, irradiated by low-LET radiation. The risk estimates for lung derived in Appendix A are almost entirely dependent on the epidemiological studies of miners exposed to high-LET radiation in the form of alpha radiation from radon inhalation and on the lung cancer incidence in Hiroshima. The assumption of the rather low RBE values of 10 for alpha irradiation and 5 for fast neutron irradiation exaggerates the effect of low levels of low-LET radiation. More importantly the lung section fails to note that the Nagasaki mortality data (low-LET radiation) show a *deficit* of lung cancer cases at doses up to 100 rads and this is also reflected in the Tumor Registry incidence data for low gamma ray doses.[5] The risk estimate for lung cancer from low-LET radiation is almost wholly dependent on the high dose (200 rad) ankylosing spondylitis study and is likely to be considerably less at low doses.

Additional comment by Dr. Mays: "I support the thoughtful comments of Dr. Edward W. Webster, and am particularly concerned that the risk coefficient derived from the sum of individual site risks exceeds by a factor of about 13 that derived directly from the A-bomb life-span mortality data. I feel that the latter is more likely to be appropriate and that future efforts by the Scientific Community should be directed toward resolving this discrepancy."

REFERENCES

1. Ishimaru, T., Otake, M. and Ishimaru, M.: Dose-response relationship of neutrons and gamma rays to leukemia incidence among atomic bomb survivors in Hiroshima and Nagasaki by type of leukemia, 1950-71. Rad. Res. 77:377-394, 1979.
2. Rossi, H. H.: The effects of small doses of ionizing radiation. Phys. Med. Biol. 15: 255-262, 1970.
3. Ullrich, R. L., et al.: The influence of dose and dose rate on the incidence of neoplastic disease in RFM mice after neutron irradiation. Rad. Res. 68:115-131, 1976.
4. Stewart, A. and Kneale, G. W.: Radiation dose effects in relation to obstetric x-rays and childhood cancers. Lancet 1:1185-1188, 1970.
5. Beebe, G. W., Kato, H., and Land, C. E.: Life Span Study Report 8: Mortality experience of atomic bomb survivors 1950-74. Technical Report RERF TR 1—77. Radiation Effects Research Foundation, Hiroshima, Japan, 1978, pages A202 and A350.

APPENDIX A
TO CHAPTER V:

SITE-SPECIFIC DATA CONCERNING
RADIATION-INDUCED CANCERS

The epidemiologic data concerning radiation-induced cancers now available permit an assessment of the relative importance of cancers by site in the total risk from radiation. It is evident from the evaluations discussed below that different tissues in the body respond differently to radiation, with some tissues highly sensitive to development of cancer and some evidently very resistant. In defining the total cancer risk from radiation exposure, we may distinguish four main categories:

• The major sites that are now well-documented as sensitive to radiation and contribute a large part of the total risk.
• Sites where radiation cancer induction is well-documented, but that contribute to a lesser degree to the total cancer risk.
• Sites for which an increased cancer risk in irradiated populations remains equivocal or has not been quantitatively assessed.
• Sites or tissues in the body in which radiation-induced cancer has not been observed.

Table A-1 summarizes these four categories. The table includes qualitative assessments of the spontaneous incidence of cancer in various tissues, based on the Third National Cancer Survey carried out in 1969–1971 in the United States, and an evaluation of their relative sensitivity to cancer induction by radiation, based on the current data contained in this report. Both aspects of the particular type of cancer determine its importance: a type of cancer that is normally rare may be less im-

TABLE A-1 Sensitivity of Various Tissues to Oncogenic Influence of Radiation

Site or Type of Cancer	Spontaneous Incidence of Cancer	Relative Sensitivity to Radiation Induction of Cancer	Remarks
Major radiation-induced cancers			
Female breast	Very high	High	Puberty increases sensitivity
Thyroid	Low	Very high, especially females	Low mortality rate
Lung (bronchus)	Very high	Moderate	Quantitative effect of smoking uncertain
Leukemia	Moderate	Very high	Especially myeloid leukemia
Alimentary tract	High	Moderate to low	Occurs especially in colon
Minor radiation-induced cancers			
Pharynx	Low	Moderate	—
Liver and biliary tract	Low	Moderate	—
Pancreas	Moderate	Moderate	—

Lymphomas	Moderate	Moderate	Lymphosarcoma and multiple myeloma, but not Hodgkin's disease
Kidney and bladder	Moderate	Low	—
Brain and nervous system	Low	Low	—
Salivary glands	Very low	Low	—
Bone	Very low	Low	—
Skin	High	Low	Low mortality. High dose necessary?
Sites or tissues in which magnitude of radiation-induced cancer is uncertain			
Larynx	Moderate	Moderate	—
Nasal sinuses	Very low	Low	—
Parathyroid	Very low	Low	—
Ovary	Moderate	Low	—
Connective tissues	Very low	Low	—
Sites or tissues in which radiation-induced cancer has not been observed			
Prostate	Very high	Absent?	—
Uterus and cervix	Very high	Absent?	—
Testis	Low	Absent?	—
Mesentery and mesothelium	Very low	Absent?	—
Chronic lymphatic leukemia	Low	Absent?	—

portant—even if it has a high fractional increase because of radiation—than a common cancer that has only a moderate fractional increase because of radiation.

The relative importance of various types of cancer is changing as the followup of irradiated human populations is extended. For example, in the early studies, leukemia emerged as the major cancer type because of the high sensitivity of the cells of origin to radiation and the short latent period. Now it is apparent that, in terms of lifetime risk, other cancers collectively, and some even individually, are more important than leukemia in assessing the late effects of radiation exposure. In part, this is because induction of excess leukemia cases by radiation essentially ceases after about 25 or 30 yr. None of the other cancers shows this decline in effect except for bone cancers induced by brief irradiation.

A number of important principles concerning radiation-induced cancers are now evident. For example, sensitivity to cancer induction is not proportional to the rate of division of stem cells; if it were, the small intestine would be as sensitive as the bone marrow, but that is not the case. Nor is sensitivity to radiation necessarily related to the influence of hormones. The female breast and the thyroid are quite sensitive to radiation, but the uterus and prostate are not, although pituitary or sex hormones are important factors in cancer sensitivity in all these tissues.

Some groups in the general population appear to be at increased risk of induction of cancer by radiation. Noteworthy is the evidence of greater risk of radiation-induced thyroid cancer in Jewish children than in other ethnic groups. Cigarette-smoking appears to lead to greater excess risk of lung cancer from radiation exposure, when smokers and nonsmokers are compared, even though the data no longer support the view that radiation and cigarette-smoking act in a multiplicative fashion in defining the cancer risk. The special influence of age on the risk of radiation-induced cancer has already been discussed.

The data on leukemia induction among the Japanese atomic-bomb survivors support the view that the gamma-ray dose-response curves in both cities are curvilinear upward—that is, the effects per unit dose are lower at low doses than at high doses. In the analysis of risks at low doses of low-LET radiation, therefore, the Subcommittee has adopted a linear-quadratic model (see Chapter II) for induction of leukemia. Some evidence is available that strongly suggests that this type of model is applicable to other types of cancer, especially bone cancer, but the Subcommittee believes that different types of cancer in man may have individual radiation dose-response relationships.

BREAST

The female breast is one of the organs most susceptible to radiation carcinogenesis. The 1972 BEIR report[40] considered evidence of radiation-induced breast cancer among female patients exposed to multiple fluoroscopic chest examinations during treatment for tuberculosis in a Nova Scotia sanatorium,[32,37] the members of the Atomic Bomb Casualty Commission (ABCC) Adult Health Study sample,[60] and women given localized x-ray treatment for acute postpartum mastitis in a Rochester, New York, hospital.[34] Breast-cancer mortality in the Japan National Institute of Health (JNIH)-ABCC Life Span Study (LSS) sample[24] was also considered. Since then, a number of studies have added substantial new information on radiogenic breast cancer in women. Evidence concerning male breast cancer is confined to case reports of cancers in men exposed to therapeutic radiation for benign conditions.[17,31,41]

There is good evidence that in female rats radiation-induced mammary tumors (fibroadenomas and adenocarcinomas) are caused by irradiation only of mammary tissue,[50] and not, as in the mouse, by irradiation of other tissue as well.[10] The rat mammary adenocarcinoma bears some morphologic resemblance to its human counterpart,[62] but differs in that it only very rarely exhibits metastasis; the fibroadenoma that appears in rats in response to radiation is another factor that complicates the interpretation of experimental data with respect to human risk.[48] The available epidemiologic data make it unnecessary to base major conclusions about the risk of radiogenic breast cancer in women on evidence from animals; nevertheless, the experimental evidence is highly relevant to the interpretation of human data. Excellent reviews of the experimental literature have been prepared by the first BEIR Committee[40] and UNSCEAR[56,57] and by committees of the National Cancer Institute's Breast Cancer Task Force[19] and the NCRP,[38] which specifically addressed the problem of radiogenic risk associated with mammography.

A case-control study of 37 breast-cancer deaths and 37 matched controls chosen from among former patients of tuberculosis sanatoria in Ontario[16] found 15 discordant matched pairs in each of which only one member received pneumothorax therapy with associated multiple chest fluoroscopies; pneumothorax therapy and breast cancer were associated in 11 of these pairs (RR* = 2.8, p = 0.07). Nine of 12 cancers among unilaterally exposed patients were on the exposed side (p = 0.07). Although dose estimates were not given, the method of exposure (subjects

---

* Relative risk.

faced away from the x-ray tube) was such as to suggest that doses were much lower than the average 600–1,200 rads estimated for the Nova Scotia series.

A followup study of former patients of two Massachusetts tuberculosis sanatoria[6] found 41 breast cancers (versus 23.3 expected, according to population rates) among 1,047 pneumothorax patients, compared with 15 (versus 14.1 expected) among 717 nonexposed women, for an age-adjusted relative risk of 1.7 ($p = 0.06$). Among the 578 women with average doses over 100 rads, 31 cancers were observed, versus 13.7 expected (RR = 2.1, $p = 0.01$). About 75% of the exposures were made with the patients' backs to the x-ray source, and in most cases the exposures were made with the shutters open or included a scan of the opposite lung. The dose per examination averaged 1.5 rads to both breasts, and the cumulative average breast dose was 150 rads. [6,7]

A study of Delarue et al. [18] found no difference in breast-cancer incidence between exposed and nonexposed former tuberculosis sanatorium patients. The average dose for exposed women was only 17 rads, however, and the sample sizes (358 exposed and 332 nonexposed women) were far too small for this study to have had much chance of detecting an increase of the magnitude to be expected at this dose.

A recent followup of the New York acute postpartum mastitis patients[53] included three control groups of nonexposed women: age-matched, nonirradiated mastitis patients treated at another hospital, sisters of the irradiated patients, and sisters of the nonirradiated patients. The three control groups did not differ with respect to breast-cancer incidence, but had substantially lower incidences than the exposed group. Following the fifth year after treatment, there were 36 breast cancers among the 571 exposed women, compared with 32 cancers among the 993 controls, for an age-adjusted RR of 2.0 ($p < 0.001$). The irradiated women received one to 10 exposures, within a few days or weeks; the average cumulative mean dose to both breasts was 247 rads.

Two surveys of breast-cancer incidence in the LSS sample have confirmed that female breast cancer is a major late effect of ionizing radiation from the Hiroshima and Nagasaki atomic bombs. The first survey[33] found 231 breast cancers diagnosed during the period 1950–1969, including 82 among women with 10 rads kerma or more and 144 among the nonexposed and the exposed with smaller doses (RR = 1.8, $p < 0.0001$). The second[55] found 360 cancers diagnosed during 1950–1974, including 108 among women with breast-tissue exposure of 10 rads or more and 243 among low-dose and nonexposed women (RR = 1.7, $p < 0.00001$). Most of the cases in the first survey were included in the second; however, the two series were independently ascertained, and the second survey should not be considered as merely an update of the first.

A followup study of Swedish women who had received radiation therapy[1] for benign breast disease included 855 patients treated for fibroadenomatosis, 120 for acute mastitis, and 49 for chronic mastitis and 13 women irradiated as young girls for unilateral breast hypertrophy. In 1,168 irradiated breasts, 115 breast cancers were observed 5 yr or more after the initial therapeutic exposure, compared with 28.7 expected according to population rates. There were 20 cancers in the nonirradiated breasts, compared with 19.9 expected. The data for the exposed patients yielded a large SMR (standard mortality ratio), 4.01, and an extremely small *p* value, but the possibility of a relationship between breast cancer and the treated conditions, fibroadenomatosis and chronic mastitis in particular, cannot be ruled out.

The data from the Massachusetts fluoroscopy series,[6] the New York mastitis series,[53] and the two LSS series[33,55] are sufficiently numerous, with sufficient dose information, to support specific inferences with respect to the shape of the dose-response curve, the influence of age at exposure on radiogenic tissue response, the latent period for radiogenic breast cancer, and estimation of risk. This report relies heavily on a parallel analysis of the raw data from those studies.[5] The other human studies and experimental studies of radiogenic breast cancer in animals provide information on particular questions. This report also relies heavily on a recent report by an NCI working group on risks associated with mammography[59] and on reviews of experimental studies.[19,38,56,57]

A few caveats are appropriate. Migration of atomic-bomb survivors, especially the younger ones, from Hiroshima and Nagasaki since 1950, although apparently unrelated to dose,[3] is likely to have caused overall underreporting of incidence. This is particularly true of women exposed in Nagasaki. The estimated dose response may therefore be biased downward with respect to absolute risk, especially in Nagasaki. The bomb survivors received whole-body irradiation, and it is possible that other effects interacted with the effects of radiation on breast tissue. The breast-cancer incidence patterns of Japanese and American women are very different, both in absolute incidence and in distribution by age at diagnosis.[21]

The tuberculosis patients' disease and associated nutritional and immunologic factors could have affected the carcinogenic response to ionizing radiation. Although the experience of the three control groups for the mastitis series argues strongly against interpreting the observed radiation dose response as an artifact of the treated condition, it is possible that lactation or inflammation of the breast tissue influenced that response. Data on older women, particularly on women exposed at ages above 50, are limited to a relatively few atomic-bomb survivors and Swedish radiation-therapy patients.

Dose estimates are more reliable for the patients given radiation therapy than for the patients given multiple fluoroscopic examinations and for the bomb survivors. Dose estimation for both pneumothorax patients and bomb survivors had to be based on reconstructions of their exposures. [7,35] Jablon[23] has estimated the standard errors of individual kerma estimates for the LSS sample to be ±30% and has also suggested that high kerma estimates (>80 rads in Hiroshima, >320 rads in Nagasaki) tend to be biased upward, whereas lower estimates are probably biased slightly downward. Kerma estimates of over 600 rads are adjusted downward to 600 rads in most studies of the LSS sample, and that custom is followed here.

Table A-2 contains summaries of the numerator and denominator information obtained from the Massachusetts fluoroscopy, New York mastitis, and 1950–1974 LSS series, by age at initial exposure and breast-tissue dose. The data are limited to breast cancers and woman-years (WY) of followup occurring more than 5 yr after initial exposure. A

TABLE A-2   Breast-Cancer Cases and Woman-Years (WY) at Risk for Three Radiation Studies, by Dose and Age at First Treatment[a]

		No. Cases/WY		
Series	Age at First Exposure, yr	0-Rad Dose	<100-Rad Dose	>100-Rad Dose
Atomic-bomb survivors[55]	10–19	31/180,742	26/96,011	17/19,579
	20–29	44/154,764	26/81,555	12/13,184
	30–39	49/144,282	22/77,402	12/7,523
	40–49	38/116,794	23/66,163	1/7,232
	50+	26/82,190	16/46,477	3/3,494
New York mastitis patients[6]	15–19	0/718	0/51	2/490
	20–29	14/12,818	0/588	18/7,187
	30–39	17/6,719	1/290	13/3,013
	40–44	1/395	0/22	3/231
Massachusetts tuberculosis fluoroscopy patients[53]	10–19	4/7,602	3/2,203	12/5,077
	20–29	6/4,053	4/4,157	14/6,297
	30–39	3/2,758	3/1,806	3/1,903
	40–45	1/670	0/585	1/286

[a] Reprinted with permission from Boice et al.[5] First 5 yr of followup excluded.

5-yr period was selected because the LSS data were necessarily limited and because 5 yr seemed to be a conservative lower limit for the minimal latent period for radiogenic breast cancer.[6,33,53,55]

EFFECT OF DOSE

A major difficulty with the analogy between the breast-cancer risk experience of Japanese and American exposed populations is the 13–30% neutron component of breast-tissue dose among women exposed to the Hiroshima bomb. Theoretical considerations based on microdosimetric principles suggest a general form for carcinogenic dose response that is linear in neutron dose (D_N), but has both linear and quadratic components in gamma dose (D_γ).[25] This is supported by experimental evidence on a number of cell systems[39] and by the leukemia-incidence data from the Hiroshima and Nagasaki bomb survivors.[3,21,59] The epidemiologic data do not, however, rule out the possibility of a linear dose response to gamma radiation.[2,22,43,44]

Breast-cancer data offer little support for a dose-response model with strong upward curvature in D_γ. The dose-response curves for mammary tumors in female rats given total-body x and gamma irradiation tend to be linear.[9,36,47,49] Functions of D and D_N fitted to the breast-cancer incidence rates for Hiroshima and Nagasaki, standardized to the age distribution of the combined cities, suggested a relationship linear in both D_γ and D_N.[26] Specifically, the best-fitting function linear in both D_γ and D_N corresponded to a chi-square statistic for lack of fit that was half as large as that obtained from the best-fitting model linear in $D_\gamma{}^2$ and D_N. Furthermore, the best-fitting regression on D_γ, $D_\gamma{}^2$, and D_N, subject to the constraint that all coefficients be nonnegative, had a zero coefficient for $D_\gamma{}^2$. Linear-model coefficients for D_γ and D_N did not differ significantly, and the RBE values most consistent with the data gave linear-model risk estimates that differed only slightly from those obtained with the assumption of an RBE of 1. Accordingly, the following analyses of the LSS sample data do not distinguish between the gamma and neutron components of breast-tissue dose. If comparability between gamma and x irradiation is assumed, the Japanese-American analogy also rests on the (testable) assumption that any differences between the two should not involve the shape of the dose-response function.

The dependence of breast-cancer risk on radiation dose has been shown to vary by age at exposure.[6,33,53,55] But the age-specific data are generally too sparse for fitting any but the simplest dose-response functions. As a way around this dilemma, it was adopted as a working assumption that within a given population the shape (but not necessarily the magnitude) of the breast-cancer dose-response function was independent of age at exposure. Given this assumption, the shape of the dose-response curve for each population should be obtainable from an investigation of summary rates, standardized for age at exposure to permit adjusting for possible confounding of age with radiation dose. Figure A-1 shows the observed adjusted rates, plotted against breast-tissue dose in rads, for the 1950–1974 LSS series[55] (also adjusted for city), the Massachusetts fluoroscopy series,[7] and the New York mastitis series;[53] each is standardized with respect to its own age distribution. For comparison, the crude rates from the Nova Scotia fluoroscopy series, plotted against number of fluoroscopic examinations as shown in the 1972 BEIR report,[40] are also presented.

The shape of the dose-response curve was investigated by using several

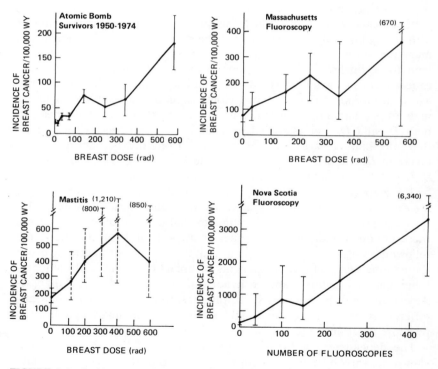

FIGURE A-1   Incidence of breast cancer in relation to radiation dose. Reprinted with permission from Boice et al.[5]

functional forms of dose: a linear form, $F_1(D) = \alpha_0 + \alpha_1 D$; a quadratic form, $F_2(D) = \alpha_0 + \alpha_1 D + \alpha_2 D^2$; a linear form with an exponential multiplier to account for the competing effect of cell-killing at high doses, $F_3(D) = (\alpha_0 + \alpha_1 D)e^{-\beta_2 D^2}$; and a quadratic form modified for cell-killing, $F_4(D) = \alpha_0 + (\alpha_1 D + \alpha_2 D^2)e^{-\beta_2 D^2}$.* All parameters except α_0 and α_1 were constrained to be nonnegative, so the coefficient α_2 corresponds to upward curvature at the low end of the dose range, and β_2 corresponds to downward curvature at the higher doses. The linear coefficient α_1 represents the excess risk per rad at low doses. Form F_4 is a simplified version of a form suggested by experimental data and radiobiologic theory, [13,58,59] but has more parameters than can be used with the available human data. [27]

Radiobiologic considerations suggest the existence, generally, of upward curvature at low doses of radiation for carcinogenesis dose-response functions, at least for low-LET radiation. [13,22,25,39,43,44,58] Breast cancer may be a special case, however, as suggested by some animal studies [9,36,47,49] and by the similarity in breast-cancer dose response between Hiroshima and Nagasaki bomb survivors. [26] Of more importance is the fact that the present analyses did not suggest the need for a quadratic term. None of the more complicated functional forms provided a significantly better fit to any of the data sets than the simple linear form F_1. For the Japanese data, the constraint that all parameters be nonnegative reduced each form to F_1, whereas, in the case of the fluoroscopy series, there was a slight but nonsignificant suggestion of upward curvature ($p = 0.36$ for α_2 in F_2). More substantial, but still nonsignificant, evidence of deviation from linearity was provided by the mastitis series, for which the dose response at high doses suggested a cell-killing component ($p = 0.08$ for β_2 in F_3). There was also a slight suggestion in the mastitis series of low-dose upward curvature ($p = 0.18$ for α_2 in F_4) as an adjustment to the linear form with cell-killing.

Because of the large number of women in the mastitis series with unilateral irradiation, an analysis was also performed by dose to single breasts. The linear model with cell-killing gave a significantly better fit to these data than the simple linear model ($p = 0.01$ for β_2 in F_3); this suggested that cell-killing at high doses (400–1,400 rads) may be important. Interestingly, it is only the single-breast exposure data of the mastitis patients that strongly suggest a turndown at the high doses. For comparable and even larger doses (average breast dose, 1,215 rads), no turndown is apparent among Nova Scotia patients who received

---

* See Chapter IV for explanation of symbols.

high-fractionated exposures delivered over several years (Figure A-1). For completeness, it should be mentioned that other functional forms for the dose-response relationship in radiogenic breast cancer have been suggested[12] and have been commented on.[4,28,52]

The evidence of a radiation effect among Japanese women exposed to low doses does not depend on extrapolation between breast-cancer incidence rates at zero and high (100+ rads) doses. In both the 1950–1969 and the 1950–1974 series, trend tests based on the lower end of the dose range rejected the null hypothesis of no dose effect: $p = 0.07$ for 0–49 rads kerma in the 1950–1969 series[27] and $p = 0.06$ for 0–49 rads breast-tissue dose in the 1950–1974 series.[55] Also, linear-regression estimates of the increment in risk per rad based on truncated data did not decrease with decreasing maximal dose down to these levels. Thus, "the actual observations at low levels themselves contribute strongly to the evidence for a linearly increasing dose response at low dose levels."[59]

DOSE FRACTIONATION

Dose-protraction effects could not be examined, because all the available studies had high dose rates. However, the effects of dose fractionation could be examined, at least indirectly, by comparing the risk estimates derived from the two multiple-fluoroscopy series, in which doses were highly fractionated and single doses were small (mean, 1.5 rads in the Massachusetts series and about 7.5 rads in the Nova Scotia series), with the other series of western women, which had little dose fractionation. In the published reports, which used a fixed minimal latent period of 10 yr, the absolute risk values in the high-fractionated irradiation series were 6.2 per 10^6 wy-rad (6.2 excess breast cancers per million women per year per rad) in the Massachusetts fluoroscopy series[7] and 8.4 per 10^6 wy-rad in the Nova Scotia series.[40] For comparison, the two radiotherapy series of patients treated for benign disease involved little fractionation of total dose (one to 10 exposures within a few days or weeks in the Rochester series), and the absolute risk values were 8.3 per 10^6 wy-rad in the Rochester postpartum mastitis series[53] and approximately 6.8 per 10^6 wy-rad in the Swedish radiotherapy study (calculations based on Table 2 of Baral et al.[1]). Thus, the risk estimates were approximately 6–8.5 per 10^6 wy-rad in both the fractionated- and unfractionated-exposure series. The fact that multiple low-dose exposures did not produce fewer cancers per unit dose than a single exposure suggests that radiation damage is cumulative and that highly fractionated x radiation is approximately as effective in inducing breast cancer as unfractionated radiation.

AGE AT EXPOSURE AND OTHER HOST FACTORS

Case reports of breast cancers in young women with histories of therapeutic high-dose radiation of the chest area during infancy have been treated by their authors as examples of radiogenic cancer, because of the high doses of radiation involved and because breast cancer is so rare in young women. [20,42] Substantial evidence from controlled studies of increased breast-cancer risk in women exposed to ionizing radiation before the age of 10 is lacking. Only one (nonexposed) breast cancer was found in the 1950–1969 LSS series among women 0–9 yr old at the time of the bombing. [33] Five cancers in the same age group were found in the 1950–1974 series, including one with a breast-tissue dose of 57 rads and four with doses of less than 20 rads. [55] However, this cohort is only now reaching the ages at which the radiation-related excess in the cohort who were 10–19 yr old at the time of the bombing became apparent. Another 5–10 yr of followup should determine the extent to which radiation exposure has affected breast-cancer incidence in the youngest cohort.

For women first exposed between the ages of 10 and 39, the data from the 1950–1974 LSS, Massachusetts fluoroscopy, and New York mastitis series all suggest a substantial radiation risk. Although the mastitis data for women exposed at ages under 20 and the fluoroscopy data for women first exposed at ages 30–39 are weak, contrasts of women exposed at 100+ rads versus women with zero rads are mutually supportive for first-exposure ages of 10–19, 20–29, and 30–39. The relative risks for exposure ages 40–44 in the mastitis series and ages 40–49 in the fluoroscopy series are high, but they are based on very small numbers, and neither series contains information about risk from exposures at higher ages. [6,53] The numerically strong LSS data are contradictory, in that there is a high-dose excess, based on small numbers, among women 50 or older at the time of the bombing, but a deficit in the 40–49 cohort. [33,55] The deficit could conceivably be due to the effects of radiation on the ovaries at ages associated with marked changes in ovarian function. On the one hand, women treated with x radiation for metropathia haemorrhagica at Scottish radiotherapy centers between 1940 and 1960, who were mostly in their 40s when irradiated and whose ovaries received fairly high radiation doses, later had less than half the breast-cancer mortality expected according to population rates. [54] On the other hand, no such reduction in breast-cancer incidence was found among a somewhat older group of women in whom artificial menopause was induced by x radiation.[11] At any rate, the findings with respect to the LSS series women exposed between the ages of 40 and 49 considerably complicate the problem of risk estimation for women with breast-tissue exposure at these ages.

Hormonal and other host factors appear to modify the carcinogenic effects of radiation on mammary tissue in rats. Postpubertal irradiation is apparently more effective than prepubertal irradiation of Sprague-Dawley rats, whereas no differences were observed in adult animals of the same age that were either virgin or lactating at the time of exposure to x rays.[46] Oophorectomy before or soon after x-ray exposure reduced the incidence of breast cancer, whereas chronic diethylstilbestrol (DES) treatment of irradiated AxC rats (but not Sprague-Dawley rats) caused a significant increase in breast-neoplasm incidence over that of rats treated with x rays or DES alone.[45,46,51] Other experiments have shown a synergism between ionizing radiation and pituitary factors[61] and additional interactions between radiation and hormones in mammary carcinogenesis.[14,15]

There is suggestive evidence from human studies that breast tissue may be more sensitive to radiation carcinogenesis if irradiation occurs at times of breast proliferation, as at menarche or during pregnancy.[8,33]

There is little information on the role played by different cell types of breast cancer, which have different age distributions and which could conceivably vary in their relationship to prior radiation. In the atomic-bomb survivors, all histologic types of infiltrating breast carcinoma appear to be involved in the excess attributable to radiation (D. H. McGregor, personal communication). But Tokunaga et al. found a statistically nonsignificant suggestion of an increase in the proportion of medullary tubular cancers and a corresponding decrease in the proportion of papillotubular cancers among the more heavily irradiated.[55]

LATENT PERIOD

The elapsed time between radiation exposure and the diagnosis of a breast cancer caused by that exposure appears not to depend on dose, but it does depend strongly on age at exposure. This follows from comparisons of the temporal distributions of date of diagnosis for breast cancers among high-dose and low-dose atomic-bomb survivors of the same ages[29,30] and similar comparisons based on the two main medical series.[26] After a probable minimal latent period, the age distribution of radiation-induced breast cancer appears to be identical with that of other breast cancers. There is probably a built-in bias toward over-estimating the minimal latent period, in that breast-cancer incidence normally increases with age and the evidence of increased risk in women exposed at ages over 40 is either nonexistent (e.g., in the LSS cohort who were 40–49 at the time of bombing) or based on small numbers (the oldest LSS cohort and the women over 40 at first medical exposure). The two LSS series suggest an excess risk within 5–9 yr after exposure, and, although the two medical series do not, they are based on relatively young samples. In view of these results and the suggestions of possible upward bias, it seems reasonable to assume a minimal latent period of 5 yr for women 25 yr old or older at exposure. But it appears that a further period may be required before there is substantial expression of the excess risk.

The existence or nonexistence of a maximal latent period (and there-fore a risk "plateau") cannot be determined from the available data—except that, if one exists, it must be greater than 30 yr.

RELATIVE- VERSUS ABSOLUTE-RISK MODELS

Breast-cancer risk depends on age. For women with histories of radiation exposure, the risk may also depend on the age at which the exposure occurred. Available data are far too sparse to permit reliance on risk estimates calculated separately for specific ages at exposure and at the time of observation. It is therefore necessary to assume that a woman's excess risk at one age has a simple relationship to her risk at another age, provided that she received a given radiation dose at a given age. An absolute-risk model implies that the risk of breast cancer at a given age is the sum of the natural risk at that age plus a dose-dependent increment that may be related to age at exposure but not to age at the time of observation. The difference between the risk for exposed women and the risk for otherwise similar nonexposed women remains constant over time. A relative-risk model expresses risk at a given age as the product of age-specific natural risk and a factor that depends on dose and age at exposure. If incidence data based on a relatively short followup of women irradiated at early ages are used to estimate lifetime risk of breast cancer, and if the natural incidence of breast cancer increases with age throughout a woman's lifetime, then lifetime-risk estimates based on relative-risk models will tend to be greater than estimates based on similar absolute-risk models. The correctness of either approach depends, of course, on the degree to which it represents the action of the unknown carcinogenic mechanism.

Breast-cancer rates observed in high-dose (100+ rads) women in

TABLE A-3 Linear-Model Risk Estimates for Breast Cancer, by Series and Age at First Exposure[a]

Series	Age at First Exposure, yr	No. Breast-Cancer Cases Among Exposed Women with 1 Rad[b]	Estimated Absolute Risk/per Rad ± SD[c]	Estimated Increase in Relative Risk/per Rad, % ± SD[d]
Atomic-bomb survivors[55][e]	10–19	40	9.0 ± 2.2	3.0 ± 0.98
	20–29	36	2.9 ± 0.88	0.92 ± 0.31
	30–39	28	4.9 ± 2.5	1.5 ± 0.85
	40–49	15	−1.0 ± 0.45	−0.30 ± 0.14
	50+	15	3.3 ± 2.2	0.97 ± 0.68
New York mastitis patients[6]	15–19	2[f]	27.9 ± 19.8[f]	not obtainable
	20–29	18	6.3 ± 2.0	0.43 ± 0.18
	30–39	13	9.4 ± 3.4	0.35 ± 0.16
	40–44	3[f]	52.1 ± 21.0[f]	1.57 ± 1.21[f]
Massachusetts tuberculosis fluoroscopy patients[53]	10–19	13	8.9 ± 3.1	0.84 ± 0.45
	20–29	18	3.8 ± 2.1	0.23 ± 0.16
	30–39	4[f,g]	6.9 ± 4.5[f]	2.3 ± 3.1[f]
	40–49	1[f,h]	6.4 ± 15.6[f]	0.54 ± 1.7[f]

[a] Reprinted with permission from Land et al.[26]

[b] Excluding the first 10, 15, and 20 yr of followup for women aged 20+, 15–19, and 10–14 yr at exposure, respectively.

[c] Excess cases per 10^6 women per rad per year of life after 10 yr after exposure or age 30, whichever is later.

[d] Excess risk per rad, as a percentage of age-specific natural breast-cancer risk.

[e] Only women with 1 rad or more to breast tissue are included.

[f] Estimate based on small numbers; normal theory inference based on the estimate and its standard deviation may be misleading.

[g] 4.8 breast cancers would have been expected if Connecticut Tumor Registry rates applied.[34]

[h] 1.6 breast cancers would have been expected if Connecticut Tumor Registry rates applied.[34]

the three series and in their appropriate low-dose comparison group were compared by age at the time of observation or years of followup for different ages at exposure.[26] Rate ratios appeared to be at least as stable over time as the rate differences; this suggests that lifetime-risk estimates based on the relative-risk versions of each of the dose-response models should be considered, as well as estimates based on the absolute-risk versions.

AGE-SPECIFIC RISK ESTIMATES

Linear-model (F_1) absolute- and relative-risk estimates were calculated for each series and each age at exposure represented in Table A-2, and they are shown in Table A-3. The estimates are for risk after 10 yr for women 20 or older at first exposure, and after 15 and 20 yr for women 15–19 and 10–14 yr old at first exposure, respectively.

It is remarkable that the absolute-risk estimates for women exposed at ages 10–19, 20–29, and 30–39 are so similar among the three studies. It appears that younger Japanese women may be as sensitive to radiation as western women, with respect to absolute risk of radiogenic breast cancer. As for relative risk, the effect on Japanese women is, of course, greater, given approximately the same absolute risk as western women and a much lower natural breast-cancer risk.

The risk estimates for ages 10–39 at exposure also indicate approximate equivalence of effect per rad between the fluoroscopy exposures delivered in small doses over extended periods and the concentrated exposures of the atomic-bomb survivors and the mastitis patients. Experimental results consistent with this interpretation were obtained by Shellabarger *et al.*,[48] who found that fractionation and protraction, as opposed to a single dose of 500 rads of whole-body x rays, did not change the overall incidence of mammary tumors in rats. These results suggest that the cumulative effect of many low-dose exposures is equivalent to that of a single, high-dose exposure and that low-dose exposures thus have a carcinogenic effect. They also suggest that the dose-response function is linear; otherwise, the breast-cancer effect of, for example, a single 100-rad exposure would be different from that of 100 separate 1-rad exposures. The experimental data indicate a possibly greater effect of fractionated dose, in that the proportion of adenocarcinomas, as opposed to fibroadenomas, increased with increasing fractionation and protraction. However, this result is difficult to interpret, because of competing mortality in the groups with fewer fractions and because of the presence of fibroadenomas.[19]

The strongly negative risk coefficients for the LSS cohort 40–49 yr old at the time of bombing underline the complete absence of a dose-response relationship in this group. The positive coefficients ($p = 0.059$ for absolute risk) for the cohort 50 + yr old confuse the situation further. The extremely high coefficients for the New York mastitis patients 40–44 yr old at treatment are based on only four breast cancers and therefore do not strongly suggest that the breast tissues of older women in this series were more sensitive to radiation than those of younger women. They do, however, suggest that sensitivity to radiogenic breast cancer did not markedly decrease in the mastitis patients with increasing age at exposure. Unless some unknown artifact (such as radiation effects on ovarian function) is the reason for the lack of a response in the LSS cohort 40–49 yr old at the time of bombing (and the strongly negative risk coefficient suggests the existence of such an artifact), it may be that the Japanese and American populations covered by these studies differ in their breast-cancer response to radiation received after the age of 40. However, the hormonal state of the breast in postpartum women may mask age-specific variations in radiation sensitivity that would otherwise apply. Estrogen-replacement therapy for postmenopausal women is another unknown factor that might affect risk assessment of radiogenic breast cancer for U.S. women exposed after the age of 40.

The Swedish radiation-therapy study reported a decreasing excess risk per rad, compared with population rates, with increasing age at treatment. Dose was highly correlated with age at treatment, however, and average doses were very high (285, 437, 667, 886, and 995 rads for women treated at ages 10–19, 20–29, 30–39, 40–49, and 50+, respectively). It is not possible to tell whether the variation in risk per rad by age at treatment was due to differences in sensitivity, to a high-dose cell-killing effect like that suggested by the analysis of the single-breast mastitis data, or even to age-related variations in the diseases treated.

The relatively small numbers in each of the two American series and the apparent Japanese-American differences with respect to naturally occurring and, possibly, radiogenic breast cancer suggest that risk estimates for American women should be based on pooled data from the two American series. Aside from the LSS cohort 40–49 yr old at the time of bombing, the only age difference suggested by the coefficients in Table A-3 is a greater radiation sensitivity for breast tissue of women exposed between the ages of 10 and 19. This suggests basing estimates for older women on the pooled American data for ages 20+ and higher. The analysis of single-breast data from the New York mastitis series suggests that a dose-response model incorporating cell-killing at high doses (F_3) should be used, as well as a linear model (F_1). Estimates of risk appropriate for low-dose exposures of normal breast tissue of American women are presented in Table A-4. As expected, the estimates differ greatly between the two age groups, but the effect of adding to the model a parameter for cell-killing at high doses, so that the estimated risks at low doses are less affected by downturn in risk at high doses, is confined to the younger group ($p < 0.001$ for β_2 in the 10–19 group, but $p = 0.39$ for women over 20 at exposure). The absolute-risk estimates for the older group are not greatly different from the previous 1972 BEIR estimate of six excess cases per year per 10^6 women, after a fixed 10-yr latent period.[40]

Life-table estimates of lifetime risk of radiogenic breast cancer due to a single 1-rad exposure are given in Table A-5 for different ages at exposure. The absolute-risk model estimates decrease with age at exposure, reflecting the decreasing expected number of years of life after exposure. The relative-risk model estimates decrease more slowly with age at first exposure, reflecting the increasing average natural breast-cancer risk per year over these remaining years.

The estimates in Table A-5 are considered to be the best estimates of risk, given the assumptions with respect to dose-response functions, absolute- or relative-risk model, and specificity for age at exposure in obtaining the corresponding risk coefficients in Table A-4. Perhaps the greatest uncertainty pertains to postmenopausal exposures. The decision to assume a uniform risk for all exposures after age 20 was made in

TABLE A-4 Estimated Risk of Radiogenic Breast Cancer among American Women[a]

| | | Risk Estimates | |
| | | Absolute Risk per Rad (excess cancers per 10^6 WY per rad) $\pm$ SD | Increase in Relative Risk per Rad, % $\pm$ SD |
Dose-Response Model	Age at Exposure, yr		
Linear risk = $\alpha_0 + \alpha_1$	10–19	10.4 $\pm$ 3.8	1.03 $\pm$ 0.64
	20+	6.6 $\pm$ 1.9	0.42 $\pm$ 0.15
Linear with cell-killing: risk = $(\alpha_0 + \alpha_1 D)e^{-\beta_2 D^2}$	10–19	22.4 $\pm$ 5.3	2.7 $\pm$ 1.30
	20+	8.7 $\pm$ 3.6	0.57 $\pm$ 0.29

[a] Reprinted with permission from Boice *et al.*[5] Estimates computed by pooling data from the Massachusetts[53] and New York[6] series after adjusting for study and age. The estimates are for risk after a minimal latent period of 10 yr for women aged 20 or over at first exposure and a correspondingly longer latent period for younger women (20 and 15 yr for women aged 10–14 and 15–19 at irradiation, respectively).

TABLE A-5   Life-Table Estimates of Lifetime Risk of Breast Cancer
Induced by a Single Exposure Resulting in a Breast Dose of 1 Rad,
by Age at Exposure

| | | Estimated Lifetime Risk of Radiogenic Breast Cancer per 10⁶ Women[a] | | | |
| | | Linear Dose Response[b] | | Linear Dose Response With Cell-Killing at High Doses[c] | |
Age at Exposure, yr	Expected No. Breast Cancers Incident in Year of Exposure per 10⁶ Women	Absolute-Risk Model	Relative-Risk Model	Absolute-Risk Model	Relative-Risk Model
35	524	234	312	307	425
40	1,036	202	288	266	391
45	1,590	172	257	226	350
50	1,713	143	226	187	307
55	1,911	115	191	151	259
60	2,251	88	154	116	208
65	2,324	64	117	84	158
70	2,566	42	79	55	108

[a] Data from Boice et al. [5]

[b] An absolute risk of 6.6 cancers per 10⁶ WY per rad and a 0.42% increase in relative risk per
rad were used in the computation.

[c] An absolute risk of 8.7 cancers per 10⁶ WY per rad and a 0.57% increase in relative risk per
rad were used in the computation.

the absence of data pertaining to U.S. women exposed after age 50 and
with only equivocal data from the LSS sample and the Swedish radiotherapy series. It is unlikely that the true risks are greater than twice
those given in the last column of Table A-5. They could be as low as
one-third of the risks corresponding to the linear absolute-risk model,
and could conceivably be zero at 1 rad if the models used for Tables
A-4 and A-5 are not applicable at very low doses.

REFERENCES

1. Baral, E., L.-E. Larsson, and B. Mattsson. Breast cancer following irradiation of
   the breast. Cancer 40:2905–2910, 1977.
2. Beebe, G. W., H. Kato, and C. E. Land. Life Span Study Report 8. Mortality
   Experience of Atomic Bomb Survivors 1950–74. Radiation Effects Research Foundation Technical Report TR 1-77. Hiroshima: Radiation Effects Research Foundation,
   1978.

3. Belsky, J. L., K. Tachikawa, and S. Jablon. The health of atomic bomb survivors. A decade of examination in a fixed population. Yale J. Biol. Med. 46:284-296, 1973.
4. Boice, J. D., Jr. Dose-response relationship in radiogenic breast cancer. J. Natl. Cancer Inst. 60:729-730, 1978. (reply)
5. Boice, J. D., Jr., C. E. Land, R. E. Shore, J. E. Norman, and M. Tokunaga. Risk of breast cancer following low-dose exposure. Radiology 131:589-597, 1979.
6. Boice, J. D., Jr., and R. A. Monson. Breast cancer in women after repeated fluoroscopic examinations of the chest. J. Natl. Cancer Inst. 59:823-832, 1977.
7. Boice, J. D., Jr., M. Rosenstein, and E. D. Trout. Estimation of breast doses and breast cancer risk associated with repeated fluoroscopic chest examinations of women with tuberculosis. Radiat. Res. 73:373-390, 1978.
8. Boice, J. D., Jr., and B. J. Stone. Interaction between radiation and other breast cancer risk factors, pp. 231-249. In Late Biological Effects of Ionizing Radiation. Vol. 1. Vienna: International Atomic Energy Agency, 1978.
9. Bond, V. P., E. P. Cronkite, S. W. Lippincott, and C. J. Shellabarger. Studies on radiation-induced mammary gland neoplasia in the rat. III. Relation of the neoplastic response to dose of total body radiation. Radiat. Res. 12:276-285, 1960.
10. Boot, L. M., P. Bentvelzen, J. Calafat, G. Röpcke, and A. Timmermans. Interaction of x-ray treatment, a chemical carcinogen, hormones, and viruses in mammary gland carcinogenesis, pp. 434-440. In R. L. Clark, R. W. Cumley, J. E. McCay, and M. M. Copeland, Eds. Oncology 1970. Vol. I. Chicago: Year Book Medical Publishers, 1971.
11. Brinkley, D., and J. L. Haybittle. The late effects of artificial menopause by x-radiation. Brit. J. Radiol. 42:519-521, 1969.
12. Bross, I. D. J. Dose-response relationship in radiogenic breast cancer. J. Natl. Cancer Inst. 60:727-728, 1978. (letter to the editor)
13. Brown J. M. Linearity vs nonlinearity of dose response for radiation carcinogenesis. Health Phys. 31:231-245, 1976.
14. Clifton, K. H., E. B. Douple, and B. N. Sridharan. Effects of grafts of single anterior pituitary glands on the incidence and type of mammary neoplasm in neutron- or gamma-irradiated Fischer female rats. Cancer Res. 36:3732-3735, 1976.
15. Clifton, K. H., B. N. Sridharan, and E. R. Douple. Mammary carcinogenesis-enhancing effect of adrenalectomy in irradiated rats with pituitary tumor MtT-F4. J. Natl. Cancer Inst. 55:485-487, 1975.
16. Cook, D. C., O. Dent, and D. Hewitt. Breast cancer following multiple chest fluoroscopy. The Ontario experience. Can. Med. Assoc. J. 111:406-409, 1974.
17. Curtin, C. T., B. McHeffy, and A. J. Kolarsick. Thyroid and breast cancer following childhood radiation. Cancer 40:2911-2913, 1977.
18. Delarue, N. C., G. Gale, and A. Ronald. Multiple fluoroscopy of the chest. Carcinogenicity for the female breast and implications for breast cancer screening programs. Can. Med. Assoc. J. 112:1405-1410, 1413, 1975.
19. Dethlefsen, L. A., J. M. Brown, A. V. Carrano, and S. Nandi. Report of the BCTF Ad Hoc Committee on X-Ray Mammography Screening for Human Breast Cancer. Can animal and in vitro studies give new, relevant answers? J. Natl. Cancer Inst. 61:1537-1545, 1978.
20. Iknayan, H. F. Carcinoma associated with irradiation of the immature breast. Radiology 114:431-434, 1975.
21. International Union Against Cancer. Cancer Incidence in Five Continents. Vol. II. R. Doll, C. Muir, and J. Waterhouse, Eds. New York: Springer Verlag, 1973.
22. Ishimaru, T., M. Otake, and M. Ichimaru. Incidence of leukemia among atomic

bomb survivors in relation to neutron and gamma dose, Hiroshima and Nagasaki, 1950-71. Radiat. Res. 77:377-394, 1979.

23. Jablon, S. Atomic Bomb Radiation Dose Estimation at ABCC. Atomic Bomb Casualty Commission Technical Report TR 23-71. Hiroshima: Atomic Bomb Casualty Commission, 1971.

24. Jablon, S., and H. Kato. Studies of the mortality of A-bomb survivors. 5. Radiation dose and mortality, 1950-1970. Radiat. Res. 50:649-698, 1972.

25. Kellerer, A. M., and H. H. Rossi. The theory of dual radiation action. Curr. Top. Radiat. Res. Q. 8:85-158, 1972.

26. Land, C. E., J. D. Boice, Jr., R. E. Shore, J. E. Norman, and M. Tokunaga. Breast cancer risk from low-dose exposure to ionizing radiation. Results of an analysis in parallel of three different exposed populations. J. Natl. Cancer Inst. (in press)

27. Land, C. E., and D. H. McGregor. Breast cancer incidence among atomic bomb survivors. Implications for radiobiologic risk at low doses. J. Natl. Cancer Inst. 62:17-21, 1979.

28. Land, C. E., and D. H. McGregor. Dose-response relationship in radiogenic breast cancer. J. Natl. Cancer Inst. 60:728-729, 1978. (reply)

29. Land, C. E., and D. H. McGregor. Temporal distribution of risk after exposure, pp. 831-843. In H. E. Nieburgs, Ed. Prevention and Detection of Cancer. Part 1. Prevention. Vol. 1. Etiology. New York: Marcel Dekker, 1977.

30. Land, C. E., and J. E. Norman. Latent periods of radiogenic cancers occurring among Japanese A-bomb survivors, pp. 29-47. In Late Biological Effects of Ionizing Radiation. Vol. 1. Vienna: International Atomic Energy Agency, 1978.

31. Lowell, D. M., R. G. Martineau, and S. B. Luria. Carcinoma of the male breast following radiation. Report of a case occurring 35 years after radiation therapy of unilateral prepubertal gynecomastia. Cancer 22:581-586, 1968.

32. MacKenzie, I. Breast cancer following multiple fluoroscopies. Brit. J. Cancer 19: 1-8, 1965.

33. McGregor, D. H., C. E. Land, K. Choi, et al. Breast cancer incidence among atomic bomb survivors, Hiroshima and Nagasaki, 1950-69. J. Natl. Cancer Inst. 59:799-811, 1977.

34. Mettler, F. A., Jr., L. H. Hempelmann, A. M. Dutton, et al. Breast neoplasms in women treated with x-rays for acute postpartum mastitis. A pilot study. J. Natl. Cancer Inst. 43:803-811, 1969.

35. Milton, R. C., and T. Shohoji. Tentative 1965 Radiation Dose (T65D) Estimation for Atomic Bomb Survivors, Hiroshima and Nagasaki. Atomic Bomb Casualty Commission Technical Report TR 1-68. Hiroshima: Atomic Bomb Casualty Commission, 1968.

36. Moskalev, Y. Remote After Effects of Radiation Damage, pp. 203-204. U.S. AEC Technical Report 7387. Washington, D.C.: U.S. Atomic Energy Commission, 1972.

37. Myrden, J. A., and J. E. Hiltz. Breast cancer following multiple fluoroscopies during artificial pneumothorax treatment of pulmonary tuberculosis. Can. Med. Assoc. J. 100:1032-1034, 1969.

38. National Council on Radiation Protection and Measurements. Report of NCRP Committee No. 56. (in preparation)

39. National Council on Radiation Protection and Measurements. Review of the Current State of Radiation Protection Philosophy. NCRP Report No. 43. Washington, D.C.: National Council on Radiation Protection and Measurements, 1975.

40. National Research Council, Advisory Committee on the Biological Effects of Ionizing Radiations. The Effects on Populations of Exposure to Low Levels of Ionizing Radiation. Washington, D.C.: National Academy of Sciences, 1972.

41. Orine, S. K., R. W. Chambers, and R. H. Johnson. Postradiation carcinoma of male breast bilaterally. J. Amer. Med. Assoc. 201:707, 1967.

42. Reimer, R. R., J. F. Fraumeni, Jr., R. Reddick, and E. L. Moorhead. Breast carcinoma following radiotherapy of metastatic Wilms' tumor. Cancer 40:1450-1452, 1977.

43. Rossi, H. H. and A. M. Kellerer. The validity of risk estimates of leukemia incidence based on Japanese data. Radiat. Res. 58:131-140, 1974.

44. Rossi, H. H., and C. W. Mays. Leukemia risk from neutrons. Health Phys. 34: 353-360, 1978.

45. Segaloff, A., and W. S. Maxfield. The synergism between radiation and estrogen in the production of mammary cancer in the rat. Cancer Res. 31:166-168, 1971.

46. Shellabarger, C. J. Modifying factors in rat mammary gland carcinogenesis, pp. 31-43. In J. M. Yuhas, R. W. Tennant, and J. D. Regan, Eds. Biology of Radiation Carcinogenesis. New York: Raven Press, 1976.

47. Shellabarger, C. J., V. P. Bond, E. P. Cronkite, and G. E. Aponte. Relationship of dose to incidence of mammary neoplasia in female rats, pp. 161-172. In Radiation Induced Cancer. Vienna: International Atomic Energy Agency, 1969.

48. Shellabarger, C. J., V. P. Bond, G. E. Aponte, and E. P. Cronkite. Results of fractionation and protraction of total-body radiation on rat mammary neoplasia. Cancer Res. 26(Part I):509-513, 1966.

49. Shellabarger, C. J., R. D. Brown, A. R. Rao, J. P. Shanley, V. P. Bond, A. M. Kellerer, H. H. Rossi, L. J. Goodman, and R. E. Mills. Rat mammary carcinogenesis following neutron or x-radiation, pp. 391-401. In Biological Effects of Neutron Irradiation. Vienna: International Atomic Energy Agency, 1974.

50. Shellabarger, C. J., and R. W. Schmidt. Mammary neoplasia in the rat as related to dose of partial-body irradiation. Radiat. Res. 30:497-506, 1967.

51. Shellabarger, C. J., J. P. Stone, and S. Holtzman. Synergism between neutron radiation and diethylstilbestrol in the production of mammary adenocarcinomas in the rat. Cancer Res. 36:1019-1022, 1976.

52. Shore, R. E. Dose-response relationships in radiogenic breast cancer. J. Natl. Cancer Inst. 60:728, 1978. (reply)

53. Shore, R. E., L. H. Hempelmann, E. Kowaluk, *et al.* Breast neoplasms in women treated with x-rays for acute postpartum mastitis. J. Natl. Cancer Inst. 59:813-822, 1977.

54. Smith, P. G., and R. Doll. Late effects of x irradiation in patients treated for metropathia haemorrhagica. Brit. J. Radiol. 49:224-232, 1976.

55. Tokunaga, M., J. E. Norman, Jr., M. Asano, S. Tokuoka, H. Ezaki, I. Nishimori, and Y. Tsuji. Malignant breast tumors among atomic bomb survivors, Hiroshima and Nagasaki, 1950-1974. J. Natl. Cancer Inst. 62:1347-1359, 1979.

56. United Nations Scientific Committee on the Effects of Atomic Radiation. Ionizing Radiation, Levels and Effects. A report to the General Assembly. New York: United Nations, 1972.

57. United Nations Scientific Committee on the Effects of Atomic Radiation. Sources and Effects of Ionizing Radiation. 1977 Report to the General Assembly, with Annexes. New York: United Nations, 1977.

58. Upton, A. C. Radiobiological effects of low dose. Implications for radiobiological protection. Radiat. Res. 71:51-74, 1977.

59. Upton, A. C., G. W. Beebe, J. M. Brown *et al.* Report of NCI ad hoc working group on the risks associated with mammography in mass screening for the detection of breast cancer. J. Natl. Cancer Inst. 59:481-493, 1977.

60. Wanebo, C. K., K. G. Johnson, K. Sato, *et al.* Breast cancer after exposure to the

atomic bombings of Hiroshima and Nagasaki. N. Engl. J. Med. 279:667–671, 1968.
61. Yokoro, K., and J. Furth. Relation of mammotropes to mammary tumors. V. Role of mammotropes in radiation carcinogenesis. Proc. Soc. Exp. Biol. Med. 107:921–924, 1961.
62. Young, S., and R. C. Hallowes. Tumors of the mammary gland, pp. 31–73. In V. S. Turosov, Ed. Pathology of Tumors in Laboratory Animals. Vol. I. Tumors of the Rat. Part I. Lyon, France: IARC, 1973.

THYROID

Since the 1972 BEIR report,[35] there has been a considerable resurgence of interest in radiation-induced thyroid disease. Several large populations, many of which were exposed during childhood, have come to light in this interval, and this has resulted in a rash of new and updated reports.[5,30,34,36,46] Indeed, radiation-induced thyroid disease has been described as "endemic" in some populations.[9] In the United States especially, a number of large populations have had medical head and neck irradiation during childhood for a variety of indications. Unfortunately, the long period between the irradiation and the recognition of induced abnormalities has resulted in a lack of availability of detailed records. Where records are available, specific populations have been studied in detail—e.g., at the Michael Reese Hospital in Chicago, Illinois.[15,19,46] Because of the magnitude of the problem, the National Cancer Institute has broadly disseminated information to physicians to inform them of the problem and to help identify people who were irradiated.[50] There is a continuing program to reach members of the public at large and alert them to potential dangers if they were irradiated. Despite the availability of a considerable amount of additional information on irradiated populations, the overall absolute-risk estimate for induction of malignant neoplasia previously reported in BEIR I, 1.6–9.3 cases per 10^6 person-years (PY) per rad, does not seem to have changed appreciably. But the additional information does appear to have increased the understanding of and improved information on modifying factors and on the kinds of thyroid disease induced.

EXPERIMENTAL INDUCTION OF THYROID NEOPLASIA WITH RADIATION

Numerous experiments have been performed, primarily in rats, to demonstrate the induction of thyroid neoplasia with radiation.[4,13,14,16,47] The adult rat thyroid, weighing only 15 mg, requires doses higher than 1,000 rads of external low-LET radiation or administration of 40 μCi (about 8,000–10,000 rads) of iodine-131 to induce measurable increases in thyroid carcinoma consistently. [13] Doniach has reported a very significant increase in incidence of both thyroid adenoma and carcinoma after administration of goitrogens that secondarily increase thyroid-stimulating hormone (TSH) with or without radiation. [13] Only one carcinoma was observed with 500 rads of external radiation in 17 thyroid-irradiated animals, whereas the same dose in 10 animals receiving goitrogens produced an excess of five follicular carcinomas over the control, goitrogen-only animals. [13] Other causes of increased TSH, such as iodine deficiency, have also increased the incidence of thyroid neoplasia. [16] "It is thought that the subsequent development of benign and malignant tumors results from summation of the TSH-induced hyperplasia with neoplastic transformation initiated by the radiation." [13] Conversely, administration of thyroid hormone with suppression of TSH decreases the incidence of tumors. [16]

Interestingly, the histology of radiation-induced carcinomas in animals is essentially of the papillary and follicular type, with apparent suppression of carcinoma of the alveolar type. [16] This finding is similar to observations in irradiated humans, in whom the carcinomas are of the papillary and follicular type, with no anaplastic carcinomas, which usually constitute 15% of thyroid cancers. Indeed, in the population of Japanese atomic-bomb survivors reported by Parker *et al.*, there were no cases of anaplastic carcinoma, although several would have been expected. [36] Thus, the highly malignant anaplastic carcinoma that is responsible for most deaths from thyroid cancer appears to be unaffected by radiation, which results primarily in an increased incidence of cancer, rather than of mortality.

KINDS OF POPULATIONS IRRADIATED

The bulk of reports of radiation-induced thyroid disease have resulted from the use of a variety of therapeutic medical procedures. Such procedures were used beginning about 1925 and extending to about 1955, with peak use probably in the 1930s. The radiation therapeutic procedures included scalp irradiation for ringworm, [1,32,48] chest irradiation for enlarged thymus, [27] chest irradiation for pertussis, [51] head and neck irradiation for various lymph node abnormalities (such as enlarged tonsils and adenoids), [7] skin irradiation for acne and hemangiomas, [21] and the use of radioiodine (principally iodine-131) for ablation of the thyroid gland. [12] The large population of Japanese atomic-bomb survivors in the cities of Hiroshima and Nagasaki has now been followed for 30 yr, and additional data on the development of thyroid tumors have been reported, so the previously available information has been refined. [36,37]

A smaller group of Marshallese accidentally exposed to nuclear fallout have also been studied in more detail. [5,6] Low-level radiation from fallout iodine-131 has been studied in groups of western U.S. schoolchildren. [41]

TYPES OF IRRADIATION IMPLICATED IN THE INDUCTION OF THYROID DISEASE

Recent evidence has continued to implicate external photon irradiation of the thyroid gland as the prime cause of radiation-induced thyroid neoplasia. In addition, more information has been obtained since 1972 on the effect on the thyroid gland of internally administered particulate radiation, primarily in the form of beta particles from iodine-131. Detailed studies of large populations treated therapeutically with radio-iodine have been reported with no evidence of resulting radiation-induced neoplasia; however, a very significant incidence of other radiation-induced thyroid disease, such as hypothyroidism, has been reported. [10,30] External photon radiation, per unit of radiation dose, primarily in the form of x rays, has been found to be considerably more efficient than internal beta radiation in inducing neoplasia. [30] Some reports have estimated that the ability of external photon radiation to induce thyroid neoplasia is some 10–80 times that of internally administered particulate beta radiation. [30] These estimates were derived mainly from animal experiments, very little information being available on human populations.

Microdosimetry must be taken into account, as well as macrodosimetry, in assessing the effects of ionizing radiation and may in great part explain differences in the effectiveness of radiation quality. Microdosimetry studies of the thyroid reported by Anspaugh, assuming average-sized follicles 300 μm in diameter (most of which represented "inert" colloid), suggested that the iodine-131 particulate radiation dose to the thyroid may be homogeneous over long periods, but there probably are dose inhomogeneities over short periods, especially in a thyroid that is not normal. [2] Inhomogeneities could lead to intensive irradiation of individual follicles that are functioning, but spare nonfunctioning follicles. This would approach an all-or-none effect and result in a microdosimetric inhomogeneous dose distribution. Such inhomogeneities could also partially explain the greater biologic effectiveness of iodine-125, which is much longer-lived than iodine-131. [10] In contrast, external photon irradiation does not depend on thyroid function for its effect and results in a more homogeneous distribution of dose.

THYROID RADIATION DOSES

The induction of thyroid disease by radiation has been studied over a wide range of radiation doses. Therapeutic external radiation has ranged in most series from approximately 100 to 1,500 rads. Larger doses are likely to ablate the thyroid. Radiation doses as low as 6.5 rads to the thyroid may have induced thyroid neoplasia. [34,48] Radiation doses in populations irradiated therapeutically with beta particles from internally administered radionuclides are considerably higher, usually above 10,000 rads.

EFFECTS OTHER THAN NEOPLASIA

With respect to radiation-induced thyroid disease, the greatest attention has been focused on thyroid neoplasia. But other important thyroid diseases, including acute thyroiditis and hypothyroidism, are associated with the use of higher doses of ionizing radiation than those associated with neoplasia.

The highest absorbed radiation doses are associated with the development of acute thyroiditis at threshold doses over 20,000 rads, which are possible only with internally administered beta radiation from radionuclides, such as iodine-131. [30]

Primary hypothyroidism has been observed after external irradiation at about 2,000 rads. The thyroid is thus ablated and probably has less potential for neoplastic degeneration. Internally administered beta-radiation doses perhaps as low as 5,000 rads in routine clinical use of iodine-131 have been associated with the development of hypothyroidism. [30] According to recent evidence, hypothyroidism occurs with a two-phase response after radioiodine treatment of patients for hyperthyroidism. [20] Thus, there appears to be, as a second phase, an inherent incidence of hypothyroidism associated with the prior hyperthyroidism, and for approximately the first 2 yr after treatment with radioiodine the induction of hypothyroidism is proportional to the amount of irradiation. Thus, both external photon and internal particulate radiation can induce hypothyroidism, but with thresholds of approximately 2,000 rads and 5,000 rads, respectively. [30]

CLASSIFICATION OF THYROID NEOPLASIA

Universally accepted criteria for definition of various types of thyroid neoplasia have been developed and promulgated by the World Health Organization. [8] This classification divides thyroid cancer into follicular, papillary, squamous-cell, undifferentiated (anaplastic), and medullary types. [8] A clinical subcategory of mixed papillary-follicular cancer has been included in the category of papillary carcinoma. It must be recognized that there is not universal agreement among experts on the cell type of some thyroid neoplasms in studies thus far reported. Recent well-documented studies of radiation-induced thyroid cancers have indicated that only the papillary and follicular types appear to be related to radiation induction, with perhaps slightly more than the usual preponderance of the papillary type. [22,46] Furthermore, the radiation-induced papillary tumors that have been observed may have somewhat less malignant potential than those arising spontaneously. [46] Papillary carcinoma accounts for approximately 80% of all spontaneous thyroid cancers, but about 89% of radiation-induced cancers. [22,46] The mortality from well-managed spontaneous papillary carcinoma is now less than 5% and appears to be the same with papillary carcinoma induced by radiation. [31]

Besides more uniform terminology, there is increased recognition of the entity of "minimal or occult microscopic thyroid cancer," defined as a tumor of 1 cm or less in diameter. It is thought that such a lesion has essentially no malignant potential and should not be considered cancer. [45] This lesion has been recognized in the Japanese population at necropsy in up to 28% of patients and in the United States in up to 15%. [45] Furthermore, there is minimal or no evidence that this lesion is induced by radiation. In any series of reports of radiation-induced thyroid carcinoma, it is imperative that the occult carcinoma not be lumped with clinical disease as being radiation-induced. Because some series reporting radiation-associated thyroid carcinoma include some cases of occult carcinoma, incidence figures for clinically significant cancers may not be correct. It is critical that thyroid pathology sections from patients reported to have thyroid carcinoma associated with radiation in these series be reviewed by panels of experts to establish the true incidence.

There is no evidence that benign thyroid neoplasia have malignant potential. Rather, the processes appear to be independent, parallel phenomena. No instance has been reported of progress of a benign nodule to malignancy in the thyroid, despite many years of widespread needle biopsy of such benign lesions. However, there is ample clinical evidence in some instances that papillary or follicular carcinoma may advance to an undifferentiated thyroid cancer. [8]

THERAPEUTIC RADIATION EXPOSURES

University of Rochester Followup of Population with Thymus Irradiation during Infancy [24-28,38-40]

This study compares 2,872 people who were irradiated during the first year of life for presumed enlargement of the thymus gland with 5,055 untreated siblings; there have been four mail surveys over the last 20 yr. A subgroup (Group C) of the population, 261 persons, has been identified as having received higher radiation doses and is thought to be at greater risk than the rest of the subjects. This subgroup has also been followed longer (i.e., it is an older population), and it has a high proportion of Jewish subjects. Although it is small, it has contributed 13 of the 24 cancers found. Furthermore, 11 of the 24 cases of thyroid cancer have developed in the 8% of the population that is Jewish.

The overall ratio of observed-to-expected cases of thyroid cancer was 24:0.29, indicating a relative risk of nearly 100. For Group C, the relative risk was over 300. Thyroid neoplasms diagnosed at surgery have not shown an increase in incidence over the period of followup, which now exceeds 35 yr.

A plot of the incidence of thyroid carcinomas against absorbed dose suggests a linear proportionality, but it may be curvilinear, with slopes of 3 cases per 10^6 PY per rad for the entire group and 4.8 cases per 10^6 PY per rad for Group C. Benign neoplasia incidence is approximately 3 times higher. Hempelmann raised some questions as to strict dependence on dose measurements, because of the uncertainty of position of the thyroid in the primary beam. [26]

Sex is seen to be an additional risk factor, females having 2.3 times the incidence of males. The combination of being female and having a Jewish ethnic background results in a 17-fold increase over the rest of the study population. There appears to be no relation of longer latent period with lower doses.

University of Chicago Head and Neck Irradiation Sample[9,43]

Of 100 patients with a history of head and neck irradiation at about 4.5 yr of age, 26 were found to have nodular thyroid disease, with seven cancers found at operation; i.e., there was a 7% prevalence. Both the base population and the occurrence of abnormalities were evenly divided between sexes. Five cancers were basically papillary, and two follicular. No occult carcinomas were reported, but one cancer was found incidentally in the opposite lobe from that of the benign lesion for which the operation was performed. Radiation dose ranged from 180 to 1,500 rads, with an estimated average of 750 rads. Peak incidence, as judged from discovery and surgery, was at about 24.5 yr of age. Cancer patients appeared to be in the higher-dose groups. An absolute risk of approximately 4 cancers per 10^6 PY per rad can be estimated from these data.

Michael Reese Hospital Head and Neck Irradiation Sample[3,15,17-19,46]

Of a population of 5,226 known to have received radiation to the head, neck, or chest during infancy, childhood, or adolescence (about 90%

less than 10 yr of age), 49% were contacted and 28% (1,476) were examined. A nearly constant 55:45 male-to-female ratio was present throughout the group contacted and the group examined. About 80% of the total population received 750 rads. Characteristics of age, radiation dose, and year of first therapy were similar between the contacted-only and a demographically selected sample of the examined groups, suggesting that the examined group was nearly representative of the overall irradiated population. Of the followup group of 2,189 subjects actually contacted on whom adequate data could be obtained, 32.6% have been found to have nodular thyroid disease. Approximately one-third of those with nodular disease have been found to have cancer—an 11.7% prevalence. The benign nodular disease has a prevalence of 20.9%,[2] or about double that of cancer. An absolute risk of 5 cancers per 10^6 PY per rad and a lower limit of 2.1 cancers are estimated, assuming a linear dose response. A minimal latency of 10 yr after irradiation was observed, with an apparent peak incidence of about 19 yr for thyroid cancer.[46] A slight but significant inverse relationship was seen between age at treatment and latency, i.e., shorter latent periods in older persons. About 91% of the lesions were papillary. Some 35% were less than 5 mm in diameter and were found incidentally at surgery; another 47% were 6–15 mm in diameter. Thus, a total of perhaps 82% were within the occult-carcinoma category.[45] Although it was not reported, it is interesting that this population was predominantly of Jewish origin.

Scalp Irradiation for Tinea Capitis in Israel[32-34,52]

A total of 10,902 Jewish children immigrating into Israel were studied after having received scalp irradiation for ringworm in one of three medical facilities. An estimated thyroid dose of 6–9 rads was received. All but 60 of the patients were successfully traced and matched against an equal number of nonirradiated controls with tinea capitis and a nonirradiated sibling group of half the size. A sixfold increase in malignant thyroid tumors was found in the irradiated group, compared with the controls. Nine of the 12 thyroid cancers in the irradiated group occurred in females, most of them of the papillary-cell type. Ten of the tumors occurred between 9 and 16 yr after therapy. In the most recent revised reports, two of the patients with cancer were found to have received more than one course of radiation. Thus, 10 of the patients who developed cancer had an estimated dose of about 6–9 rads to the thyroid, and the other two received 12 and 18 rads.[33,34] Only two cases of thyroid cancer would have been expected in this study, so the excess was 10 cases. On the basis of this revision, the absolute-risk estimate is 6.3 cases per 10^6 PY per rad. Most of the cancers were papillary carcinoma, and there were no cases of anaplastic carcinoma. No data were given on the occurrence of occult carcinoma. Detailed dosimetry studies have been repeated by the authors and have confirmed the low radiation doses measured in phantoms that reproduced the circumstances of the scalp irradiation. However, as discussed by the authors, if a small amount of movement or misalignment occurred during exposure and the thyroid gland came into the primary therapy beam for only a few seconds, the thyroid radiation dose might have been considerably higher than calculated.[32] The case-finding techniques in these investigations from the Israeli Tumor Registry did not permit the identification of benign thyroid tumors.

Scalp Irradiation for Tinea Capitis in New York [1,23,48]

Shore, Albert, and Pasternak reported on the second survey of a population of 2,215 irradiated and 1,395 nonirradiated control subjects with tinea capitis.[48] Scalp epilation was accomplished with essentially the same technique as in the Israeli population just discussed; the authors produced almost exactly the dosimetry estimates of 6–10 rads to the thyroid. The average age at irradiation was about 8 yr, and the average interval of followup was about 20 yr after irradiation. No thyroid cancers were observed, although eight patients with benign adenomas were identified. The variance of this study from that of Modan *et al.*[34] may be due to the much smaller size of the population.

National Thyrotoxicosis Followup of Patients Treated with Surgery, Antithyroid Drugs, or Radioiodine [11,12]

In A. U.S. Public Health Service thyrotoxicosis followup, 21,714 adult patients treated with radioiodine were matched against 11,732 patients treated with surgery and 1,144 treated with antithyroid drugs; 667 patients were operated on more than a year after one of the forms of therapy, and 27 thyroid cancers were found in the patients operated on. Sixteen of the cancers were in patients who had previously received iodine-131 for therapy. Thus, no clear-cut increase in the incidence of thyroid cancer due to radioiodine was found in this study, compared with the incidental malignant lesions found in patients with thyrotoxicosis who were treated primarily with surgical thyroidectomy.

ACCIDENTAL RADIATION EXPOSURES

Marshall Islands Population Exposed to Fallout[5,6]

In March 1954, 64 inhabitants of Rongelap Island (105 nautical miles from detonation site), 28 Americans on a nearby island, 18 Rongelap natives who happened to be on Ailingnae (also a nearby island), and 157 islanders on Uterik (about 200 miles farther east in the Marshall Islands) were accidentally exposed to "fresh" fallout from a thermo-nuclear detonation. Thyroid radiation doses could only be approximated from urine collections for Rongelap people assayed for iodine-131, although the shorter-lived radioiodine isotopes to which the inhabitants must have been exposed during the early period after the explosion delivered 2–3 times the dose of iodine-131. The approximate adult thyroid dose was estimated at 220–450 rads, and that of a 4-yr-old child, 700–1,400 rads. Thyroid-function studies, even on control "normal" Marshallese, suggested unusual function, with excess iodinated organic products in serum—perhaps evidence of underlying dyshormonogenesis. Within 22 yr, 40 had developed thyroid nodules, and seven, thyroid cancer, all in the nodule of concern in clinical examination. The latent period varied between 11 and 22 yr. There is some evidence that glands that received lower doses developed tumors later. An estimate of thyroid-cancer absolute risk of 3.5 cases per 10^6 PY per rad is similar to that found in series of people who developed cancer as a result of therapeutic external photon radiation. The higher energy of the short-lived iodine-132, -133, and -135—resulting in higher dose rates and more uniform exposure than iodine-131—may explain the similarity of risks to those from therapeutic external photon radiation. All the tumors occurred in females, and the data do not support significant differences in the risk of cancer between exposed children and exposed adults; the majority of the children (15 of 19) had most of their thyroid tissue removed surgically. Followup of the Americans has not been reported.

Atomic-Bomb Survivors[36,37,49,53]

By 1961, 16 yr after the detonation of atomic weapons over Hiroshima and Nagasaki, an excess of thyroid neoplasms had developed. Parker *et al.* published new data in 1973[36] and 1974[37] on the occurrence of thyroid cancer in the Japanese atomic-bomb survivors followed to 1971, 26 yr after whole-body irradiation in 1945. In the approximately 17,000 members of the Adult Health Study sample resident in or near Hiroshima or Nagasaki in the period 1958–1971, 40 clinically diagnosed and micro-scopically confirmed cases of thyroid cancer were found (28 in Hiroshima, 12 in Nagasaki). In autopsies performed in the same period, 34 clinically silent cases were discovered by routine procedures (27 in Hiroshima, seven in Nagasaki). The clinically evident and clinically silent cases differed markedly as to cell type, clinical cases being predominantly papillary (27 of 40), and autopsy cases being mainly papillary sclerosing (23 of 34)—the usual cell type of occult papillary carcinoma. There were 11 follicular cases among the 40 clinical cases, and seven among the 34 autopsy cases. The clinically silent tumors were usually less than 1.5 cm in diameter and appeared to be of little clinical significance. They did not include cases detected only in the special study of Sampson *et al.*, in which serial sections were examined and the prevalence of occult thyroid carcinoma at autopsy among zero-dose survivors was estimated at 28%.[45]

The 34 autopsy cases cannot be used in incidence calculations and do not, by themselves, provide strong evidence of the carcinogenic effect of radiation on thyroid tissue; but for females only the excess was statistically significant at the 0.02 level. The 40 clinical cases may perhaps be considered incidence cases—not for the period 1958–1971, as reported by Parker *et al.*,[36] but for the period 1950–1971, because the Adult Health Study sample was defined by schedules filled out at the time of the 1950 census. Nine of the 40 cases were reported to have had their onset before 1958, but most subjects were not examined at the Atomic Bomb Casualty Commission before 1958, when the first cycle of examinations in the Adult Health Study began. Multiplication of the exposure-years reported by Parker *et al.* for 1958–1971 by the factor 21/13 provides an approximate adjustment consistent with the view that the 40 cases are best regarded as incidence cases for 1950–1971.

Although examinations during the first two cycles (1958–1962) of the Adult Health Study included a strong emphasis on the detection of thyroid disease,[49] only about half (70 of 131) of the recommended surgical biopsies were taken, and the succeeding 2-yr cycles were charac-terized by less zealous case-finding. The report of cases by Parker *et al.* underestimated the true risk of thyroid cancer in the atomic-bomb survivors.

The clinical series, with person-years counted from 1950, is summarized in Table A-6 in terms of excess cases per 10^6 PY per rad of tissue dose, derived from the contrast between those with essentially zero dose and those exposed to 50+ rads kerma. For both cities combined, the estimate is 1.89, and the rates of 2.2 for Hiroshima and 1.5 for Nagasaki do not

TABLE A-6 Estimation of Absolute Risk of Thyroid Cancer Attributable to Ionizing Radiation, Atomic-Bomb Survivors, Hiroshima and Nagasaki, 1950–1971[a]

City	Sex	Statistic	Rads (Kerma)				Absolute-Risk Tissue Dose[b]
			Not in City, <1	1-49	50+	Total	
Total	Total	Person-years, thousands	162.7	65.3	83.7	311.7	
		Cases observed	9	6	25	40	
		Cases expected	9	3.61	4.63	—	
		Tissue dose, rads[c]	—	—	129	—	1.89
Total	Male	Person-years, thousands	60.6	22.0	32.2	114.8	
		Cases observed	1	2	5	8	
		Cases expected	—	3	1.17	4.17	
		Tissue dose, rads[c]	—	—	129	—	0.92

Female	Person-years, thousands	102.1	43.3	51.5	196.9	
	Cases observed	8	4	20	32	
	Cases expected[d]	8	3.39	4.04	15.4	
	Tissue dose, rads[c]	—	—	129	—	2.40
Hiroshima	Total					
	Person-years, thousands	115.1	57.3	50.2	222.6	
	Cases observed	6	6	16	28	
	Cases expected[d]	6.36	3.17	2.78	—	
	Tissue dose, rads[c]	—	—	119	—	2.21
Nagasaki	Total					
	Person-years, thousands	47.6	8.0	33.5	89.1	
	Cases observed	3	0	9	12	
	Cases expected[d]	2.63	0.44	1.85	—	
	Tissue dose, rads[c]	—	—	145	—	1.47

[a] Modified from Parker et al.[36]

[b] Excess cases per 10[6] persons per year per rad of tissue dose, 50+ rads.

[c] Kerma-to-tissue rad factors from Kerr.[29] For the 50+ rads (kerma) group, the gamma and neutron components are, respectively, 117 and 12 for both cities, 99 and 20 for Hiroshima, and 144 and 1 for Nagasaki.

[d] Based on rate for not in city + <1 rad for both cities combined.

differ significantly. Regression estimates, based on Tumor-Registry ascertainment in the much larger Life Span Study sample for 1959–1970 and adjusted for age, are considerably lower: 1.3 for each city when converted from kerma to tissue dose. Both coefficients are significant at the 0.01 level.

In every dose group of the clinical sample, the incidence for females exceeds that for males, and the absolute-risk estimates are 2.4 for females and 0.9 for males. The baseline data for males are too few, however, to determine whether the absolute risk is reliably different between the two sexes. Differences in susceptibility by age are also difficult to explore in the clinical sample, for the same reason: the baseline rate for those under age 20 in 1945 is based on only two cases. Although, therefore, the calculated absolute-risk estimate of 2.85 excess cancers per 10^6 PY per rad to thyroid tissue of persons who were under age 20 in 1945 exceeds that of 1.29 for those who were 20 or older, the sampling errors are so large that the difference is no more than suggestive of the greater susceptibility of thyroid tissue of younger people. In their original analysis, Parker *et al.* concluded that, among female subjects, the relative risk of thyroid cancer was significantly higher in persons who were under 20 in 1945.

As reported by Parker *et al.*, the year-of-onset distribution (5 before 1955, 12 during 1955–1959, 20 during 1960–1964, and 3 during 1965–1971) suggests a peaking of incidence about 15 yr after the bombing, with a sharp subsidence thereafter. In view of the pattern characterizing the ascertainment effort, however, it would be unwise to conclude that incidence has declined in this fashion.

Children Potentially Exposed to Fallout Radioiodine

Two groups of children—one group of 2,691 residing in a relatively high-fallout area of Utah and Nevada, and another group of 2,140 in a minimal-fallout area of Arizona during their infancy and early childhood—were compared by Rallison and co-workers[41,42] for evidence of thyroid disease. Benign neoplasms were observed in 6 exposed and 10 nonexposed children. Two carcinomas were found, but only in the nonexposed children, 15–20 yr after the fallout period. The estimated radiation dose, primarily from iodine-131, was approximately 120 rads to the exposed group (C. Mays, personal communication). An average radiation dose of 18 rads quoted by the authors is misleading, in that it is based on the sum of the exposed and nonexposed populations. Radiation doses may actually have been higher, ranging from 30 to 240 rads (Mays, personal communication).

DISCUSSION OF IRRADIATED POPULATIONS

Ranges of external radiation dose of 6.5–1,500 rads have been associated with the induction of thyroid carcinoma, as noted. The data presented may suggest that thyroid-carcinoma induction by external photon irradiation at high dose rate is a nonthreshold, linear phenomenon. Maxon has plotted the estimated dose response for this range of observed doses and has found an apparent linear dose-incidence response.[30] Such amounts of radiation may well be in the range of some currently used diagnostic radiographic studies.

Minimal evidence is available to establish a relationship between induction of thyroid carcinoma and beta particulate radiation. What little evidence is available from children treated with iodine-131 for hyperthyroidism does not demonstrate the carcinogenic effect seen with *external radiation.*[44]

Observations of the Marshallese are difficult to analyze, because their radiation exposures were to a mixture of high-dose-rate external and internal gamma photons, as well as to beta radiation. A mixture of fission iodine radioisotopes is considerably different from iodine-131 in radiation characteristics. The radiation-induced thyroid disease in these populations, particularly thyroid carcinoma, is probably more analogous to the results of external radiation than to the results of the therapeutic use of iodine-131. Indeed, tumor incidence in these populations seems to approach that in those exposed to external photon irradiation.

THYROID ADENOMAS

Radiation-induced thyroid adenomas have been observed in all the populations in whom thyroid carcinoma has been induced, where data sources permit the detection of benign disease. Thyroid adenoma may have a higher incidence than thyroid carcinoma with smaller amounts of radiation. In all the population series studied, the relative increase in thyroid nodularity was significantly higher than that in thyroid carcinoma—12 cases per 10^6 PY per rad, or about 3 times that of thyroid carcinoma.[30]

MODIFYING FACTORS IN RADIATION-ASSOCIATED THYROID NEOPLASIA

Many of the classically known modifying factors in radiation effect have been identified in the study of radiation-induced thyroid disease. The influence of some of these factors has been more clearly delineated since the 1972 BEIR report.

Age may be a weak factor in influencing the effect of radiation on the thyroid, at least with external high-dose-rate irradiation. This is particularly true of thyroid neoplasia for both malignant and benign lesions. There may be some increased risk under the age of 20, but this suggestion is based on minimal data (G. W. Beebe, personal communication). The body of data available on neoplastic induction in the thyroid of adults is very small. Unfortunately, extensive therapeutic use, particularly involving the head and neck region early in childhood, has resulted in numerous reports of the incontrovertible association of neoplastic induction in the immature thyroid. Hypothyroidism and thyroiditis may be induced at any age, given a large enough absorbed radiation dose. The "apparent" inverse relation of radiation-induced thyroid neoplasia with age is probably mistakenly assumed, inasmuch as the nonmalignant conditions for which medical irradiation was used occurred primarily in infants and children.

There is a greater predominance of thyroid neoplasia in females, as is the case with almost all thyroid disease.[18] There may be as much as a fourfold difference in induction of thyroid neoplasia between sexes. This is probably related to the fluctuating hormonal status in females, with significantly greater variations in the pituitary-thyroid axis and in secretion of thyroid-stimulating hormone than in males. Other hormonal interdependences of the thyroid in the endocrine system may also be involved. Animal data clearly indicate the increased efficacy of induction of thyroid neoplasia in the presence of thyroid stimulation increased by TSH.[4]

Some questions have been raised as to the dependence of thyroid neoplasia induction on ethnic background—specifically, whether there is an increase in susceptibility to induction in those of Jewish descent, particularly females. Reevaluation of the Hempelmann data by R. E. Shore (personal communication) has confirmed the increased risk in the Jewish component of the population study. The Michael Reese series[46] of patients with head and neck irradiation and the Israeli tinea capitis series[34] were predominantly Jewish. Thus, hereditary-familial background may be a moderating factor of some significance.

Many, if not most, series have suggested that there is a peak incidence of thyroid carcinoma induction 15–25 yr after irradiation. This is observed in the recent data from the Michael Reese series,[46] suggesting a peak incidence at approximately 25 yr, and in the more recent reports from Japan, with a peak incidence at about 15 yr.[36] Other series have reported a peak incidence at approximately 20 yr, which appears reasonable.[46] In contrast, a more recent evaluation of the data from Hempelmann's series suggested no definite peak incidence, but a continuingly increasing incidence after 35 yr, although the case numbers are small.[26] An artifact in the cumulative incidence due to changing amounts of radiation given in the last few years of treatment of the group may obscure a peaking of incidence. It is not clear, however, to what extent latency is associated with the amount of radiation received. Another peak-incidence artifact could be introduced by variation in the intensity of patient followup.

At this point, the effect of fractionation of irradiation on the induction of thyroid neoplasia is not established. This influence is extremely difficult to evaluate, because detailed data are not available on fractionation in all series. Where such data are available, particularly in the Hempelmann study,[26] a difference in effect with fractionation is not evident.

Related to fractionation is dose rate, which may be a significant factor accounting in part for the difference observed between external radiation and internally absorbed, longer-lived beta radiation. Because the bulk of the experience with the latter stems from iodine-131, which has a relatively long half-life, the low dose rate may lead to a significant cellular recovery rate. In the special instances of the Marshallese, where radioisotopes of much shorter half-life may have contributed to a large fraction of the dose, the results of internal irradiation are much closer to those of external irradiation.

A further influence related to dose rate and fractionation is the degree of homogeneity of delivery of radiation. It is most likely that the delivery of external radiation is considerably more homogeneous than that of absorbed beta particles in the thyroid gland. Furthermore, most of the radioiodine incorporated into thyroid hormone resides in colloid of the follicles of approximately 300 μm, delivering variable beta radiation to the cellular component of the follicle. A considerable amount of the energy deposition is inconsequential, because it is deposited in the biologically unimportant colloid within the follicles. These factors probably are mostly responsible for the marked differences between the results of external photon irradiation and of internal particulate irradiation.

RISK ESTIMATES

Risks are derived from the numerous reported series in the literature.[5,27,34,36,43,46] Risks are given for incidence, and not mortality, because the mortality rate from thyroid cancer is extremely low. Inasmuch as insufficient data are available, it is not possible to subtract the minimal-latent-period years from the years at risk in all these reports. These results, of course, must be qualified by the conditions that limited the studies. Risk estimates are given for the various types of radiation-induced thyroid disease in adults and children where the data are available. The risk estimates continue to approximate four cases of thyroid malignancy per 10^6

PY per rad for doses up to 1,000 rads and perhaps down to 6.5 rads. For benign thyroid adenoma or nodule induction, this figure appears to be approximately 12 cases per 10^6 PY per rad. In the more recent series, it appears that the relative risk of development of thyroid carcinoma in persons with radiation-induced thyroid nodular disease is approximately twice that in persons with spontaneously occurring nodular disease—i.e., a thyroid nodule in an irradiated person is twice as likely to be carcinoma as is the usual clinical nodule.[46] It must be recognized that both these absolute- and relative-risk factors are contingent on the number of person-years in each series and also depend on whether a peak or equilibrium incidence is reached in each population. If a peak or equilibrium incidence is not reached, then, with increasing numbers of years of observation, the tumor incidence would be expected to continue to increase in the population. However, if a peak or wave of incidence is experienced, as may be the case in the Japanese and Michael Reese populations, then, with increasing numbers of years of observation, the incidence would actually be reduced. Thus, final absolute-risk factors will be available only when the entirety of an irradiated population has been observed for the total length of its life span. It would therefore be expected that the absolute- and relative-risk factors will continue to be modified through the years, and any present estimates are necessarily tentative.

SUMMARY

The effect of thyroid irradiation is primarily an increase in the incidence, and not the mortality, of thyroid neoplasms. The malignancies induced by radiation—namely, of the papillary-follicular type—are usually associated with a normal life span. Indeed, there appears to be a lack of the lethal form of thyroid malignancy—the anaplastic type—in the irradiated populations reported thus far. Furthermore, a distinction must be made as to the types of observed tumors, because it has recently been recognized that the occult type may be of little significance and may account for one-third to one-half the incidence of thyroid carcinoma reported in various irradiated populations. A minimal latent period of 10 yr seems to be reasonable, paralleling other radiation-induced solid tumors. A peak incidence perhaps 20 yr after exposure is suggested by some studies. A consistent three-fold increase in incidence is seen in women, compared with men. Jewish ethnic background may predispose to a higher incidence of development of thyroid cancer. There are no significant data to substantiate an age-at-irradiation effect, and the best estimate of risk for all ages appears to be approximately four carcinomas per 10^6 PY per rad, which includes incidence of occult carcinomas in some series. Benign adenomas are also induced by radiation, with an absolute risk of 12 adenomas per 10^6 PY per rad.

REFERENCES

1. Albert, R. E., and A. R. Omran. Follow-up study of patients treated by x-ray epilation for tinea capitis. Arch. Environ. Health 17:899–918, 1968.
2. Anspaugh, L. R. Special problems of thyroid dosimetry: Considerations of I^{131} dose as a function of gross size and inhomogeneous distribution. UCRL-12492. Biology and Medicine, UC-48. TID-4500. 39th Ed. 1965.
3. Arnold, J., S. Pinsky, U. Ryo, *et al.* ^{99m}Tc-Pertechnetate thyroid scintigraphy in patients predisposed to thyroid neoplasms by prior radiotherapy to the head and neck. Radiology 115:653–667, 1975.
4. Christov, K. Thyroid cell proliferation in rats and induction of tumors by x-rays. Cancer Res. 35:1256–1262, 1975.
5. Conard, R. A. Summary of thyroid findings in Marshallese 22 years after exposure to radioactive fallout, pp. 241–257. In L. J. DeGroot, L. A. Frohman, E. L. Kaplan, and S. R. Refetoff, Eds. Radiation-Associated Thyroid Carcinoma. New York: Grune & Stratton, 1977.
6. Conard, R. A., B. M. Dobyns, and W. W. Sutow. Thyroid neoplasia as late effect of exposure to radioactive iodine in fallout. JAMA 214:316–324, 1970.
7. Crile, G., Jr. Carcinoma of the thyroid after radiation to the neck. Surg. Gynecol. Obstet. 141:602–603, 1975.
8. DeGroot, L. J., L. A. Frohman, E. L. Kaplan, and S. R. Refetoff, Eds. Radiation-Associated Thyroid Carcinoma. New York: Grune & Stratton, 1977. 539 pp.
9. DeGroot, L. J., and E. Paloyan. Thyroid carcinoma and radiation. A Chicago endemic. JAMA 225:487–491, 1973.
10. De Ruiter, J., C. F. Hollander, G. A. Boorman, G. Hennemann, R. Docter, and L. M. Van Putten. Comparison of carcinogenicity of I 131 and I 125 in thyroid gland of the rat, pp. 21–33. In Biological and Environmental Effects of Low-Level Radiation. Vol. 1. Vienna: International Atomic Energy Agency, 1975.
11. Dobyns, B. M. Radiation hazard. Experience with therapeutic and diagnostic ^{131}I, pp. 459–483. In L. J. DeGroot, L. A. Frohman, E. L. Kaplan, and S. R. Refetoff, Eds. Radiation-Associated Thyroid Carcinoma. New York: Grune & Stratton, 1977.
12. Dobyns, B. M., G. E. Sheline, J. B. Workman, *et al.* Malignant and benign neoplasms of the thyroid in patients treated for hyperthyroidism. A report of the Cooperative Thyrotoxicosis Therapy Follow-up Study. J. Clin. Endocrinol. Metab. 38:976–998, 1974.
13. Doniach, I. Carcinogenic effect of 100, 250, and 500 rad x-rays on the rat thyroid gland. Brit. J. Cancer 30:487–495, 1974.
14. Doniach, I. Experimental evidence of etiology of thyroid cancer. Proc. R. Soc. Med. 67:1103, 1974.
15. Favus, M. J., A. B. Schneider, M. E. Stachura, *et al.* Thyroid cancer occurring as a late consequence of head-and-neck irradiation. N. Engl. J. Med. 294:1019–1025, 1976.
16. Foster, R. S., Jr. Thyroid irradiation and carcinogenesis. Review with assessment of clinical implications. Amer. J. Surg. 130:608–611, 1975.

17. Frohman, L. A. Irradiation and thyroid carcinoma: Legacy and controversy. J. Chronic Dis. 29:609-612, 1976.
18. Frohman, L. A., A. B. Schneider, M. J. Favus, M. E. Stachura, J. Arnold, and M. Arnold. Risk factors associated with the development of thyroid carcinoma and of nodular thyroid disease following head and neck irradiation, pp. 231-240. In L. J. DeGroot, L. A. Frohman, E. L. Kaplan, and S. R. Refetoff, Eds. Radiation-Associated Thyroid Carcinoma. New York: Grune & Stratton, 1977.
19. Frohman, L. A., A. B. Schneider, M. J. Favus, M. E. Stachura, J. Arnold, and M. Arnold. Thyroid carcinoma after head and neck irradiation. Evaluation of 1476 patients, pp. 5-15. In L. J. DeGroot, L. A. Frohman, E. L. Kaplan, and S. R. Refetoff, Eds. Radiation-Associated Thyroid Carcinoma. New York: Grune & Stratton, 1977.
20. Glennon, J. A., E. S. Gordon, and C. T. Sawin. Hypothyroidism after low dose I-131 treatment of hyperthyroidism. Ann. Int. Med. 76:721-723, 1972.
21. Goldschmidt, H. Dermatologic radiotherapy and thyroid cancer. Arch. Dermatol. 113:362-364, 1977.
22. Greenspan, F. S. Radiation exposure and thyroid cancer. JAMA 237:2089-2091, 1977.
23. Harley, N. H., R. E. Albert, R. E. Shore, et al. Follow-up of patients treated by x-ray epilation for tinea capitis. Estimate of the dose to the thyroid and pituitary glands and other structures of the head and neck. Phys. Med. Biol. 21:631-642, 1976.
24. Hazen, R. W., J. W. Pifer, E. T. Toyooka, et al. Neoplasms following irradiation of the head. Cancer Res. 26:305-311, 1966.
25. Hempelmann, L. H. Risk of thyroid neoplasms after irradiation in childhood. Science 160:159-163, 1968.
26. Hempelmann, L. H. Thyroid neoplasms following irradiation in infancy, pp. 221-229. In L. J. DeGroot, L. A. Frohman, E. L. Kaplan, and S. R. Refetoff, Eds. Radiation-Associated Thyroid Carcinoma. New York: Grune & Stratton, 1977.
27. Hempelmann, L. H., W. J. Hall, M. Phillips, et al. Neoplasms in persons treated with x-rays in infancy. Fourth survey in 20 years. J. Natl. Cancer Inst. 55:519-530, 1975.
28. Hempelmann, L. H., J. W. Pifer, G. J. Burke, et al. Neoplasms in persons treated with x rays in infancy for thymic enlargement. A report of the Third Follow-up Survey. J. Natl. Cancer Inst. 38:317-341, 1967.
29. Kerr, G. D. Organ Dose Estimates for the Japanese Atomic Bomb Survivors. Oak Ridge National Laboratory Report 5436, 1978. Springfield, Va.: National Technical Information Service. 46 pp.
30. Maxon, H. R., S. R. Thomas, E. L. Saenger, C. R. Buncher, and J. G. Kereiakes. Ionizing irradiation and the induction of clinically significant disease in the human thyroid gland. Amer. J. Med. 63:967-978, 1977.
31. Mazzaferri, E. L., R. L. Young, J. E. Oertel, W. T. Kemmerer, and C. P. Page. Papillary thyroid carcinoma. The impact of therapy in 576 patients. Medicine 56:171-196, 1977.
32. Modan, B., D. Baidatz, H. Mart, et al. Radiation-induced head and neck tumours. Lancet 1:277-279, 1974.
33. Modan, B., E. Ron, and A. Werner. Thyroid cancer following scalp irradiation. Radiology 123:741-744, 1977.
34. Modan, B., E. Ron, and A. Werner. Thyroid neoplasms in a population irradiated for scalp tinea in childhood, pp. 449-457. In L. J. DeGroot, L. A. Frohman, E. L. Kaplan, and S. R. Refetoff, Eds. Radiation-Associated Thyroid Carcinoma. New York: Grune & Stratton, 1977.
35. National Research Council, Advisory Committee on the Biological Effects of Ionizing

Radiation. The Effects on Populations of Exposure to Low Levels of Ionizing Radiation. Washington, D.C.: National Academy of Sciences, 1972.

36. Parker, L. N., J. L. Belsky, T. Yamamoto, *et al.* Thyroid Carcinoma Diagnosed between 13 and 26 Years after Exposure to Atomic Radiation. A Study of the ABCC-JNIH Adult Health Study Population, Hiroshima and Nagasaki 1958–71. Atomic Bomb Casualty Commission Technical Report TR 5-73. Hiroshima: Atomic Bomb Casualty Commission, 1973.

37. Parker, L., J. L. Belsky, T. Yamamoto, S. Kawamoto, and R. J. Keehn. Thyroid carcinoma after exposure to atomic radiation. Ann. Int. Med. 80:600–604, 1974.

38. Pifer, J. W., L. H. Hempelmann, H. J. Dodge, *et al.* Neoplasms in the Ann Arbor series of thymus-irradiated children. Amer. Roentgenol. Radium Ther. Nucl. Med. 53:13–18, 1968.

39. Pifer, J. W., E. T. Toyooka, R. W. Murray, *et al.* Neoplasms in children treated with x rays for thymic enlargement. I. Neoplasms and mortality. J. Natl. Cancer Inst. 31:1333–1356, 1963.

40. Pincus, R. A., S. Reichlin, and L. H. Hempelmann. Thyroid abnormalities after radiation exposure in infancy. Ann. Int. Med. 66:1154–1164, 1967.

41. Rallison, M. L., B. M. Dobyns, F. R. Keating, Jr., J. E. Rall, and F. H. Tyler. Thyroid disease in children. A survey of subjects potentially exposed to fallout radiation. Amer. J. Med. 56:457–463, 1974.

42. Rallison, M. L., B. M. Dobyns, F. R. Keating, Jr., J. E. Rall, and F. H. Tyler. Thyroid nodularity in children. JAMA 233:1069–1072, 1975.

43. Refetoff, S., J. Harrison, B. T. Karanfilski, *et al.* Continuing occurrence of thyroid carcinoma after irradiation to the neck in infancy and childhood. N. Engl. J. Med. 292:171–175, 1975.

44. Safa, A. M., O. P. Schumacher, and A. Rodriguez-Antunez. Long-term follow-up results in children and adolescents treated with radioactive iodine (^{131}I) for hyperthyroidism. N. Engl. J. Med. 292:167–171, 1975.

45. Sampson, R. J., C. R. Key, C. R. Buncher, *et al.* Thyroid carcinoma in Hiroshima and Nagasaki. I. Prevalence of thyroid carcinoma at autopsy. JAMA 209:65–70, 1969.

46. Schneider, A. B., M. J. Favus, M. E. Stachura, J. Arnold, M. J. Arnold, and L. A. Frohman. Incidence, prevalence and characteristics of radiation-induced thyroid tumors. Amer. J. Med. 64:243–252, 1978.

47. Shellabarger, C. J. Radiation carcinogenesis. Cancer 37:1090–1096, 1976.

48. Shore, R. E., R. E. Albert, and B. S. Pasternak. Follow-up study of patients treated by x-ray epilation for tinea capitis. Arch. Environ. Health 31:17–28, 1976.

49. Socolow, E. L., A. Hashizume, S. Neriishi, and R. Niitani. Thyroid carcinoma in man after exposure to ionizing radiation. A summary of the findings in Hiroshima and Nagasaki. N. Engl. J. Med. 268:406–410, 1963.

50. U.S. Department of Health, Education, and Welfare. National Cancer Institute, Division of Cancer Control and Rehabilitation. Information for Physicians: Irradiation-Related Thyroid Cancer. DHEW Publ. No. (NIH) 77-1120, 1977.

51. Webber, B. M. Radiation therapy for pertussis. A possible etiologic factor in thyroid carcinoma. Ann. Int. Med. 86:449–450, 1977.

52. Werner, A., B. Modan, and E. Ron. Thyroid Dosimetry Re-evaluation after Treatment for Scalp Tinea. Presented at 14th International Conference on Medical Physics, July 1976, Ottawa.

53. Wood, J. W., H. Tamagaki, S. Neriishi, *et al.* Thyroid carcinoma in atomic bomb survivors, Hiroshima and Nagasaki. Amer. J. Epidemiol. 89:4–14, 1969.

LUNG

Lung cancer—or, more properly, bronchial cancer—was the first internal cancer of which exposure to ionizing radiation was implicated as a cause (in Bohemian miners). As followup investigations of radiation-exposed groups have been extended, bronchial cancer has emerged as one of the most important radiation-induced cancers. Since the 1972 BEIR report,[31] our understanding of radiation induction of bronchial cancer in man and lung tumors in animals has advanced considerably. Moreover, further information about radiation dosimetry related to lung cancer associated with inhalation of radionuclides has become available. There are also new experimental and epidemiologic data on the role of cigarette-smoking in relation to radiation exposure in lung-cancer induction.

EXPERIMENTAL STUDIES

The experimental production of cancers of the respiratory tract in animals by ionizing radiation has recently been reviewed;[22,32] this brief summary stresses evidence from the animal data most pertinent to human experience. In experimental studies of lung cancer, the origin of tumors in rodents and dogs commonly is found to be bronchoalveolar; they arise from regions adjacent to the respiratory bronchioles. In contrast, human cancers induced by cigarette-smoking or exposure to environmental agents nearly always arise from epithelium in proximal regions of the bronchial tree (down to the first few generations of branching). This difference in site of origin has raised important questions about the applicability of animal data to the human disease.

Animals exposed to aerosols of beta- or gamma-emitting isotopes or to x rays develop primarily bronchoalveolar tumors, but may have tracheal or bronchial tumors;[7,35] the radiation exposure of all the tissues is generally uniform. Bronchoalveolar tumors are also characteristic in animals that have inhaled alpha-emitting elements.[32] Beta- or gamma-emitting nuclides implanted in the upper airways give rise to tumors near the site of implantation.[13,24]

Kennedy and co-workers observed in hamsters that bronchoalveolar tumors induced by intratracheal instillation of polonium-210 arose from the Clara cells in the terminal regions of bronchial epithelium near the respiratory bronchioles.[23] These tumors occurred whether the polonium was absorbed on iron oxide particles or in solution.[26] At equivalent mean lung doses, the tumor yield was similar in the two cases. For particle-absorbed polonium, aggregation of activity occurred in the terminal bronchial region where the tumors arose; for free polonium, there was rather uniform distribution in the terminal bronchiolar and alveolar tissues. Soon after instillation of polonium-210, the free polonium could be found in the Clara cells.[21] Little and O'Toole[29] demonstrated that tracheal instillation of benzo[a]pyrene in hamsters induced primarily epidermoid cancers in the trachea and large bronchi and concluded that the distribution and kinetics of the carcinogenic agent in pulmonary tissues are important determinants of the site of cancer induction.

Hamsters were given zirconium oxide (ZrO_2) microspheres containing plutonium-238 or plutonium-239 intravenously; the microspheres lodged in the pulmonary capillaries and produced only adenomatoid changes in the bronchiolar region in a few animals, but few frank tumors.[43] This change was not dose-related, nor were the few cancers that occurred. Intratracheally administered plutonium-239 microspheres, however, have been reported to induce "lung cancers."[22] These data suggest that the response depends on the extent to which alpha particles reach cells that are sensitive to cancer induction.

Lung cancer produced in animals by inhalation of radon daughters has now provided an experimental model of cancer production from this source in man. Chameaud et al.[8] exposed rats to radon daughters at various concentrations; the cumulative dose was a function mainly of the number of exposures. Both bronchogenic epidermoid and bronchoalveolar cancers occurred at a dose-related frequency, but the cell type was independent of total dose. In 26 rats given 300–500 working-level months* of exposure (the lowest-dose group), one bronchogenic cancer and one bronchoalveolar cancer were found. Change of the exposure regimen from 2,500 WL for 5 h/d, or about 74 WLM/d, for 20–60 d (cumulative doses, 1,500–4,500 WLM), to 3,000 WL for 16 h/d, or about 282 WLM/d for 7–20 d (cumulative doses, 2,000–5,500 WLM), yielded a higher cancer percentage (at equivalent doses) from the more protracted dose. The occurrence of induced tumors as a function of cumulative radon-daughter exposure was in good agreement with similar data in man, although the dose rate was much higher in the animal experiments.

Filipy and co-workers, in a long-term study,[12] exposed three groups of 20 beagles to radon daughters (600 WL) and uranium-ore dust (15 mg/m³), 4 h/d and 5 d/wk, with and without concomitant smoking of cigarettes (10 cigarettes/d, 5 d/wk). One other group was exposed to smoking alone.

---

*The "working level" (WL) is a concentration of radon daughters equivalent to equilibrium with 100 pCi of radon per liter of air. It is defined as the activity in air that gives 1.3×10^5 MeV of alpha radiation per liter from ultimate decay of the short-lived daughters. The working-level month (WLM) is defined as exposure at 1 WL for 170 h.

Thus far, after exposure periods of 4–5 yr and cumulative doses of more than 11,000 WLM, seven dogs among the groups exposed to radon and uranium have developed lung cancer: three bronchoalveolar cancers (without cigarette-smoking), three epidermoid cancers (smoking status not stated, but one evidently nonsmoking), and one fibrosarcoma of the peripheral lung (nonsmoking). In animals of these groups that were sacrificed after 40 mo or more, extensive adenomatosis was found at the bronchoalveolar junction, as well as granulomas and bullous emphysema.[47] Three dogs have had squamous carcinomas of the nasal mucosa (two nonsmoking and one smoking). No cancer at any site has appeared among the smoking dogs not exposed to radon and uranium, and at 4–5 yr pulmonary changes have been minimal in this group.

The Hanford group has exposed hamsters and rats to radon daughters (900 and 1,200 WL) with and without uranium-ore dust (15 mg/m^3) for 5 mo, with total cumulative doses of about 10,000–12,000 WLM.[14,48] Squamous metaplasia of the nasopharynx was a very common observation in both hamsters and rats exposed to radon daughters only, and there were a few squamous-cell cancers of the nasal epithelium. The groups exposed to dust and radon daughters had changes in the deep lung, instead of nasal mucosal metaplasia. Hamsters thus exposed had no lung neoplasia, but did have fibrosis, emphysema, and adenomatosis. However, rats exposed to radon daughters and dust had a high proportion of bronchoalveolar squamous carcinomas and occasional adenocarcinomas. It is evident that hamsters were more resistant than rats to development of lower respiratory tract neoplasia. Moreover, the contrast in results between those exposed and not exposed to dust indicates that radiation exposure without dust was chiefly to the upper airway, presumably owing to absorption there of the free-ion fraction of radon daughters.

Little and colleagues have demonstrated bronchoalveolar cancer induction from polonium-210 instilled intratracheally in hamsters at mean doses as low as 15 rads.[27] A single dose of 100 nCi in 0.2 ml of saline produced tumors in fewer animals (two of 31) than a slightly lower cumulative dose given in 15 instillations each of 5.6 nCi each (14 of 59 animals).[25] A most important observation in these experiments has been that instillation of salt solution alone increased the effects of polonium instillation. For example, if, under the same conditions of exposure, the single instillation (100 nCi) was followed by 14 weekly intratracheal instillations of 0.2 ml of normal saline, the proportion of animals with tumors was approximately the same as in the group given 15 instillations of 5.6 nCi. A number of such experiments have shown the enhancement of lung-cancer production in hamsters from alpha-radiation exposure by this potential stimulus to bronchial cell proliferation.[28] Saline alone was as effective in increasing the carcinogenic action of polonium as was the instillation of benzo[a]pyrene in saline; thus, the increase in cancer production from polonium instillation by benzo[a]pyrene can be ascribed to the effects of the saline vehicle.

These results may explain the relative resistance to lung-cancer induction observed in hamsters, compared with other rodents.[41] Hamsters are resistant to chronic respiratory infection,[11,34] whereas rats and mice commonly develop chronic peribronchial inflammation, even if they are bred to minimize maternally transmitted infection during weaning. An endemic factor associated with a bronchial cell-proliferation stimulus may render the epithelium more susceptible to the development of cancer in response to an initiating agent, in this case alpha radiation. In man, factors acting as proliferative stimuli to the respiratory epithelium are more widespread than in animals maintained under controlled laboratory environments.

These considerations may be important in the relationship of cigarette-smoking to radiation-induced cancers. The results in the Hanford beagle experiments do not suggest, at least so far, a marked increase in radon-daughter induction of bronchial cancers when regular cigarette-smoking is added. In rats exposed to plutonium or americium aerosols by inhalation,[30] exposure to cigarette smoke added to exposure to americium-241 aerosol substantially increased the incidence of pulmonary and extrapulmonary cancers and reduced the latent period for their appearance.[33]

Patrick and Stirling[36] have shown in rats that about 1% of the retained activity of a radiolabeled aerosol given intratracheally could be found in the bronchial wall; this activity did not clear rapidly, as it would if it were associated with the mucociliary blanket, but remained for 30 d or more. In the case of nonpenetrating radiation, such as alpha radiation, if particles were retained for relatively long times at this location adjacent to the proliferative epithelial cells, the local dose to this cell population could differ substantially from that inferred from the average lung concentration. Radford and Martell[37] estimated from preliminary data obtained from human lung tissues that about 1% of the lead-210 activity inhaled in cigarette smoke is found in bronchial epithelium—in good agreement with the experimental results of Patrick and Stirling. The residence time of the insoluble particles containing lead-210 in this location may be as long as several months. These results suggest that classical models of retention of radionuclides in the human respiratory tract may require some modification to take account of local exposure of insoluble particles resident near the proliferating-cell layer of the bronchial epithelium.

In summary, results obtained in experimental animals support the following general conclusions:

• Respiratory tract tumors develop in animals exposed to radiation at sites where the local radiation exposure is greatest.

• Bronchial and nasal sinus tumors have been produced in animals by exposure to radon and its daughters.

• The effect of cigarette-smoking on the development of bronchial cancers in the latter experiments remains equivocal.

• The sensitivity of the respiratory tract of animals to cancer induction by radiation may be increased by irritant or other proliferative stimuli given after the radiation exposure.

• The bronchial tissue in the lungs is itself a separate compartment whose uptake and release of inhaled materials may play an important role in diseases, such as bronchogenic carcinoma, arising in the bronchial epithelium.

These results are significant new additions to our ability to relate human bronchial cancer to the experimental models that have been studied.

HUMAN STUDIES

Substantial progress since the 1972 BEIR report has been due to longer followup observations of several human populations. The latent period for lung-cancer induction by radiation is relatively long, and thus many more cases are being added with further observation.

Patients with Radiotherapy for Ankylosing Spondylitis

A followup of over 14,000 patients, up to January 1, 1970, has been reported.[10,44] Treatments were given in 1935–1954; currently, the study includes the patients who received one course of x-ray therapy, originally about half the total group, plus the experience of the remainder up to a second treatment. On the average, the patients received about 10 treatments over 4–6 wk, directed to the whole spine and sacroiliac area or to more restricted fields. Of these patients, 83% were men; so far, there is no analysis of sex-specific cancer risks in this group.

Of the patients followed to 1970, 124 had died of lung cancer, compared with 87.3 expected from age-, sex-, and year-specific rates for England and Wales. This difference is highly significant statistically. Analysis of lung-cancer deaths by time since x-ray treatment showed that for 0–5 yr there were 23 observed versus 17.8 expected; this excess generally occurred within a year after treatment. Smith and Doll[44] attributed this "early" excess of deaths to lung cancer that existed at the time of treatment, with metastatic disease attributed to reactivated spondylitis. For 6 yr or more after treatment, there were 101 observed lung-cancer deaths versus 69.5 expected; this is also a highly significant difference. Because there was no significant excess of cancers of the heavily irradiated sites, excluding leukemia, 6–8 yr after treatment, the excess can be considered to have arisen 9 yr or more after treatment. Smoking histories were not available for these patients; for purposes of comparison, it was assumed that the smoking experience of these patients did not differ from that of the general population (for which the expected rates were calculated).

The mean radiation dose to the bronchial tree has been estimated from exposure data (see Chapter III). From the analysis presented we estimate that the total bronchial dose may have been about 197 rads in this treatment group, with each exposure yielding an average bronchial dose of about 20 rads. On this basis, the absolute-risk estimate is 2.8 lung-cancer deaths per 10^6 PY per rad for the period 9 yr or more after treatment. At the time of the last followup, these patients had an average age of about 55 yr. A preliminary analysis (Radford, unpublished data) of those who died of lung cancer 9 yr or more after irradiation indicates that there was only a slight effect of age at exposure on the average latent period from x-ray treatment to death. This observation must be considered tentative until analysis is available of the distribution of the expected lung-cancer deaths in time after exposure.

Japanese Atomic-Bomb Survivors

Results of the mortality followup of survivors in Hiroshima and Nagasaki are now complete through 1974.[5] Autopsy studies indicate that lung cancer was underdiagnosed on death certificates by about one-third;[46] thus, mortality data based on Japanese death certificates seriously underestimate the risk of this kind of cancer. The types of lung tumors observed in the bomb survivors have been investigated by Cihak *et al.*[9] Only small-cell anaplastic cancers were observed significantly more frequently in the survivors than in nonirradiated persons; but, from the small numbers reported, one cannot exclude a general effect involving all cell types.

Through 1974, the Life Span Study (LSS) yielded little or no evidence of an excess of lung cancers in Nagasaki, except possibly in the highest dose groups. In contrast, a significant trend of increasing lung-cancer deaths per unit population at risk with increasing dose ($p = 0.002$) has been observed in Hiroshima. Absolute-risk estimates are 0.54 (0.22 to 0.86*) excess death per million persons per year per rad (kerma) for Hiroshima and 0.12 (−0.23 to 0.47) for Nagasaki. It is important to note that, when the LSS death-certificate data are augmented by information from the

---

*Numbers in parentheses are 90% confidence intervals.

Tumor Registry in each city for the period 1959–1970, the rate for Nagasaki is significantly above zero, the difference between Nagasaki and Hiroshima largely disappears, and the absolute risks are higher than for the death-certificate data[5]—1.06 ± 0.41 for Hiroshima and 1.02 ± 0.41 for Nagasaki.

We know that deaths certified as resulting from lung cancer are about 83% accurate according to autopsy data, but that 45% of the lung cancers seen at autopsy have not been certified as lung cancer.[46] These two sources of error may be sufficient, in the Nagasaki mortality results, to dilute a real effect that is better demonstrated in Tumor-Registry data. Ascertainment of cancer incidence in the Nagasaki Tumor Registry is believed to be good, and incidence data are less subject to errors of misdiagnosis. The fact that the association between lung cancer and radiation dose among the atomic-bomb survivors was not demonstrated until 1967 is an argument against the likelihood of a dose-related reporting bias in the Tumor-Registry data. For lung cancer, the Tumor-Registry data do not indicate a marked difference in risk per rad for the two cities, but the Tumor-Registry data are believed to be more nearly complete in Nagasaki than in Hiroshima.

The induction of lung cancer depends heavily on age at exposure and duration of observation. Those who were over 50 yr old at the time of the bombing had a mortality excess above expectation beginning 10 yr after the bombing; those who were 20–34 yr old are just beginning to show an excess above expectation 33 yr later. Among those who were 35–49 yr old, the onset of lung-cancer excess occurred about 15 yr later. Beebe *et al.*[5] calculated the following age-specific mortality risks for cancer of the trachea, bronchus, and lung for the period 1950–1974 for the two cities combined, and with no adjustment for incomplete reporting:

Age in 1945, yr	No. Deaths, 100+ Rad Group	Relative Risk, 100+ vs. 0–9 Rads	Absolute Risk— Excess Deaths per 10^6 PY per Rad (Kerma)
0–9	0	—	—
10–19	1	—	—
20–34	4	—	0.19
35–49	19	1.45	0.84
>49	16	2.54	1.91
ALL AGES	40	1.79	0.35

If one takes into account the marked underdiagnosis of lung cancer in Japan, converts kerma dose to absorbed dose, and applies an arbitrary RBE of 5 for neutrons, absolute-risk estimates increase to 2–3 times the absolute-risk estimates given above, 1.25 for Hiroshima and 0.35 for

Nagasaki, for all ages combined. Age-specific estimates are not available separately for each city, unfortunately; but, for both cities combined, the regression estimates, corrected as above, are 2.1 excess deaths per million persons per year per rem for those aged 35–49 in 1945, and 4.9 for those aged 50 or older. If the coefficients had been calculated, not for 1950–1974, but for 1955–1974, to allow for a more reasonable latent period and to provide estimates more comparable with that for the ankylosing-spondylitis patients, and if the age-specific estimates were weighted by the age distribution for the ankylosing-spondylitis patients, one would obtain an estimate of 2.0. This is not significantly below the 2.8 for the ankylosing-spondylitis patients with a latent period of 9 yr. Application of an RBE of 10 for neutrons would increase this difference slightly.

Ishimaru *et al.*[18] examined the smoking histories of persons found to have lung cancer at autopsy. For the nonirradiated group (less than 1 rad), the lung-cancer risk of smokers was 6.2 times that of nonsmokers. For patients exposed to greater than 200 rads kerma, the relative risks were 8.6 and 3.0 for smokers and nonsmokers, respectively, compared with nonirradiated nonsmokers. These preliminary results are consistent with the conclusion that exposure to external radiation does not interact strongly with smoking in increasing lung-cancer risk; the separate risks from smoking and external radiation were nearly additive in this study.

The Japanese analysis is important also because it is the only one available that allows comparison of the risks in women and men. The radiation-exposed women had a somewhat greater relative risk of lung cancer than the men (2.33 and 1.57, respectively, for 100+ rads versus 0–9 rads), but the absolute risk was somewhat less in the women than in the men (0.28 and 0.43, respectively, excess death per million persons per year per rad kerma). The discrepancy between these two measures reflects the much lower natural lung-cancer risk in Japanese women than in men because of the sex difference in prevalence of cigarette-smoking. The relatively small and statistically insignificant difference in absolute risk between the two sexes is further indication that cigarette-smoking does not influence strongly the cancer-induction process related to exposure to external radiation.

Because excess lung-cancer risk continues beyond 25 yr after exposure and because the lung-cancer effect is only now being expressed in persons exposed before the age of 35, it is evident that future estimates are likely to be somewhat higher than those available now.

Miners Exposed to Radon Daughters in Underground Mines

Additional followup data are now available on several of the groups of miners discussed in the 1972 BEIR report and exposed occupationally to alpha radiation from short-lived radon daughters. The Czechoslovakian and U.S. uranium miners and Newfoundland fluorspar miners mentioned in BEIR I have additional followup data; there have also been new investigations of Canadian uranium miners and Swedish metal miners. In some of the studies, the concentrations of radon and daughters in the mines are now well established.* Results are for males only, and generally they are not analyzed according to cigarette-smoking experience. For all groups except the Swedish miners, smoking is common; e.g., about 70% of the U.S. and Czechoslovakian uranium miners have been smokers. This proportion was stated by the Czechoslovakian investigators to be fairly close to national rates.[42] Smoking has never been as prevalent in Sweden as in other western countries, and in the miners it has probably not exceeded 50%—not much greater than national statistics for the proportion who have ever smoked.[6]

Czechoslovakian Uranium Miners Ševc *et al.*[42] recently reported followup of a group of Czechoslovakian uranium miners who began underground mining in the period 1948–1952 (Group A). Lung cancer in these men has been investigated through December 31, 1973. The exact number of miners was not stated, but evidently was comparable with the number of U.S. uranium miners. More than 98% of these miners had had no previous hard-rock mining experience.

Over 120,000 radon-gas measurements were made in the mines. Before 1960, the degree of equilibrium of radon daughters was estimated from ventilation and other data; from 1960 on, some measurements of radon daughters were made directly. Thus, the mean WLM estimates of exposure to radon daughters in the mine atmosphere for the different dose groups have standard deviations of about 1%. A random sample of smoking histories in 700 miners indicated that the proportion of cigarette-smokers

---

*Because the polonium-214 alpha particle has high energy, enabling it to reach the basal-cell layer of the bronchi more readily than the polonium-218 alpha particle, the principal biologic effects of radon daughters in man are from the polonium-214 daughter. Thus, the WL as defined is not entirely satisfactory in characterizing health risks in all cases. The degree of equilibrium of lead-214 (RaB) and bismuth-214 (RaC) with polonium-218 (RaA) may vary, and thus the proportion of polonium-214 alpha decays to total alpha decays will be variable. In mine atmospheres, the extent of equilibrium will depend on the relative ventilation, and in practice the degree of disequilibrium does not greatly affect the ratio of polonium-214 to total alpha activity. In other atmospheres, however, such as homes or buildings, the degree of disequilibrium may be substantial. Another variable factor affecting the health significance of a given WL is the fraction of free daughter ions unattached to dust particles[16]—not a major problem in mines, but potentially important in relatively clean spaces, such as homes.

(70%) was the same as that in the general male population in Czechoslovakia.

The exposure in the Czechoslovakian mines was relatively slight: if the underground work experience was 20 yr or more and the average cumulative exposure was about 300 WLM, then the concentrations of radon daughters were about 1 WL—much lower than in the U.S. uranium mines before 1960.

An important observation in this group of miners was an assessment of excess lung cancers by age at which underground exposure began. Three groups were separated: miners who began work at less than 30 yr of age, those who began at age 30–39, and those who began at 40 or older. The excess risk per cumulative dose for the two lower age groups showed a reasonable straightline fit to the data; but for the oldest age group, the three highest dose groups (cumulative dose greater than 300 WLM) showed a relatively constant excess risk that was independent of dose.

The lung-cancer risk estimates were given by the authors simply as excess cases per 1,000 miners and included all years since the start of mining (on the average, about 23 yr). Obviously, the excess risk was zero for several of these years and then increased.

To convert the data to risk per person-year with the 10-yr latent period excluded, the years at risk are taken to be 13. Although some slight excess may have occurred earlier than 10 yr after mining began, it is apparent from other mine populations that the full lung-cancer excess is not reached until 15 yr or more from initial exposure; thus, exclusion of 10 yr on the assumption of zero risk may underestimate the risk slightly. This procedure permits comparison with the other studies in this report.

The authors' calculation of excess lung cancers per 1,000 miners per WLM is converted to excess risk per 10^6 PY per WLM by multiplying by 1,000/13, or 77. To eliminate the smaller effect of higher doses on cumulative risk per dose in the older miners, at doses greater than 300 WLM, risk estimates have been calculated only below that cumulative dose. Moreover, each of the two lowest dose groups (less than 100 WLM) were stated to have only half as many men as the higher dose groups; therefore, the data for less than 100 WLM have been combined into one group, and no weighting by dose category is required.

Thus adjusted, the total excess risk is found to be 19.0 excess cases per 10^6 PY per WLM. Precise correction of the published relative-risk estimates to eliminate the latent-period years is not possible; but, on the assumption that the expected lung-cancer deaths per year during the first 10 yr were one-third the expected during the succeeding years of followup, an approximate value is obtained of 1.8% excess lung-cancer risk per WLM over the period under study. This value indicates a doubling dose of about 56 WLM.

Similar calculation of the absolute risk per WLM for the three age subgroups and for exposures less than 300 WLM (from Ševc et al., [42] Table 2) gives 8.8 cases per 10^6 PY per WLM for the group that began mining before the age of 30, 13.3 cases per 10^6 per WLM for those who began at 30–39, and 46.7 cases per 10^6 PY per WLM for those who began at 40 or older. Although these risk estimates obviously are subject to the statistical uncertainties of the data presented, they show a marked effect of age at first exposure or of age at risk and are consistent in that regard with the results obtained in the Japanese bomb survivors.

Horáček et al. [17] studied the histologic type of bronchial cancers in 115 cases among these miners by comparing the frequency of cancer cell type (according to the WHO classification as modified by Yesner et al. [50]) with data from 326 nonminers matched for smoking experience. The results indicated that the frequencies of epidermoid and small-cell anaplastic cancers (WHO types 1 and 2) were about equally increased and dose-dependent, with only a small excess of adenocarcinomas (type 3). The excess of "other" cancers (types 4–6) may have been due to the inclusion of mixed epidermoid cancers and adenocarcinomas in this category. The authors concluded that their results confirmed those of Archer et al. [1] and indicate that radiation-induced cancers are not limited to the small-cell anaplastic types.

Of those 115 miners, five were nonsmokers and three were pipe-smokers; this indicates that a substantial excess of lung cancer has already begun to occur in the nonsmokers among these miners. The expected cases for comparison with the 115 miners would be 24.5. On the assumptions that the relative risk of lung cancer among Czechoslovakian smokers compared with nonsmokers is 10, that nonsmokers constitute 30% of the total miner population at risk, and that the age distributions of the smokers and nonsmokers are similar, the expected deaths for nonsmokers would be 1.0 and for smokers 23.5. On this basis, the relative risk among the smokers, 110/23.5 = 4.7, is approximately the same as that among the nonsmokers, 5/1.0 = 5.0. Because the latent period for lung-cancer induction in nonsmokers is longer than that for smokers,[2,4] with further followup the relative risk would be expected to rise more rapidly for nonsmokers than for smokers.

U.S. Uranium Miners   The group of men under study was identified by medical examinations carried out between 1950 and 1960 at the uranium mines in the Colorado Plateau region. All miners included had at least 1 mo of underground employment before December 31, 1963; 3,366 white and 780 nonwhite miners had adequate records of age, race, and mining experience and met the above criterion for inclusion. The nonwhite miners were nearly all American Indians. Followup of these miners began at the time of their first medical examination and has continued to the present. Data on mortality are complete through September 30, 1974 (V. E. Archer *et al.,* unpublished manuscript).

The following table gives the results of analysis for the white miners by cumulative dose categories. All lung-cancer deaths and person-years less than 10 yr after the start of mining have been excluded. Expected rates have been calculated from age- and year-specific rates for white males in Colorado, Utah, New Mexico, and Arizona, with an upward correction of 10% to account for the inclusion of some lung-cancer cases diagnosed among miners who are still living. The U.S. uranium miners had exposures to high concentrations of radon daughters; at least before 1960, the radon-daughter concentrations ranged generally from 10 to 100 or more WL. This explains the fact that the average cumulative exposure, 1,180 WLM, is well above that of most of the other mining populations studied. The estimates of risk are therefore heavily weighted by experience associated with high cumulative doses and at relatively high dose rates. The table indicates that, except for the lowest dose group, in whom no lung-cancer excess has been observed, the lower exposure groups have risk estimates 2–3 times those for the higher dose groups.

Cumulative WLM		Person-Years	Lung Cancers		Absolute Risk, cases per 10^6 PY per WLM	Relative Risk, % increased risk/WLM
Range	Midpoint		Observed	Expected		
0–119	60	5,183	3	3.96	—	—
120–239	180	3,308	7	2.24	8.0	1.2
240–359	300	2,891	9	2.24	7.8	1.0
360–599	480	4,171	19	3.33	7.8	1.0
600–839	720	3,294	9	2.62	2.7	0.3
840–1,799	1,320	6,591	40	5.38	4.0	0.5
1,800–3,719	2,760	5,690	49	4.56	2.8	0.4
>3,719	7,000 (est.)	1,068	23	0.91	3.0	0.3
ALL	1,180 (mean)	32,196	159	25.24	3.52	0.45

If we consider only the data for miners exposed to less than 360 WLM, the absolute-risk estimate is 6.0 cases per 10^6 PY per WLM, and the relative risk is 0.8% per WLM. These values indicate a risk well below the results for the Czechoslovakian miners with comparable total cumulative doses.

This difference cannot be explained by smoking experience, and the American miners have had about the same followup as the Czechoslovakian miners. A possible explanation for the lower risk in the U.S. miners is the high dose rate at which exposure occurred. An increased bone-cancer effect from a reduced dose rate of alpha-radiation exposure from radium-224 has also been observed.

Archer *et al.*[1] analyzed the histologic types among 107 bronchial cancers in 104 miners. In three cases, two different types of cancer were present at autopsy: in two there were simultaneous epidermoid and small-cell undifferentiated cancers (WHO types 1 and 2), and in one there were simultaneous small-cell cancer and adenocarcinoma (WHO types 2 and 3). Nearly all tissue sections were reviewed independently by a panel of two or three pathologists with long experience in evaluating lung-cancer cell types. The frequency of cancer by type was compared with the frequency in a group of 121 lung cancers in nonminers matched for smoking history.

The authors concluded that small-cell anaplastic type 2 cancers were in greater excess than other types, but that epidermoid cancer, adenocarcinoma, and mixed epidermoid cancer and adenocarcinomas (types 1, 3, and 5) were also present in greater numbers than expected. Only the large-cell undifferentiated cancers (type 4), carcinoids, and bronchoalveolar tumors (among type 6) were not in excess among the miners. The proportion of excess tumors by type was not dose-related. These observations are important in refuting the earlier conclusion[39] that only small-cell anaplastic tumors are the result of exposure to radon daughters.

No detailed comparison of risk by cigarette-smoking category is available for the U.S. miners, but it is possible to make an estimate similar to that given above for the Czechoslovakian miners. Among the 159 lung-cancer cases, nine were in men who had never smoked or who had given up smoking 15 yr or more before death. On the basis that U.S. cigarette-smokers have 12 times the risk of lung cancer as nonsmokers (those who never smoked or ex-smokers of long standing), that 30% of the miners were nonsmokers, and that the age distributions of the smokers and nonsmokers were similar, the expected cases among the nonsmokers would be 0.87, and among the smokers, 24.37. Thus, the relative risk for nonsmokers is $9/0.87 = 10.3$, and for smokers, $150/24.37 = 6.2$. The somewhat higher relative risk for nonsmokers is consistent with the conclusion given above for the Czechoslovakian miners: the excess cases among the nonsmokers may rise proportionately more rapidly with further followup, because the latent period for nonsmokers is longer. Because the assumptions used to derive these data are relatively crude, however, further information will be needed to settle the question of whether exposure of these miners to radon daughters simply adds to the effect of cigarette-smoking, or whether the effects are greater than additive when both are present.

Canadian Uranium Miners A Royal Commission study of Canadian miners who worked in the Ontario mines during 1955–1974 has been published.[15] In the uranium mines, radon-daughter measurements have been routine since 1957. Among 15,094 persons who worked underground in the uranium mines for at least 1 mo during that period, 81 deaths from lung cancer were certified. For these cases, the median year of starting mining was 1957, and the followup was 17 yr. It is evident from the published data that many of these miners worked underground only relatively short times, and it is not possible to determine which should be excluded on the grounds of having been followed for less than 10 yr after beginning mining (when the excess risk would be essentially zero). For these reasons, one cannot derive risk estimates with any confidence, but it is evident that the lung-cancer data in this group of miners have unusual potential for defining low risks, if studied adequately, as the report recommended.[15] Not only is the population larger than that of other mining groups under study, but exposures have generally been low and there has been reasonably good monitoring of these mines since they were opened. An evaluation of the effect of cigarette-smoking should also be possible.

Despite the limitations of this aspect of the Royal Commission's report, several important points can be made from the data presented. First, although exposures were below 1 WL, except in a few mines, a significant excess risk of lung cancer has been observed. From age- and year-specific data for lung cancer for Ontario males, the expected number of lung-cancer deaths was 45.1 for the entire roll of more than 15,000 miners; the relative risk of 81/45.1, or 1.8, is undoubtedly an underestimate, because of incomplete ascertainment of cases and because of the inclusion of years at low risk during the latent period in calculating the expected deaths. Second, miners who began underground work after the age of 35 were at somewhat higher relative risk than miners who began work when they were younger, so the absolute risk would be substantially greater in the older miners, as was found in the Czechoslovakian study. Third, a plot of lung-cancer deaths, as an estimated proportion of the population born before 1933, versus cumulative exposure in WLM gives a reasonably linear relationship, the slope being such that the crude doubling dose is about 12 WLM. This latter figure is not an accurate indication of the relative risk, because important factors, such as age and smoking, may have varied by exposure dose category; but this observation suggests that a more complete analysis may well show this group of miners to be at high risk.

The lowest cumulative dose category in this analysis was 1–30 WLM, in which 29 lung-cancer deaths were recorded. Of these, eight occurred less than 10 yr after mining was begun and may be considered to be un-related to the mining experience, or in other words to represent the ex-pected cases during this interval. If we assume that during the 8 yr of further followup (10–17 yr) the expected cases per year were about twice the rate per year during the first 10 yr (or about 0.8/yr), the expected deaths would total about 12.8 during this latter followup period. This would give a relative risk of 21/12.8, or 1.64, for this group in this interval. The mean cumulative dose for miners who died 10 yr or more after starting mining was 10.9 wlm; thus, on this basis, the doubling dose for this low dose group would be 17 wlm, in reasonable agreement with the analysis discussed above. Although this assessment is tentative, the data suggest an excess risk for these miners at this very low cumulative-dose range. The importance of an adequate epidemiologic followup of this mining popu-lation is obvious.

Newfoundland Fluorspar Miners   A. J. deVilliers and D. T. Wigle (un-published manuscript) have continued the followup study of these miners through 1971. Underground mining in these mines in St. Lawrence, New-foundland, began in 1936. The total employed population, both under-ground and on the surface, was 2,414 men, whose work records and mortality experience have been determined for the period 1933–1971. The number of miners who worked underground before 1960 was 1,118, with 16,845 py of followup more than 10 yr after the start of underground mining. The average followup has been 25.3 yr, and the average age at the start of mining was 28.

Radon-daughter measurements were begun in 1959, and the concen-trations before then have been estimated, with mining methods, ventila-tion history of the mines, and work locations taken into account. Esti-mates of radon-daughter concentrations varied from 2 to 8 wl, according to the type of work during the period up to 1960, when with improved ventilation they decreased to below 0.5 wl. Exposures in these mines were therefore substantially lower than in the U.S. uranium mines, but some-what higher than in the Czechoslovakian or Canadian uranium mines.

Sixty-five deaths from lung cancer have occurred among the under-ground miners (lung cancer was the cause of 27% of all deaths up to 1971) and six among the surface workers (4% of all deaths). The first lung-cancer death in the underground workers occurred in 1949, and 64 oc-curred after 1952; the number continued to rise sharply up to 1971. No lung-cancer deaths were observed less than 10 yr after the start of under-ground work, and the average latent period was 22.6 yr. The risk of lung cancer had a highly significant correlation with cumulative dose and with age at the start of underground mining. The risk per WLM has not been analyzed by age at the start of mining.

For the entire group of underground miners during the years under study, the expected number of deaths was 3.76, on the basis of age- and year-specific lung-cancer death rates for Newfoundland males. The average cumulative exposure weighted for person-years at risk was 204 WLM; thus, the absolute risk was 17.7 deaths per 10^6 PY per WLM. In this group, the relative risk was 8.0% per WLM, which yields a doubling dose of 12.5 WLM. These results illustrate the problem of determining relative-risk data for lung cancer, whose incidence is so strongly influenced by cigarette-smoking. These miners were nearly all smokers (86% according to a 1960 survey), and many smoked heavily. Thus, the expected rate is probably too low—a bias that affects the relative risk substantially, but has little effect on the absolute risk in this instance. Another consequence of the high proportion of smokers is that the number of nonsmokers was probably too small in this study group to provide an adequate comparison of risks between smokers and nonsmokers.

From the data presented on the 65 lung-cancer cases, it is possible to analyze the latent period as a function of age at start of mining, on the assumption that the expected cases are so few that they can be neglected. The results of this analysis are as follows:

Age When Began Mining, yr	No. Deaths	Time to First Death, yr after start	Mean Latent Period, yr, + SD
< 20	18	13	22.9 ± 5.0
20–24	13	12	22.9 ± 5.9
25–29	10	14	24.1 ± 6.2
30–34	7	17	23.7 ± 4.0
35–39	8	11	21.9 ± 6.4
> 39	9	12	21.4 ± 7.7

In these miners, there was virtually no effect of age on the latent period, at least up to now. This finding is in sharp contrast with the data on the Japanese atomic-bomb survivors, but is in reasonable agreement with the tentative results in the British spondylitics. The explanation for the difference is not clear, but one possibility is that the Japanese were light smokers,[18] in comparison with the miners. Smokers have a shorter latent period and one that may be less influenced by age-specific factors than would be the case for an exposed group containing a high proportion of nonsmokers and light smokers, as the Japanese survivors appear to have been in the postwar period.

Wright and Couves[49] reported on the cytology of sputum obtained from 29 fluorspar miners in whom the diagnosis of bronchogenic cancer was made. Twenty-six had squamous-cell carcinoma, two oat-cell (or small-cell) cancer, and one adenocarcinoma. This marked preponderance of well-differentiated tumors diagnosed during life has also been found in a smaller number of U.S. uranium miners.[40] The distribution is in contrast with that found at autopsy among the U.S. and Czechoslovakian miners. This observation is consistent with the view that the degree of differentiation of these cancers may be a function of the stage of progression of the disease.

Swedish Metal Miners   Several reports of lung-cancer excess among Swedish metal miners have been published. A number of these reports[3,20,38,45] are preliminary and include incomplete followup or material on only active miners. Therefore, it is not possible to determine risk estimates from them. Axelson and Sundell[4] have recently published data on a group of zinc miners studied for the period 1956–1976, with deaths ascertained by local parish records. The zinc mine in question has been in operation since before 1900. The number of miners is small; about 100 have worked underground at any one time, with relatively slight turnover until recently. Radon concentrations have been extensively measured in the shafts since 1969 and, according to analyses before institution of new ventilation methods, have been found to be equivalent to 0.3–1 WL. Expected lung-cancer-mortality rates were obtained from Swedish national statistics by an indirect method.

Twenty lung-cancer deaths have been observed, compared with 2.32 expected, with 2,154 PY at risk. The mean cumulative exposure is estimated at 270 WLM, and the risk estimate is 30.4 deaths per 10^6 PY per WLM. This group of miners has had very long followup into retirement, so the approximate value obtained is applicable to a relatively old age, probably equivalent to the oldest age group of Czechoslovakian uranium miners. The long followup of these men is indicated by the mean times from beginning mining to death of 34 yr for smokers and 43 yr for nonsmokers (median, 37 and 49 yr, respectively). Of the 20 cases, smoking histories have been obtained on 19. Nine of the miners were nonsmokers and 10 were smokers. If this mining population had smoking experience similar to that of other Swedish laboring groups[6] and if the work experience of smokers and nonsmokers were the same, these data would indicate little difference in radiation risk between smokers and nonsmokers.

Forty-five lung-cancer deaths have been observed between 1953 and 1976 in Swedish iron miners at Malmberget—a larger group than the zinc miners, but also with very long followup (E. P. Radford and K. G. St. C.

Renard, unpublished data). Smoking histories were obtained in these cases from some miners before death, and the rest from the families or from co-workers. Seven lung-cancer deaths have occurred in men who never smoked, nine in men who had stopped smoking 15 yr or more before death, and 29 in current smokers. Smoking surveys among active and retired miners indicate that about two-thirds of the miners have smoked, but only one-third are current smokers. The study is not yet completed, but these proportions indicate that the excess risk for smokers may not be markedly greater than that for nonsmokers. The very long followup of these Swedish groups is an important factor in determining risk estimates for nonsmokers, because of the long latent period that may be observed in these cases.

Summary of Risk Estimates in Underground Miners   From the data presented above, the risk estimates for lung cancer from exposure to radon daughters now range from about 6 to 47 cases per 10^6 PY per WLM; the range reflects in large part the effect of age at exposure or at onset of the cancer. The Newfoundland fluorspar miners and Czechoslovakian uranium miners have risk estimates very comparable with those for the entire population; the Swedish zinc miners have higher estimates, even with less evidence than for the other groups of cigarette-smoking as a factor, apparently because they have been followed to a greater age. The U.S. uranium miners have risk estimates well below those of all the other groups. Only two explanations seem reasonable to account for this latter difference: either the radon-daughter measurements in the U.S. mines have overestimated exposures by as much as a factor of 3 (not likely, in view of the great efforts made to obtain this information) or the much higher dose rate (working levels in the mines) has led to less risk per unit of cumulative exposure than the lower working levels in the other mines. There is no evidence that the age distribution of the U.S. miners differs significantly from that of the Czechoslovakian or Newfoundland miners.

The most likely risk estimates, at exposure of about 1 WL and with characteristic smoking experience, are about 10 cases per 10^6 PY per WLM for the age group 35–49, 20 cases per 10^6 PY per WLM for the age group 50–65, and about 50 cases per 10^6 PY per WLM for those over 65. These values apply to the age at diagnosis and are consistent with available followup data.

CONCLUSIONS

The more proximal regions of the human bronchial tree are most sensitive to induction of bronchogenic cancer by radiation or other environmental agents. Experimental studies have indicated that cancer of these proximal regions of the bronchi can be induced in animals with radiation, provided that the dose delivered to those regions is sufficient. The data of Little and colleagues have drawn attention to the possible role of nonspecific proliferative stimuli to bronchial tissues in the induction of bronchial cancers. Such stimuli may be widespread in human populations, but it is of interest that emerging data in man indicate that cigarette-smoking does not contribute as strongly to the risk of bronchial cancer induced by radiation as has previously been thought, although smoking shortens the latent period.

The possible influence of "hot spots" of insoluble radioactive particles deposited in pulmonary tissues on cancer risk has been evaluated in a previous report.[32] The evidence is still insufficient to determine whether aggregates of radioactivity that remain localized in specific regions of the lungs give a greater or smaller risk of lung cancer per average lung dose than uniformly deposited radiation. Preliminary experimental data indicate that a small fraction of inhaled insoluble particles may remain in the bronchial epithelial layer for long periods, but the significance of this local exposure on lung-cancer risk is still uncertain.

Risk estimates for lung cancer depend on the age of the subject at the time of radiation exposure, as well as the age at the time of appearance of the cancer. There is little evidence of an increased risk before the age of 35, regardless of the age at exposure, but the risk at later ages rises steeply, as does the risk in the general population. The latent period from radiation exposure to death from lung cancer is generally 10 yr or more, with excess cases appearing in some populations 50 yr or more after the beginning of exposure. Among the Japanese atomic-bomb survivors, latent periods are much longer for those exposed when younger, compared with those exposed when older, but there is little evidence of this effect in other groups studied.

To compare the risk estimates obtained from the Japanese atomic-bomb survivors or the British patients with spondylitis with those from underground miners, the assumed minimal latent period, as well as the age distribution over which risk estimates are compared, must be approximately the same. For the British spondylitis patients, the estimate is 2.8 cases per 10^6 py per rad (x ray) versus 2.0 for the Japanese survivors of both cities (quality factor of 5 for neutrons), age-adjusted to the age distribution of the spondylitics. The miners are older, and the most reliable population estimates are for the Newfoundland fluorspar miners and the Czechoslovakian uranium miners; on the average, these yield about 18 cases per 10^6 py per wLM.

Conversion of wLM to a rad dose to the basal-cell layer of the proximal bronchial segments has recently been reevaluated.[16,19] Several factors influence these estimates, such as the fraction of free ions compared with the fraction inhaled and bound to dust particles, breathing pattern and whether the subject is mouth- or nose-breathing, and the thickness of the epithelium. On the basis of average data presented in these reports, the conversion is taken as 1 WLM = 0.4-0.8 rad, the range reflecting the variables mentioned above.

Application of this range for the miners yields a risk estimate of 22–45 cases per 10^6 PY per rad of alpha-radiation exposure to the bronchial epithelium. The Japanese bomb survivors and British spondylitics have risk estimates in the equivalent (older) age groups of about 3 cases per 10^6 PY per rem. From these data, the RBE for alpha irradiation for induction of this cancer is between 8 and 15. The uncertainties in this comparison are substantial, but the results fit conventional views of the relative effectiveness of experimental alpha irradiation. Regardless of the conversion factor applied for WLM to rad, comparison of the empirical data for risks per WLM with risks per rem yields about 6 rem/WLM—very close to the value of 5 assumed in the 1972 BEIR report.

Expression of the age-specific lung-cancer risk estimates in rems yields the following values, based on available data and on the assumption that the smoking experience of the exposed population is typical of the whole population of which it is a segment:

Age at Diagnosis of Cancer, yr	Excess Risk, cases per 10^6 PY per rem
<35	0
35–49	1.5
50–65	3.0
>65	7.0

These values are based on the combined estimates for the miners, but are reasonably consistent with the data on all the groups studied, except for the Japanese data, which are anomalous in terms of the long latent periods observed in the younger groups. No available data indicate whether these values may be used for groups irradiated in childhood. At the least, the minimal latent period applicable to irradiation before the age of 15 would be long, i.e., about 25 yr. Above that age, the minimal latent periods are approximately 15–20 yr for those irradiated at the age of 15–34 and 10 yr for those irradiated at the age of 35 or above.

The effect of cigarette-smoking on these risk estimates cannot be finally evaluated. If smoking and radiation risks are merely additive, then the risk estimates presented above apply to either smokers or nonsmokers. But, as the Japanese data suggest, the risks could develop among nonsmokers at higher ages than indicated in the table. If the lung-cancer risk after radiation exposure is proportional to the usual age-specific rates for smokers and nonsmokers (as in the relative-risk concept, consistent with a multiplicative effect of radiation on cigarette-induced cancer), then the estimates of excess risk should be increased by about 50% to apply to smokers and reduced by a factor of about 6 for nonsmokers, as well as delayed in time. Some evidence now available is consistent with both positions, and it is probable that the truth is somewhere in between. The evidence now indicates, however, that a purely multiplicative effect on lung-cancer risk related to radiation exposure and cigarette-smoking is highly unlikely.

The risk estimates given above have been derived on the assumption of a linear relationship between dose and effect. The new information since the 1972 BEIR report is consistent with this procedure for high-LET radiation, at least. The lowest dose at which the lung-cancer rate is increased has been lowered. Experimentally, an excess of neoplasms has been found in hamsters at 15 rads for polonium alpha radiation.[27] The Canadian uranium miners appear to have an excess of lung cancer in the lowest dose group (1–30 WLM; mean, 10.9 WLM), at a cumulative dose to the bronchi of 4–9 rads. In Hiroshima survivors, a significant excess of lung cancers has been observed in the dose range of 10–49 rads kerma; this group has a mean absorbed dose to the lungs of 8.8 rads from gamma radiation and 0.95 rad from neutrons. For the British patients with spondylitis who were given x-ray therapy, the mean bronchial dose is estimated at 197 rads on the average given in doses of about 20 rads each.

REFERENCES

1. Archer, V. E., G. Saccomanno, and J. H. Jones. Frequency of different histologic types of bronchogenic carcinoma as related to radiation exposure. Cancer 34:2056–2060, 1974.
2. Archer, V. E., J. K. Wagoner, and F. E. Lundin, Jr. Uranium mining and cigarette smoking effects on man. J. Occup. Med. 15:204–211, 1973.
3. Axelson, O., H. Josefson, M. Rehn, and L. Sundell. Svensk pilotstudie over lung cancer hos gruvarbetere. Lakartidningen 68:5687–5694, 1971.
4. Axelson, O., and L. Sundell. Mining, lung cancer and smoking. Scand. J. Work Environ. Health 4:46–52, 1978.
5. Beebe, G. W., H. Kato, and C. E. Land. Life Span Study Report 8. Mortality Experience of Atomic Bomb Survivors 1950–74. Radiation Effects Research Foundation Technical Report TR 1-77. Hiroshima: Radiation Effects Research Foundation, 1978.
6. Cederlof, R., L. Friberg, Z. Hrubec, and U. Lorich. The relationship of smoking and some social covariables to mortality and cancer morbidity. A ten-year follow-up in a probability sample of 55,000 Swedish subjects age 18–69. Report of the Department of Environmental Hygiene. Stockholm: Karolinska Institute, 1975.

7. Cember, H. Radiogenic lung cancer. Prog. Exp. Tumor Res. 4:251–303, 1964.

8. Chameaud, J., R. Perraud, R. Masse, J. C. Nenot, and J. Lafuma. Cancers du Poumon Provoques chez le Rat par le Radon et Ses Descendants a Diverses Concentrations. Paper No. IAEA-SM-202/410. Proceedings of International Atomic Energy Agency, Chicago, 1977.

9. Cihak, R. W., T. Ishimaru, A. Steer, and A. Yamada. Lung cancer at autopsy in A-bomb survivors and controls, Hiroshima and Nagasaki, 1961–1970. I. Autopsy findings and relation to radiation. Cancer 33:1580–1588, 1974.

10. Doll, R., and P. G. Smith. Mortality from cancer and other causes after radiotherapy for ankylosing spondylitis. Further observations. Cited in Sources and Effects of Ionizing Radiation. New York: United Nations, 1977.

11. Erlich, R. Effect of nitrogen dioxide on resistance to respiratory infection. Bacteriol. Rev. 30:604–614, 1966.

12. Filipy, R. E., G. E. Dagle, R. F. Palmer, and B. O. Stuart. Carcinogenesis of inhaled radon daughters with uranium ore dust in beagle dogs, pp. 49–55. Pacific Northwest Lab. Annu. Rep. BNWL-2100. Part I. May 1977.

13. Gates, O., and S. Warren. The production of bronchial carcinomas in mice. Amer. J. Pathol. 36:653–672, 1960.

14. Gaven, J. C., R. F. Palmer, K. E. McDonald, J. E. Lund, and B. O. Stuart. Comparative toxicity in rats vs. hamsters of inhaled radon daughters with and without uranium ore dust, pp. 56–60. Pacific Northwest Lab. Annu. Rep. BNWL-2100. May 1977.

15. Ham, J. M. Report of the Royal Commission on the Health and Safety of Workers in Mines. Toronto: Ministry of the Attorney General, Province of Ontario, 1976.

16. Harley, N. H., and B. S. Pasternack. Alpha absorption measurements applied to lung dose from radon daughters. Health Phys. 23:771–782, 1972.

17. Horáček, J., V. Plaček, and J. Ševc. Histologic types of bronchogenic cancer in relation to different conditions of radiation exposure. Cancer 40:832–835, 1977.

18. Ishimaru, T., R. W. Cihak, C. E. Land, A. Steer, and A. Yamada. Lung cancer at autopsy in A-bomb survivors and controls, Hiroshima and Nagasaki, 1961–1970. II. Smoking, occupation, and A-bomb exposure. Cancer 36:1723–1728, 1975.

19. Jacobi, W. Relations between the inhaled potential alpha-energy of ^{222}Rn- and ^{220}Rn-daughters and the absorbed alpha-energy in the bronchial and pulmonary region. Health Phys. 23:3–11, 1972.

20. Jorgensen, H. S. A study of mortality from lung cancer among miners in Kiruna 1950–1970. Work Environ. Health 10:125–133, 1973.

21. Kennedy, A. R., and J. B. Little. Cellular localization of intratracheally administered ^{210}Po in the hamster lung using autoradiography of this section from plastic embedded tissue, pp. 475–484. In E. Karbe and J. F. Park, Eds. Experimental Lung Cancer. Carcinogenesis and Bioassays. Berlin: Springer Verlag, 1974.

22. Kennedy, A. R., and J. B. Little. Radiation carcinogenesis in the respiratory tract, pp. 189–261. In C. C. Harris, Ed. Pathogenesis and Therapy of Lung Cancer. New York: Marcel Dekker, Inc., 1978.

23. Kennedy, A. R., R. B. McGandy, and J. B. Little. Histochemical, light and electron microscopic study of polonium-210 induced peripheral tumors in hamster lungs. Evidence implicating the Clara cell as the cell of origin. Eur. J. Cancer 13:1325–1340, 1977.

24. Laskin, S., M. Kuschner, and R. T. Drew. Studies in pulmonary carcinogenesis, pp. 321–351. In M. G. Hanna, Jr., P. Nettesheim, and J. R. Gilbert, Eds. Inhalation Carcinogenesis. U.S. Atomic Energy Commission Symposium Series 18, 1970.

25. Little, J. B., and A. R. Kennedy. Evaluation of radiation-induced respiratory carcinogenesis in Syrian hamsters. Total dose and dose-rate. Prog. Exp. Tumor Res. (in press)

26. Little, J. B., A. R. Kennedy, and R. B. McGandy. Effect of dose distribution on the induction of experimental lung cancer by alpha radiation. Health Phys. 35:595–606.

27. Little, J. B., A. R. Kennedy, and R. B. McGandy. Lung cancer induced in hamsters by low doses of alpha radiation from polonium-210. Science 188:737–738, 1975.

28. Little, J. B., R. B. McGandy, and A. R. Kennedy. Interactions between polonium-210, α-radiation, benzo(a)pyrene, and 0.9% NaCl solution instillations in the induction of experimental lung cancer. Cancer Res. 38:1929–1935, 1978.

29. Little, J. B., and W. F. O'Toole. Respiratory tract tumors in hamsters induced by benzo(a)pyrene and ^{210}Po alpha radiation. Cancer Res. 34:3026–3032, 1974.

30. Morin, M., J. C. Nenot, R. Masse, D. Nolibe, H. Metivier, and J. Lefuma. Induction de Cancers chez le Rat apres Inhalation de Radioelements Emetteurs Alpha. Paper No. IAEA-SM-202/404. Proceedings of International Atomic Energy Agency, Chicago, 1977.

31. National Research Council, Advisory Committee on the Biological Effects of Ionizing Radiations. The Effects on Populations of Exposure to Low Levels of Ionizing Radiation. Washington, D.C.: National Academy of Sciences, 1972.

32. National Research Council, Advisory Committee on the Biological Effects of Ionizing Radiations, ad hoc Committee on "Hot Particles." Health Effects of Alpha-Emitting Particles in the Respiratory Tract. Washington, D.C.: National Academy of Sciences, 1976.

33. Nenot, J. C. Discussion of paper IAEA-SM-202/402, p. 228. In Proceedings of International Atomic Energy Agency, Chicago, 1977.

34. Nettesheim, P. Respiratory carcinogenesis studies with the Syrian golden hamster. A review. Prog. Exp. Tumor Res. 16:185–200, 1972.

35. Nettesheim, P., M. G. Hanna, Jr., D. G. Doherty, R. F. Newell, and A. Hellman. Effects of chronic exposure to artificial smog and chromium oxide dust on the incidence of lung tumors in mice, pp. 305–317. In M. G. Hanna, Jr., P. Nettesheim, and J. R. Gilbert, Eds. Inhalation Carcinogenesis. U.S. Atomic Energy Commission Symposium Series 18, 1970.

36. Patrick, G., and C. Stirling. The retention of particles in large airways of the respiratory tract. Proc. R. Soc. Lond. B 198:455–462, 1977.

37. Radford, E. P., and E. A. Martell. Polonium-210:lead-210 ratios as an index of residence times of insoluble particles from cigarette smoke in bronchial epithelium, pp. 567–582. In W. H. Walton, Ed. Oxford: Pergamon, 1977.

38. Renard, K. G. St. C. Lungcancerdodlighet vid LKAB's gruvar ei Malmberget. Lakartidningen 71:158–163, 1974.

39. Saccomanno, G., V. E. Archer, O. Auerbach, M. Kuschner, R. P. Saunders, and M. G. Klein. Histologic types of lung cancer among uranium miners. Cancer 27:515–523, 1971.

40. Saccomanno, G., V. E. Archer, R. P. Saunders, O. Auerbach, and M. G. Klein. Early indices of cancer risk among uranium miners with reference to modifying factors. Ann. N.Y. Acad. Sci. 271:377–383, 1976.

41. Sanders, C. L., and R. C. Dagle. Inhalation toxicology of ^{238}PuO$_2$ in Syrian golden hamsters. Radiat. Res. 70:334–344, 1977.

42. Ševc, J., E. Kunz, and V. Plaček. Lung cancer in uranium miners and long-term exposure to radon daughter products. Health Phys. 30:433–437, 1976.

43. Smith, D. M., E. C. Anderson, J. R. Prine, L. M. Holland, and C. R. Richmond. Biological Effect of Focal Alpha Radiation on the Hamster Lung. Paper No. IAEA-SM-202/410. Proceedings of International Atomic Energy Agency, Chicago, 1977.

44. Smith, P. G., and R. Doll. Age- and time-dependent changes in the rates of radiation-induced cancers in patients with ankylosing spondylitis following a single course of

x-ray treatment, pp. 205–214. In Late Biological Effects of Ionizing Radiation. Vienna: International Atomic Energy Agency, 1978.

45. Snihs, J. O. The approach to radon problems in nonuranium mines in Sweden, pp. 900–912. In Proceedings of Third International Congress of the International Radiation Protection Association. Oak Ridge: U.S. Atomic Energy Commission, 1974.

46. Steer, A., C. E. Land, I. M. Moriyama, T. Yamamoto, M. Asano, and H. Sanefuji. Accuracy of diagnosis of cancer among autopsy cases. JNIH-ABCC population for Hiroshima and Nagasaki. GANN 67:625–632, 1976.

47. Stuart, B. O., R. F. Palmer, R. E. Filipy, and G. E. Dagle. Biological effects of inhaled cigarette smoke in beagle dogs, pp. 3.65–3.69. Pacific Northwest Lab. Annu. Rep. BNWL-2500. Part I. February 1978.

48. Stuart, B. O., R. F. Palmer, R. E. Filipy, and J. Gaven. Inhaled radon daughters and uranium ore dust in rodents, pp. 3.70–3.72. Pacific Northwest Lab. Annu. Rep. BNWL-2500. Part I. February 1978.

49. Wright, E. S., and C. M. Couves. Radiation-induced carcinoma of the lung. The St. Lawrence tragedy. J. Thorac. Cardiov. Surg. 74:495–498, 1977.

50. Yesner, R., B. Gerstl, and O. Auerbach. Application of the WHO classification of lung cancer to biopsy material. Ann. Thorac. Surg. 1:33–49, 1965.

LEUKEMIA

The 1972 BEIR report summarized knowledge available up to 1972 regarding radiation-induced leukemia. Knowledge regarding mechanisms was derived largely from animal studies, and estimates of risk were obtained from human epidemiologic investigations.

Experiments with many species led to the conclusion that the leukemogenic effect of ionizing radiation varied with radiation quality (LET), dose rate, and total accumulated dose. Particular attention was paid to myeloid leukemia induced in RF mice (inasmuch as myeloid leukemia was known to predominate in the Japanese atomic-bomb survivors and the British ankylosing-spondylitis patients). A strong dose-rate dependence was observed in many laboratory animal species after gamma-ray exposures. In RF mice, the leukemogenic effect of a single dose of x or gamma rays was noted to be 5 or more times that of the same total dose accumulated through daily exposures. The observed reduced effectiveness of low dose rates from low-LET radiation was attributed to repair of incipient injury and accounted for the corresponding departure from linearity of the dose-effect curve seen at high dose rates.[31] No dose-rate dependence was observed after high-LET radiation exposures.

Despite extensive studies in irradiated mice and numerous studies in rats, guinea pigs, dogs, cats, swine, and monkeys, it was not possible to establish dose-effect relationships at low doses or to characterize the process of leukemogenesis in detail. The 1972 BEIR report therefore placed greatest reliance on the data from epidemiologic studies of irradiated human beings.

The most extensively studied group was that of the Japanese atomic-bomb survivors. All forms of leukemia, except chronic lymphocytic leukemia, were increased in incidence in atomic-bomb survivors of Hiroshima and Nagasaki. The excess risk increased with dose. On the assumption that a linear relationship held at all doses, the average leukemia risk was estimated to be about one excess case of leukemia per million exposed persons per year per rad (kerma) in Nagasaki. Excess risk after high doses was evident within 3–4 yr after irradiation and declined within 15 yr, but persisted for 25 yr after exposure. The estimate of the RBE, for neutrons in Hiroshima, at high doses and dose rates was between 1 and 5, on the basis of kerma. An RBE of 5 was claimed to give the best fit[31] and yielded an excess leukemia incidence (for all ages combined) in Hiroshima survivors of 1.7 cases per million persons per year per rem; for an RBE of 1, the excess was 3.1 cases per million persons per year per rem. The excess rate of leukemia for Nagasaki was about one case per million persons per year per rem and was almost unaffected by neutron RBE. These are average rates computed for 16 yr of followup (1950–1966). Absolute and relative risks were higher for those under 10 yr old at the time of the bombing in both cities.

The data from British ankylosing-spondylitis patients treated with x rays revealed 0.88 (0.71–1.0) excess leukemia deaths per million persons per year per rem averaged over the total red marrow (for a followup period of 5–25 yr). An increased risk of leukemia was observed in cancer patients receiving pelvic radiotherapy for ovarian sterilization, but not for cervical cancer. Leukemia incidence was increased after radiation therapy for thymic enlargement and for scalp ringworm in childhood, and the excess risk was approximately 3.0 cases per million persons per year per rem averaged over the total red marrow. No increased leukemia incidence was observed after iodine-131 therapy for hyperthyroidism or in occupationally exposed workers, other than radiologists. Susceptibility to leukemia induction was higher in the fetus irradiated *in utero* in patients exposed to diagnostic x rays, but not in the atomic-bomb populations irradiated with x rays and neutrons. Some studies reported an association between leukemia in adults and prior diagnostic exposure to x rays. Leukemia risk was increased in patients who had received Thorotrast as an x-ray contrast agent. Chronic lymphocytic leukemia was not found to be increased in any study of irradiated people.

ANIMAL STUDIES

Experiments have been conducted in many species to investigate the factors that influence radiation induction of leukemia.[40,41] Additional confidence in the conclusions drawn from leukemia induction in irradiated RF mice was provided by recent studies in CBA mice. This strain has a negligible spontaneous incidence of leukemia and is subject to an increasing incidence of myeloid leukemia with increasing dose, followed by a plateau and then a decrease in incidence. These effects are seen promptly after radiation exposure, before increased mortality from competing risks, which simplifies analysis and interpretation.[24]

In general, the following observations appear to be relevant to the human situation:

- Incidence rises less rapidly in the low-dose region after low-LET than after high-LET radiation. This is presumed to reflect differences in induction and repair rates—greater induction by high-LET radiation and greater repair with low-LET radiation.
- There is a dose above which a decreasing incidence of leukemia occurs; this is presumed to reflect killing (or mitotic inactivation) of cells in which malignant transformations have been induced. Protraction and dose-rate effects indicate that repair mechanisms operate more effectively after low-dose-rate, low-LET radiation than after high-dose-rate, high-LET radiation.
- Despite the availability of information on the importance of these different variables for leukemia induction, statistics associated with the small samples that are practical in such studies are too limited to permit the assessment of risk at very low doses.
- Differences in types of leukemia and the importance of viruses, particularly in murine leukemias, make it difficult to extrapolate directly from animal studies to man, but the general principles outlined above are widely accepted.

HUMAN STUDIES

Since the 1972 BEIR report, new information has become available from several important epidemiologic studies.

Leukemia in Japanese Atomic-Bomb Survivors

Newly available information is derived from:

- Additional years of followup, allowing for the documentation of the duration of increased leukemia risk after a single whole-body radiation exposure.
- New information on organ dose distribution from gamma and

TABLE A-7 Deaths from Leukemia[a] in the Life-Span Study Sample, 1950–1974

T65 Dose, rads (kerma)	Average Kerma, rads			Person-Years	No. Leukemias				Relative Risk
	Total	Gamma	Neutron		Observed	Expected[b]	O/E		
Hiroshima									
0	—	—	—	630,094	28	54.0	0.52		1.0
1–9	3.7	2.9	0.8	291,890	14	25.5	0.55		1.7
10–49	21.9	17.6	4.3	227,467	18	19.5	0.92		2.7
50–99	70.2	56.8	13.4	56,939	7	4.9	1.42		7.7
100–199	138.9	108.6	30.3	35,861	13	3.2	4.06		12.2
200–299	243.0	186.2	56.8	14,028	8	1.2	6.45		28.6
300–399	346.4	254.3	92.1	7,830	10	0.7	15.08		24
>399	524.6	380.5	144.1	10,601	12	0.9	12.65		
TOTAL				1,274,710	110	110			
Nagasaki									
0	—	—	—	101,165	3	8.6	0.35		1.0
1–9	3.9	3.9	0	142,845	9	11.4	0.79		0.5
10–49	21.5	21.5	0	79,734	2	6.7	0.30		0
50–99	70.8	70.6	0.2	28,591	0	2.4	0		2.0
100–199	144.2	142.9	1.3	30,481	3	2.5	1.22		8.9
200–299	241.4	238.0	3.4	16,431	7	1.3	5.31		6.6
300–399	340.0	334.6	5.4	6,060	2	0.5	3.94		20.7
>399	524.9	514.4	10.5	7,792	8	0.6	12.40		
TOTAL				413,099	34	34			

[a] Death-certificate diagnoses. Data from Beebe et al.[4] (Tables 2, 3, and 8).
[b] Normalized to number of person-years at risk.

neutron exposures (which permits a more careful analysis of the dose-response curve).

• New calculations of the neutron RBE according to these new data. These have been based on death certificates from the Life Span Study sample* and on the Leukemia Registry (hematologically verified cases).

The total number of deaths certified as due to leukemia in the Hiroshima and Nagasaki LSS atomic-bomb survivors for the period 1950–1974 was 144 (Table A-7). During the years 1971–1974, 14 of these deaths occurred. The distribution of these 14 with dose was similar to that observed in the earlier years; hence, the shape of the leukemia dose-response curve remains essentially unchanged.

The absolute-risk estimates are shown in Table A-8 for Hiroshima and Nagasaki, for sequential intervals. The new leukemia cases were found mostly in Hiroshima,[4] with only one new case in Nagasaki survivors. The Hiroshima cases included four cases in the 100+ rads exposure group, and a significant correlation with radiation dose persisted through 1974. In the period 1971–1974, however, the mortality rate was below one case per million persons per year per rad for the first time in Hiroshima and declined to below expected in Nagasaki.

The temporal changes in leukemia frequency differed not only by city, but also by age at the time of bombing. Table A-9 shows observed and expected deaths from leukemia in survivors exposed to 100+ rads in both cities combined, compared with the number expected on the basis of all Japan death rates, by broad age classes. In recent years, 1963–1974, less

TABLE A-8   Excess Leukemia Deaths per Million Persons per Year per Rad (Kerma),[a] by City

	Period	No. Deaths	
		Hiroshima	Nagasaki
	1950–1954	4.11	4.27
	1955–1958	3.67	0.30
	1959–1962	1.27	2.49
	1963–1966	1.48	0.21
	1967–1970	1.73	0.43
	1971–1974	0.90	−0.08
TOTAL	1950–1974	2.33	1.46

[a] Linear-regression estimates based on death-certificate data.

*89% detection rate, 83% confirmation rate.[39]

TABLE A-9   Observed and Expected Deaths from Leukemia at All
Japan Death Rates, in Those Exposed to 100+ Rads, by Age at Time of
Bombing and by Calendar Period[a]

| | Age at Time of Bombing, yr | | | | | | | | | |
| | 0–9 | | 10–19 | | 20–34 | | 35–49 | | 50+ | |
Period	Obs.	Exp.	Obs.	Exp.	Obs.	Exp.	Obs.	Exp.	Obs.	Exp.
1950–1954	7	0	5	0	6	0	4	0.1	1	0
1955–1958	2	0	2	0.1	2	0.1	4	0.1	3	0
1959–1962	5	0	2	0.1	1	0.1	1	0.1	1	0
1963–1966	0	0	1	0.1	2	0.1	1	0.1	2	0
1967–1970	1	0	0	0.2	3	0.1	2	0.2	1	0
1971–1974	0	0	1	0.1	3	0.1	0	0.2	0	0

[a] Reprinted with permission from Beebe et al. [4]

effect is seen in those under age 20 at the time of bombing. The most recent cases were largely in those aged 20–34 at the time of bombing (at approximately the same relative risk that persisted since the earlier years in
the same age group).

Figure A-2 shows the relative- and absolute-risk models for leukemia induction in both cities combined for different ages at the time of bombing.
Both models show the maximal increment in leukemia risk in the youngest
(0–9) and oldest (50+) survivors. Age-specific mortality trends since 1950
are shown in Figure A-3, where the number of deaths from leukemia per
1,000 living in 1950 is plotted against time, separately for those with 100+
rads and those with 0–9 rads. The rate of increase in the 0–9 age group
from 1950 to 1954 exceeded that experienced by any other group. Only in
the 50+ age group did the initial rate of mortality persist through 1970,
after which the effect may have disappeared. Cases in the youngest people
tended to occur in the earliest years (before 1957), whereas the incidence
in older survivors increased more gradually.

The first case of leukemia reported in survivors occurred over 2 yr after
exposure. The maximal incidence occurred in the mid-1950s, and the
most recent mortality data from Japan suggest that the effect may have
persisted in Hiroshima, although at a very diminished rate, as late as
1974.

Age at exposure, time after exposure, and type of leukemia are important variables influencing the risk of radiation-induced leukemia. These
factors have been examined by Bizzozero et al.[6] and by Ishimaru et al.[13] in

the experience of the atomic-bomb survivors, and the latter have provided a schematic representation of the influence of these factors that is reproduced as Figure A-4 with specific risk estimates drawn from the Ishimaru *et al.* report (Table A-10). Although sampling errors are large, the major patterns seem clear enough: Over the entire period of observation from 1950 to 1971 (Table A-11), acute forms of leukemia predominate, but in the early years, 1950–1955, acute leukemia cases do not greatly outnumber cases of chronic granulocytic leukemia; after that, the latter virtually disappears, especially among the younger members of the sample. Risks are correlated inversely with age in the earliest interval in each subtype, as well as the total for all forms of leukemia. With the passage of time, the early age-risk relation is reversed, as the excess cases disappear among those who were youngest in 1945 and continue to be substantial in those in the older groups; thus, in the last interval, the rates increase with age in 1945, most clearly and emphatically for acute leukemia, which dominates the later experience. The available data constitute firm evidence of what has been claimed previously: radiation-induced leukemia in the atomic-bomb survivors has followed a characteristic latency pattern that distinguishes these cases of leukemia from those due to other causes. Furthermore, these data provide another reason to believe that chronic granulocytic leukemia is induced by

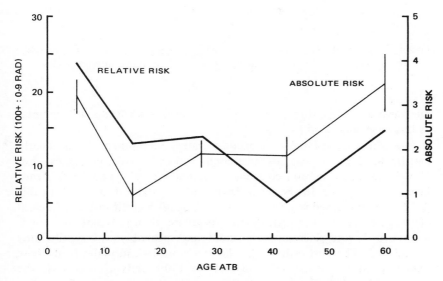

FIGURE A-2   Age-specific relative estimates and absolute-risk estimates (excess deaths per 10^6 PYR) with 90% confidence intervals. Reprinted with permission from Beebe *et al.*[5]

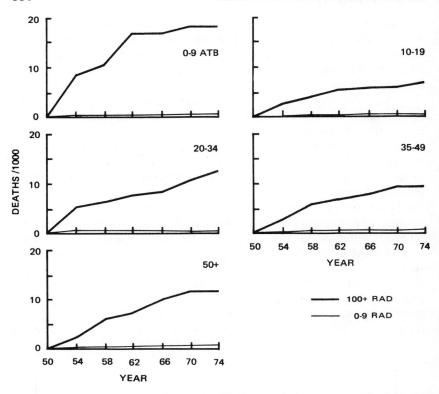

FIGURE A-3   Deaths from leukemia per 1,000 persons alive October 1, 1950, by age at time of bombing, and T65 dose, cumulative 1950–1974. Reprinted with permission from Beebe *et al.*[5]

mechanisms different from those which induce other forms of leukemia, as has been suggested by Mole.[28]

Age-adjusted leukemia mortality rates, computed from death-certificate data, reveal an increased leukemia risk in males. For Hiroshima and Nagasaki combined, the male:female ratio for 0–9 rads kerma is 1.36; for 10–99 rads, 2.86; and for 100+ rads, 1.56 (Table A-12). The increased risk in males is statistically significant; i.e., in eight of 10 age groups (>10 rads), the ratio is greater than 1. Where the ratio is less than 1, the decrement is small and probably not significantly diminished. The best estimate of the increased risk in males from these data is presumed to be 56%, in that the 100+ rads exposed have the smallest admixture of nonradiogenic cases. The factors responsible for the apparently heightened sensitivity of males are not understood.

Previous estimates of T65 kerma (dose) were computed for approximately 79,000 of the 82,200 atomic-bomb survivors included in the LSS.[15,30] New estimates of attenuation by self-shielding yield absorbed doses to active bone marrow of survivors that vary significantly from the T65 kerma estimates.[3,16-18,20] For an adult Japanese survivor, recent data show that the ratio of the low-LET absorbed dose in active marrow to kerma from gamma rays is 0.56, whereas the ratios of high-LET absorbed dose and neutron-capture gammas to kerma from neutrons are 0.28 and 0.067, respectively.[19]

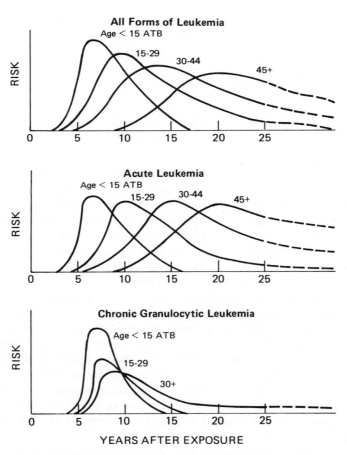

FIGURE A-4   Schematic model of influence of age at time of bombing and calendar time on leukemogenic effect of radiation (heavily exposed survivors). Reprinted with permission from Okada *et al.*[33]

TABLE A-10   Excess Mortality from Leukemia among Atomic-Bomb Survivors Exposed to 100+ Rads, by Age in 1945, Calendar Time, and Type of Leukemia[a]

Type of Leukemia	Period	No. Excess Leukemia Deaths per 100,000 per Year by Age in 1945[b]			
		< 15 yr	15–29 yr	30–44 yr	> 44 yr
Total	1950–1955	176	78	84	0*
	1955–1960	53*	20*	57*	87*
	1960–1971	0*	20	46	64*
Acute	1950–1955	101	76	24*	0*
	1955–1960	52*	38*	57*	62*
	1960–1971	0*	16*	39*	61*
Chronic	1950–1955	72	56*		26*
granulocytic	1955–1960	0*	0*		10*
	1960–1971	0*	5*		6*

[a] Data from Ishimaru *et al.*[13] (Tables 5, 6, and 7).
[b] Difference between rate for 100+ rads and for <1 rad; rates adjusted for city and sex. Asterisk indicates ratio based on fewer than five cases among 100+ rads group.

Figure A-5 compares the leukemia dose-response curve for mortality in the LSS sample with a parallel estimate based on the dose distribution of all cases in the Leukemia Registry and the dose distribution of survivors enumerated at the time of the 1950 census. The sparsity of leukemia cases in the Nagasaki LSS sample below 100 rads kerma results in apparent curvilinearity in the low-dose region, which is much less marked when all the Registry cases are used in relation to the total population enumerated in Nagasaki in 1950. In the Nagasaki LSS group, the incidence at all doses is less than observed in Hiroshima survivors, with the exception of the 0–9 rads group. The increased incidence in Hiroshima implies that neutrons are more leukemogenic than gamma rays—i.e., the neutron RBE is greater than 1. The curvilinearity requires that the neutron RBE also increase as the dose diminishes. This is thought to be due to greater repair of effects of low-dose, low-LET radiation, rather than increased damage per unit of higher-LET radiation, such as neutrons (mainly from proton recoil interactions in tissue). These concepts are based on radiobiologic evidence from many species. The current question is the strength of the evidence that similar effects apply to man.

Analyses of leukemia in the LSS samples are complicated by the fact that there were few Nagasaki survivors in the low-dose group; hence, statistical confidence in the estimation of rates in survivors who received less than 100 rads is low. Thus, Beebe *et al.*[4] were led to analyze the data for all members of the Leukemia Registry, regardless of membership in the defined samples ordinarily relied on. The use of the Leukemia Registry increases the number of cases available for analysis: in Hiroshima from 120 to 323, and in Nagasaki from 46 to 231. The population base for rate calculations is uncertain for the Leukemia Registry; however, assuming that it has the same dose distribution as survivors enumerated at the time of the 1950 population census, Beebe *et al.*[4] computed relative risks based on death certificates for survivors in the LSS sample and for the total Leukemia Registry in the entire city populations from which the Registry cases are drawn. Figure A-5 shows good correspondence in Hiroshima between the compared groups, whereas the Nagasaki curve is much less curvilinear when the total Leukemia Registry is used.

The shape of the Nagasaki curve is a strong determinant of the value for

TABLE A-11   Incidence of Leukemia in Japanese Atomic-Bomb Survivors in Life Span Study, Hiroshima and Nagasaki, 1950–1971 (Cases from Leukemia Registry)[a]

T65 Dose, rads (kerma)	Average Kerma Gamma	Average Kerma Neutron	No. Cases[b] Person-Years	No. Cases[b] AL	No. Cases[b] CGL	No. Cases per 100,000 PY[b] AL	No. Cases per 100,000 PY[b] CGL	All
Hiroshima								
400–600	381	144	9,535	10	2	104.9	21.0	125.9
200–399	211	70	19,614	8	7	40.8	35.7	76.5
100–199	109	30	32,384	9	3	27.8	9.3	37.1
50–99	57	13	51,456	3	4	5.8	7.8	13.6
1–49	9	2	469,060	11	14	2.3	3.0	5.3
<1	0	0	569,266	16	4	2.8	0.7	3.5
TOTAL	—	—	1,151,315	57	34	5.0	3.0	7.9
Nagasaki								
400–600	514	11	6,981	6	1	85.9	14.3	100.3
200–399	264	4	20,151	7	1	34.7	5.0	39.7
100–199	143	1	27,355	4	0	14.6	0.0	14.6
50–99	71	0	25,643	0	0	0.0	0.0	0.0
1–49	10	0	200,417	6	3	3.0	1.5	4.5
<1	0	0	90,944	2	0	2.2	0.0	3.3[c]
TOTAL			371,491	25	5	6.7	1.3	8.3[c]

[a] Reprinted with permission from Ishimaru *et al.*[13] (Table 1).
[b] AL = acute leukemia; CGL = chronic granulocytic leukemia.
[c] 1 case of chronic lymphocytic leukemia was included.

TABLE A-12   Leukemia Deaths per Million Atomic-Bomb Survivors per Year, Hiroshima and Nagasaki, by Age, Kerma, and Sex, 1950–1974

Age, yr	Leukemia Deaths								
	0–9 Rads (Kerma)			10–99 Rads (Kerma)			100+ Rads (Kerma)		
	M	F	M/F	M	F	M/F	M	F	M/F
0–9	25	33	0.76	40	21	1.90	1,118	301	3.71
10–19	20	19	1.05	196	19	10.32	268	319	0.84
20–34	73	24	3.04	33	39	0.85	769	499	1.54
35–39	84	54	1.56	92	19	4.84	479	228	2.10
>39	8	35	0.23	27	21	1.29	389	229	1.70
TOTAL	57	42	1.36	120	42	2.86	736	473	1.56

the RBE for neutrons derived from the Hiroshima (neutron-rich) and Nagasaki (neutron-deficient) exposures. Several analyses of neutron RBE have been reported.[13,14,19,35]

Rossi and Kellerer[35] analyzed published data and concluded from a comparison based on kerma that the biologic effectiveness of the radiation in Hiroshima relative to that in Nagasaki varied smoothly from a value of 2 at high kerma to between 5 and 25 at a kerma of 10 rads. To derive neutron RBEs, they made allowances for the fact that only about one-fifth of the kerma at Hiroshima was due to neutrons and for a twofold difference of absorption by tissues overlying the bone marrow for gamma rays and neutrons. In this case, an RBE between 50 and 250 was obtained for a marrow neutron dose of 0.5 rad, the most likely value being about 70.[35] (Rossi—personal communication—now believes that the best estimate for the variation of neutron RBE with kerma for leukemia induction in the Japanese atomic-bomb survivors is approximately $45/D_n^{1/2}$.)

To estimate the neutron RBE, Ishimaru et al.[13] considered two models, one that is linear with gamma-ray dose and one that varies with the square of the gamma-ray dose, and tested these against data for the period 1950–1971 (Table A-10). In the first model, the incidence depends linearly on both gamma-ray dose and neutron dose:

$$P = \alpha_0 + \alpha_1 D_\gamma + \beta_1 D_n,$$

where P = leukemia incidence, α_0 = baseline incidence in each city, D_γ = gamma dose, D_n = neutron dose, and α_1 and β_1 are fitted constants. In the second model, the incidence depends linearly on neutron dose and on the square of the gamma-ray dose:

$$P = \alpha_0 + \alpha_2 D_\gamma^2 + \beta_1 D_n.$$

Both models fit the leukemia-incidence data (all leukemias) in the LSS (1950–1971) equally well, and the best model cannot be discriminated statistically. For acute leukemia, the gamma-ray component (α term) is statistically most significant in each model; for chronic granulocytic leukemia, the neutron component (β_1 term) predominates.[13] (This is in agreement with earlier observations made by Mole.[28]) In fact, the gamma-ray term (α_1 or α_2 coefficient) was not significantly different from zero for induction of chronic granulocytic leukemia in either model, and the β_1 term was significant at the 0.01 level for chronic granulocytic leukemia in both models. With regard to acute leukemia, both the α_2 and β_1 terms were significantly different from zero with model II, but only α_1 was significant with the linear model. These findings held whether kerma or marrow dose was used.

The neutron RBE was computed as D_γ/D_n for those values of D_γ and D_n which produced the same effect. For model I, the RBE was computed as β_1/α_1; for model II, the RBE varies with dose and was computed as $[\beta_1/(\alpha_2 D_n)]^{1/2}$.

For acute leukemia and for all forms of leukemia, the RBEs for kerma and marrow dose derived from model II are listed in Table A-13. The

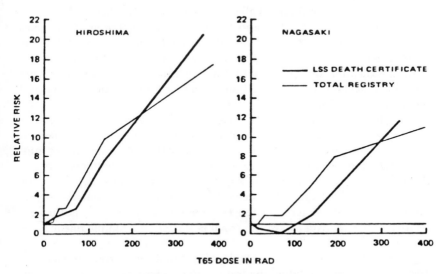

FIGURE A-5 Dose response for leukemia, LSS sample and death certificates, 1950–1974, versus atomic-bomb survivors' survey and total Leukemia Registry, 1946–1974. Reprinted with permission from Beebe *et al.*[5]

TABLE A-13 RBE for Neutrons for Leukemia in Hiroshima and Nagasaki, Model II[a]

Neutron Dose, rads	Kerma, rads		Marrow Dose, rads	
	Acute Leukemia	All Forms	Acute Leukemia	All Forms
1	30	45	32	48
10	9.6	14	10.2	15
100	3	4	3.2	4.8
500	1.3	2	1.4	2.1

[a] Data from Ishimaru *et al.* [13]

results obtained by using kerma and marrow dose are almost identical. The RBE was not computed for chronic granulocytic leukemia, because the gamma coefficient (α_1) was not significantly different from zero. The data from Hiroshima and Nagasaki are consistent with the observations that neutrons were more important than gamma rays for the induction of chronic granulocytic leukemia and that the neutron RBE for induction of chronic granulocytic leukemia greatly exceeds the values for induction of acute leukemia.

With current information on absorbed dose and published leukemia incidence data (1950–1971), Land *et al.*[21] have compared linear and quadratic dose models for the two cities on the basis of the LSS data. The three models used included the following relationships for the gamma and neutron terms: linear gamma and linear neutron (*L-L*), quadratic gamma and linear neutron (*Q-L*), and both linear and quadratic gamma and linear neutron (*LQ-L*).

Age-adjusted leukemia risk coefficients for gamma and neutron radiation (bone-marrow dose) are presented in Table A-14 by type of leukemia based on Leukemia Registry incidence data (1950–1971) from the LSS. RBE and χ^2 (goodness-of-fit) values are computed for the different model types. The value of α_1 indicates the age-adjusted risk per rad of gamma rays for the linear term in the *LQ-L* model; α_2 is the analogous parameter for the gamma-dose-squared term in the two nonlinear models; β_1 is the risk coefficient for neutron exposure, which is treated as a simple linear process.

The best-fitting model for all leukemias (total) and for acute leukemia (the *LQ-L* model) contains quadratic gamma-dose and linear neutron-dose terms. The linear model (in both gamma and neutron doses) and the

Q-L model fit the chronic granulocytic leukemia equally well, but not significantly better than the LQ-L model. The RBE for neutrons for acute leukemia, which best fits the linear model, is 5.4 $\pm$ 5.6, and the RBE for the Q-L model is $30.6/D_n^{1/2} \pm 10.7/D_n^{1/2}$.

The wide confidence intervals on RBE reflect the statistical fluctuations in the data, especially in the low-dose region, and in the Nagasaki data in particular. Analyses on larger samples, such as the Leukemia Registry,[21] provide RBE estimates that are compatible with or higher than those shown in Table A-13.

When the appropriate risk coefficients calculated for the different types of leukemia (from Table A-14) are multiplied by dose to the bone marrow, the increased risk to a population exposed to that dose can be estimated. The results of applying age-specific risk estimates to a hypothetical (but reasonable) population distribution are shown in Table A-15. If a million persons received 1 rad of gamma radiation to the bone marrow, we would

TABLE A-14   Age-Adjusted Leukemia Risk Coefficient for Gamma and Neutron Dose to Marrow, by Type of Leukemia, Hiroshima and Nagasaki Data, 1950–1971[a]

Type of Leukemia	Model[b]	α_1	α_2	β_1	RBE	χ^2	df	p
Total	L-L	2.24	—	25.4	11.3	11.5	12	0.5
	Q-L	—	0.0137	31.1	$47.6/D_n^{1/2}$	12.3	12	0.4
	LQ-L	0.99	0.0085	27.5	≤ 27.8	10.4	11	0.5
Acute	L-L	2.02	—	10.9	5.4	20.6	12	0.07
	Q-L	—	0.0147	13.7	$30.6/D_n^{1/2}$	17.9	12	0.12
	LQ-L	0.19	0.0136	13.0	≤ 68.6	17.6	11	0.09
Chronic	L-L	0.013	—	16.5	[c]	20.6	12	0.07
granulocytic	Q-L	—	0.00004	16.6	[c]	20.6	12	0.07
	LQ-L	0.012	0.0	16.5	[c]	20.6	11	0.05

[a] Leukemia Registry incidence cases in the Life Span Study sample from Ishimaru *et al.*;[11] regression coefficients and RBE estimates from Land *et al.*[21]

[b] L-$L = \alpha_0 + \alpha_1 D_\gamma + \beta_1 D_n$

Q-$L = \alpha_0 + \alpha_2 D_\gamma^2 + \beta_1 D_n$

LQ-$L = \alpha_0 + \alpha_1 D_\gamma + \alpha_2 D_\gamma^2 + \beta_1 D_n$

with α_0, the risk at zero dose

with α_1, the risk coefficient per rad of gamma dose

with α_2, the risk coefficient per (rad of gamma dose) squared

with β_1, the risk coefficient per rad of neutron dose

[c] Not calculated.

TABLE A-15   Leukemia Risk per 10^6 PY[a]

Type of Leukemia	Model	Leukemia Risk					
		Gamma-Ray Dose to Marrow, rads			Neutron Dose to Marrow, rads		
		1	10	100	1	10	100
All types	L-L	2.2	22.4	224	25	254	2,540
	LQ-L	1.0	10.8	184	28	275	2,750
	Q-L	0.01	1.4	137	31	311	3,110
Acute	L-L	2.0	20.2	202	11	109	1,090
	LQ-L	0.2	3.3	155	13	130	1,300
	Q-L	0.01	1.5	147	14	137	1,370
Chronic granulocytic	L-L	0.01	0.13	1.3	17	165	1,650
	LQ-L	0.01	0.1	1.2	17	165	1,650
	Q-L	0.	0.	0.4	17	166	1,660

[a] Data based on Table A-14.

expect to see on the average 2.2 cases of leukemia per year, starting 2 yr after exposure and continuing for 25 yr thereafter, if a linear model is correct. If the linear-quadratic model is correct (it fits the data slightly better than the linear-linear model), the expected number falls to 1.0; and, if the quadratic-linear dose model is used, only 0.01 additional case per year would be expected. The differences between the various model predictions after gamma-ray exposure can be appreciable at the lowest doses, but the differences at the higher doses are considerably smaller. The risk of chronic granulocytic leukemia after gamma-ray exposure is much smaller than the risk of acute leukemia at all doses.

The possible effect of dose rate has not been included in these models. Thus, if sparing due to low dose rate or dose fractionation occurs, as it does in lower mammals, then the gamma-dose risk estimates are too high. Differences in leukemia risk between males and females have not been taken into account explicitly.

The risk estimates presented are based on leukemia-incidence data for the period from October 1, 1950, through December 31, 1971, and not for the period of presumed excess risk (mid-1947 through 1970 for Nagasaki and through 1974 for Hiroshima). In the excluded periods, the numbers of cases do not exceed (on an average annual basis) those occurring during the study period. The exclusion of the early cases is of interest with respect to neutron RBE, in that a high proportion (9 of 13) of the early cases had chronic granulocytic leukemia and were excluded from rate calculations. The addition of these chronic granulocytic leukemia cases would further increase the values estimated for neutron RBE.

Leukemia in Ankylosing-Spondylitis Patients

Since 1957, important information on leukemia risk after human exposure to ionizing radiation has been derived from a series of investigations of British ankylosing-spondylitis patients treated with radiation therapy directed toward the spine. Reservations concerning the significance of excess risk observed in patients after radiation therapy for a serious medical condition arise when the incidence of associated disease in nonirradiated patients with the same primary disease is not known. A recent review of cancer mortality in 859 British ankylosing-spondylitis patients diagnosed between 1935 and 1957 who were not treated with x-ray therapy and were followed for over 13 yr through 1967 has gone far to allay some of these reservations.[37] Leukemia, lymphoma, and aplastic anemia, which were increased after radiation therapy, were not observed to be increased in nonirradiated patients. Although it is still possible that more seriously ill patients received radiation therapy and that radiation in severely ill patients may potentiate a leukemogenic effect, this study suggests that ankylosing-spondylitis patients do not have a high spontaneous incidence of malignant disease.

The leukemia-mortality patterns following a single course of radiation therapy (average, 10 x-ray treatments over a period of 5–6 wk) were determined through 1969—an average followup of 16.2 yr after treatment.[36] A mean whole-marrow dose of 321 rads and a 2-yr minimal latent period were estimated. (Recent unpublished data from Fabrikant and Lyman suggest that the average marrow dose was about 214 rads.) There were 29.15 excess leukemia deaths in 112,970 PY, or an estimated absolute risk of 0.8 leukemia death per 10^6 PY per rad, according to R. Doll and P. G. Smith (personal communication)—1.2, according to Fabrikant and Lyman's dose estimate. The leukemia induced in these patients was acute, and the increment in incidence was most marked in the older irradiated patients. The smaller increase in chronic granulocytic leukemia is consistent with expectation, owing to the absence of neutron exposure in x-ray therapy.

The time course of the increased risk can be assessed most readily from the followup of patients given a single course of x-ray treatment.[36] Excess risk first appeared 2 yr after the start of treatment and persisted for 20 yr. This, too, is in good agreement with observations on atomic-bomb survivors.

Leukemia in Other Irradiated Human Populations

An increased incidence of leukemia has been observed in Portuguese, Danish, and German patients who received Thorotrast (thorium dioxide) as a contrast agent for diagnostic radiography.[29] In 3,772 patients, many followed for over 30 yr, after an average dose of 25 ml of Thorotrast, 44 cases of leukemia were observed.[29,40] Bone-marrow exposures were estimated to average 270 rads by 30 yr after administration.[29]

The first leukemia cases were seen about 8 yr after Thorotrast injection;[8] cases are continuing to appear in excess.[29] Only seven of the 44 cases were chronic granulocytic leukemia in the persons exposed to high-LET radiation. A possible explanation for the smaller number of such patients than expected, if chronic granulocytic leukemia is induced preferentially by high-LET radiation, may be cell inactivation or death of transformed cells at the high doses received by these patients. The excess risk was estimated as 40 cases per million persons per rad, when the dose was calculated 30 yr after the dose was administered.[29]

Two investigators have reported leukemia after therapeutic scalp irradiation for tinea capitis in children. The followup of 2,043 children who received an estimated 30 rads (average marrow dose) revealed four leukemia cases versus 0.9 expected, for a relative risk of 4.4.[1] The absolute incidence was estimated at 3.4 per million persons per year per rem.[23] More recently, Modan *et al.*[26] surveyed a large group (10,902 irradiated, 10,902 population controls, and 5,496 sibling controls) over a period of 12–23 yr and found a smaller increment in risk: seven cases in the irradiated (0.6 case/10^3), five in the population controls (0.5 case/10^3), and two in the sibling control (0.4 case/10^3). The relative risk in this study thus was 1.4–1.8, and this was not judged by the authors to be a demonstrated effect in this study group, in which the cranial marrow dose was estimated to be several hundred rads.[42]

The mortality experience among radiologists in the United States includes a significant increase in leukemia mortality among those in medical practice during the period 1920–1939; no excess was observed in radiologists who began practice thereafter.[25] Uncertainty of the dose for the leukemia cases and the population at risk make it impossible to establish a dose-response curve for these persons. Estimates of lifetime radiation exposures to the bone marrow for radiologists practicing in the 1920s and 1930s are 600 and 240 rads, respectively (BEIR I). This equates to a lifetime risk of 20–50 excess cases per million persons per rad, which is close to the value for low-LET irradiation of atomic-bomb survivors and ankylosing-spondylitis patients. Taken at face value, this observation suggests that protraction of exposure over many years does not diminish the leukemia rates below that observed after single high-dose, high-dose-rate exposures. Lack of knowledge of the doses received by radiologists who developed leukemia and those who did not makes such a conclusion highly suspect.

Two new studies have shown that leukemia is increased in persons given radium intravenously for medical therapy and in radium-dial-painters.[34,38]

The incidence of bone cancer in German patients given repeated intravenous injections of radium-224 is summarized later in this appendix. Among the 816 traced patients of known dose and injection span in the Spiess *et al.* series,[38] two cases of leukemia occurred. Paramyeloblastic leukemia was diagnosed in a 44-yr-old man and lymphatic leukemia (presumed to be acute) was diagnosed in a 29-yr-old woman; they had received radium-224 2 and 3 yr previously, respectively. In a population in which 0.8 case was expected, three cases of panmyelosis were diagnosed (two in adults, one in an adolescent). All cases had onset 2–4 yr after the start of intravenous injections. The authors noted that juveniles received higher doses, but no cases of malignant hematologic conditions occurred in that age group. The red marrow dose was estimated to be approximately 60% of the average skeletal dose from radium-224. The average skeletal dose was computed as being 282 rads in adult females and 186 rads in adult males, from high-LET, alpha-particle radiation. The short, well-defined latent period and the significantly increased incidence suggest that these effects are real.

Mortality in a cohort of 634 women who worked in the United States as painters of radium dials between 1915 and 1929 has recently been reviewed.[34] In addition to large excesses in bone cancer (22 observed versus 0.3 expected), diseases of blood and blood-forming organs were increased significantly (4 observed versus 1.0 expected). The number of deaths from leukemia and related conditions was significantly greater than expected in the period before 1945, whereas bone cancers were observed to peak later. These observations are consistent with data from the Japanese atomic-bomb survivors. Unfortunately, dosimetry studies are lacking; hence, risk coefficients cannot be computed and compared between series. The findings, however, closely parallel observations described above on patients given radium-224 intravenously.

Knowledge regarding the incidence of neoplasms after therapeutic irradiation of infants for enlarged thymic glands has been extended by a fourth survey of a well-studied sample.[9] This represents a 20-yr followup of nearly 3,000 treated infants and approximately 5,000 nonirradiated siblings. One additional leukemia case was identified in the fourth survey, bringing the total to seven (2.27 expected). In the most heavily irradiated subgroup, two cases were observed in the earlier surveys (0.25 expected). Fewer cases than expected were observed in the sibling comparison groups. Dose information does not permit an accurate assessment of leukemia risk coefficients, but it appears likely that leukemia risk was increased in these subjects, especially in the earlier years after therapy.[9]

Other Hematologic Cancers in Man

Aplastic Anemia   An increased incidence of aplastic anemia was reported in the Japanese atomic-bomb survivors in the early years after exposure.[22] It is not certain whether this effect was real and quickly attenuated or was a secondary manifestation of other diseases. More recent surveys have revealed that, in 56 patients among the atomic-bomb survivors who had confirmed diagnoses of aplastic anemia between 5 and 28 yr after exposure, there was no temporal trend in incidence; and aplastic anemia did not increase in frequency at the time that the leukemia incidence increased. The incidence of unconfirmed aplastic anemia (death-certificate diagnoses) was 10 times higher in survivors exposed to 100 rads or more than in those exposed to less than 1 rad, but analysis showed no association between exposure and incidence in the confirmed cases.[10]

Aplastic anemia was reported in the British ankylosing-spondylitis patients who received radiotherapy to the spine, but it was finally concluded that this was aleukemic leukemia in many of the patients.[7]

In addition to the increased mortality from leukemia in American radiologists, 17 excess deaths occurred from aplastic anemia, compared with none in the cohort (1930–1939), with an average followup of 30 yr. No reliable estimates of radiation dose are available, and it was not possible to locate and review the histologic evidence on classification. Lewis also reviewed the disease experience of American radiologists and suggested that an excess of aplastic anemia had been observed during the years 1948–1961.[23]

Malignant Lymphoma   Evidence from human studies, primarily those of the atomic-bomb survivors and the British ankylosing-spondylitis patients, indicates that lymphoma incidence after radiation exposure is increased, but to a lesser extent than the incidence of leukemia.[4,7] Additional evidence is available from studies of American radiologists,[25] uranium-mill workers,[2] and infants irradiated for thymic enlargement.[9]

• *Japanese Atomic-Bomb Survivors*: Pathology data (autopsy and biopsy-proven cases) from the LSS (Nishiyama *et al.*[32] and A. Steer *et al.*, in preparation) demonstrate an increased incidence of malignant lymphoma in heavily exposed (over 100 rads kerma) Hiroshima survivors, but for Nagasaki survivors the data are too few for a judgment to be made. The induction interval was shortest in the youngest heavily exposed survivors.

In the two cities, 10 malignant lymphomas were observed in persons exposed to kerma greater than 100 rads, and 2.3 were expected—i.e., a greater than fourfold increase—and the greatest excess occurred in Hiroshima (Table A-16). In this group, among 3,128 survivors, there were four cases of lymphosarcoma (0.4 expected), one case of reticulum cell sarcoma (0.5 expected), and three cases of Hodgkin's disease (0.2 expected). In the LSS, there were 75 deaths attributed on death certificates to cancer of the lymphatic and hematopoietic tissue, other than leukemia in the ABCC LSS (1950-1974).[4] All patients exposed to more than 100 rads in this group had confirmed pathologic diagnoses. Thus, the effect was not due to the inclusion of misdiagnosed cases of leukemia, for example. There were too few cases to support definite conclusions with respect to trends over time, but four of the 10 cases observed in subjects exposed to 100 rads or more occurred in 1971-1974, and only one in the 1950-1954 period—a distribution completely different from that observed in leukemia cases. There were too few cases to draw conclusions about variations with age, but the major effect was not in young persons (those under 20 at the time of bombing). The ratio of excess leukemia to excess lymphoma deaths in the Japanese survivors above 100 rads kerma (from

TABLE A-16   Expected and Observed Frequencies of Malignant Lymphoma in Japanese Atomic-Bomb Survivors[a]

City	Dose, rads (kerma)	No. Cases Malignant Lymphoma	
		Observed	Expected[b]
Hiroshima	1-99	7	11.0
	≥ 100	8	1.3
Nagasaki	1-99	6	4.7
	≥ 100	2	1.0
TOTAL	—	23	18.0
	Excess	5.0	

[a] Derived from Nishiyama *et al.*[32]
[b] Based on rate observed in 37,675 survivors,[29] who received < 1 rad (kerma).

Tables A-9 and A-16) is $61.1/7.7 = 8$. It will be important to determine whether different types of irradiation induce similar relative ratios of risk.

• *British Ankylosing-Spondylitis Patients*: Recent reviews of mortality in ankylosing spondylitis (Doll and Smith, personal communication) have shown that 13 lymphoma deaths were observed after radiotherapy, whereas 6.59 were expected—6.41 excess deaths in a period of 6 yr or more. For leukemia, the excess is $31 - 6.47 = 24.5$ for 0+ yr, 19.53 for 3+ yr, and 13.11 for 6+ yr. Doll and Smith prefer to use the 3+-yr latent period for calculation of the leukemia risk. The ratio of excess leukemia to excess lymphoma is thus $19.53/6.41 = 3.0$. Fabrikant and Lyman have reanalyzed the dosimetry data and, assuming that the disease originates in mediastinal lymph nodes, computed an absolute risk of 0.3 cases per million persons per year per rad absorbed dose to the mediastinal lymph nodes.

• *Other Human Populations*: Increased mortality from malignant lymphoma has also been observed in two groups of workers occupationally exposed to ionizing radiation. These are radiologists[25] and uranium-mill workers,[2] but dosimetry is inadequate for quantitative risk estimation in both situations.

In addition to the increased leukemia mortality noted in radiologists (American College of Radiologists), in comparison with pathologists (College of American Pathologists), from 1920 to 1939, increased mortality from neoplasms involving the lymphatic and hematopoietic system (excluding leukemia) was noted from 1930 to 1949. The standard mortality ratios (SMRs) computed were 3.57 and 1.61 for the two specialties for 1930–1939 and 5.71 and 1.04 for 1940–1949. The lymphoma increase was observed later than the leukemia effect; this corresponds to the experience in the Japanese atomic-bomb survivors.

Following 662 men employed in U.S. uranium mills from 1950 through 1967, Archer *et al.*[2] observed 104 deaths in the workers, compared with 105 expected. Although the numbers in the study are small, the one condition that showed a significant increase above expected was malignant neoplasia involving the lymphatic system (excluding leukemia): four observed versus 1.02 expected. The radioactive nuclides identified in air in aerosol form included uranium-238, uranium-234, thorium-230, radium-228, radium-226, and lead-210. Urine analyses (1950–1953) showed average uranium concentrations of 8.2 μg/L, with 10% of the workers excreting uranium at concentrations up to 20 times as high. Air concentrations were thought to be approximately 10 times higher, on the average, than ICRP standards, as defined in 1959.

In addition to thyroid cancer and leukemia, an increased incidence of malignant lymphoma has been identified in the followup of infants given radiation therapy for thymic enlargement.[9] Two new cases were observed in the treated infants in the fourth survey (bringing the total to eight, compared with 3.97 expected). The highest dose group contained two cases, with 0.49 expected. No increase was observed in the sibling control groups. Thus, from all the evidence reviewed, it seems highly probable that malignant lymphoma is increased in children after high levels of radiation exposure. The appearance of increased risk comes later than for leukemia and may last longer.

Multiple Myeloma   Since the 1973 study,[32] a more extensive study of multiple myeloma[12] has identified 22 confirmed cases among survivors in the LSS sample: 14 in Hiroshima and eight in Nagasaki. With the data of the two cities pooled for analysis, those exposed to 100+ rads kerma have a significantly increased risk relative to that of those exposed to less than 1 rad (relative risk, 4–7). Excess mortality was estimated at 0.45 death per million persons per rad of marrow dose for the mixed neutron and gamma radiation.[12]

Summary

Extensive information has been accumulated on hematologic malignancies (especially leukemia) after whole-body radiation exposure. Because of the low natural occurrence of leukemia, the high radiosensitivity of stem cells, and the short minimal latent period (2–3 yr), leukemia was recognized early as a potential consequence of high-level radiation exposure in man.[31] To understand the pathophysiology of these diseases and the mechanisms by which radiation exerts its effects, many experimental studies have been conducted in a large number of animal strains. These have led to several important conclusions that bear on the interpretation of dose-response curves derived from epidemiologic studies in man. Given the body of knowledge derived from controlled animal studies and parallel (but uncontrolled) studies in persons exposed to high doses (where statistically valid observations can be made), we have been led to fit the observed data to models that are consistent with animal and human data.

Leukemia induction after low doses of low-LET radiation in experimental studies in mammals is curvilinear, with decreasing effectiveness per rad at the lowest doses. Also, a decrease in effect is seen at progressively diminished dose rates. The data indicate that small amounts of radiation-induced damage can be repaired to a substantial extent. Similar repair mechanisms are believed to operate in man, inasmuch as they occur in so many different organisms.

Several models consistent with the data in man have been discussed. Linear and quadratic models have been considered with different estimates of RBE from the Japanese atomic-bomb survivors. Because this group includes all ages and both sexes, findings in these survivors provide a basis for estimation of risk in man after a single exposure at any time in life.

Assuming that a million persons are exposed to 1 rad of gamma radiation, the total expected number of radiation-induced cases of leukemia that would appear during the 25-yr period starting 2 yr later would vary between 0.01×25 and 2.2×25, or between 0.25 and 55, depending on which proposed model is correct. These would include primarily cases of acute leukemia, with fewer cases of chronic myeloid leukemia, but no chronic lymphocytic leukemia.

The two large populations whose followup has provided the most useful information are the Japanese atomic-bomb survivors and the British ankylosing-spondylitis patients. The most recent age-specific estimates of excess leukemia mortality in the British patients are those of Doll and Smith (personal communication) for the period after a single course of x-ray therapy. The risk increases rapidly with age at irradiation, from 13 per 10^5 PY at risk for those aged 25–34 at the time of therapy to 52 for those aged 55 or older. If these estimates are transformed to excess deaths per million persons per year per rad, on the assumption that the average marrow dose was 214 rads, and compared with those obtained for the atomic-bomb survivors in Tables V-16 through V-18, the agreement is fairly close for the *L-L* and *LQ-L* models. The comparison in Figure A-5, it should be pointed out, which focuses on the influence of age, makes use of marrow dose for the atomic-bomb survivors in terms of rads. Conversion to rems, with an RBE of 10–15, would bring the two curves much closer together. All the indications are that the incidences of radiation-induced leukemia observed in these two well-studied irradiated human populations are in very close accord. The increased incidence of leukemia in irradiated human populations is a well-documented effect. An increased incidence of lymphoma has also been detected in several populations, although the effect is less striking. The ratio of leukemia mortality to lymphoma mortality observed in the atomic-bomb survivors is 8, and in the ankylosing-spondylitis patients, 3.0. The concordance of these diseases in surveys may give additional confidence in establishing a cause-effect relationship.

Because data are available for age and sex groups only for the atomic-bomb survivors, we have used the risk estimates obtained from the Japanese survivors for estimating risks for other groups exposed to high-dose-rate radiation at relatively high doses, i.e., upwards of 1 rad. Until we know the radiobiologic basis for leukemia induction, we cannot be confident regarding the choice of model or parameter values for use in risk calculations at even lower doses. It should be recognized that risk estimated at a selected point in the high-dose region may overestimate the magnitude of hazards of low-dose exposures by a factor of 2–10, depending on the type of radiation, its rate of delivery, and the high-dose point at which the observations were made. It seems reasonable to estimate risk from the most similar set of exposure circumstances for which specific assessments have been made, rather than to try to establish a single risk figure for radiation-induced leukemia. The estimated risks from the atomic-bomb and ankylosing-spondylitis treatment represent upper limits, in that both are derived from high-dose-rate exposures.

REFERENCES

1. Albert, R. E., and A. R. Omran. Follow-up study of patients treated by x-ray epilation for tinea capitis. Arch. Environ. Health 17:899–918, 1968.
2. Archer, V. E., J. K. Wagoner, and F. E. Lundin, Jr. Cancer mortality among uranium mill workers. J. Occup. Med. 15:11–14, 1973.
3. Auxier, J. A. Physical dose estimates for A-bomb survivors. Studies at Oak Ridge, U.S.A. J. Radiat. Res. (Tokyo) 16(Suppl.):1–11, 1975.
4. Beebe, G. W., H. Kato, and C. E. Land. Life Span Study Report 8. Mortality Experience of Atomic Bomb Survivors, 1950–74, pp. 64–67. Radiation Effects Research Foundation Technical Report TR 1-77. Hiroshima: Radiation Effects Research Foundation, 1978.
5. Beebe, G. W., H. Kato, and C. E. Land. Studies of the mortality of A-bomb survivors. 6. Mortality and radiation dose, 1950–1974. Radiat. Res. 75:138–201, 1978.
6. Bizzozero, O. J., Jr., K. G. Johnson, A. Ciocco, et al. Radiation-related leukemia in Hiroshima and Nagasaki 1946–64. 2. Observations on type specific leukemia, survivorship and clinical behavior. Ann. Int. Med. 66:522–530, 1967.
7. Court Brown, W. M., and R. Doll. Mortality from cancer and other causes after radiotherapy for ankylosing spondylitis. Brit. Med. J. 2:1327–1332, 1965.
8. Faber, M. Malignancies in Danish Thorotrast patients. Health Phys. 35:153–158, 1978.
9. Hempelmann, L. H., W. J. Hall, M. Phillips, et al. Neoplasms in persons treated with x-rays in infancy. Fourth survey in 20 years. J. Natl. Cancer Inst. 55:519–530, 1975.
10. Ichimaru, M., and T. Ishimaru. Aplastic Anemia and Atypical Leukemia in the A-bomb Survivors and Control in the Fixed Cohort of Hiroshima and Nagasaki, 1950–1973. Radiation Effects Research Foundation Technical Report Draft Document, February 2, 1977.
11. Ichimaru, M., T. Ishimaru, J. Belsky, T. Tomiyasu, N. Sadamori, T. Hoshino, M. Tomonaga, N. Shimizu, and H. Okada. Incidence of Leukemia in Atomic Bomb Survivors, Hiroshima and Nagasaki, 1950–71, by Radiation Dose, Years after Exposure, Age, and Type of Leukemia. Radiation Effects Research Foundation Technical Report TR 10-76. Hiroshima: Radiation Effects Research Foundation, 1976.
12. Ichimaru, M., T. Ishimaru, M. Mikami, and M. Matsunaga. Incidence of multiple myeloma among atomic bomb survivors, Hiroshima and Nagasaki, by dose, 1950–1976. (in preparation)

13. Ishimaru, T., M. Otake, and M. Ichimaru. Dose-response relationship of neutrons and γ rays to leukemia incidence among atomic-bomb survivors in Hiroshima and Nagasaki by type of leukemia, 1950–1971. Radiat. Res. 77:377–394, 1979.

14. Jablon, S. Environmental Factors in Cancer Induction. Appraisal of Epidemiologic Evidence. Leukemia, Lymphoma and Radiation, pp. 239–243. In Cancer Epidemiology Environmental Factors (Proceedings of 11th International Cancer Congress, Florence, 1974). Vol. 3. Amsterdam: Excerpta Medica ics 351, 1975.

15. Jablon, S., and H. Kato. Studies of the mortality of A-bomb survivors. 5. Radiation dose and mortality. Radiat. Res. 50:649–698, 1972.

16. Jones, T. D. A Chord Simulation for Insult Assessment to the Red Bone Marrow. ORNL Report 5191. Oak Ridge, Tenn.: Oak Ridge National Laboratory, 1976. 44 pp.

17. Jones, T. D. Radiation Insult to the Active Bone Marrow as Predicted by a Method of Chords. ORNL TM/5337. Oak Ridge, Tenn.: Oak Ridge National Laboratory, 1976. 36 pp.

18. Jones, T. D., J. A. Auxier, J. S. Cheka, and G. D. Kerr. *In vivo* dose estimates for A-bomb survivors shielded by typical Japanese houses. Health Phys. 28:367–381, 1975.

19. Kerr, G. D. Organ Dose Estimates for the Japanese Atomic-Bomb Survivors. ORNL-5436. Springfield, Va.: National Technical Information Service, 1978. 46 pp.

20. Kerr. G. D., *et al.* An analysis of leukemia data from studies of atomic-bomb survivors based on estimates of absorbed dose to active bone marrow, pp. 714–718. In Proceedings of IVth International Congress of International Radiation Protection Association, Vol. 3, Paris, April 24, 1977.

21. Land, C. E., G. W. Beebe, and S. Jablon. Role of Neutrons in Late Effects of Radiation among A-Bomb Survivors. Presented at Health Physics Society, Minneapolis, Minnesota, June 21, 1978.

22. Lange, R. D., S. W. Wright, M. Tomonaga, *et al.* Refractory anemia occurring in survivors of the atomic bombing in Nagasaki, Japan. Blood 10:312–324, 1955 (Atomic Bomb Casualty Commission Technical Report TR 23-59).

23. Lewis, E. B. Leukemia, multiple myeloma, and aplastic anemia in American radiologists. Science 142:1492–1494, 1963.

24. Major, I. R., and R. H. Mole. Myeloid leukaemia in x-ray irradiated CBA mice. Nature 272:455–456, 1978.

25. Matanoski, G. N., *et al.* The current mortality rates of radiologists and other physician specialists. Specific causes of death. Amer. J. Epidemiol. 101:199–210, 1975.

26. Modan, B., *et al.* Radiation-induced head and neck tumours. Lancet 1:277–279, 1974.

27. Mole, R. H. Cancer production by chronic exposure to penetrating gamma radiation. Natl. Cancer Inst. Monogr. Ser. 14:217, 1964.

28. Mole, R. H. Ionizing radiation as a carcinogen. Practical questions, and academic pursuits. Brit. J. Radiol. 48:157–169, 1975.

29. Mole, R. H. The radiobiological significance of the studies with ^{224}Ra and Thorotrast. Health Phys. 35:167–174, 1978.

30. Moriyama, I. M., and H. Kato. Life Span Study Report 7. Mortality Experience of A-Bomb Survivors, 1970–1972, 1950–1972. Atomic Bomb Casualty Commission Technical Report TR 15-73. Hiroshima: Atomic Bomb Casualty Commission, 1973.

31. National Research Council, Advisory Committee on the Biological Effects of Ionizing Radiations. The Effects on Populations of Exposure to Low Levels of Ionizing Radiation, pp. 103, 114. Washington, D.C.: National Academy of Sciences, 1972.

32. Nishiyama, H., R. E. Anderson, T. Ishimaru, *et al.* The incidence of malignant lymphoma and multiple myeloma in Hiroshima and Nagasaki atomic bomb survivors, 1945–1965. Cancer 32:1301–1309, 1973.

33. Okada, S., H. B. Hamilton, N. Egami, S. Okajima, W. J. Russell, and K. Takeshita, Eds. A review of thirty years study of Hiroshima and Nagasaki atomic bomb survivors. J. Radiat. Res. (Tokyo) 16 (Suppl.):1–164, 1975.
34. Polednak, A. P., A. F. Stehney, and R. E. Rowland. Mortality among women first employed before 1930 in the U.S. radium dial-painting industry. Amer. J. Epidemiol. 107:179–195, 1978.
35. Rossi, H. H., and A. M. Kellerer. The validity of risk estimates of leukemia incidence based on Japanese data. Radiat. Res. 58:131–140, 1974.
36. Smith, P. G., and R. Doll. Age- and Time-Dependent Changes in the Rates of Radiation-Induced Cancers in Patients with Ankylosing Spondylitis Following a Single Course of X-Ray Treatment, pp. 205–218. IAEA Report SM-224/711. Vienna: International Atomic Energy Agency, 1978.
37. Smith, P. G., R. Doll, and E. P. Radford. Cancer mortality among patients with ankylosing spondylitis not given x-ray therapy. Brit. J. Radiol. 50:728–734, 1977.
38. Spiess, H., A. Gerspach, and C. W. Mays. Soft-tissue effects following ^{224}Ra injections into humans. Health Phys. 35:61–81, 1978.
39. Steer, A., I. M. Moriyama, and K. Shimizu. ABCC-JNIH Pathology Studies, Hiroshima and Nagasaki. Report 3. The Autopsy Program and the Life Span Study: Jan. 1951–Dec. 1970. Atomic Bomb Casualty Commission Technical Report TR 16-73. Hiroshima: Atomic Bomb Casualty Commission, 1973.
40. U.N. Scientific Committee on the Effects of Atomic Radiation. Sources and Effects of Ionizing Radiation, p. 376. 1977 Report to the General Assembly, with Annexes. New York: United Nations, 1977.
41. Upton, A. C. Radiobiological effects of low doses. Implications for radiological protection. Radiat. Res. 71:51–74, 1977.
42. Weyner, A., B. Modan, and D. Davidoff. Doses to brain, skull and thyroid, following x-ray therapy for tinea capitis. Phys. Med. Biol. 13:247–258, 1968.

ESOPHAGUS

Esophageal tissue is not thought to be especially sensitive to the carcinogenic action of ionizing radiation,[8] but its precise sensitivity remains to be established.[9] Experimental data relevant to the esophagus in man are primarily those of Warren and Gates[5,11] derived from continuous gamma irradiation of the esophagus and some incidental findings reported by Upton *et al.*[10] and Cosgrove *et al.*[3] on the forestomach of mice; but there are data on man based on x-ray therapy for ankylosing spondylitis (Court Brown and Doll[4] and R. Doll, personal communication) and on exposure to the atomic-bomb explosions in Japan.[1,2]

Warren and Gates,[5,11] reporting on experiments in which rodents of five different species were continuously exposed to gamma radiation from a cobalt-60 source, concluded that esophageal tissue was generally sensitive to the carcinogenic action of radiation, with little or no variation among eight strains of mice, but some variation among species. The incidence in mice was about half that in other species. Total doses varied from 8,500 to

1,900,000 rads. Incidence depended on both dose rate and total dose, and a dose rate in the range of 751–1,000 rads/d appeared to be maximally effective for all species; both higher and lower dose rates were less effective. In their analysis of late effects of atomic-bomb irradiation of mice, Upton *et al.*[10] noted that squamous-cell carcinoma of the forestomach was the most common form of stomach tumor observed in their animals, but that no significant excess was seen in either gamma- or neutron-irradiated mice. Cosgrove *et al.*,[3] reporting on a large series of experiments on laboratory rodents of several kinds, recorded a relatively large increase in tumors of the stomach, and especially in squamous-cell carcinomas of the forestomach, after irradiation. They also stated that neutron irradiation appeared to be more effective in one strain of mouse (LAF₁), but not in another (RF).

In the 1965 report on the mortality of patients with ankylosing spondylitis treated with x rays, Court Brown and Doll found no evidence of excess mortality from cancer of the esophagus, but both observed and expected frequencies were very low.[4] For the period beginning 6 yr after therapy, there were three observed versus 2.25 expected in the complete followup

TABLE A-17   Observed and Expected Deaths from Esophageal Cancer among Atomic-Bomb Survivors, RERF Mortality Sample, 1950–1974, by City[a]

| T65 Dose, rads (kerma) | No. Deaths from Esophageal Cancer | | | |
| | Hiroshima | | Nagasaki | |
	Observed	Expected	Observed	Expected
0	57	50.9	10	7.4
1–9	12	24.2	5	8.8
10–49	18	18.8	7	5.5
50–99	6	5.1	3	2.1
100–199	2	3.5	2	2.1
200–299	7	1.2	1	1.1
300–399	0	0.5	0	0.5
>399	3	0.9	0	0.5
TOTAL	105	105	28	28
Test *p*s: Homogeneity	0.01		0.50	
Trend	0.01		0.50	
Excess deaths per 10⁶ PY per rad	0.39		−0.09	
90% confidence limits	0.18, 0.60		−0.31, 0.13	

[a] Data from Beebe *et al.*[2]

experience through 1959, and three versus 3.37 in the incomplete experience through 1962. In the recent report on causes of death among patients with ankylosing spondylitis, after a single treatment course with x rays, Doll and Smith found nine deaths versus 4.27 expected in the more than 10-yr interval beginning 6 yr after therapy and ending January 1, 1970 (Doll, personal communication). The difference was significant at about the 0.04 level, but no precise dose information is available for this heavily irradiated site. Patients with ankylosing spondylitis *not* treated with x rays have recently been followed up and found to have no excess risk of such cancer.[6]

For the atomic-bomb survivors, the mortality data for the period 1950–1974 provide more definite evidence that esophageal tissue is sensitive to the carcinogenic influence of ionizing radiation.[2] Table A-17 gives the experience of 60,470 Hiroshima survivors and 19,255 somewhat younger Nagasaki survivors. Only for exposure to the Hiroshima bomb did the observed deaths exceed expectation in dose-related fashion ($p <$ 0.01). Because 19–27% of the total kerma dose in Hiroshima, but much less of the unknown tissue dose, is attributable to neutrons, it is quite possible that some of the observed excess is attributable to the neutron dose, which constituted only 1–2% of the total kerma dose in Nagasaki. The observations are too few to testify in support of any specific dose-response function. The effect is apparent in both sexes and is thus far seen only in those who were 35 or older when exposed in 1945. Age-specific regression estimates of excess deaths per 10^6 PY per rad are not available for the Hiroshima experience alone; but, for both cities combined, the estimates are as follows:[1]

	All Ages	0–9	10–19	20–34	35–49	50+
H + N	0.19	0	0.06	−0.04	0.21	1.80
H	0.39			------(not available)------		
N	−0.09			------(not available)------		

The 90% confidence interval for those aged 50 or older in 1945 is 0.48–3.12. The effect is not especially concentrated in time.

In summary, human esophageal tissue may be sensitive to the carcinogenic action of ionizing radiation, but the evidence is neither extensive nor strong. The small experience of the Nagasaki survivors is entirely negative as to the effect of gamma radiation, but followup of the ankylosing-spondylitis patients treated with x rays is more than suggestive. The strongest evidence of the effect, and the only basis for estimating its magnitude, is derived from the followup of the Hiroshima survivors ($p <$ 0.01), but applies to a mix of neutron and gamma radiation. Because esophageal cancer has a 70% detection rate by death certificate and 70% of such death-certificate diagnoses are confirmed by autopsy in the experience of the ABCC,[7] the linear estimate of 0.39 excess cancer per 10^6 PY per rad (kerma) is the best obtainable from the experience of the Hiroshima atomic-bomb survivors of all ages in the absence of estimated tissue doses. For persons who were 35 or older in 1945, however, the estimate would be considerably higher.

REFERENCES

1. Beebe, G. W., H. Kato, and C. E. Land. Life Span Study Report 8. Mortality Experience of Atomic Bomb Survivors, 1950–74. Radiation Effects Research Foundation Technical Report TR 1-77. Hiroshima: Radiation Effects Research Foundation, 1978.
2. Beebe, G. W., H. Kato, and C. E. Land. Studies of the mortality of A-bomb survivors. 6. Mortality and radiation dose, 1950–1974. Radiat. Res. 75:138–201, 1978.
3. Cosgrove, G. E., H. E. Walburg, Jr., and A. C. Upton. Gastrointestinal lesions in aged conventional and germfree mice exposed to radiation as young adults, pp. 303–312. In M. F. Sullivan, Ed. Gastrointestinal Radiation Injury. Excerpta Medica Found. Monogr. Nucl. Med. Biol. No. 1. Amsterdam: Excerpta Medica Foundation, 1968.
4. Court Brown, W. M., and R. Doll. Mortality from cancer and other causes after radiotherapy for ankylosing spondylitis. Brit. Med. J. 2:1327–1332, 1965.
5. Gates, O., and S. Warren. Radiation-induced experimental cancer of the esophagus. Amer. J. Pathol. 53:667–685, 1968.
6. Radford, E. P., R. Doll, and P. G. Smith. Mortality among patients with ankylosing spondylitis not given x-ray therapy. N. Engl. J. Med. 297:572–576, 1977.
7. Steer, A., I. M. Moriyama, and K. Shimizu. ABCC-JNIH Pathology Studies, Hiroshima and Nagasaki, Report 3. The Autopsy Program and the Life Span Study: January 1951–December 1970. Atomic Bomb Casualty Commission Technical Report TR 16-73. Hiroshima: Atomic Bomb Casualty Commission, 1973.
8. Storer, J. B. Radiation carcinogenesis, pp. 453–484. In F. F. Becker, Ed. Cancer. A Comprehensive Treatise. Vol. 1. New York: Plenum Press, 1975.
9. United Nations Scientific Committee on the Effects of Atomic Radiation. Report to the General Assembly. 32nd Session. Suppl. No. 40 (A/32/40). Annexes G (Radiation Carcinogenesis) and I (Experimental Carcinogenesis). New York: United Nations, 1977.
10. Upton, A. C., A. W. Kimball, J. Furth, K. W. Christenberry, and W. N. Benedict. Some delayed effects of atom-bomb radiations in mice. Cancer Res. 20(8, Part 2): 1–60, 1960.
11. Warren, S., and O. Gates. Cancers induced in different species by continuous irradiation. Arch. Environ. Health 17:697–704, 1968.

STOMACH

In the 1972 BEIR report,[14] the stomach was mentioned as an organ possibly sensitive to the carcinogenic effects of ionizing radiation; Court Brown and Doll had reported[4] a quite significant excess of stomach cancer among patients given x-ray therapy for ankylosing spondylitis. On the assumption that the average tissue dose ranged from 250 to 500 rads, the absolute risk was estimated at 0.32–0.64 death per 10^6 PY per rad.[14] It was then uncertain, however, whether the excess was attributable to radiation or to selection factors associated with the disease process or its treatment. Moreover, there was no substantial evidence of excess stomach cancer in the Japanese atomic-bomb survivors.[11,12] By 1977, however, as noted in the most recent UNSCEAR report,[18] doubts about the role of radiation in the reported excess incidence of stomach cancer among patients with ankylosing spondylitis treated with x rays had been dissipated by the failure of Radford *et al.*[17] to find any excess cancer among similarly studied ankylosing-spondylitis patients treated by means other than x rays. Moreover, in 1977, Nakamura reported[13] the first solid evidence of excess stomach cancer among high-dose survivors of the Hiroshima atomic bomb.

In experimental animals, the natural incidence of stomach cancer is low,[21,22] and experimental observations suggest that the glandular stomach is less susceptible to radiation-induced neoplasia than many other organs of mice of the strains thus far investigated.[3,19] Several investigators have, however, shown definite evidence that ionizing radiation will cause gastric cancer in laboratory animals.[8,9,15,16,20] In experiments with localized x-ray exposure of both mice and rats, Hirose[8,9] has found x radiation to be effective in inducing gastric cancer. Nowell *et al.*[15,16] observed some increase in gastric cancer in mice irradiated with x rays and with fast neutrons, and Upton *et al.*[20] reported an uncertain excess of gastric cancer in mice exposed to atomic-bomb radiation. In their 1968–1969 reviews of available data on radiation-induced carcinoma of the glandular stomach, Cosgrove *et al.*[3] and Upton *et al.*[19] concluded that the likelihood of radiogenic stomach cancer in mice is a function of the dose and quality of radiation, with neutrons being more effective than x rays or gamma rays in LAF mice, but not in RF mice. In their discussion, Cosgrove *et al.*[3] remarked on the difficulty of conducting an adequately thorough necropsy examination of the gastrointestinal tract of mice and suggested that all reported counts may be underestimates.

In the 1965 report on their followup of about 14,000 patients with ankylosing spondylitis treated with x rays in the interval 1935–1954, Court Brown and Doll[4] tallied 28 deaths from stomach cancer versus 16.0 expected from a point 6 yr after treatment through 1960, and 38 versus 23.6 expected through 1963, when followup was still incomplete. The latter comparison corresponds to an excess of 14.4 deaths in 89,432 PY of followup. Although Dolphin and Eve[5] estimated the average tissue dose to the stomach at 60 rads, the BEIR Committee used the much higher range of 250–500 rads and estimated the excess as 0.32–0.64 death from stomach cancer per 10^6 PY per rad.[14] The most recent report on the mortality followup of the ankylosing-spondylitis patients was for the period through 1969 and for patients given a single course of x-ray therapy (R. Doll, personal communication). If the observed and expected counts of 36 and 24.57 deaths over the period 6 yr after therapy through 1969 are combined with the lower BEIR estimate of 250 rads and 77,494 observed person-years, the result is 0.59 excess death per 10^6 PY per rad, with approximate 90% confidence limits of 0.11 and 1.19. There is no information with which to examine the form of the dose-response function.

Among the atomic-bomb survivors, it is only in the survivors of the Hiroshima bombing that mortality from stomach cancer is clearly related to radiation dose.[2] Table A-18 gives the observations for 1950–1974 by city and total dose in rads kerma. Further observation will be necessary to determine whether the excess above 400 rads in Nagasaki is reliable.

TABLE A-18   Observed and Expected Deaths from Stomach Cancer among Atomic-Bomb Survivors by T65 Dose and by City, 1950–1974[a]

T65 Dose, rads (kerma)	No. Deaths from Stomach Cancer			
	Hiroshima		Nagasaki	
	Observed	Expected[b]	Observed	Expected[b]
0	610	601.3	55	63.2
1–9	253	276.0	94	84.0
10–49	228	223.9	50	50.5
50–99	55	59.7	18	18.2
100–199	42	37.7	13	17.6
200–299	18	13.3	7	8.9
300–399	8	6.3	4	3.9
>399	14	9.7	10	4.8
TOTAL	1,228	1,228	251	251
p (homogeneity)	0.43		0.21	
p (linear trend)	0.03		0.16	
Excess deaths per 10^6 PY per rad	0.81		0.40	
90% confidence limits	0.12, 1.50		−0.25, 1.05	

[a] Data from Beebe et al.[1]

[b] Calculated on the assumption of independence of risk and dose, and adjusted for age and sex; the Nagasaki sample is appreciably younger than the Hiroshima sample and has a very different dose distribution.

Nagasaki Tumor Registry data for 1959–1970, covering 231 cases among survivors of known dose (versus 251 deaths for the period 1950–1974), also do not show a significant linear trend with dose ($p = 0.18$), whereas in the Hiroshima Tumor Registry data the test for a linearly increasing incidence with dose yields $p = 0.02$. For Hiroshima, a regression analysis yields the average (linear) estimate of 0.81 excess cancer death per 10^6 PY per rad (kerma). If the mean dose of those exposed to $10+$ rads kerma is converted to an approximate tissue dose by means of the attenuation factors published by Hashizume and Maruyama,[7] one obtains a mean dose to stomach tissue of 37 rads. Recalculation of the average absolute risk in terms of tissue dose yields 1.57 excess deaths per 10^6 PY per rad, for which the 90% confidence limits are 0.23 and 2.90. That this estimate depends heavily on the age structure of the sample of atomic-bomb survivors and on length of followup is suggested by the following age-specific estimates for Hiroshima only:[1]

	Excess Deaths per 10^6 PY per Rad (Kerma)	
Age in 1945, yr	Estimate	90% Confidence Limits
0–9	0.56	0.20, 0.91
10–19	0.00	−0.44, 0.42
20–34	0.71	−0.22, 1.65
35–49	2.22	0.30, 4.15
>49	0.23	−4.72, 5.19
ALL AGES	0.81	0.12, 1.50

Although the values for ages 0–9 and >49 seem atypical, there is otherwise a suggestion of an increasing absolute risk with age. In the absence of age-specific data on the ankylosing-spondylitis patients, and in the absence of an effect in Nagasaki, it seems pointless to attempt an RBE estimate for the effect of neutrons, but it is of some interest that the rough estimate of absolute risk after x radiation, 0.59 excess death per 10^6 PY per rad, is of the same order as that of 1.6 for the mixed gamma and neutron radiation released by the Hiroshima bomb. Apart from age at exposure, there is no information on the influence of host factors. There is no clear evidence of an effect in Nagasaki, but the data are so few that it cannot be said that they are inconsistent with those for the ankylosing-spondylitis patients. The 90% confidence limits on the Nagasaki estimate of absolute risk are −0.25 and 1.05 excess deaths per 10^6 PY per rad.

Over the 24-yr period of followup, the time-specific estimates of absolute risk for Hiroshima atomic-bomb survivors have been as follows:[1]

Excess Deaths per 10^6 PY per Rad (Kerma)

Period	Regression Estimate	90% Confidence Limits
1950–1974	0.81	0.12, 1.50
1950–1954	0.01	−1.20, 1.24
1955–1958	−0.87	−2.42, 0.67
1959–1962	1.48	−0.18, 3.16
1962–1966	1.70	−0.09, 3.49
1967–1970	1.67	−0.22, 3.58
1971–1974	1.32	−0.78, 3.43

The distribution of the excess over time does not easily lend itself to the estimation of a latent period. But, if the end of the latent period is taken as the beginning of the 4-yr interval after which there is a consistent cumulative excess in Hiroshima, then the estimate becomes 14 yr after exposure in 1945. Perhaps because the excess mortality attributable to radiation is small in relation to natural incidence, the effect has been difficult to establish. Early suggestions of a possible effect seen in Tumor Registry data[6] and in an early report on the LSS[10] were discounted, because the evidence was inconsistent or did not build up in time until Nakamura's analysis.[13] On the linear hypothesis, the excess in Hiroshima by the end of 1974 amounted to only about 20 deaths (0.81 × 25.2 per 10^6 PY per rad) among 1,228 from stomach cancer.

That Nagasaki atomic-bomb survivors should not exhibit excess mortality from stomach cancer seems inconsistent with the observations on the ankylosing-spondylitis patients, but the data on Nagasaki survivors are so few as to be not inconsistent with the latter. Perhaps the apparent excess (10 observed versus 4.8 expected) at 400+ rads (kerma) is reliable and in time the evidence of an effect will grow stronger.

The dose-response plot for Hiroshima (Figure A-6) suggests linearity for this mixture of gamma and neutron radiation, but the excess cases are so few that a variety of functional forms would fit well enough.

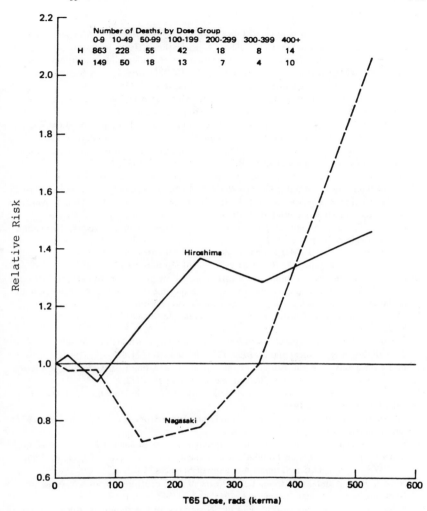

FIGURE A-6 Relative risk of death from stomach cancer, atomic-bomb survivors, by T65 dose and by city, 1950-1974.

REFERENCES

1. Beebe, G. W., H. Kato, and C. E. Land. Life Span Study Report 8. Mortality Experience of Atomic Bomb Survivors, 1950-74. Radiation Effects Research Foundation Technical Report TR 1-77. Hiroshima: Radiation Effects Research Foundation, 1978.
2. Beebe, G. W., H. Kato, and C. E. Land. Studies of the mortality of A-bomb survivors. 6. Mortality and radiation dose, 1950-1974. Radiat. Res. 75:138-201, 1978.
3. Cosgrove, G. E., H. E. Walburg, Jr., and A. C. Upton. Gastrointestinal lesions in aged conventional and germfree mice exposed to radiation as young adults, pp. 303-312. In M. F. Sullivan, Ed. Gastrointestinal Radiation Injury. Excerpta Medica Found. Monogr. Nucl. Med. Biol. No. 1. Amsterdam: Excerpta Medica Foundation, 1968.
4. Court Brown, W. M., and R. Doll. Mortality from cancer and other causes after radiotherapy for ankylosing spondylitis. Brit. Med. J. 2:1327-1332, 1965.
5. Dolphin, G. W., and I. S. Eve. Some aspects of the radiological protection and dosimetry of the gastrointestinal tract, pp. 465-476. In M. F. Sullivan, Ed. Gastrointestinal Radiation Injury. Excerpta Medica Found. Monogr. Nucl. Med. Biol. No. 1. Amsterdam: Excerpta Medica Foundation, 1968.
6. Harada, T., and M. Ishida. Neoplasms among atomic bomb survivors, Hiroshima. Tumor Registry Report 1. J. Natl. Cancer Inst. 25:1253-1264, 1960.
7. Hashizume, T., and T. Maruyama. Physical dose estimates for atomic-bomb survivors. Studies at Chiba, Japan. J. Radiat. Res. (Tokyo) 16(Suppl.):12-23, 1975.
8. Hirose, F. Experimental induction of carcinoma in the glandular stomach by localized x-irradiation of gastric region, pp. 75-113. In Experimental Carcinoma of the Glandular Stomach. Japanese Cancer Association GANN Monograph No. 8. Tokyo, 1969.
9. Hirose, F. Induction of gastric adenocarcinoma in mice by localized x-irradiation. GANN 60:253-260, 1969.
10. Jablon, S., M. Ishida, and G. W. Beebe. Studies of the mortality of A-bomb survivors. 2. Mortality in selections I and II, 1950-1959. Radiat. Res. 21:423-445, 1964.
11. Jablon, S., and H. Kato. Studies of the mortality of A-bomb survivors. 5. Radiation dose and mortality, 1950-1970. Radiat. Res. 50:649-698, 1972.
12. Murphy, E. S., and A. Yasuda. Carcinoma of the stomach in Hiroshima, Japan. Amer. J. Pathol. 34:531-542, 1958.
13. Nakamura, K. Stomach cancer in atomic-bomb survivors. Lancet 2:866-867, 1977. (letter to the editor)
14. National Research Council, Advisory Committee on the Biological Effects of Ionizing Radiations. The Effects on Populations of Exposure to Low Levels of Ionizing Radiation. Washington, D.C.: National Academy of Sciences, 1972.
15. Nowell, P. C., and L. J. Cole. Late effects of fast neutrons versus x-rays in mice. Nephrosclerosis, tumors, longevity. Radiat. Res. 11:545-550, 1959.
16. Nowell, P. C., L. J. Cole, and M. E. Ellis. Neoplasms of the glandular stomach in mice irradiated with x-rays or fast neutrons. Cancer Res. 18:257-260, 1958.
17. Radford, E. P., R. Doll, and P. B. Smith. Mortality among patients with ankylosing spondylitis not given x-ray therapy. N. Engl. J. Med. 297:572-576, 1977.
18. United Nations Scientific Committee on the Effects of Atomic Radiation. Ionizing Radiation. Levels and Effects. Vol. II. Effects. Publ. E721X18. New York: United Nations, 1977.
19. Upton, A. C., G. E. Cosgrove, and C. C. Lushbaugh. Induction of Carcinoma of the Glandular Stomach in Mice by Whole-Body Irradiation, pp. 63-74. In Experimental Carcinoma of the Glandular Stomach. Japanese Cancer Association GANN Monograph No. 8. Tokyo, 1969.

20. Upton, A. C., A. W. Kimball, J. Furth, K. W. Christenberry, and W. H. Benedict. Some delayed effects of atom-bomb radiations in mice. Cancer Res. 20:1–60, 1960.
21. Wells, H. G., M. Slye, and H. F. Holmes. Comparative pathology of cancer of the alimentary canal with report of cases in mice. Amer. J. Cancer 33:223–238, 1938.
22. Willis, R. A. Pathology of Tumors. London: Butterworth and Company, 1948.

INTESTINE AND RECTUM

Cancers of the large and small intestine and rectum were not specifically mentioned in the 1972 BEIR report,[13] although gastrointestinal cancers as a group, excluding those of the stomach, were discussed with respect to mortality on the basis of the experience of the Japanese atomic-bomb survivors[12] and British ankylosing-spondylitis patients treated with x irradiation.[7] Other information available at that time included apparent excesses of colon and rectal cancers in women whose ovaries had been irradiated by external x rays or radium implant to produce artificial menopause.[3,10,15] ICRP Publication 14, in 1969,[11] listed the colon as an organ of apparent, but uncertain, sensitivity to radiation carcinogenesis and the small intestine as an organ of low sensitivity.

Experimental studies have suggested that intestinal cancers can be induced by whole-body irradiation in mice, although not consistently enough or in large enough numbers to establish induction by radiation.[5,14,22] Carcinogenesis by irradiation of temporarily exteriorized intestinal tissue in the rat appears to be a standard treatment in studies of intestinal cancer not directly concerned with radiation carcinogenesis.[6,21]

The Japanese atomic-bomb survivor mortality experience based on death-certificate information has so far failed to provide unequivocal evidence of a relationship between radiation dose and intestinal or rectal cancer.[2] Only the Hiroshima female data show a statistically significant increasing trend in colon-cancer mortality with increasing dose for the period 1955–1974. The estimated risk for combined cities, sexes, and ages at the time of bombing is 0.1 ± 0.1 excess death per 10^6 PY per rem, assuming a linear dose response and a neutron RBE of 15, with no increased risk for males and 0.3 ± 0.2 excess death per 10^6 PY per rem for females. The age-specific estimates for combined cities and sexes are unstable, but tend to increase with increasing age at the time of bombing. However, Tumor-Registry data for the period 1959–1970, which are not sex-specific, show a statistically significant increasing trend in incidence of colon cancer with increasing dose ($p < 0.02$), which, moreover, is found in both Hiroshima ($p < 0.02$) and Nagasaki ($p < 0.05$). Regression estimates based on these data are 1.45 ± 0.67 excess cases per 10^6 PY per rad (intestinal dose) for Hiroshima and 0.60 ± 0.45 for Nagasaki. The Hiroshima Tumor Registry is known to be incomplete; furthermore, the Tumor Registries have not been subjected to critical analysis for ascertainment bias, although the Nagasaki Tumor-Registry data seem reliable enough (I. M. Moriyama, personal communication). It is interesting that autopsy studies based on the LSS sample have not demonstrated any relationship between radiation dose and benign or malignant tumors of the intestine.[25]

Neither the death-certificate data nor the Tumor-Registry data for the LSS sample suggest a dose-response relationship for rectal cancer.

Earlier studies of women irradiated for benign pelvic disorders included a mail survey by Palmer and Spratt[15] of 731 women with an average followup of 16.1 yr; seven rectal cancers were reported versus 2.1 expected ($p = 0.006$). Doll and Smith[10] reported results of a 13.6-yr (average) followup of 2,068 metropathia haemorrhagica patients treated with irradiation; there were 11 deaths from intestinal cancer versus 5.84 expected ($p = 0.04$), and five from rectal cancer versus 2.24 expected ($p = 0.08$). Another series of 267 patients followed for an average of 16.1 yr by Brinkley and Haybittle[3] yielded four deaths from intestinal cancer versus 1.0 expected ($p = 0.02$) and three rectal cancers versus 0.5 expected ($p = 0.014$).

However, no significant excess mortality from pelvic cancer (26 versus 21.5 expected; $p = 0.19$) was found in another followup of 2,049 metropathia haemorrhagica patients treated with internal radium (14%) or external x irradiation (86%).[1] Wagoner[23] observed one cancer of the small intestine versus 0.78 expected, 32 colon cancers versus 28.77 expected, and 16 rectal cancers versus 13.63 expected among Connecticut women who received radiotherapy for benign gynecologic disorders between 1935 and 1966; these excesses are statistically nonsignificant, both individually and in total. An incomplete followup of women treated with radium for benign uterine hemorrhage between 1926 and 1966 also found no excess mortality from pelvic cancers, or indeed from leukemia;[9] but this may simply reflect inadequate tracing of patients.[9,19]

A study of women given radium treatment for cancer of the cervix found 13 deaths versus 9.94 expected ($p = 0.20$) from colon cancer among 923 women who survived more than 5 yr after treatment, a nonsignificant excess.[8] In the same study, an excess of rectal cancers was found: 12 observed versus 4.35 expected ($p < 0.01$). Castro et al.[4] have reported circumstantial evidence of radiation involvement in a substantial fraction of 26 colon or rectal cancers in women previously irradiated for carcinoma of the cervix and uterus.

A recent followup of the 2,068 patients in the Doll-Smith metropathia haemorrhagica series found 32 deaths from intestinal and rectal cancers versus 19.1 expected ($p = 0.004$) in 28,857 PY 5 yr or more after irradiation.[19] These included three cancers of the small intestine versus 0.4 expected ($p = 0.01$), 21 colon cancers versus 13.5 expected ($p = 0.035$), and eight rectal cancers versus 5.2 expected ($p = 0.16$).[18,19]

The earlier followup study by Court Brown and Doll of irradiated ankylosing-spondylitis patients found a significant excess mortality from colon cancer (25 observed versus 14.8 expected), but it seemed possible that this might arise from the known associations between the treated condition and ulcerative colitis and between ulcerative colitis and colon cancer.[7] The observation of only one colon cancer versus 1.6 expected in a series of ankylosing-spondylitis patients not given x-ray therapy[17] does not make an especially convincing counterargument, because this is easily consistent with an underlying risk twice as high as that expected; but the observation of 21 versus 21.5 expected total cancers is somewhat more convincing.

The most recent followup of the irradiated ankylosing-spondylitis patients covered 14,109 patients who entered the study after a single treatment course (R. Doll and P. G. Smith, personal communication). These patients were followed until the end of the year after their second treatment, if any, or until lost to followup, or until January 1, 1970. Of the 7,453 patients who were re-treated before January 1, 1970, the average followup was 3.5 yr. The average followup for the remainder, who received only one treatment, was 16.2 yr. Twenty-eight deaths were attributed to cancer of the colon, compared with 17.30 expected ($p = 0.011$). Of these, six occurred in the first 3 yr of followup (versus 2.52 expected), and four in the next 3 yr (versus 2.22 expected). Two colon-cancer deaths occurred in the next 3 yr (versus 2.17 expected); this is consistent with the inference that the early excess was not caused by radiation, but was related to the treated disease. Although the group not treated with radiation showed no such early excess,[17,20] it is possible that this group was less severely affected by the underlying disease. There were 16 colon-cancer deaths 9 yr or more after the first radiation treatment, versus 10.39 expected ($p = 0.164$), with a total of 58,014 PY of followup, or 1.7 ± 1.0 excess deaths per 10^6 PY per rad, assuming an average dose of 57 rads to the colon (J. I. Fabrikant and J. T. Lyman, personal communication).

Cancers of the small intestine and rectum were not tabulated separately in the most recent report on the ankylosing-spondylitis patients (Doll and Smith, personal communication).

Polednak, Stehney, and Rowland, in a mortality followup through 1976 of a cohort of 634 women ascertained from employment lists as having been employed before 1930 in the U.S. radium-dial industry, found a statistically significant excess of deaths from colon cancer (10 observed versus 4.96 expected).[16] The incidence of other cancers of the digestive organs and peritoneum was only slightly different from expectation (five observed versus 7.57 expected). The excess colon cancer occurred mainly in women who were first employed before 1925, that is, before the practice of pointing brush tips with the lips was banned. There were five colon-cancer deaths versus 2.11 expected, among 360 women measured alive for radium body burden in 1954 or later; all five occurred among the 302 with body burdens of less than 50 μCi (1.72 expected). The role of the colon in the excretion of radium in man suggests that the relation between colon cancer and radium ingestion may be causal.

Although there is evidence of a causal relationship between ionizing radiation and cancers of the small intestine and rectum, there is no information on which to base estimates of excess risk per rad. The LSS sample data suggest that these sites are relatively minor in terms of the overall excess cancer risk.

Risk estimates for radiation-induced colon cancer vary from 0.1 to 1.7 excess deaths per 10^6 PY per rad of low-LET radiation and include the value 0.6 per 10^6 PY per rad for excess incidence. The incidence estimate based on the Nagasaki Tumor-Registry data for the LSS sample is probably the most reliable, given its completeness of ascertainment of disease (I. M. Moriyama, personal communication) and the good individual dosimetry and unselected nature of the exposed population. In the case of the ankylosing-spondylitis series, the possibility that the treated disease itself, which was more severe among those irradiated, may have contributed to the excess of colon-cancer deaths is cause for discounting the high risk estimate obtained. The different population rates—colon cancer being 4 times more frequent in the United Kingdom, and 7 times more frequent in the United States, than in Japan[24]—offer an alternative explanation of the discrepancy, but data for cancers of other organs tend to support comparability of absolute, rather than relative, estimates of risk among irradiated populations having different underlying cancer rates (see sections on leukemia, breast cancer, and lung cancer). Accordingly, the most reasonable estimate of risk is 0.6 excess colon-cancer case per 10^6 PY per rad, 15–25 yr after exposure for a population exposed at ages similar to those of the Japanese atomic-bomb survivors. In terms of mortality, between 60% and 70% of the incidence cases might be expected eventually to result in death from colon cancer (B. F. Hankey, personal communication), which corresponds to 0.4 excess colon-cancer death per 10^6 PY per rad.

REFERENCES

1. Alderson, M. R., and S. M. Jackson. Long term follow-up of patients with menorrhagia treated by irradiation. Brit. J. Radiol. 44:295-298, 1971.
2. Beebe, G. W., H. Kato, and C. E. Land. Life Span Study Report 8. Mortality Experience of Atomic-Bomb Survivors, 1950-74. Radiation Effects Research Foundation Technical Report TR 1-77. Hiroshima: Radiation Effects Research Foundation, 1978.
3. Brinkley, D., and J. L. Haybittle. The late effects of artificial menopause by x radiation. Brit. J. Radiol. 42:519-521, 1969.
4. Castro, E. B., P. P. Rosen, and S. H. Q. Quan. Carcinoma of large intestine in patients irradiated for carcinoma of cervix and uterus. Cancer 31:45-52, 1973.
5. Cole, L. J., P. C. Nowell, and J. S. Arnold. Late effects of x radiation. The influence of dose fractionization on life span, leukemogenesis, and nephrosclerosis incidence in mice. Radiat. Res. 12:173-185, 1960.
6. Coop, K. L., J. G. Sharp, J. W. Osborne, and G. R. Zimmerman. An animal model for the study of small-bowel tumors. Cancer Res. 34:1487-1494, 1974.
7. Court Brown, W. M., and R. Doll. Mortality from cancer and other causes after radiotherapy for ankylosing spondylitis. Brit. Med. J. 2:1327-1332, 1965.
8. Dickson, R. J. Late results of radium treatment of carcinoma of the cervix. Clin. Radiol. 23:528-535, 1972.
9. Dickson, R. J. The late results of radium treatment for benign uterine haemorrhage. Brit. J. Radiol. 42:582-594, 1969.
10. Doll, R., and P. G. Smith. The long term effects of x irradiation in patients treated for metropathia haemorrhagica. Brit. J. Radiol. 41:362-368, 1968.
11. International Commission on Radiological Protection. Radiosensitivity and Spatial Distribution of Dose. ICRP Publication 14. Oxford: Pergamon Press, 1969.
12. Jablon, S., and H. Kato. Life Span Study Report 6. Mortality among A-bomb Survivors 1950-71. Atomic Bomb Casualty Commission Technical Report TR 10-71. Hiroshima: Atomic Bomb Casualty Commission, 1971.
13. National Research Council, Advisory Committee on the Biological Effects of Ionizing Radiations. The Effects on Populations of Exposure to Low Levels of Ionizing Radiation. Washington, D.C.: National Academy of Sciences, 1972.
14. Nowell, P. C., and L. J. Cole. Late effects of fast neutrons vs x-rays in mice. Nephrosclerosis, tumors, longevity. Radiat. Res. 11:545-556, 1959.
15. Palmer, J. P., and D. W. Spratt. Pelvic carcinoma following irradiation for benign gynecological diseases. Amer. J. Obstet. Gynecol. 72:497-505, 1956.
16. Polednak, A. P., A. F. Stehney, and R. E. Rowland. Mortality among women first employed before 1930 in the U.S. radium dial-painting industry. A group ascertained from employment lists. Amer. J. Epidemiol. 107:179-195, 1978.
17. Radford, E. P., R. Doll, and P. G. Smith. Mortality among patients with ankylosing spondylitis not given x-ray therapy. N. Engl. J. Med. 297:572-576, 1977.
18. Smith, P. G. In United Nations Scientific Committee on the Effects of Atomic Radiation. Ionizing Radiation. Levels and Effects. Vol. II. Effects, p. 408. New York: United Nations, 1977.
19. Smith, P. G., and R. Doll. Late effects of x irradiation in patients treated for metropathia haemorrhagica. Brit. J. Radiol. 49:224-232, 1976.
20. Smith, P. G., R. Doll, and E. P. Radford. Cancer mortality among patients with ankylosing spondylitis not given x-ray therapy. Brit. J. Radiol. 50:728-734, 1977.
21. Stevens, R. H., G. P. Brooks, J. W. Osborne, C. W. Englund, and D. W. White. Lymphocyte cytotoxicity in x-irradiation-induced adenocarcinoma of the rat small

bowel. I. Measurement of target cell destruction by release of radioiodinated membrane proteins. Brief communication. J. Natl. Cancer Inst. 59:1315-1319, 1977.

22. Upton, A. C., M. L. Randolph, and J. W. Conklin. Late effects of fast neutrons and gamma-rays in mice as influenced by the dose rate of irradiation. Induction of neoplasia. Radiat. Res. 41:467-491, 1970.

23. Wagoner, J. K. Leukemia and Other Malignancies Following Radiation Therapy for Gynecological Disease. Unpublished Doctoral Thesis, Harvard School of Public Health, Boston, 1970.

24. Waterhouse, J., C. Muir, P. Correa, and J. Powell, Eds. Cancer Incidence in Five Continents. Vol. III. International Agency for Research on Cancer Scientific Publication No. 15. Lyon, France: International Agency for Research on Cancer, 1976.

25. Yamamoto, T., H. Kato, and G. S. Smith. Benign tumors of the digestive tract among atomic bomb survivors, 1961-1970, Hiroshima. GANN 66:623-630, 1975.

LIVER

Until recently, the risk of liver cancer from radiation has been largely overlooked. The liver was regarded as being relatively radioresistant, owing to the lack of acute radiation damage and the long latency of radiation-induced liver cancer. Furthermore, much of the early radiobiologic research on plutonium and other actinide elements was done in laboratory mice and rats, which rapidly excrete these elements from the liver. However, in man, 45% of the plutonium reaching the bloodstream is now thought to be deposited in the liver and to remain there with a biologic half-time of 40 yr.[12] Thus, the dose to the human liver from the maximal permissible body burden of 0.04 μCi of plutonium-239 is much higher than previously calculated by the internal-dose committees of the ICRP[11] and NCRP.[27] In addition, an increasing number of patients are receiving radiopharmaceuticals for diagnostic liver scans, and many radiotherapy patients are surviving long enough to be at later risk from cancers induced by the therapy. No longer can the risk of radiation-induced liver cancer be ignored.

THOROTRAST PATIENTS

Patients given Thorotrast (colloidal [^{232}Th]thorium dioxide) injections provide by far the most significant evidence of liver-cancer induction in man by a radioactive material. Thorotrast was injected intravascularly as an x-ray contrast medium, primarily for the diagnosis of suspected brain diseases, starting in 1928.[24] Its use was stopped around 1955, after the discovery that Thorotrast causes liver cancer.[16] The three most common types of liver cancer in the Thorotrast patients are angiosarcomas, bile-duct carcinomas, and hepatic-cell carcinomas (Table A-19).

TABLE A-19 Liver Cancers in Thorotrast Patients (at Latest Detailed Tumor Classification)

Types of Liver Cancer	Germany[29] as of 1975	Denmark[7] as of 1977	Portugal[5] as of 1974	Japan[19] as of 1975	TOTAL[a]
Angiosarcomas	37	20	19	20	96
Bile-duct carcinomas	48	15	11	57	131
Hepatic-cell carcinomas	21	15	2	10	48
"Liver" carcinomas[b]	20	0	0	0	20
Unspecified[c]	14	0	43	6	63
TOTAL	—	—	—	—	358

[a] In the general population, the most frequent types of primary liver cancer are hepatic-cell carcinomas and bile-duct carcinomas.[30] Angiosarcomas are extremely rare, except in persons exposed to Thorotrast, arsenic, or vinyl chloride.

[b] Either bile-duct carcinoma or hepatic-cell carcinoma.

[c] Fatal, but not classified histologically, except for one case of reticulosarcoma in the Portuguese series.

Thorotrast was used in many countries, but the followup studies in Germany,[28,29] Denmark,[7,8] and Portugal[5] are of special value, because large populations at risk have been identified and investigated systematically. The status of these studies is shown in Table A-20. Among the European Thorotrast patients in the followup series, only one of the 301 cases of liver cancer had a reported appearance time shorter than 18 yr.[5] However, an appearance time of 12 yr was reported for the first liver cancer ascribed to Thorotrast in the United States.[16] Thus, virtually no

TABLE A-20 Thorotrast Patients Surviving at Least 10 Yr After Intravascular Injection

Country and Year of Last Followup	No. Cases Liver Cancer	Traced Patients Surviving at Least 10 Yr	Person-Years at Risk from 10 Yr After Injection to Death or Last Contact
Germany, 1977[28]	176	1,733	28,424
Denmark, 1977[7,8]	50	646	12,274
Portugal, 1974[5]	75	667	12,673[a]
TOTAL	301	3,046	53,371

[a] The fraction of the Portuguese patients surviving at least 10 yr and their average time to death or last contact were considered similar to those documented for the Danish patients. In both countries, suspected brain diseases were the main reason for the intravascular injection of Thorotrast (80% in Portugal and nearly 100% in Denmark).

radiation-induced liver cancers are expected to appear within at least the first 10 yr after injection. Similarly, the last 10 yr of irradiation can be considered "wasted," with respect to inducing observable liver cancers— although, alternatively, a shorter span could be disregarded, such as the last 5 yr of irradiation. About 3,046 traced persons in Germany, Denmark, and Portugal have lived at least 10 yr after Thorotrast injection (Table A-20). Their average time at risk, from the 10-yr minimal latent period to death or latest contact, is now about 18 yr, with a collective 53,371 PY at risk.

The natural yearly incidence of cancers of the liver (including gall bladder), per 100,000 persons of all ages, is eight in Denmark, 15 in Hamburg, Germany, and 14 in the Saarland, Germany, but unavailable for Portugal.[30] Multiplying the average for these known rates, 12 liver cancers per 10^5 PY, by the 53,371 PY at risk, yields a prediction of about six naturally occurring liver cancers. Thus, of the 301 observed cases of liver cancer in Table A-20, almost all (295 cases) are attributed to Thorotrast.

Intravascular injections of Thorotrast mostly ranged between 10 and 100 ml, averaging about 25 ml in Germany,[28] 23 ml in Denmark,[8] and 26 ml in Portugal.[5] The average injection of 25 ml contained about 5 g of thorium (0.6 μCi of thorium-232 with additional radioactivity from its daughters). When Thorotrast is injected intravascularly, whether by artery or by vein, about 60% is deposited in the liver.[14] For the injection of 25 ml of Thorotrast, the alpha-particle dose rate to the liver of a standard 70-kg man (assuming that 65% of the alpha energy escapes from the Thorotrast aggregates and becomes absorbed in tissue) is about 25 rads/yr.[14] Multiplying 25 rads/yr by the 53,371 PY at risk in Table A-20 gives a collective population dose of 1,334,275 person-rads. Thus, the risk coefficient, *up to the time of the latest followup,* is:

$$\frac{295 \text{ liver cancers}}{1,334,275 \text{ person-rads}} = \frac{221 \text{ liver cancers}}{10^6 \text{ person-rads of alpha radiation}}.$$

However, this does not include the additional risk during the remaining life spans of the surviving patients.

About one-fourth of the recent deaths among the Thorotrast patients are from liver cancer.[7,28,29] Thus, an estimated 256 additional liver cancers are expected among the total 1,026 surviving traced Thorotrast patients. There were 591 surviving patients as of 1977 in Germany,[28] 294 as of 1977 in Denmark,[7] and 141 as of 1974 in Portugal.[5] The age of these survivors averaged about 60 yr at the last followup. Allowing for the toxicity of Thorotrast, a mean additional survival time of about 15 yr is anticipated, with individual survival times ranging from zero to over 30 yr. Therefore, when all these patients have died, an additional 1,026 persons $\times$ 15 yr = 15,390 PY and an additional 15,390 PY $\times$ 25 rads/yr = 384,750 person-rads are predicted. The total combined values to the end of the life span become 557 liver cancers (of which about eight would be expected naturally, leaving 549 as Thorotrast-induced), 68,761 PY at risk, and 1,719,025 person-rads. The projected risk coefficient *to the end of the life span* becomes:

$$\frac{549 \text{ liver cancers}}{1,719,025 \text{ person-rads}} = \frac{319 \text{ liver cancers}}{10^6 \text{ person-rads of alpha radiation}}.$$

As an alternative to the assumption that the last 10 yr of irradiation are "wasted," with respect to producing an observable liver tumor, one may assume that only the last 5 yr of irradiation are wasted, in which the collective radiation dose would be increased by 3,046 persons $\times$ 5 yr $\times$ 25 rads/yr = 380,750 person-rads. The total collective dose would then be 380,750 + 1,719,025 = 2,099,775 person-rads, and the projected risk coefficient to the end of the life span would become 549 liver cancers per 2,099,775 person-rads = 261 liver cancers per 10^6 person-rads of alpha radiation.

These evaluations are consistent with a rounded risk coefficient projected to the end of the life span of:

300 liver cancers per 10^6 person-rads of alpha radiation.

This is 3 and 4 times greater than estimates previously derived from the Thorotrast data by Mays[17] and by Mole,[18] respectively, because these authors did not include the future risk to the surviving patients, they did not exclude the patients who died before the minimal latent period of 10 yr, and Mole did not exclude the "wasted" radiation received too late to produce an observable tumor. If one corrects for these effects, the risk coefficients of Mays and of Mole would become about the same as derived here.

There are uncertainties, however, in applying the Thorotrast risk to liver irradiation from other alpha-emitters, such as plutonium-239. First, colloidal Thorotrast in the liver is taken up mainly by phagocytic cells,[24] whereas plutonium is deposited more uniformly throughout all cells of the liver.[9] Thus, the distribution of alpha-particle radiation among the various types of liver cells differs initially for plutonium and Thorotrast, although both are later concentrated in phagocytic cells. Second, the Thorotrast patients have several grams of thorium dioxide in their livers, whereas the permissible total-body burden of plutonium-239 for radiation workers is only 0.6 μg (0.04 μCi). Therefore, Thorotrast might (or might not) involve a chemical toxicity that would certainly be insignificant for plutonium and other actinide elements. When (or if) the toxicity ratio of plutonium to Thorotrast is evaluated for liver-cancer induction in suitable laboratory animals, an upward or downward revision may be necessary to obtain the most appropriate liver-risk coefficient for plutonium and the other actinide elements. Until this information can be obtained, the Thorotrast risk coefficient probably should be used as the best available estimate. However, as will now be shown, it is unlikely that the risk coefficient for plutonium could be over 10 times that derived for Thorotrast.

PERSONS WHO RECEIVED PLUTONIUM INJECTIONS

To evaluate the relationship between urinary excretion and plutonium body content, 17 persons of presumed short life expectancy received intravenous injections of plutonium in 1945 or 1946.[6,23] Unexpectedly, six of these patients survived at least 10 yr, and two were still alive as of August 1, 1978 (Table A-21). No cancers of liver or bone have appeared. Furthermore, it seems unlikely that any plutonium-induced cancers will appear in the two present survivors, because of their advanced ages and the size of their doses.

TABLE A-21 Patients of Short Life Expectancy Who Lived at Least 10 Yr After the Intravenous Injection of Plutonium[a]

Patient	Injected Amount, Ci/kg	Age at Injection, yr	Time from Injection to Death or Aug. 1, 1978, yr	Average Dose to Liver, rads		
				At Death or 1978	5 Yr Before	10 Yr Before
Cal-1	0.0608[b]	58	20.67	1,460	1,173	848
HP-1	0.0040	67	14.25	85	58	28
HP-3	0.0043	48	32.70[c]	226	199	170
HP-6	0.0044	44	32.52[c]	194	171	145
HP-8	0.0073	41	29.73	282	244	203
HP-10	0.0053	52	10.89	91	51	8
TOTAL	—	—	—	2,338	1,896	1,402

[a] Data from Rowland and Durbin.[23]
[b] Cal-1 received plutonium-238 (VI) nitrate; the other patients, plutonium-239 (IV) citrate.
[c] Patients HP-3 and HP-6 still alive as of August 1, 1978 (R. E. Rowland, personal communication).

Disregarding the last 10 yr of irradiation as "wasted," these six patients had a collective 1,402 person-rads of alpha dose to the liver. Multiplying this by the Thorotrast risk coefficient of 300 liver cancers per 10^6 person-rads yields an expectation of 0.4 case of liver cancer—in good agreement with the zero cases observed. However, if the risk coefficient were 10 times higher for plutonium than for Thorotrast, then four liver cancers would have been predicted, and the chance of having no liver cancers would have been very small ($p = 0.02$). Therefore, it seems very unlikely that the risk coefficient for plutonium could exceed 10 times that for Thorotrast. This conclusion is very important, because it is based on experience with human irradiation from plutonium.

ATOMIC-BOMB SURVIVORS

Liver cancers from the Tumor-Registry data at Hiroshima and Nagasaki are shown in Table A-22. Among the atomic-bomb survivors followed from 1959 to 1970, those at Nagasaki exposed to an air kerma above 10 rads showed the same incidence of liver cancer as those exposed to under 9 rads. Because nearly all the dose at Nagasaki was from gamma rays, this suggests that sparsely ionizing radiation, up to a few hundred rads, is not very effective in the induction of liver cancer. However, the Hiroshima radiation contained a biologically significant component of fast neutrons.[2,21,22] These neutrons, in colliding with atoms in tissue, produced densely ionizing tracks, somewhat similar to those from alpha particles. At Hiroshima, 31 liver cancers were reported in persons exposed to an air kerma above 10 rads, compared with 23.34 cases expected on the basis of the incidence in those exposed to under 10 rads, thus yielding an excess of 7.66 cases.

In the persons at Hiroshima exposed to a kerma of 10–600 rads, the absorbed neutron dose to the liver gave a collective 518,693 PY-rads. That would be the relevant dosage figure if, on the basis of the Nagasaki liver data, the gamma-ray dose were considered ineffective. Alternatively, if the relative effectiveness of gamma rays averaged one-tenth of that of neutrons,[10] the neutron "equivalent" would be (0.1) $(4,746,430)$ + $518,693 = 993,336$ PY-rads of neutron dose. The corresponding risk coefficients for a "plateau period" of 30 yr (starting at 10 and ending at 40 yr after an abrupt irradiation) would be: for negligible effectiveness of gamma rays,

$$\frac{7.66 \text{ excess liver cancers (30 yr)}}{518,693 \text{ PY-rads}} = \frac{443 \text{ liver cancers}}{10^6 \text{ person-rads of neutron dose}};$$

TABLE A-22 Liver Cancers, Including Those of the Gall Bladder and Bile Ducts, from the Tumor-Registry Data, 1959-1970, of the Atomic-Bomb Survivors[a]

Kerma, rads Range	Liver Dose[15]				Person-Years (1959-1970)	Person-Year-Rads		Liver and Biliary Cancers		
	Gamma	Neutron	Gamma[b]	Neutron[c]		Gamma	Neutron	Observed	Expected[d]	Net
Hiroshima										
200-600	266.7	94.1	132.4	16.9	15,837	2,096,819	267,645	4	2.15	1.85
100-199	108.6	30.3	53.3	5.5	17,458	930,511	96,019	5	2.37	2.63
50-99	56.8	13.4	27.7	2.4	27,653	765,988	66,367	6	3.76	2.24
10-49	17.6	4.3	8.6	0.8	110,827	953,112	88,662	16	15.06	0.94
TOTAL	—	—	—	—	171,775	4,746,430	518,693	31	23.34	7.66
0-9	0.9	0.3	0.4	0.1	448,945			61	61.00	0.00
Nagasaki										
200-600	331.5	5.7	156.2	1.0	14,859	2,320,976	14,859	5	3.74	1.26
100-199	142.9	1.3	67.3	0.2	14,966	1,007,212	2,993	2	3.77	(—)1.77
50-99	70.6	0.2	33.2	0.0	14,034	465,929	0	5	3.53	1.47
10-49	21.5	0.0	10.1	0.0	38,908	392,971	0	9	9.79	(—)0.79
TOTAL	—	—	—	—	82,767	4,187,088	17,852	21	20.83	0.17
0-9	2.3	0.0	1.1	0.0	119,203			30	30.00	0.00

[a] Data from Beebe et al.[3]
[b] Liver gamma dose = 0.47 gamma kerma + 0.075 neutron kerma.
[c] Liver neutron dose = 0.18 neutron kerma.
[d] Based on rate for 0-9 rads.

and for gamma rays one-tenth as effective as neutrons,

$$\frac{7.66 \text{ excess liver cancers (30 yr)}}{993,336 \text{ PY-rads}} = \frac{231 \text{ liver cancers}}{10^6 \text{ person-rads of neutron dose}}.$$

These results support the risk coefficient of 300 liver cancers per 10^6 person-rads of densely ionizing alpha radiation as derived from the Thorotrast experience for which the statistical significance is very much better. The atomic-bomb results have the advantage that they represent relatively uniform irradiation of liver tissue (in contrast with the focal deposition of Thorotrast) and are not complicated by the possibility of chemical toxicity (which might exist from the 3 g of thorium in the liver of a typical Thorotrast patient). However, the atomic-bomb data have severe statistical limitations: at Hiroshima, the liver-tumor excess of 7.66 cases has a standard deviation of ± 6.32 cases; furthermore, whereas the accuracy of diagnosis is generally good in the Tumor-Registry data, the data are incomplete, especially for Hiroshima, and have not been investigated for possible bias in reporting and dose assignment.

The most important conclusion from the atomic-bomb results is that the true risk coefficient for radiation-induced liver cancer seems unlikely to exceed, by a large factor, that derived from the Thorotrast data.

RECOMMENDED RISK COEFFICIENTS

For a population of mixed ages at the start of liver irradiation, the best estimate of the cumulative risk during the remaining life span is regarded as that derived from alpha-emitting Thorotrast:

risk from densely ionizing (alpha and fast-neutron) radiation =

$$\frac{300 \text{ liver cancers}}{10^6 \text{ person-rads of high-LET radiation}}.$$

Dividing the above by the quality factor of 20 currently recommended in ICRP Report 26 for alpha radiation:[10]

risk from sparsely ionizing (x-ray, gamma, and beta) radiation =

$$\frac{15 \text{ liver cancers}}{10^6 \text{ person-rads of low-LET radiation}}.$$

There are 3,046 traced European Thorotrast patients who lived at least 10 yr after injection. When all these patients have died, they are projected to have about 68,761 PY at risk beyond the first 10 yr, or an average of about 23 yr at risk, from 10 yr after injection to their end of life. Dividing the preceding life-span risk coefficients by 23 yr yields the following risk *rate* coefficients:

risk *rate* from densely ionizing radiation

$$= \frac{13 \text{ liver cancers per YEAR}}{10^6 \text{ person-rads of high-LET radiation}},$$

and

risk *rate* from sparsely ionizing radiation

$$= \frac{0.7 \text{ liver cancer per YEAR}}{10^6 \text{ person-rads of low-LET radiation}}.$$

DISCUSSION

Because nearly all human liver cancers are fatal, the mortality is approximately equal to the incidence.[1]

It is unknown whether children or adults are the most susceptible to radiation-induced liver cancer. Suitable life-table analyses of the Thorotrast patients, grouped by age at injection, might resolve this uncertainty. Relevant information on susceptibility versus age should be available within several years from experiments in progress at the University of Utah on the effects of plutonium injected into beagle puppies, young adults, and older adults.[13] Until definite information is obtained, the same risk-rate coefficient is recommended for all ages at irradiation.

The possible chemical toxicity of Thorotrast and the difference between the distributions of plutonium and Thorotrast in liver tissue raise an important question as to how reliably the liver risk coefficient for Thorotrast represents that for plutonium and the other actinide elements. This uncertainty is likely to remain unresolved until the ratio of plutonium toxicity to Thorotrast toxicity is evaluated in suitable laboratory animals, or until definitive results occur among the increasing number of plutonium-contaminated persons. In the absence of better data, the Thorotrast coefficient is used, with the realization that the true risk from plutonium could be greater or less. However, the lack of liver cancers in a small group of patients who received plutonium injections indicates that the risk to the liver from plutonium is very unlikely to exceed 10 times that from Thorotrast.

From data on the long-term German Thorotrast patients living to 1968 or later,[28] the incidence of liver cancer seems to be somewhat linearly proportional to dose, although alternative dose-response relationships cannot be excluded. Inasmuch as over 90% of the radiation dose from Thorotrast is from alpha particles, it seems reasonable in the light of present knowledge to assume a linear dose-response relationship for the induction of liver cancer by other alpha-emitters, such as plutonium-239. However, the dose-response relationship for the induction of liver cancer by beta-emitting cerium-144 in rats is strongly concave upward.[20] This result and the lack of excess liver cancers in the Nagasaki survivors who received gamma-ray doses of up to a few hundred rads[3] suggest that the dose-response relationship for liver-cancer induction in man from sparsely ionizing radiation may also be concave upward, rather than linear. Thus, the true risk to the liver from low doses of x rays, gamma rays, and beta particles could be considerably less than indicated from a linear risk coefficient for low-LET radiation obtained by dividing the linear risk coefficient for high-LET radiation by an assumed constant RBE of 20. There is increasing evidence from a number of different biologic systems that the relative biologic effectiveness of high- versus low-LET radiation is not constant, but increases as the dose decreases.[21,22]

The indicated risk coefficient for liver exceeds that for the endosteal layer in bone. For protracted irradiation from repeated injections of alpha-emitting radium-224 in persons, the risk coefficient in terms of average skeletal dose is about 200 bone sarcomas per 10^6 person-rads.[17] Dividing this by 7.5 (which is the ratio of endosteal dose to average skeletal dose for radium-224 and its daughters decaying half on bone surfaces and half within bone volume) yields a *risk coefficient in terms of endosteal dose* of about 27 bone sarcomas per 10^6 person-rads. By comparison, the liver risk coefficient from the Thorotrast patients is about 300 liver cancers per 10^6 person-rads.

If a 70-kg "reference man" inhales 1 μCi of plutonium-239 in particles having a median aerodynamic diameter of about 1 μm, the average organ doses 50 yr later would be about 40 rads to the 1,800-g liver and 13 rads to the 7,000-g skeleton without marrow.[17] The corresponding endosteal dose for plutonium-239 is about 9.27 times as high as the average skeletal dose of 13 rads and is 120 rads, assuming that half the skeletal plutonium decays on bone surfaces, and the other half decays randomly throughout bone volume.

The predicted lifetime cancer incidences per inhaled microcurie of plutonium-239 would be:

$$\text{liver-cancer incidence} = \frac{300 \text{ liver cancers}}{10^6 \text{ person-rads}} (40 \text{ rads}) = 1.2\%$$

and

$$\text{bone-sarcoma incidence} = \frac{27 \text{ bone sarcomas}}{10^6 \text{ person-rads}} (120 \text{ rads}) = 0.3\%.$$

These predictions indicate that the risk from plutonium intake by man might be 4 times as high for liver cancer as for bone cancer. Because of uncertainties in the risk coefficients for both liver and the endosteum, as applied to plutonium in man, the ratio of liver cancers to bone sarcomas could be either larger or smaller than 4.

Liver tumors have generally been less common than bone sarcomas in laboratory animals exposed to plutonium, but there is uncertainty in extrapolating this result to man for the following reasons: The relative sensitivity to the induction of cancers of liver versus bone may be inherently different in man and other animals. Alcohol, solvents, and toxic chemicals may increase the susceptibility to radiation-induced liver cancer. People are often exposed to these potential liver toxins, whereas this has rarely occurred in laboratory animals in plutonium experiments. Studies of plutonium toxicity have often been done in laboratory rats and mice, which quickly excrete plutonium from their livers. Thus, the infrequency of liver tumors in these rodents could be due more to the loss of plutonium than to an absence of its toxicity. The beagle liver tenaciously retains plutonium, but the beagle skeleton seems about 25 times more sensitive than the human skeleton to the radiation induction of bone sarcomas.[17] The short latent period and high induction of bone sarcomas in beagles prevent the adequate expression of liver tumors after long latency, except at lower dosages. Whereas some of the induced liver tumors in beagles have been malignant (usually fibrosarcomas or bile-duct carcinomas), most appear to be bile-duct adenomas.[26] It is unknown whether these small adenomas would have progressed into carcinomas if a longer normal life span, such as occurs in man, had been available. In the Thorotrast patients, virtually all the liver tumors have been malignant at clinical recognition.

However, part of the carcinogenicity of the Thorotrast may be due to its "foreign-body effect."

To estimate more reliably the risk of plutonium-induced liver cancer in man, the risk to the Thorotrast patients should be multiplied by the plutonium-to-Thorotrast toxicity ratio. It is hoped that this ratio can be established in suitable animal species having a prolonged retention of actinide elements in the liver. Suitable species might be beagles,[26] Chinese hamsters (*Cricetulus griseus*),[4] deer mice (*Peromyscus maniculatus*),[25] and grasshopper mice (*Onychomys leucogaster*).[25]

REFERENCES

1. American Cancer Society. Cancer Facts and Figures. 1978.
2. Auxier, J. A., J. S. Cheka, F. F. Haywood, T. D. Jones, and J. H. Thorngate. Field-free radiation-dose distributions from the Hiroshima and Nagasaki bombings. Health Phys. 12:425–429, 1966.
3. Beebe, G. W., H. Kato, and C. E. Land. Life Span Study Report 8. Mortality Experience of Atomic Bomb Survivors 1950–74, pp. A6, A9, A12, and A350. Radiation Effects Research Foundation Technical Report TR 1-77. Hiroshima: Radiation Effects Research Foundation, 1978.
4. Benjamin, S. A., A. L. Brooks, and R. O. McClellan. The biological effectiveness of Pu-239, Ce-144, and Sr-90 citrate in producing chromosome damage, bone-related tumors, liver tumors, and life shortening in the Chinese hamster, pp. 241–244. In Inhalation Toxicology Research Institute Annual Report LF-52, 1975.
5. da Silva Horta, J., M. E. da Silva Horta, L. C. da Motta, and M. H. Tavares. Malignancies in Portuguese Thorotrast patients. Health Phys. 35:137–152, 1978.
6. Durbin, P. W. Plutonium in man. A new look at the old data, pp. 469–530. In B. J. Stover and W. S. S. Jee, Eds. Radiobiology of Plutonium. Salt Lake City: J. W. Press, University of Utah, 1972.
7. Faber, M. Epidemiology of Thorotrast Malignancies in Man. Review paper prepared for the World Health Organization Scientific Group on the Long Term Effects of Radium and Thorium in Man. WHO Working Paper 12, Geneva, Sept. 12–16, 1977. (Copy obtainable from Prof. Mogens Faber, Finsen Institute, Strandboulevard 49, Copenhagen, Denmark.)
8. Faber, M. Malignancies in Danish Thorotrast patients. Health Phys. 35:153–158, 1978.
9. Grube, B. J., W. Stevens, and D. R. Atherton. The retention of plutonium in hepatocytes and sinusoidal lining cells isolated from rat liver. Radiat. Res. 73:168–179, 1978.
10. International Commission on Radiological Protection. Recommendations, p. 5. ICRP Publication 26. Oxford: Pergamon Press, 1977.
11. International Commission on Radiological Protection. Report of Committee II on Permissible Dose for Internal Radiation, p. 227. ICRP Publication 2. Oxford: Pergamon Press, 1959.
12. International Commission on Radiological Protection. The Metabolism of Compounds of Plutonium and Other Actinides, p. 50. ICRP Publication 19. Oxford: Pergamon Press, 1972.
13. Jee, W. S. S., D. R. Atherton, F. W. Bruenger, J. H. Dougherty, R. D. Lloyd, C. W. Mays, C. J. Nabors, Jr., W. Stevens, B. J. Stover, G. N. Taylor, and L. A. Woodbury. Current status of Utah long-term [239]Pu studies, pp. 79–94. In M. Lewis, Ed. Biological and Environmental Effects of Low-Level Radiation. Vienna: International Atomic Energy Agency, 1976.
14. Kaul, A., and W. Noffz. Tissue dose in Thorotrast patients. Health Phys. 35:113–122, 1978.
15. Kerr, G. D. Organ Dose Estimates for the Japanese Atomic-Bomb Survivors. Health Phys. 37:487–508, 1979.

16. MacMahon, H. E., A. S. Murphy, and M. I. Bates. Endothelial-cell sarcoma of liver following Thorotrast injections. Amer. J. Pathol. 23:585–611, 1947.

17. Mays, C. W. Estimated risk from [239]Pu to human bone, liver and lung, pp. 373–384. In M. Lewis, Ed. Biological and Environmental Effects of Low-Level Radiation. Vol. 2. Vienna: International Atomic Energy Agency, 1976.

18. Mole, R. H. The radiobiological significance of the studies with [224]Ra and Thorotrast. Health Phys. 35:167–174, 1978.

19. Mori, T., Y. Kato, T. Shimamine, and S. Watanabe. Statistical analysis of Japanese Thorotrast-administered autopsy cases. In International Meeting on the Toxicity of Thorotrast and Other Alpha-Emitting Heavy Elements, Lisbon. Environ. Res. 18:231–244, 1979.

20. Moskalev, Y. I., V. N. Streltsova, and L. A. Buldakov. Late effects of radionuclide damage, p. 505. In C. W. Mays et al., Eds. Delayed Effects of Bone-Seeking Radionuclides. Salt Lake City: University of Utah Press, 1969.

21. Rossi, H. H., and A. M. Kellerer. The validity of risk estimates of leukemia incidence based on Japanese data. Radiat. Res. 58:131–140, 1974.

22. Rossi, H. H., and C. W. Mays. Leukemia risk from neutrons. Health Phys. 35:353–360, 1978.

23. Rowland, R. E., and P. W. Durbin. Survival, causes of death, and estimated tissue doses in a group of human beings injected with plutonium, pp. 329–341. In W. S. S. Jee, Ed. Health Effects of Plutonium and Radium. Salt Lake City: J. W. Press, University of Utah, 1976.

24. Swarm, R. L., Ed. Distribution, retention, and late effects of thorium dioxide. Ann. N.Y. Acad. Sci. 145:523–858, 1967.

25. Taylor, G. N., P. A. Gardner, C. W. Jones, and R. D. Lloyd. Prolonged liver retention of [241]Am in deer mice and grasshopper mice, pp. 127–134. In Research in Radiobiology. University of Utah Report COO-119-254, March 1979.

26. Taylor, G. N., W. S. S. Jee, J. L. Williams, and L. Shabestari. Hepatic changes induced by [239]Pu, pp. 105–127. In B. J. Stover and W. S. S. Jee, Eds. Radiobiology of Plutonium. Salt Lake City: J. W. Press, University of Utah, 1972.

27. U.S. Department of Commerce. National Bureau of Standards Handbook 69. Maximum permissible body burdens and maximum permissible concentrations of radionuclides in air and in water for occupational exposure, p. 87. Washington, D.C.: U.S. Government Printing Office, 1959.

28. van Kaick, G., A. Kaul, D. Lorenz, H. Muth, K. Wegener, and H. Wesch. Late effects and tissue dose in Thorotrast patients. Recent results of the German Thorotrast study, pp. 263–276. In Late Biological Effects of Ionizing Radiation. Vienna: International Atomic Energy Agency, 1978.

29. van Kaick, G., H. Muth, D. Lorenz, and A. Kaul. Malignancies in German Thorotrast patients and estimated tissue dose. Health Phys. 35:127–136, 1978.

30. Waterhouse, J., C. Muir, P. Correa, and J. Powell. Cancer.

PANCREAS

The 1972 BEIR report[14] referred to cancer of the pancreas as one of a number of cancers reported to occur in excess in persons exposed to ionizing radiation; information on incidence in human populations was relatively limited, however. The 1969 ICRP Publication 14[7] listed the pancreas among organs of apparent, but uncertain, sensitivity to radiation carcinogenesis. This observation was based on preliminary data on reported excess mortality from pancreatic cancer among British radiologists who entered the practice of radiology before 1921[2] and among ankylosing-spondylitis patients treated with radiation.[3] The pancreas was one of the heavily irradiated organs considered by Court Brown and Doll in their survey of the ankylosing-spondylitis patients treated with radiation.[3, 19, 20] The risk of pancreatic cancer at moderate radiation doses was difficult to assess, although it seemed likely that it was relatively low, for example, relative to the risk of leukemia induced under the same conditions of irradiation.[8] There are no definitive experimental studies in animals on the radiation induction of cancer of the pancreas or other pancreatic tumors. However, pancreatic cancer was found in mice exposed to gamma irradiation, although not in sufficient numbers to establish its induction by radiation.[21]

RADIOTHERAPY FOR BENIGN DISEASE

The initial study of 14,554 ankylosing-spondylitis patients treated with irradiation[3] found nine deaths from pancreatic cancer, compared with 3.78 expected according to population rates, or 5.2 (0.9, 11.9*) excess cancer deaths, in patients followed from 6 yr after the first irradiation treatment until January 1, 1960. Furthermore, there were 12 observed deaths due to pancreatic cancer, compared with 5.71 expected, or an excess of 6.3 (1.2, 13.7), in an incomplete followup of treated patients to January 1, 1963.

The most recent analysis by Doll and Smith (Smith and Doll[16] and R. Doll and P. G. Smith, personal communication) of 14,109 ankylosing-spondylitis patients who had received radiotherapy and who were later followed from the date of their first treatment until the year after their second treatment, if any, or until January 1, 1970, found a significant increase in deaths from cancer of the pancreas. The excess was 8.5 (2.1, 17.2) deaths from panceatic cancer, or 18 deaths observed versus 9.49 expected (mean followup, 9.5 yr (133,881 PY). In a subset of 6,838 patients observed for more than 5 yr after radiotherapy, with 77,494 PY followup after the fifth year, this estimate fell to 4.5 (−0.5, 12.0) excess deaths from pancreatic cancer (12 deaths observed versus 7.47 expected). This is to be compared with no excess deaths due to cancer of the pancreas observed in 836 patients with ankylosing spondylitis not given x-ray therapy, with an average followup of 7.9 yr to January 1, 1968.[15,17] The values for this control series were one case of cancer of the pancreas observed versus 0.8 expected.

---

*Numbers in parentheses are equitailed 90% confidence limits, assuming Poisson variation for the observed numbers of deaths.

There were six deaths observed versus 2.02 expected during the first 6 yr after treatment; this suggested either that the minimal latent period for radiation-induced pancreatic cancer is less than 6 yr or that the treated disease had an associated risk of cancer of the pancreas. Complete follow-up of the patient group with ankylosing spondylitis not treated with radiation[15,17] does not clarify this, in that one death from pancreatic cancer was observed versus 0.8 expected from the second year after enrollment in the patient series until January 1, 1968. However, this figure is nevertheless consistent with an underlying risk some 2–3 times that expected according to population rates. Nevertheless, the pancreas was not prominent among tissues associated with excess mortality in the series of nonirradiated patients; the overall finding was 21 cancer deaths from all cancers versus 21.51 expected from the population rates.

If it is assumed that the mean radiation dose to the pancreas was approximately 90 rads for the treatment group receiving only one course of radiotherapy, these data suggest an absolute risk of 0.7 (0.2, 1.4) excess death from pancreatic cancer per million patients exposed per year per rad (PYR), assuming no minimal latent period, and an excess of 0.6 (−0.1, 1.7) death per 10^6 PYR beginning 6 yr after treatment.

ATOMIC-BOMB SURVIVORS

The most recent survey of the atomic-bomb Life Span Study[1] mortality data on survivors in Hiroshima and Nagasaki contains little suggestion of a firm relationship between radiation dose and the induction of pancreatic cancer. However, cancer of the pancreas is often poorly diagnosed on death certificates in Japan,[18] and death certificates are commonly completed before autopsy findings become known. A search of the Tumor Registries maintained by the city medical associations of Hiroshima and Nagasaki has, nevertheless, revealed suggestive evidence of an increasing trend in the induction of pancreatic cancer with increasing radiation dose, among members of the LSS during the period 1959–1970.[1] For the two cities combined, the estimated linear trend was 0.18 ± 0.15 excess case of pancreatic cancer per 10^6 PYR kerma. For a ratio of organ dose to kerma dose of 0.37 for the two cities combined, for an RBE of 1, the estimated excess cancer risk was 0.49 ± 0.41 case per 10^6 PYR. For the two cities separately, the trend toward an increased incidence of cancer of the pancreas appeared in the Nagasaki survivors, but not in the Hiroshima survivors; the Hiroshima Tumor Registry is known to be incomplete, however. The risk estimate for the Nagasaki survivors was 0.33 ± 0.21 excess cancer case per 10^6 PYR kerma. For a ratio of organ dose to kerma dose of 0.40 for Nagasaki exposure, and a neutron RBE of 1, the estimate was 0.83 ± 0.53 excess cancer death per 10^6 PYR. For the incomplete Hiroshima Tumor Registry, it was 0.04 ± 0.22 excess cancer case per 10^6 PYR. These estimates of the excess cancer-induction rate are subject to bias, in that the known atomic-bomb survivors, or survivors known to have been heavily exposed, may have received more thorough diagnostic medical attention than would other persons under normal circumstances. The conclusion may be drawn, however, that, although the LSS data for the 24-yr followup period from October 1, 1950, to September 30, 1974, do not by themselves suggest a firm radiation effect, the data do lend plausibility to the epidemiologic evidence from the ankylosing-spondylitis patient series.

OCCUPATIONAL EXPOSURE

Pancreatic cancer is one of the two cancers reported by Mancuso, Stewart, and Kneale[10,12] and confirmed by other analysts of the same material[5,6,13] to be associated with cumulative radiation-badge dose among nuclear workers at the Hanford Works. Doubling-dose estimates based on proportional-mortality analyses and assuming a linear dose response are extremely low, from 7 to 13 rems,[5,10,12] and a population-based data analysis yields a formal absolute-risk estimate of about 10 excess deaths per million persons exposed per year per rem.[6] However, although the various analyses of these data confirm that the observed association of pancreatic-cancer mortality with cumulative badge dose is unlikely to be an artifact of the original analysis, there remains considerable doubt that these data give an accurate representation of the relationship between radiation dose and pancreatic cancer. Both exposed and nonexposed workers showed higher-than-expected SMRs with respect to cancer of the pancreas,[5] which has been linked to chemical exposures.[11] Further, and more complete, studies of cancer risk and exposures to radiation and other potential carcinogens among nuclear workers are needed. The preliminary findings from the Hanford study suggest the existence of an increased risk of pancreatic cancer among nuclear workers that may or may not be causally related to radiation, but these data appear to offer only limited information about the dose-response relation between this cancer and radiation.

RADIOTHERAPY FOR MALIGNANT DISEASE

Excess mortality from pancreatic cancer (seven cases observed versus 2.85 expected) has been reported in 923 patients who survived 5 yr or more after radiotherapy for carcinoma of the cervix.[4] Pancreatic carcinoma has been reported in patients treated with radiation for lymphoma.[9] Dose estimates are not available for the reliable assessment of excess risk in these radiotherapy patients.

CONCLUSIONS

New data from the British survey of ankylosing-spondylitis patients treated with radiation tend to confirm and refine the earlier observations of an increased radiation risk of cancer of the pancreas. The most recent report of the LSS on Japanese atomic-bomb survivors has suggested a radiation dose-response relationship for pancreatic cancer, but this is not apparent from death-certificate information. A recent study of proportional mortality among workers at the Hanford nuclear plant suggested that workers in the nuclear industry may be at increased risk of pancreatic cancer. Nevertheless, pancreatic cancer continues to be an especially difficult malignancy to study for possible radiation carcinogenesis. Thus far, the only three positive studies have given widely varied risk estimates. This may be explained, at present, on the basis of inaccuracy of death-certificate diagnoses, ascertainment bias, inaccurate or incomplete dosimetry, and the possible association of radiation with other carcinogens and environmental pollutants. It appears, primarily from the series of ankylosing-spondylitis patients and the LSS data on atomic-bomb survivors, that the increase in pancreatic-cancer induction rate may be attributable to exposure to radiation. The induction rate per rad appears to be low, but this is not known with certainty.

REFERENCES

1. Beebe, G. W., H. Kato, and C. E. Land. Life Span Study Report 8. Mortality Experience of Atomic-Bomb Survivors, 1950–74. Radiation Effects Research Foundation Technical Report TR 1-77. Hiroshima: Radiation Effects Research Foundation, 1978.
2. Court Brown, W. M., and R. Doll. Expectation of life and mortality from cancer among British radiologists. Brit. Med. J. 2:181–187, 1958.
3. Court Brown, W. M., and R. Doll. Mortality from cancer and other causes after radiotherapy for ankylosing spondylitis. Brit. Med. J. 2:1327–1332, 1965.
4. Dickson, R. J. Late results of radium treatment of carcinoma of the cervix. Clin. Radiol. 23:528–535, 1972.
5. Gilbert, E. S., and S. Marks. Cancer mortality in Hanford workers. Radiat. Res. 79:122–148, 1979.
6. Hutchison, G. B., B. MacMahon, S. Jablon, and C. E. Land. Review of report by Mancuso, Stewart, and Kneale of radiation exposure of Hanford workers. Health Phys. 37:207–220, 1979.
7. International Commission on Radiological Protection. Radiosensitivity and Spatial Distribution of Dose. Reports prepared by two Task Groups of ICRP Committee 1. Publication 14. Oxford: Pergamon Press, 1969.

8. International Commission on Radiological Protection. Recommendations of the International Commission on Radiological Protection. ICRP Publication No. 24. Annals of the ICRP. Vol. 1. No. 3. New York: Pergamon Press, 1977.

9. Jochimsen, P. R., N. W. Pearlman, and R. L. Lawton. Pancreatic carcinoma as a sequel to therapy of lymphoma. J. Surg. Oncol. 8:461–464, 1976.

10. Kneale, G. W., A. M. Stewart, and T. F. Mancuso. Reanalysis of data relating to the Hanford study of the cancer risks of radiation workers, pp. 386–412. In Late Biological Effects of Ionizing Radiation. Vol. 1. Vienna: International Atomic Energy Agency, 1978.

11. Li, F. P., J. F. Fraumeni, Jr., N. Mantel, and R. W. Miller. Cancer mortality among chemists. J. Natl. Cancer Inst. 43:1159–1164, 1969.

12. Mancuso, T. F., A. Stewart, and G. Kneale. Radiation exposures of Hanford workers dying from cancer and other causes. Health Phys. 33:369–385, 1977.

13. Marks, S., W. S. Gilbert, and B. D. Breitenstein. Cancer mortality in Hanford workers, pp. 369–386. In Late Biological Effects of Ionizing Radiation. Vol. 1. Vienna: International Atomic Energy Agency, 1978.

14. National Research Council, Advisory Committee on the Biological Effects of Ionizing Radiations. The Effects on Populations of Exposure to Low Levels of Ionizing Radiation. Washington, D.C.: National Academy of Sciences, 1972.

15. Radford, E. P., R. Doll, and P. G. Smith. Mortality among patients with ankylosing spondylitis not given x-ray therapy. N. Engl. J. Med. 297:572–576, 1977.

16. Smith, P. G., and R. Doll. Age- and time-dependent changes in the rates of radiation-induced cancers in patients with ankylosing spondylitis following a single course of x-ray treatment, pp. 205–214. In Late Biological Effects of Ionizing Radiation. Vienna: International Atomic Energy Agency, 1978.

17. Smith, P. G., R. Doll, and E. P. Radford. Cancer mortality among patients with ankylosing spondylitis not given x-ray therapy. Brit. J. Radiol. 50:728–734, 1977.

18. Steer, A., C. E. Land, I. M. Moriyama, T. Yamamoto, M. Asano, and H. Sanefuji. Accuracy of diagnosis of cancer among autopsy cases. JNIH-ABCC population for Hiroshima and Nagasaki. GANN 67:625–632, 1976.

19. Upton, A. C. Effects of radiation on man. Annu. Rev. Nucl. Sci. 18:495–528, 1968.

20. Upton, A. C. Somatic and genetic effects of low-level radiation, pp. 1–40. In J. H. Lawrence, Ed. Recent Advances in Nuclear Medicine. New York: Grune & Stratton, 1974.

21. Upton, A. C., M. L. Randolph, and J. W. Conklin. Late effects of fast neutrons and gamma rays in mice as influenced by the dose rate of irradiation. Induction of neoplasia. Radiat. Res. 41:467–491, 1970.

PHARYNX, HYPOPHARYNX, AND LARYNX

The 1972 BEIR report referred to the excess occurrence of carcinoma of the pharynx in man after therapeutic irradiation of regions of the head and neck.[7] The 1977 recommendations of the ICRP[5] do not indicate that the pharynx and hypopharynx would be human tissues at risk in radiation carcinogenesis. However, in a category including all other tissues and organs of the digestive tract, the evidence suggests that in these tissues there is a carcinogenic risk at moderate radiation doses. No experimental radiation carcinogenesis of the pharynx or hypopharynx has thus far been reported in animals. This is of particular interest, in view of the extensive studies on thyroid neoplasia in mice and rats.[7,11]

RADIOTHERAPY PATIENTS

Cancer of the pharynx and hypopharynx in man has been observed after therapeutic irradiation for benign or malignant conditions in adjacent tissues—frequently the esophagus, the larynx, the thyroid, and the spine. Goolden[3] reviewed a series of 37 patients who had previously received radiotherapy for thyrotoxicosis or other lesions of the neck; the latent periods were extremely long (mean, 23.8 yr), the radiation doses were high, and exposure was continuous or fractionated (external radiotherapy fractionated doses of some 3,000–6,000 rads delivered over 3–6 wk). Raven and Levinson[10] have reported 10 patients with cancer of the pharynx after radiotherapy; the mean latent period was 25.0 yr, and doses were in the therapeutic range. Yoshizawa and Takeuchi[13] reviewed 130 cases of pharynx and larynx radiation neoplasia; the mean latent period was 27.3 yr. Other reports of similar radiation cancers[6,8] indicate quite long latent periods, in the range of 23–24 yr.

In their study of 14,554 males treated with x-ray therapy for ankylosing spondylitis, Court Brown and Doll[2] demonstrated an excess of solid cancers, including cancer of the pharynx, in heavily irradiated sites.[4] Two groups of patients have been reviewed, each with excess cancers that occurred at least 6 yr after therapy. In the group with complete followup to January 1, 1960 (14,796 PY), there were four observed cancers of the pharynx and 0.70 expected, for an excess of 3.3 (0.7, 8.5*). In the incompletely followed group to January 1, 1963 (165,631 PY), the values were five observed and 1.05 expected, for an excess of 3.95 (0.9, 9.5; $p <$ 0.025; induction rate, 0.35 per 1,000 patients).[2,4] However, the number of excess cases of cancer of the larynx over expected was not statistically significant.[12] In the most recent followup of the ankylosing-spondylitis patients treated with one course of radiotherapy (R. Doll and P. G. Smith, personal communication), no significant increase in deaths from cancer of the pharynx over that previously observed was recorded. The ankylosing-spondylitis surveys still require precise dose estimates for radiation risk to be determined. Risk estimates may be obtainable from this population when radiation dose absorbed by the tissues of the pharynx during

---

*Numbers in parentheses are 90% confidence limits.

radiotherapy has been reliably determined. A mean radiation dose to the spinal canal of 880 rads was estimated for the thoracic spine.[2] However, the induction rate is probably not significantly greater than that observed in the atomic-bomb Life Span study—perhaps 5–10 excess cases per million exposed patients per rad over almost 20 yr of followup.

Radford *et al.*[9] reported in their survey of mortality among patients with ankylosing spondylitis who were not given x-ray therapy that the only deaths from cancer showing an apparent excess risk were from cancer of the pharynx and hypopharynx (two deaths observed, 0.13 expected; $p <$ 0.01). The authors concluded, however, that the numbers were too small to permit firm conclusions concerning a relationship between ankylosing spondylitis and cancer of the pharynx.

ATOMIC-BOMB SURVIVORS

The Japanese LSS does not specify pharynx and hypopharynx neoplasia observed in excess, but this is included in a category of cancer of other digestive organs.[1]

CONCLUSIONS

It is important to recognize that the latent period for cancers of the pharynx and larynx is unusually long; mean latent periods exceeding 25 yr have been recorded in some clinical studies. It follows, therefore, that the values observed in both the LSS and the ankylosing-spondylitis patients would be below the true values. With a mean latent period of 25 yr, the total number of cancers occurring after irradiation, provided that patients do not die from other causes, would be perhaps only half the number of all cancers induced by radiation. Thus, in the absence of more precise figures on occurrence of cancers and absorbed radiation dose in the pharynx and larynx, only the following limited conclusions can be drawn:

• There is now a significantly increased rate of induction of cancers of the pharynx in irradiated populations.
• The mean latent period probably is some 25 yr after exposure.
• The present value is an underestimate, and a large proportion of radiation cancers of the pharynx and larynx may be expected to occur in surviving populations over the next decade.
• Any radiation-risk estimates are not precise, because of underestimated values and lack of information on absorbed radiation dose.

REFERENCES

1. Beebe, G. W., H. Kato, and C. E. Land. Life Span Study Report 8. Mortality Experience of Atomic-Bomb Survivors, 1950–74. Radiation Effects Research Foundation Technical Report TR 1-77. Hiroshima: Radiation Effects Research Foundation, 1978.
2. Court Brown, W. M., and R. Doll. Mortality from cancer and other causes after radiotherapy for ankylosing spondylitis. Brit. Med. J. 2:1327–1332, 1965.
3. Goolden, A. W. G. Radiation cancer. A review with special reference to radiation tumours in the pharynx, larynx, and thyroid. Brit. J. Radiol. 30:626–640, 1957.
4. International Commission on Radiological Protection. Radiosensitivity and Spatial Distribution of Dose. Reports Prepared by Two Task Groups of Committee 1 of the International Commission on Radiological Protection. ICRP Publication 14. Oxford: Pergamon Press, 1969.
5. International Commission on Radiological Protection. Recommendations of the International Commission on Radiological Protection. ICRP Publication No. 26. Annals of the ICRP. Vol. 1. No. 3. New York: Pergamon Press, 1977.
6. Kikuchi, A., C. Watanabe, M. Abe, et al. Head and neck cancer following therapeutic irradiation and brief review on those in Japan. Nippon Acta Radiol. 34:491–501, 1974.
7. National Research Council, Advisory Committee on the Biological Effects of Ionizing Radiations. The Effects on Populations of Exposure to Low Levels of Ionizing Radiation. Washington, D.C.: National Academy of Sciences, 1972.
8. Nitze, H. R. Radiogene Tumoren in H.N.O. Bereich Arch. Klin. Ohren-Nasen-Kehlkopfheilk. 199:634–638, 1971.
9. Radford, E. P., R. Doll, and P. G. Smith. Mortality among patients with ankylosing spondylitis not given x-ray therapy. N. Engl. J. Med. 297:572–576, 1977.
10. Raven, R. W., and V. B. Levinson. Radiation cancer of the pharynx. Lancet 2:683–684, 1954.
11. United Nations Scientific Committee on the Effects of Atomic Radiation. Ionizing Radiation. Levels and Effects. Vol. II. Effects. New York: United Nations, 1977.
12. Upton, A. C. Somatic and genetic effects of low-level radiation, pp. 1–40. In J. H. Lawrence, Ed. Recent Advances in Nuclear Medicine. Vol. 4. New York: Grune & Stratton, 1974.
13. Yoshizawa, Y., and T. Takeuchi. Search for the lowest irradiation dose from literatures on radiation induced cancer in pharynx and larynx. Nippon Acta Radiol. 34:903–909, 1974.

SALIVARY GLANDS

Neoplasms of the salivary glands in man, both benign and malignant, have been reported to occur in excess after irradiation, but the data have been too sparse to provide estimates of radiation risk. The early reports[12,25,26] concerned primarily children exposed to therapeutic irradiation of the neck region at high dose rates and atomic-bomb survivors of all age groups; the 1972 BEIR report[17] mentioned salivary-gland tumors only briefly, in connection with other neoplasms of specific types. Since then, additional data have been reported from several sources. The 1977 UNSCEAR report[24] summarized briefly the major epidemiologic studies in which salivary-gland tumors have been reported after exposure to ionizing radiation, particularly after radiotherapy for benign disease. From these studies, data are emerging that may provide a preliminary estimate of radiation induction rate in relation to exposure dose.

Tumors of the salivary glands have been observed in experimental rodents exposed to irradiation.[6,7,22]

The epidemiologic and experimental literature has not demonstrated the salivary-gland tissue to be more than moderately susceptible to the induction of benign and malignant tumors, and it probably is so only at high doses. However, recent studies have suggested a much higher susceptibility in man than was previously suspected.

RADIOTHERAPY FOR BENIGN DISEASE

In their early studies of thyroid neoplasia after therapeutic irradiation of the neck and mediastinum for various benign diseases in 1,644 infants and children between 1932 and 1950, Saenger and colleagues[19] observed two excess cases of salivary-gland tumors; comparison was made with 3,777 nonirradiated sibling controls. After a followup period of 10–18 yr, they found two malignant and no benign tumors of the salivary glands in the irradiated population and no salivary-gland tumors in the control siblings. The fields of irradiation included the salivary glands in children irradiated for lymphadenopathy in the tonsils and adenoids, and to a lesser extent for cervical adenitis. Radiation-dose estimates were difficult to ascertain, but probably were less than 600 R in air. The authors reported a cumulative incidence rate for salivary-gland tumors of 0.12% in 30,254 PY, or 66.1 cases per 10^6 PY.

Hempelmann and colleagues[9-11,18,21] have reported four benign salivary-gland tumors and no malignant tumors in 2,872 irradiated patients in the Rochester series of children irradiated between 1930 and 1951 for benign thymus enlargement, with a followup of 20–40 yr until 1971. The control group of 5,055 siblings had two benign and one malignant salivary-gland tumors. The precise estimates of radiation dose are not available, but doses were less than 600 R in air; the cumulative incidence rate for benign and malignant salivary-gland tumors was 0.14% in 47,313 PY, or 84.5 cases per 10^6 PY in the irradiated patients versus 19.8 cases per 10^6 PY in the control groups. On the basis of estimates of radiation dose to the thyroid gland, however, the risk rate would be approximately 5–10 excess salivary-gland tumors per million exposed children per rad over a followup period of 20–40 yr.

Janower and Miettinen[13] observed one benign salivary-gland tumor in 466 thymus-irradiated children treated between 1924 and 1946; two tumors occurred in 3,029 controls. The air dose was less than 400 R. The incidence rate for salivary-gland tumors was 0.21% in 14,037 patient-years, or 71.2 cases per 10^6 patient-yr in the irradiated group, and approximately 0.07% in the controls taken as a whole.

The initial studies of Albert, Shore, and their colleagues[1,8,20] of 2,215 children treated in New York during 1945–1950 with x-ray epilation for tinea capitis have now demonstrated three benign and one malignant salivary-gland neoplasms in exposed patients in a 20-yr followup to 1973. No salivary-gland tumors were observed in the control group of 1,413 persons. The cumulative incidence rate for benign and malignant salivary-gland tumors was 0.18% in approximately 44,300 patient-yr, or some 90.3 cases per 10^6 patient-yr. On the basis of estimates of radiation dose to the parotid gland of 39 rads determined by Harley et al.,[8] the radiation-risk rate for the 20-yr observation period would be roughly 12 (1, 35*) excess salivary-gland tumors (benign and malignant) per million exposed children per rad.

Mole[16] and Modan and colleagues[14-15] have reported the results of detailed observations on 10,902 children in Israel treated with scalp x irradiation for tinea capitis during the 11-yr period 1949–1960. They found four malignant and three benign tumors of the salivary (parotid) glands during the 15-yr followup to 1973 in the irradiated population; one benign tumor occurred in the two control series. On the basis of phantom calculations of the mean thyroid dose in the irradiated children, a parotid-gland dose of approximately 39 rads might be estimated from the measurements of Harley et al.[8] in the New York series. These values would yield cumulative radiation-risk estimates for benign and malignant salivary-gland (parotid-gland) tumors of at least 16 excess cases per million children exposed per rad for the 15-yr followup period.

ATOMIC-BOMB SURVIVORS

The study by Belsky and colleagues[3,4] on salivary-gland tumors in Japanese atomic-bomb survivors for the period 1957–1970 has now been extended to 1975.[5] In the LSS, the cases of salivary-gland tumors reported were those indexed in Tumor Registries in both cities from 1957 to 1970 and from the ABCC-JNIH Adult Health Study index of cases. The case-incidence data from Hiroshima and Nagasaki were combined. The gamma- and neutron-radiation estimates were added in these studies. Of 1,433 exposed persons examined in a 12-yr period (16,172 PY), there was a significant excess of two cases of malignant salivary-gland tumors ob-

*Numbers in parentheses are 90% confidence limits.

served versus 0.12 expected and an excess of one case of benign tumor observed versus 0.28 expected in the over-300-rads kerma group (observed/expected = 7.5). Assuming a mean kerma of 400–500 rads, the radiation-risk estimate was approximately three (one to eight) excess salivary-gland tumors per million persons per rad over the 12-yr followup period. No excess of salivary-gland tumors was observed in the below-300-rads kerma group (observed/expected = 0.91).

Takeichi and associates[23] have now observed 17 benign and malignant salivary-gland tumors (1.7 expected) over a 25-yr period (1945–1971) in the atomic-bomb survivors in Hiroshima and nearby Kure within 5,000 m at the time of the bombing, as determined from records of hospital pathology departments. Standardized incidence rates for benign and malignant salivary-gland tumors were calculated as 1.8 cases per 100,000 exposed persons per year and 0.7 case per 100,000 unexposed persons per year. The incidence rates decreased with increasing distance from the hypocenter, from 3.8 cases per 100,000 exposed persons per year at 0-1–500 m to 1.3 cases per 100,000 exposed persons per year at 1,501–5,000 m. For malignant tumors alone, the standardized incidence was 2.2 cases per 100,000 exposed persons per year at 0–1,500 m; 0.7 case per 100,000 exposed persons per year at 1,500 m and beyond; and 0.1 case per 100,000 nonirradiated persons per year. Thus, the incidence of all benign and malignant salivary-gland tumors was some 5.4 times greater among the high-dose survivors than in the unexposed group; in the low-dose survivors, the incidence was only some 1.9 times greater than in the nonirradiated population. This increased incidence of tumors with increasing proximity to the hypocenter was statistically significant ($p < 0.001$). The radiation-risk rate for salivary-gland tumors in survivors exposed in the region less than 1,500 m from the hypocenter (assuming total air doses of 32 rads at 1,500 m and 135 rads farther in) would be approximately 21 (9, 41) excess tumors per million persons exposed per year, and possibly only one-third of that in survivors exposed at 1,500–5,000 m. The LSS dosimetry exposure determinations[2] would permit a very rough estimation of radiation risk: perhaps no more than one or two excess salivary-gland tumors per million exposed persons per rad over the 19-yr followup period.

CONCLUSIONS

Since the 1972 BEIR report,[17] additional radiation-induced benign and malignant salivary-gland tumors have been reported in significant excess in irradiated children and in Japanese atomic-bomb survivors. The numbers in each group are small, the latent period for both benign and malignant tumors is relatively long, and the diagnosis has occurred after 13-25 yr. The induction rate for both benign and malignant tumors is low, perhaps no more than 10 excess cases per million exposed children per rad over a 20-yr period of followup; the rate would be expected to increase over a longer period of observation. Exposure in adult life might result in a decreased risk, perhaps only one-third or less of that after childhood exposure. No conclusions can yet be reached about the relationship of age at the time of irradiation to the incidence of tumors; in the childhood studies, the age range was relatively narrow. Neither can conclusions be reached on sex ratios; in the childhood studies, the patients were predominantly male. Finally, as in the case of thyroid tumors, salivary-gland tumors are both benign and malignant, and the present evidence from clinical studies of salivary-gland tumors indicates that patients with radiation-induced tumors of the salivary glands should be expected to have a high survival rate in association with modern diagnosis and management.

REFERENCES

1. Albert, R. E., and A. R. Omran. Follow-up study of patients treated by x-ray epilation for tinea capitis. Arch. Environ. Health 17:899-918, 1968.
2. Beebe, G. W., H. Kato, and C. E. Land. Life Span Study Report 8. Mortality Experience of Atomic-Bomb Survivors, 1950-74. Radiation Effects Research Foundation Technical Report TR 1-77. Hiroshima: Radiation Effects Research Foundation, 1978.
3. Belsky, J. L., K. Tachikawa, R. W. Cihak, and T. Yamamoto. Salivary gland tumours in atomic bomb survivors, Hiroshima-Nagasaki, 1957 to 1970. JAMA 219:864-868, 1972.
4. Belsky, J. L., N. Takeichi, T. Yamamoto, R. W. Cihak, F. Hirose, H. Ezaki, S. Inoue, and W. J. Blot. Salivary Gland Neoplasms Following Atomic Radiation. Atomic Bomb Casualty Commission Technical Report TR 23-72. Hiroshima: Atomic Bomb Casualty Commission, 1972.
5. Belsky, J. L., N. Takeichi, T. Yamamoto, R. W. Cihak, F. Hirose, H. Ezaki, S. Inoue, and W. J. Blot. Salivary gland neoplasms following atomic radiation. Additional cases and reanalysis of combined data in a fixed population, 1957-1970. Cancer 35:555-559, 1975.
6. Glucksmann, A., and C. P. Cherry. The induction of adenomas by the irradiation of salivary glands in rats. Radiat. Res. 17:186-202, 1962.
7. Gross, L. Attempt to recover filterable agent from x-ray induced leukemia. Acta Hematol. 19:353-361, 1958.
8. Harley, N. H., R. E. Albert, R. E. Shore, and B. S. Pasternack. Follow-up study of patients treated by x-ray epilation for tinea capitis. Estimation of the dose to the thyroid and pituitary glands and other structures of the head and neck. Phys. Med. Biol. 21:631-642, 1976.
9. Hazen, R. W., J. W. Pifer, E. T. Toyooka, J. Livingood, and L. H. Hempelmann. Neoplasms following irradiation of the head. Cancer Res. 26:305-311, 1966.
10. Hempelmann, L. H., W. J. Hall, M. Phillips, R. A. Cooper, and W. R. Ames. Neoplasms in persons treated with x-rays in infancy. Fourth survey in 20 years. J. Natl. Cancer Inst. 55:519-530, 1975.

11. Hempelmann, L. H., J. W. Pifer, G. J. Burke, R. Terry, and W. R. Ames. Neoplasms in persons treated with x rays in infancy for thymic enlargement. A report of the third follow-up survey. J. Natl. Cancer Inst. 38:317–341, 1967.
12. International Commission on Radiological Protection. Radiosensitivity and Spatial Distribution of Dose. Publication 14. Reports prepared by Task Groups of Committee 1. New York: Pergamon Press, 1969.
13. Janower, M. L., and O. S. Miettinen. Neoplasms after childhood irradiation of the thymus gland. JAMA 215:753–756, 1971.
14. Modan, B., D. Baidatz, H. Mart, R. Steinitz, and S. G. Levin. Radiation-induced head and neck tumours. Lancet 1:277–279, 1974.
15. Modan, B., E. Ron, and A. Werner. Thyroid cancer following scalp irradiation. Radiology 123:741–744, 1977.
16. Mole, R. H. Antenatal irradiation and childhood cancer. Causation or coincidence? Brit. J. Cancer 30:199–208, 1974.
17. National Research Council, Advisory Committee on the Biological Effects of Ionizing Radiations. The Effects on Populations of Exposure to Low Levels of Ionizing Radiation. Washington, D.C.: National Academy of Sciences, 1972.
18. Pifer, J. W., L. H. Hempelmann, H. J. Dodge, and F. J. Hodges II. Neoplasms in the Ann Arbor series of thymus-irradiated children. A second survey. Amer. J. Roentgenol. Rad. Ther. Nucl. Med. 103:13–18, 1968.
19. Saenger, E. L., F. N. Silverman, T. D. Sterling, and M. E. Turner. Neoplasia following therapeutic radiation for benign conditions in childhood. Radiology 74:889–904, 1960.
20. Shore, R. E., R. E. Albert, and B. S. Pasternack. Follow-up study of patients treated by x-ray epilation for tinea capitis. Arch. Environ. Health 31:17–28, 1976.
21. Simpson, C. L., L. H. Hempelmann, and L. M. Fuller. Neoplasia in children treated with x-rays in infancy for thymic enlargement. Radiology 64:840–845, 1955.
22. Takeichi, N. Induction of salivary gland tumours following x-ray examination. II. Development of salivary gland tumours in long-term experiments. Med. J. Hiroshima Univ. 23:391–411, 1975. (in Japanese)
23. Takeichi, N., F. Hirose, and H. Yamamoto. Salivary gland tumors in atomic bomb survivors, Hiroshima, Japan. I. Epidemiologic observations. Cancer 38:2462–2468, 1976.
24. United Nations Scientific Committee on the Effects of Atomic Radiation. Sources and Effects of Ionizing Radiation. Report to the General Assembly with Annexes. New York: United Nations, 1977.
25. Upton, A. C. Effects of radiation on man. Annu. Rev. Nucl. Sci. 18:495–528, 1968.
26. Upton, A. C. Somatic and genetic effects of low-level radiation, pp. 1–40. In J. H. Lawrence, Ed. Recent Advances in Nuclear Medicine. Vol. 4. New York: Grune & Stratton, 1974.

PARATHYROID

Parathyroid tumors were not mentioned in the 1972 BEIR report[15] or the 1977 UNSCEAR report.[26] However, they have been produced in animals by x irradiation.[1,2,27]

The association of parathyroid adenomas with benign and malignant thyroid tumors in man has been observed.[3,6,8-10,16] However, during the last 3–4 yr, a radiation factor in the development of primary hyperparathyroidism has been suspected; this suspicion has arisen from the association of thyroid tumors with prior irradiation. Most of the radiogenic parathyroid tumors have been hyperfunctioning adenomas, but a few carcinomas with metastases have also been reported. It is not clear whether there have been true increases in the incidence of primary hyperparathyroidism and in the incidence of parathyroid adenomas associated with irradiation. The increasing frequency of measurement of serum calcium in patients has led to the recognition of many patients with high serum calcium and may have resulted in the diagnosis of mild hyperparathyroidism and parathyroid adenomas that might otherwise have gone unrecognized for many years.

A number of clinical reports linking hyperparathyroidism and parathyroid adenomas with prior irradiation of the head and neck or upper thorax area have appeared since 1975.[4,5,7,11-14,17-25] In the largest series,[18] it was found that, of 89 surgical patients with parathyroid adenomas, at least 27 (30%) had a history of prior irradiation of the head, neck, or upper thorax—"at least," because many patients did not know whether they had received radiation in childhood. The dose of radiation was not known in most of these cases, but probably ranged from 250 to 1,000 rads.

It is difficult to determine the incidence of parathyroid adenomas in the general population; benign tumors are not ordinarily entered in Cancer Registries. Because patients are usually cured by surgery, the incidence is not reflected in mortality statistics. Furthermore, radiation histories are not taken routinely, so it will be difficult to evaluate the association of simultaneous or sequential thyroid and parathyroid tumors with prior irradiation of the head, neck, or upper thorax. Nevertheless, the suspected association of parathyroid adenomas with prior irradiation requires that this be considered in persons who received radiation to the head, neck, or upper thorax in infancy through young adulthood and who are being examined for possible thyroid tumors.

REFERENCES

1. Berdjis, C. C. Parathyroid diseases and irradiation. Strahlentherapie 143:48–62, 1972.
2. Berdjis, C. C. Pathogenesis of radiation-induced endocrine tumors. Oncology 21:49, 1967.
3. Ellenberg, A. H., L. Goldman, G. S. Gordan, and S. Lindsay. Thyroid carcinoma in patients with hyperparathyroidism. Surgery 51:708, 1962.
4. Greenspan, F. S. Radiation exposure and thyroid cancer. JAMA 237:2089–2091, 1977.
5. Grinblat, J., et al. Meningioma associated with parathyroid adenoma. Amer. J. Med. Sci. 272:327–339, 1976.
6. Heimann, P., O. Nilsson, and G. Hansson. Parathyroid and thyroid disease. II. Thyroid disease connected with hyperparathyroidism. Acta. Chir. Scand. 136:143, 467, 1970.

7. Kairaluoma, M. I., E. Heikkenen, R. Mokka, R. Huttunen, and T. K. I. Larmi. Non-medullary thyroid carcinoma in patients with parathyroid adenoma. Acta Chir. Scand. 142:447–449, 1976.

8. Kaplan, L., A. D. Katz, C. Ben-Isaac, *et al.* Malignant neoplasms and parathyroid adenoma. Cancer 28:401–407, 1971.

9. Krementz, E. E., R. Yeager, W. Hawley, *et al.* The first 100 cases of parathyroid tumor from Charity Hospital of Louisiana. Ann. Surg. 173:872–883, 1971.

10. Laing, V. O., B. Frame, and M. A. Block. Associated hyperparathyroidism and thyroid lesions. Arch. Surg. 98:709–712, 1969.

11. Leichty, R. D. Dramatic increase in hyperparathyroidism detection worldwide seen in last decade. Presented to Wyoming State Medical Society. Int. Med. News, June 1, 1977.

12. LiVolsi, V. A., and C. R. Feind. Parathyroid adenoma and non-medullary thyroid carcinoma. Cancer 38:1391–1393, 1976.

13. LiVolsi, V. A., *et al.* Coexistent parathyroid adenomas and thyroid carcinoma. Can radiation be blamed? Arch. Surg. 113:285–286, 1978.

14. Lund, R. S., *et al.* Parathyroid adenoma presenting as a "cold nodule" on thyroid scan. Minn. Med. 59:1382, 1976.

15. National Research Council, Advisory Committee on the Biological Effects of Ionizing Radiations. The Effects on Populations of Exposure to Low Levels of Ionizing Radiation. Washington, D.C.: National Academy of Sciences, 1972.

16. Ogburn, P. L., and B. M. Black. Primary hyperparathyroidism and papillary adenocarcinoma of the thyroid. Proc. Staff Meet. Mayo Clin. 31:295–298, 1956.

17. Petro, A. B., and J. D. Hardy. The association of parathyroid adenoma and non-medullary carcinoma of the thyroid. Ann. Surg. 181:118, 1975.

18. Prinz, R. A., E. Paloyan, A. M. Lawrence, J. R. Pickleman, S. Braithwaite, and M. H. Brooks. Radiation-associated hyperparathyroidism. A new syndrome? Surgery 82:296–302, 1977.

19. Rosen, I. B., H. G. Strawbridge, and J. Bain. A case of hyperparathyroidism associated with radiation to the head and neck area. Cancer 36:1111–1114, 1975.

20. Russ, J. E., E. F. Scanlon, and S. F. Sener. Parathyroid adenomas following irradiation. Cancer 43:1078–1083, 1979.

21. Schachner, S. H., *et al.* Parathyroid adenoma and previous head and neck irradiation. Ann. Intern. Med. 88:804, 1978.

22. Swelstad, J. A., E. F. Scanlon, M. A. Oviedo, *et al.* Irradiation-induced polyglandular neoplasia of the head and neck. Amer. J. Surg. 135:820–824, 1978.

23. Tisell, L. E., S. Carlsson, S. Lindberg, and I. Ragnhult. Autonomous hyperparathyroidism. A possible later complication of neck radiotherapy. Acta Chir. Scand. 142:367–373, 1976.

24. Tisell, L. E., *et al.* Occurrence of previous neck radiotherapy among patients with associated non-medullary thyroid carcinoma and parathyroid adenoma or hyperplasia. Acta Chir. Scand. 144:7–11, 1978.

25. Triggs, S. M., and E. D. Williams. Irradiation of the thyroid as a cause of parathyroid adenoma. Lancet 1:593–594, 1977. (letter)

26. United Nations Scientific Committee on the Effects of Atomic Radiation. Sources and Effects of Ionizing Radiation. Report to the General Assembly with Annexes. New York: United Nations, 1977.

27. Warren, S., and R. Chute. Parathyroid carcinoma in parabiont rats. Science 135:927–928, 1962.

URINARY ORGANS

Urinary organs, especially the kidney and bladder, appear to be among those with definite but low sensitivity to the carcinogenic action of ionizing radiation. Most experimental work has been done on the rat and the mouse; and findings are usually reported for the kidney, and seldom for the urinary bladder.

Although renal tumors are uncommon in all species of animals, they can be produced experimentally by agents of many different kinds, including ionizing radiation, and this is especially true of the rat. In his recent review of renal carcinogenesis, Hamilton[8] summarized experimental evidence of radiogenic renal tumors in the rat and the mouse. The tumors included both benign and malignant types and were induced by neutron, gamma, and x radiation and by both whole- and partial-body irradiation. Strain, sex, and age differences were reported. Although incidence varies greatly from experiment to experiment, dose-response relationships are rarely estimated. A series of reports by Maldague[10-12] on experiments in which partial-body radiation was administered to rats is exceptional. Maldague reported a threshold dose for renal carcinogenesis between 570 and 850 R, an optimally effective dose of 1,710 R, and complete absence of effect at 14,250 R. Mice exposed to atomic-bomb radiation at various doses showed too few renal tumors to yield dose-response estimates or an RBE estimate for the possibly greater neutron effect observed.[21] From experiments on Sprague-Dawley rats exposed to fast neutrons or to x radiation, Rosen et al.[16] concluded that neutrons were more effective in producing renal neoplasia. In their earlier experiments comparing the effects of x rays and fast neutrons on (C57L × A)F$_1$ mice, Nowell and Cole[15] reported an excess of renal carcinoma only in neutron-irradiated mice. Several reports are of particular interest in regard to pathogenesis. From his experiments with whole- and partial-body irradiation of Sprague-Dawley and FAC-F$_1$ rats, Berdjis[6] concluded that nephrosclerotic and arteriosclerotic lesions in the kidney play a major role in the pathogenesis of these tumors. And in his partial-body x-ray experiments with rats, Maldague[12] observed that renal tumors developed from foci of regeneration within kidneys atrophied by nephrosclerosis. Finally, Rosen and Cole,[17] after a series of experiments on combined x radiation and nephrectomy in mice, concluded that renal neoplasia arose as an interaction of the specific proliferative stimulus (unilateral nephrectomy) with radiation-altered kidney cells.

There are several sources of data on man: patients given colloidal thorium dioxide in connection with various diagnostic procedures, but especially retrograde pyelography;[22] patients with various diseases treated with x rays (da Silva Horta *et al.*,[7] Smith and Doll,[19] McIntyre and Pointon,[13] and R. Doll and P. G. Smith, personal communication); and the Japanese atomic-bomb survivors.[1] In his 1967 review of Thorotrast (thorium dioxide) tumors, Wenz[22] recorded that 26 among 124 Thorotrast tumors (after retrograde pyelography with Thorotrast) reported at that time were of the kidney, but that followup studies on cohorts of patients who received injections of Thorotrast did not often yield tumors of the kidney. He noted that the most frequent use of thorium dioxide was in arteriography and intravenous hepatography—procedures that left very little of the contrast medium in the kidney. In the da Silva Horta *et al.* series of 1,230 traced persons who received injections of Thorotrast for diagnostic purposes, for example, cerebral angiography was the diagnostic procedure in 67%, but for only three was it retrograde pyelography.[7] Although he reported 104 malignant tumors among the 1,230 patients, none was of the kidney and only two were of the bladder. In retrograde pyelography, however, less than 10 ml of Thorotrast was used and, depending on the pathologic condition, enough contrast medium might be deposited in the kidney to cause either benign or malignant disease.[22] The Thorotrast experience is significant chiefly in showing that alpha particles can produce malignant tumors of the kidney. Unfortunately, information is lacking with which to assess the degree of increased risk associated with specific radiographic procedures that use thorium dioxide; clearly, such estimates would depend on the presence of pathologic kidney conditions that cause retention of the contrast medium. Wenz[22] gave an average interval of 27.5 yr from exposure to thorium dioxide to onset or diagnosis of tumor—longer than that for other sites (17.5 yr). He also cited an estimate attributed to Alken of the number and energy of alpha particles in autoradiographies; for Thorotrast conglomerates, an effective dosage of 13.2 rem/wk was calculated.

In their most recent followup of ankylosing-spondylitis patients treated with x rays, Doll and Smith (personal communication) observed seven deaths versus 3.12 expected from cancer of the kidney in the interval from 6 yr after treatment through 1969 ($p < 0.05$ in a one-tailed test). There were eight deaths from cancer of the bladder versus 5.98 expected over this same period—an insignificant increase.

Followup studies of women treated with radium, x rays, or both for cancer of the uterine cervix or corpus have not suggested that this procedure carries a significant risk of excess mortality from cancer of the urinary bladder,[1,13] but the numbers of observations are small and serve only to place rather wide limits on any possible effect. Similarly, in their

TABLE A-23　Observed and Expected Deaths from Cancer of Urinary Organs among Atomic-Bomb Survivors, by T65 Dose and City, 1950–1974[a]

| T65 Dose, rads (kerma) | No. Deaths | | | |
| | Hiroshima | | Nagasaki | |
	Observed	Expected[b]	Observed	Expected[b]
0	34	35.8	1	3.1
1–9	16	16.4	3	4.0
10–49	11	13.3	4	2.4
50–99	3	3.6	2	0.9
100–199	4	2.2	1	0.9
200–299	4	0.8	0	0.4
300–399	1	0.4	0	0.2
>399	0	0.6	1	0.2
TOTAL	73	73	12	12
p (homogeneity)	0.02		0.39	
p (linear trend)	0.06		0.14	
Excess deaths per 10⁶ PY per rad	0.15		0.09	
90% confidence limits	0, 0.32		−0.04, 0.23	

[a] Data from Beebe et al. [4]
[b] Adjusted for age and sex; the Nagasaki sample is appreciably younger than the Hiroshima sample and has a very different dose distribution.

1976 report on the followup of women treated with x rays for metropathia haemorrhagica, Smith and Doll[19] cited only three observed deaths from cancer of the bladder versus 2.15 expected in the 13.6-yr interval.

The autopsy data of the ABCC for the period 1961–1965 contained a preliminary indication of excess mortality from bladder cancer among atomic-bomb survivors,[5] but this was not confirmed by the larger death-certificate studies for 1950–1966,[3] 1950–1970,[9] and 1950–1972.[14] By 1974, however, the death-certificate data were seen to contain evidence that mortality from cancer of urinary organs was directly associated with radiation dose; but it was only for Hiroshima that the relationship was judged to be statistically significant.[4] Table A-23 contains the basic observations, by city. Although the number of deaths in Nagasaki was quite small, a test for linear trend on the data of the two cities returned a p of 0.02, in comparison with 0.06 for Hiroshima alone. This is not strong evidence by any means, but the death certificate has a very low detection rate for cancers of urinary organs in the Japanese experience.[20] The overall estimate of absolute risk is 0.13 excess death per 10^6 PY per rad for both cities combined, with 90% confidence limits of 0.02 and 0.25. For Hiroshima alone, the estimate is 0.15, with confidence limits of 0.00 and 0.32. None of the age-specific estimates is especially striking; there were no deaths in those who were under age 10 in 1945. Analysis of the material by calendar time provides no firm basis for estimating a minimal latent period; the test for linear trend first yields a significant result in 1967–1970.

Tumor-Registry data for 1959–1970 in the two cities are summarized in Table A-24. Tumor-Registry data are known to be incomplete, especially for Hiroshima, and studies have not been done to rule out the possibility of bias in ascertainment. They are, nevertheless, suggestive of an association between radiation dose and the risk of cancer of urinary organs.[2]

Stimulated by the preliminary findings in the autopsy data, H. Sanefuji *et al.* (unpublished data) have recently completed a comprehensive study

TABLE A-24 Observed and Expected Malignant Tumors of Urinary Organs Reported by City Tumor Registries among Atomic-Bomb Survivors, by T65 Dose and City, 1959–1970[a]

| T65 Dose, rads (kerma) | No. Tumors | | | |
| | Hiroshima | | Nagasaki | |
	Observed	Expected[b]	Observed	Expected[b]
0	40	35.3	2	3.5
1–9	9	16.2	3	4.7
10–49	12	13.1	2	2.8
50–99	2	3.5	4	1.0
100–199	3	2.2	1	1.0
200–299	5	0.8	1	0.5
300–399	1	0.4	0	0.2
>399	0	0.6	1	0.3
TOTAL	72	72	14	14
p values, linear trend:				
0+ rads	0.05		0.04	
10+ rads only	0.07		0.21	
Excess cases per 10^6				
PY per rad	0.34		0.32	
90% confidence limits	0, 0.68		0.01, 0.63	

[a] Reprinted from Beebe *et al.*[2]

[b] Adjusted for age and sex; the Nagasaki sample is appreciably younger than the Hiroshima sample and has a very different dose distribution.

of urinary-bladder tumors ascertained through both clinical and pathology diagnoses made in all medical-care facilities in the two cities over the period 1961–1972 and of kidney tumors in autopsy cases in the same period among members of the LSS sample (extended). They confirmed 112 cases of (mostly malignant) urinary-bladder tumor, for 77% of which histologic diagnoses were available. Only for the two cities combined was the relationship between radiation dose and the risk of bladder cancer a statistically significant one, and the whole effect was seen in those who were aged 40 or older in 1945 (Table A-25). Among the 18 subjects with malignant tumors of the kidney established by autopsy in the LSS sample in the 1961–1972 period, only three had been exposed to 100 rads kerma or more; because of ascertainment bias in the autopsy sample and small sample size, the authors felt that they could reach no conclusion as to the relationship between dose and frequency of tumor. The dose distribution of 93 benign tumors of the kidney suggested no relationship to radiation.

Overall, then, both the kidney and the urinary bladder seem susceptible to radiogenic cancer in both man and experimental animals. The degree of susceptibility is probably low, however, in comparison with that of other organs. For man, the only quantitative estimate is that obtainable from the ankylosing-spondylitis patients and the atomic-bomb survivors: 0.13 excess death from cancer of urinary organs per 10^6 PY per rad. But its confidence interval is wide and, being based on death certificates, it seems quite likely to be an underestimate. However, entirely apart from other

TABLE A-25   Observed and Expected Incidence Cases of Urinary Bladder Tumor in the Extended Life Span Study Sample Age 40 or Older in 1945, by Dose, 1961–1972[a]

| | T65 Dose, rads (kerma) | | | |
	Not in City, <1	1–99	>99	TOTAL
Observed	52	26	9	87
Expected	50.7	32.1	4.28	87
p^b	—	—	—	0.04
RR	1.0	0.8	2.0	—

[a] Unpublished data from H. Sanefuji *et al.*, Radiation Effects Research Foundation Technical Report TR 18-79; used here with permission from Radiation Effects Research Foundation.
[b] In χ^2 test on 2 df.

uncertainties, whether such linear estimates are truly applicable to the low-dose region remains moot. Furthermore, mortality from cancer of the urinary organs is notably low in Japan—less than half that in the United States.[18] The experience of the atomic-bomb survivors suggests that persons under, perhaps, age 40 may show little evidence of an effect for 25 yr or so, perhaps because the natural incidence of these tumors is high only in the later decades of life.

REFERENCES

1. Bailar, J. C., III. The incidence of independent tumors among uterine cancer patients. Cancer 16:842–853, 1963.
2. Beebe, G. W., H. Kato, and C. E. Land. Life Span Study Report 8. Mortality Experience of Atomic-Bomb Survivors, 1950–74. Radiation Effects Research Foundation Technical Report TR 1-77. Hiroshima: Radiation Effects Research Foundation, 1978.
3. Beebe, G. W., H. Kato, and C. E. Land. Studies of the mortality of A-bomb survivors. 4. Mortality and radiation dose, 1950–1966. Radiat. Res. 48:613–649, 1971.
4. Beebe, G. W., H. Kato, and C. E. Land. Studies of the mortality of A-bomb survivors. 6. Mortality and radiation dose, 1950–1974. Radiat. Res. 75:138–201, 1978.
5. Beebe, G. W., T. Yamamoto, Y. S. Matsumoto, and S. E. Gould. ABCC-JNIH Pathology Studies, Hiroshima-Nagasaki, Report 2, October 1950–December 1965. Atomic Bomb Casualty Commission Technical Report TR 8-67. Hiroshima: Atomic Bomb Casualty Commission, 1967.
6. Berdjis, C. C. Kidney tumors and irradiation pathogenesis of kidney tumors in irradiated rats. Oncologia (Basel) 16:312–324, 1963.
7. da Silva Horta, J., L. C. da Motta, and M. H. Tavares. Thorium dioxide effects in man. Epidemiological, clinical, and pathological studies (experience in Portugal). Environ. Res. 8:131–159, 1974.
8. Hamilton, J. M. Renal carcinogenesis. Adv. Cancer 22:1–56, 1975.
9. Jablon, S., and H. Kato. Studies of the mortality of A-bomb survivors. 5. Radiation dose and mortality, 1950–1970. Radiat. Res. 50:649–698, 1972.
10. Maldague, P. Neuropathies et cancers du rein par irradiation locale chez le rat. Acta Univ. Int. Contra Cancrum 19:697–703, 1963.
11. Maldague, P. Radiocanierisation experimentale du rein par les rayons x chez le rat. Pathol. Eur. 1:321–409, 1966.
12. Maldague, P. Radiocanierisation experimentale du rein par les rayons x chez le rat. Pathol. Eur. 2:1–54, 1967.
13. McIntyre, D., and R. C. S. Pointon. Vesical neoplasms occurring after radiation treatment for carcinoma of the uterine cervix. J. R. Coll. Surg. Edinb. 16:141–146, 1971.
14. Moriyama, I. M., and H. Kato. Life Span Study Report 7. Mortality Experience of A-bomb Survivors, 1970–1972, 1950–1972. Atomic Bomb Casualty Commission Technical Report TR 15-73. Hiroshima: Atomic Bomb Casualty Commission, 1973.
15. Nowell, P. C., and L. J. Cole. Late effects of fast neutrons vs. x-rays in mice. Nephrosclerosis, tumors, longevity. Radiat. Res. 11:545–556, 1959.
16. Rosen, V. J., Jr., T. J. Castanera, D. J. Kimeldorf, and D. C. Jones. Pancreatic islet cell tumors and renal tumors in the male rat following neutron exposure. Lab. Invest. 11:204–210, 1962.

17. Rosen, V. J., and L. J. Cole. Accelerated induction of kidney neoplasms in mice after x-irradiation (690 rad) and unilateral nephrectomy. J. Natl. Cancer Inst. 28:1031–1041, 1962.
18. Segi, M., and M. Kurihara. Cancer Mortality for Selected Sites in 24 Countries. No. 3, 1960–1961. Sendai, Japan: Tohoku University School of Medicine, 1964.
19. Smith, P. G., and R. Doll. Late effects of x irradiation in patients treated for metropathia haemorrhagica. Brit. J. Radiol. 49:224–232, 1976.
20. Steer, A., I. M. Moriyama, and K. Shimizu. ABCC-JNIH Pathology Studies, Hiroshima and Nagasaki Report 3. The Autopsy Program and the Life Span Study. Atomic Bomb Casualty Commission Technical Report TR 16-73. Hiroshima: Atomic Bomb Casualty Commission, 1973.
21. Upton, A. C., A. W. Kimball, J. Furth, K. W. Christenberry, and W. H. Benedict. Some delayed effects of atom-bomb radiations in mice. Cancer Res. 20:1–60, 1960.
22. Wenz, W. Tumors of the kidney following retrograde pyelography with colloidal thorium dioxide. Ann. N. Y. Acad. Sci. 145:806–810, 1967.

OVARY

No specific category of radiation-induced ovarian neoplasms in humans was discussed in the 1972 BEIR report,[9] although it did refer to a group of miscellaneous neoplasms of other types and sites that reportedly occur in excess after irradiation.[6,12] However, the human reproductive cells appear to have a relatively low sensitivity to the induction of radiation cancer, compared with other tissues. The 1977 ICRP report[7] stated that no carcinogenic effects in these organs after irradiation had yet been documented conclusively in humans. However, there have been confirmed reports of carcinoma of the ovary in women who had received radiotherapy for benign conditions of the pelvic organs[11] and in atomic-bomb survivors of Hiroshima and Nagasaki.[1]

EXPERIMENTAL STUDIES

There is now reliable experimental evidence of the radiation induction of ovarian tumors in mice.[8,10,16] A number of important observations have been related to the dose-response relationship and the effects of total dose, of dose rate, and of LET. The dose-response relationship for the induction of ovarian tumors shows no apparent increase over the high spontaneous incidence of ovarian adenomas in mice[8,10,16] (e.g., 11% in LAF$_1$ mouse controls, 5–15% in RFM mice). The dose-response curve for acute exposure has a very steep curvilinear or sigmoidal rise for low radiation doses (about 50 rads) and shows a high susceptibility to radiation neoplasia.[10,15,16] In general, the induction of tumors results from single acute doses, reaches a maximum at 100 rads,[4,15,17] and may maintain a high-level plateau to 200 rads and then decrease. The plateau is maintained at 500–600 rads with higher LET,[2,17] but tends to decline slowly with x irradiation. At low dose rates (e.g., less than 2 rads/d) of continuous gamma irradiation, there is a small increase in incidence, but the response appears to be only slightly curvilinear or sigmoidal.[18] At higher dose rates, 112–390 rads/d, the curvilinear or sigmoidal dose response demonstrated was marked, with a plateau after 390 rads. It is of interest that, provided that the exposure time was held constant in these experiments, the incidence varied with the square of the dose.[18]

The experimental radiation studies on ovarian carcinogenesis dealt primarily with tumor induction in mice[13] and demonstrated the following: All cells, both supporting and hormone-secreting elements, that constitute the organ, but not the reproductive cells (oocytes and follicular cells), are at risk of neoplastic induction; no single element appears to be more susceptible. The ovary is relatively sensitive to the induction of radiation cancer, and as little as a 50-rad acute exposure can result in a significant increase in the tumor-induction rate.[2] At the lower doses, acute exposure to higher-LET radiation—e.g., higher-energy neutrons and protons—has no greater effectiveness. Higher doses, however, up to 400 rads, maintain a higher incidence plateau, whereas there is a falloff with x rays. The maximum is reached in the range of 100–200 rads.[4,17] There is a dose-rate effect at low doses of continuous gamma irradiation; an increase in dose rate results in an increased yield of tumors. There is a curvilinear dose-response relationship without a threshold in the range of 1.75–112 rads/d for total doses up to approximately 400 rads.[18] In general, the dose-response curve appears to be sigmoidal or curvilinear without a threshold, depending on dose rate, LET, and total dose, as well as strain and age.[14] There is a hormone-dependent relationship in ovarian neoplastic transformation after irradiation, possibly mediated by pituitary gonadotropins.[5]

HUMAN STUDIES

Radiotherapy for Benign Disease

In a retrospective study of 731 gynecologic patients treated with intracavitary radium or external x rays, primarily for uterine fibroids or other benign pelvic disorders, Palmer and Spratt[11] found an excess of 5.4 cases (eight observed versus 2.6 expected) of ovarian cancer. The mean latent period was 10.1 yr. No precise radiation-dose estimate could be determined. Air and tissue radiation doses of approximately 2,700 R and 700 R were estimated, but radium dosage was estimated in milligram-hour equivalents. The induction rate could be determined on the basis of x-ray treatment solely, but an estimate of radiation risk per rad could not be ascertained. In their review of five other clinical series, Palmer and Spratt[11] described a total of 3,968 gynecologic patients in whom eight ovarian neoplasms arose after pelvic irradiation. Precise radiation doses could not be ascertained, and followup periods were generally less than 10 yr.

ICRP Publication 14[6] assessed the data of Court Brown and Doll[3] and found, in ankylosing-spondylitis patients who developed cancer in heavily irradiated sites, that cancer of the ovary appeared in the subgroup in which the difference between the observed and expected cancer incidences was not statistically significant (four observed, two expected, and a rate of 0.8 case per 1,000 persons).

Atomic-Bomb Survivors

In their most recent report on studies of the Tumor-Registry data on the Japanese atomic-bomb survivors, Beebe and colleagues[1] indicated an increasing rate of induction of ovarian tumors in the exposed Hiroshima population, but not in the Nagasaki survivors. For the 300+ rads kerma exposure group in Hiroshima, the incidence rate per rad kerma was 0.6 ± 0.26 excess cancer per million exposed women per rad during the 12-yr followup period, 1959–1970. Considering a ratio of organ dose to kerma dose of 0.36 for an RBE of 1, the risk was 1.67 ± 0.72 excess cancers per million exposed women per rad. For a ratio of organ dose to kerma dose of 0.47 for an RBE of 5, the risk was 1.28 ± 0.55 excess cancers per million exposed women per rad for the 12-yr followup period. The induction rate for the Nagasaki cohort was 0.04 ± 0.22 excess cancer per million women per rad kerma. However, it is probable that the mean latent period for ovarian tumors is longer, and a rise in the radiation-induction rate in the atomic-bomb survivors could occur.

CONCLUSIONS

The human ovary appears to have a relatively low rate of tumor induction by radiation; both benign and malignant tumors may be induced. The only dose-response data are based on Tumor-Registry reporting, and incidence figures are unreliable. Risk estimates cannot be determined with any precision; quantitative information on dose and dose rate is lacking. In general, the radiation risk of ovarian-tumor induction is low, but is identifiable. However, aside from complex radiation variables affecting this estimate—such as dose, dose rate, duration of exposure, and LET—other biologic factors, such as age and hormonal dependence, are poorly understood.

REFERENCES

1. Beebe, G. W., H. Kato, and C. E. Land. Life Span Study Report 8. Mortality Experience of Atomic-Bomb Survivors, 1950–74. Radiation Effects Research Foundation Technical Report TR 1-77. Hiroshima: Radiation Effects Research Foundation, 1978.
2. Clapp, N. K., E. B. Darden, Jr., and M. C. Jernigan. Relative effects of whole body sublethal doses of 60 MeV protons and 300 kVP x-rays on disease incidences in RF mice. Radiat. Res. 57:158–186, 1974.
3. Court Brown, W. M., and R. Doll. Mortality from cancer and other causes after radiotherapy for ankylosing spondylitis. Brit. Med. J. 2:1327–1332, 1965.
4. Darden, E. B., G. E. Cosgrove, A. C. Upton, et al. Female RF/vn mice irradiated with single doses of 14 MeV fast neutrons. Int. J. Radiat. Biol. 12:435–452, 1967.
5. Gardner, W. V. Hormonal aspects of experimental tumorigenesis. Adv. Cancer Res. 1:173–232, 1953.
6. International Commission on Radiological Protection. Radiosensitivity and Spatial Distribution of Dose. Publication 14. Reports prepared by two Task Groups of Committee 1. New York: Pergamon Press, 1969.
7. International Commission on Radiological Protection. Recommendations of the International Commission on Radiological Protection. Annals of the ICRP. Publication 26. New York: Pergamon Press, 1977.
8. Komuro, M. Ovarian tumorigenesis after whole-body x-ray irradiation in ddY/F and C3H/Tw mice. J. Radiat. Res. 17:99–105, 1976.
9. National Research Council, Advisory Committee on the Biological Effects of Ionizing Radiations. The Effects on Populations of Exposure to Low Levels of Ionizing Radiation. Washington, D.C.: National Academy of Sciences, 1972.
10. Nowell, R. C., and L. J. Cole. Late effects of fast neutrons versus x rays in mice. Nephrosclerosis, tumors, longevity. Radiat. Res. 11:545–556, 1959.
11. Palmer, J. P., and D. W. Spratt. Pelvic carcinoma following irradiation for benign gynecological diseases. Amer. J. Obstet. Gynecol. 72:497–505, 1956.
12. Upton, A. C. Effects of radiation on man. Annu. Rev. Nucl. Sci. 18:495–528, 1968.
13. Upton, A. C. Radiation carcinogenesis, pp. 53–82. In H. Busch, Ed. Methods in Cancer Research. Vol. 4. New York: Academic Press, 1968.
14. Upton, A. C. Radiological effects of low doses. Implications for radiological protection. Radiat. Res. 71:51–74, 1977.
15. Upton, A. C., M. A. Kastenbaum, and J. W. Conklin. Age-specific death rates of mice exposed to ionizing radiation and radiomimetic agents, pp. 285–297. In R. J. C. Harris, Ed. Cellular Basis and Etiology of Late Somatic Effects of Ionizing Radiation. London: Academic Press, 1963.
16. Upton, A. C., A. W. Kimball, J. Furth, et al. Some delayed effects of atomic bomb radiations in mice. Cancer Res. 20:1–62, 1960.
17. Upton, A. C., M. L. Randolph, and J. W. Conklin. Late effects of fast neutrons and gamma-rays in mice as influenced by the dose rate of irradiation. Induction of neoplasia. Radiat. Res. 41:467–491, 1970.
18. Yuhas, J. M. Recovery from radiation-carcinogenic injury to the mouse ovary. Radiat. Res. 60:321–322, 1974.

UTERUS AND CERVIX UTERI

Cancers of the uterus (and the cervix uteri) were not considered in the 1972 BEIR report[2] to appear in excess owing to radiation exposure. However, there is ample evidence from patients who received relatively large doses of irradiation during the course of radiotherapy—either intracavitary radium in the cervix or the uterine canal or external exposure to x rays or gamma rays—that a relationship exists between radiation and cancers of the uterus and cervix. Insofar as all pelvic cancers are concerned, the experience from radiotherapy of cervical cancer suggests that radiation induction may be lower at higher doses. No animal experiments are available to support the human experience.

ATOMIC-BOMB SURVIVORS

The most recent data on the Japanese atomic-bomb survivors up to 1974 indicate that there were 282 deaths from cancer of the cervix uteri and uterus.[1] Evaluation of relative- and absolute-risk estimates according to age at the time of bombing and T65 total dose in rads from 0 to 400+ demonstrates that none of the comparisons contains any indication of a relationship with radiation dose or with age at the time of exposure. A further analysis of cancer of the cervix uteri only (58 deaths) based on death-certificate data also failed to demonstrate a relationship with radiation. The Tumor-Registry data, however, based on 297 cases of cervical cancer in both cities suggest a very slight association in Hiroshima survivors (222 cases), but none in Nagasaki (75 cases). In the 10+ and 50+ rads groups in Hiroshima, there is suggestive evidence of a linear trend ($p = 0.06$ and $p = 0.09$, respectively).

RADIOTHERAPY PATIENTS TREATED FOR BENIGN DISEASE

Palmer and Spratt[3] observed an excess of uterine fundus and cervical cancers in 651 patients treated with intracavitary radium or external x rays for benign uterine fibroids and 80 patients with other benign pelvic diseases. For uterine cancers, there were 29 observed cases versus 4.9 expected, and the mean latent period was 9.7 yr; for cervical cancers, there were 11 observed versus 6.5 expected, and the mean latent period was 8.5 yr. Dose estimates were imprecise, and dose-response relationship was difficult to ascertain. In their review of five surveys of a total of 3,968 patients, they noted 27 uterine and 10 cervical cancers; latent periods were frequently less than 10 yr, and there was no dose determination.

Smith and Doll[4] reported on 2,068 patients who had been treated with pelvic irradiation for metropathia haemorrhagica and found excess deaths occurring from uterine cancers; there were 16 observed deaths versus 10.3 expected ($p = 0.08$) 5 yr or more after irradiation. Provided that the incidence of uterine cancer was not increased in patients with this disease, an absolute-risk estimate would be approximately 7 excess deaths from uterine cancer per million exposed patients per rad for a followup period of 5–19 yr after 400 rads.

CONCLUSION

The most recent information available suggests that a relationship exists between radiation and cancers of the uterus and cervix uteri. The data on the Japanese survivors in Hiroshima and on patients treated with radiation for benign uterine bleeding are from the only reliable surveys available. The numbers are too few and the range of dose estimates is too limited to provide a dose-response relationship and an estimate of risk. In view of these new data, however, these neoplasms warrant continued study.

REFERENCES

1. Beebe, G. W., H. Kato, and C. E. Land. Life Span Study Report 8. Mortality Experience of Atomic-Bomb Survivors 1950–74. Radiation Effects Research Foundation Technical Report TR 1-77. Hiroshima: Radiation Effects Research Foundation, 1978.
2. National Research Council, Advisory Committee on the Biological Effects of Ionizing Radiations. The Effects on Populations of Exposure to Low Levels of Ionizing Radiation. Washington, D.C.: National Academy of Sciences, 1972.
3. Palmer, J. P., and D. W. Spratt. Pelvic carcinoma following irradiation for benign gynecological diseases. Amer. J. Obstet. Gynecol. 72:497–505, 1956.
4. Smith, P. G., and R. Doll. Late effects of x irradiation in patients treated for metropathia haemorrhagica. Brit. J. Radiol. 49:224–232, 1976.

BONE

Primary cancers of bone have been induced in man by internally deposited alpha-emitters[1,5-7,9,14,17,19-21,24-30] and by high doses of therapeutic x rays.[2,4,10,34] Osteosarcomas are the most common form of bone cancer (Table A-26). Bone sarcomas are usually fatal because of metastases: the 5-yr survival after diagnosis and therapy is only about 20% for osteosarcomas.[3]

The distribution of times from *brief* irradiation to clinical diagnosis is similar for bone sarcomas induced by radium-224 and leukemia in the

TABLE A-26 Primary Malignant Tumors of the Skeleton—Naturally Occurring and Radiation-Induced

Primary Malignant Tumors of Skeleton[a]	Naturally Occurring				Radiation-Induced							
	U.S.[3,32]		England and Wales[32]		X-Ray Literature[34]		X-Ray Therapy in U.S.[10]		226Ra and 228Ra in Bone in U.S.[9,32]		224Ra in Bone in Germany[21]	
	No.	%	No.	%	No.	%	No.	%	No.	%	No.	%
Osteosarcoma	652	43.0	296	60.9	155	59.2	12	44.4	42	70.0	42	84.0
Chondrosarcoma	343	22.6	73	15.0	19	7.3	3	11.1	1	1.7	3	6.0
Ewing's tumor	209	13.8	31	6.4	—	—	—	—	—	—	—	—
Chordoma	122	8.0	25	5.1	—	—	—	—	—	—	—	—
Reticulum-cell sarcoma	101	6.7	10	2.1	—	—	2	7.4	—	—	1	2.0
Fibrosarcoma	82	5.4	49	10.1	66	25.2	8	29.6	17	28.3	1	2.0
Angiosarcoma	7	0.5	2	0.4	—	—	—	—	—	—	—	—
Others	—	—	—	—	22	8.4	2	7.4	—	—	3	6.0
TOTALS	1,516	100	486	100	262	100.1	27	99.9	60	100	50	100

a Excluding leukemia, myeloma, Hodgkin's disease, neuroblastoma, adamantioma, giant-cell tumors, unspecified bone tumors, and soft-tissue tumors invading bone.

atomic-bomb survivors.[19] In German patients who received injections of short-lived radium-224, bone sarcomas started to appear at 4 yr, peaked in frequency at 6–8 yr, and seemed virtually exhausted after 22 yr.[21] Similarly, in a careful followup of patients receiving therapeutic x rays, the appearance times ranged from 4 to 27 yr, with a median of 11 yr.[10] In contrast, bone sarcomas have appeared as long as 52 yr after the start of *continuous* irradiation from long-lived radium-226 in the skeleton.[26,27] These late-appearing tumors may have been induced by radiation emitted long after the initial deposition of the radium.

The susceptibility to sarcoma induction in different regions of human bone varies widely, being highest near the knee joint and lowest in the vertebrae, for both naturally occurring and radiation-induced bone sarcomas.[3,32] Taylor *et al.*[31] have observed that sarcoma induction by bone-seeking radionuclides in humans, dogs, and mice tends to be highest in skeletal locations that have the highest natural occurrence of bone sarcoma. Therefore, the most useful risk evaluations for bone-sarcoma induction from total skeletal irradiation come from bone-seeking radionuclides that are deposited throughout the entire skeleton, rather than from localized x-ray therapy, which typically involves very high doses to a small and often poorly defined fraction of the total skeleton. For example, of the 14,000 British ankylosing-spondylitis patients receiving average doses of about 1,000 rads from x rays to the "spine," only one developed a bone sarcoma in the vertebrae, with the four remaining sarcomas occurring in the pelvic region.[4]

The best information on skeletal risk comes from patients who received radium-224 injections and persons exposed to the intake of radium-226 and radium-228.

After World War II, several thousand German patients received repeated intravenous injections of 3.6-d radium-224 as therapy for tuberculosis and ankylosing spondylitis.[17,19,21,28-30] These patients included boys and girls, as well as men and women of various ages. Two followup studies are now in progress. The original study, started in 1952, now involves 900 patients, most of whom received average skeletal doses exceeding 90 rads. As of June 1978, bone sarcomas had been identified in 54 of these patients.[21] A new study, involving about 1,000 additional patients below 90 rads, was started in 1971. So far, two skeletal sarcomas have been identified in the new series (Schales,[28] Mays,[17] and R. Wick, personal communication). The lowest average skeletal doses at which bone sarcomas have been identified are 90 rads in the original series[21] and 67 rads in the new series (Wick, personal communication).

For bone-sarcoma induction, the effectiveness of a given dose from radium-224 alpha particles *increases* as the time of irradiation is protracted, both in humans[19,21,30] and in mice.[23] The exact mathematical expression for the increase in radium-224 effectiveness with protraction is probably quite complicated and may be influenced by many factors, including the size of the dose. However, in the German patients, the observed effectiveness (bone sarcomas per million persons per rad of average skeletal dose) for an injection span of m months was well represented by the following simple equation:[22]

$$\text{radium-224 effectiveness} = 40 + 160(1 - e^{-0.09m}).$$

This equation predicts that the number of induced bone sarcomas in a million people of mixed ages, each receiving an average skeletal dose of 1 rad from radium-224 and its daughters (or 10,000 people, each receiving 100 rads, etc.), rises from 40 cases for a single injection of radium-224 to 200 cases for weekly injections extending over several years—this increase results from protraction of the alpha-particle dose.

In the original series, which contains all the traced children, the risk per rad for a given injection span is similar for children and adults[30] and for males and females.[29] More than half the patients in the original series are still alive, and the living patients have now been followed for an average of 27 yr since their first injection. The longest time from injection to sarcoma is now 22 yr, and this was the only bone sarcoma appearing between 1969 and 1978. Even if a few additional bone sarcomas develop, it seems unlikely that they would increase the present number of 54 substantially. Thus, the lifetime incidence of bone sarcoma in the radium-224 patients of the original series is unlikely to become appreciably higher than that already observed.

However, the followup times in the new series range from about 6 to 27 yr (average, about 12 yr), so a few additional bone sarcomas may appear in the important dose region below 90 rads.

The risk coefficient for radium-224 can also be expressed in terms of endosteal dose. If one uses the dosimetry of Spiess and Mays,[29] but the revised surface-to-volume ratio of Lloyd and Hodges[11] of 50 cm^2/cm^3 (rather than their preliminary value of 42 cm^2/cm^3), the ratio of endosteal dose to average skeletal dose is 7.5:1. If the risk coefficient for radium-224 of 40–200 bone sarcomas per million persons per rad of *average skeletal dose* is divided by 7.5, the risk coefficient becomes five bone sarcomas per million persons per rad of *endosteal dose* for a single intake and rises to 27 bone sarcomas per million persons per rad of *endosteal dose* for protracted irradiation continuing over several years.

The same risk coefficient in terms of endosteal dose is obtained with the dosimetry of Spiess and Mays or that of Marshall *et al.*,[12] because both groups assume the same value for radium-224 decaying on bone surfaces, and the surface deposit of radium-224 is responsible for nearly all the endosteal dose. The alpha particles resulting from radium-224 within bone volume are comparatively ineffective in irradiating the cells near the bone surface.

About 2,000 dial-painters and other persons internally contaminated with long-lived radium isotopes have been extensively studied. All these persons were exposed to 1,600-yr radium-226, and many also acquired 5.8-yr radium-228 (mesothorium). Both these long-lived radium isotopes decay mainly within bone volume. The largest exposures occurred before 1926. Bone sarcomas have occurred in 58 persons whose retained body burdens have been evaluated and in 26 others whose doses are unknown.[26] The occurrence of new sarcomas has decreased rapidly in recent years, probably because the radium-228 has virtually decayed out and remodeling and excretion are continually removing radium-226, especially from trabecular bone, where most of the radiation-induced bone sarcomas seem to originate. Only two radium subjects are known to have developed bone sarcomas since the end of 1969,[26] so it seems unlikely that the overall incidence of bone sarcoma will increase appreciably in these persons. Only one bone-sarcoma case (at 888 rads at diagnosis) is known to have occurred below 1,000 rads.[26]

Rowland[24] found that the bone-sarcoma cumulative incidence at an average skeletal dose of D rads from radium-226 and radium-228 was best described by the following dose-squared exponential:

$$\text{Cumulative incidence} = 3.7 \times 10^{-8}D^2e^{-1.4\times10^{-4}D}.$$
$$\text{(sarcomas/persons)}$$

An equation of the above form gives an excellent fit to the existing data on bone-sarcoma induction in humans by long-lived radium. However, the possibility of a linear response between zero and 1,000 rads cannot be ruled out.

Mays *et al.*[22] used the detailed tabulations of Rowland and Stehney[25] and suggested that the risk from low doses of radium-226 and radium-228 in people lies between six and 53 bone sarcomas per million persons per rad of average skeletal dose. The upper value was based on a linear fit to the 48 bone sarcomas below 10,000 rads, and the lower value on a linear fit to the one bone sarcoma below 1,000 rads. The upper value would predict a 5% chance of observing zero or one bone sarcoma in the radium subjects below 1,000 rads.

The corresponding endosteal risk is about six to 53 bone sarcomas per million persons per rad of *endosteal dose,* because, allowing for the small but efficient dose from radium-226 and radium-228 deposited on bone surfaces,[12,13] the endosteal dose is approximately equal to the average dose to the marrow-free skeleton (7 kg in reference man).

The risk in terms of endosteal dose for long-protracted alpha-particle irradiation from radium-224 (mainly a bone-surface-seeker) of 27 bone sarcomas per million persons per rad lies within the range of values obtained from radium-226 plus radium-228 (bone-volume-seekers). If a linear dose-response relation is assumed, it is suggested that the risk coefficient for radium-224 be used as a best estimate, with the coefficients for radium-226 and radium-228 serving as reasonable upper and lower limits.

Rowland *et al.*[27] have recently found that the following dose-squared exponential gives the best fit to their incidence-*rate* data:

$$\text{Incidence rate} = 9.8 \times 10^{-10}D^2e^{-1.5 \times 10^{-4}D}.$$
$$(\text{sarcomas/PY})$$

No increase in malignant bone tumors attributable to nuclear radiation was observed among atomic-bomb survivors at Hiroshima and Nagasaki,[33] although the time of observation was 5–20 yr after irradiation. Nearly all the bone sarcomas induced in the radium-224 patients[21] and in x-ray-therapy patients[10] have appeared during a corresponding interval. The 1968 ABCC report[33] stated that, "of the 25 malignant [bone] tumors found in survivors, only 5 were in persons who were within 1400 meters [of ground zero]. On the basis of the distributions by distance of the autopsy series and the surgical pathology series, the expected number with which the 5 observed can be compared is 4.67 cases. Thus, there is no indication here of an increase in the number of malignant bone tumors as a consequence of atomic bomb radiation."

RECOMMENDED RISK COEFFICIENTS

The following *provisional* risk coefficients are recommended for endosteal doses up to a few hundred rads (Table A-27). The linear coefficient for the cumulative risk (27 bone sarcomas per million persons per rad of alpha dose) is based on protracted irradiation via repeated injections of radium-224. Dividing this value by an assumed expression time of 27 yr (from 4 to 31 yr after the start of radium-224 irradiation) gives the linear risk-rate coefficient (one bone sarcoma per year per million persons per rad of alpha dose).

The dose-squared coefficients are from Rowland[24] and Rowland *et al.*[27] The exponential factor in their complete equations has been omitted, because it is approximately 1 at doses below a few hundred rads.

TABLE A-27   Risk Coefficients for Radiation-Induced Bone Sarcomas

---

Alpha particles (high-LET):

  Cumulative risk coefficients:

$$\text{Linear} = \frac{27 \times 10^{-6}\ \text{sarcoma}}{\text{person-rad}}$$

$$\text{Dose-squared} = \frac{3.7 \times 10^{-8}\ \text{sarcoma}}{\text{person-rad}^2}$$

  Risk-rate coefficients:

$$\text{Linear} = \frac{1 \times 10^{-6}\ \text{sarcoma/yr}}{\text{person-rad}}$$

$$\text{Dose-squared} = \frac{9.8 \times 10^{-10}\ \text{sarcoma/yr}}{\text{person-rad}^2}$$

Beta, gamma, and x rays (low-LET):

  Cumulative risk coefficients:

$$\text{Linear} = \frac{1.4 \times 10^{-6}\ \text{sarcoma}}{\text{person-rad}}$$

$$\text{Dose-squared} = \frac{9.2 \times 10^{-11}\ \text{sarcoma}}{\text{person-rad}^2}$$

  Risk-rate coefficients:

$$\text{Linear} = \frac{0.05 \times 10^{-6}\ \text{sarcoma/yr}}{\text{person-rad}}$$

$$\text{Dose-squared} = \frac{2.4 \times 10^{-12}\ \text{sarcoma/yr}}{\text{person-rad}^2}$$

With reference to low doses of low-LET radiation (x rays, gamma rays, and beta particles), the ICRP quality factor of 20 has been used to approximate the RBE of alpha particles.[8] Hence, the linear coefficients for alpha particles have been divided by 20, and the dose-squared coefficients have been divided by 400, the square of 20.

DISCUSSION

Linear and dose-squared risk coefficients have both been given to emphasize the uncertainty as to the true shape of the dose-response curve for bone-sarcoma induction in man. Other relationships are also possible. The existing data on the U.S. radium subjects are best fitted by a dose-squared exponential, but this is not the universal response to other sets of data. Mays has examined bone-sarcoma induction by alpha-emitters in 11 studies with data extensive enough to indicate the shape of the dose-response curve.[15,16] At the lower doses, the response appeared to be approximately linear in seven studies (radium-224 in humans, plutonium-239 in beagles, thorium-228 in beagles, plutonium-239 in rats, radium-226 in mice, thorium-227 in mice, and plutonium-239 in mice), concave downward in one (radium-224 in mice), and concave upward in three (radium-226 plus -228 in humans, radium-228 in beagles, and radium-226 in beagles). However, preliminary results from an expanded study of radium-226 in beagles indicate that the final response might be either linear, concave upward, or even concave downward.[16] It is difficult, on the basis of both experimental and theoretical considerations, to rule out a linear *component* to the dose-effect relationship for bone-sarcoma induction by alpha-emitters. This component should become dominant at very low doses.

The relative significance of a dose-squared term seems much more likely with sparsely ionizing radiation. Mays and Lloyd[18] demonstrated a response strongly concave upward for bone-sarcoma induction by beta particles from strontium-90 in mice, calcium-45 in mice, and strontium-90 in rats.

The lack of a detectable increase in bone sarcomas in the atomic-bomb survivors is supported by the very low number of cases predicted by multiplying the collective dose to marrow from gamma rays and neutrons, respectively, by the risk coefficients derived here for gamma rays and alpha particles. The risk coefficient for alpha particles should be roughly similar to that for fast neutrons, inasmuch as both result in high-LET radiation, and for fast neutrons the marrow dose should be fairly close to the endosteal dose.

REFERENCES

1. Aub, J. C., R. D. Evans, L. H. Hempelmann, and H. S. Martland. The late effects of internally-deposited radioactive materials in man. Medicine 31:221–329, 1952.
2. Cahan, W. C., H. Q. Woodard, N. L. Higinbotham, F. W. Stewart, and B. L. Coley. Sarcoma arising in irradiated bone. Cancer 1:3–29, 1948.
3. Dahlin, D. C. Bone Tumors. 2nd ed. Springfield: Charles C Thomas, 1967. 285 pp. (See pp. 11 and 174; Dr. Dahlin—personal communication—deleted the radiation-induced sarcomas.)
4. Edgar, M. A., and M. P. Robinson. Post-radiation sarcoma in ankylosing spondylitis. J. Bone Joint Surg. 55B:183–188, 1973.
5. Evans, R. D., A. T. Keane, R. J. Kolenkow, R. R. Neal, and M. M. Shanahan. Radiogenic tumors in the radium and mesothorium cases studied at M.I.T., pp.

157-194. In C. W. Mays, W. S. S. Jee, R. D. Lloyd, B. J. Stover, J. H. Dougherty, and G. N. Taylor, Eds. Delayed Effects of Bone-Seeking Radionuclides. Salt Lake City: University of Utah Press, 1969.

6. Finkel, A. J., C. E. Miller, and R. J. Hasterlik. Radium-induced malignant tumors in man, pp. 195-225. In C. W. Mays, W. S. S. Jee, R. D. Lloyd, B. J. Stover, J. H. Dougherty, and G. N. Taylor, Eds. Delayed Effects of Bone-Seeking Radionuclides. Salt Lake City: University of Utah Press, 1969.

7. Harrist, T. J., A. L. Schiller, R. L. Trelstad, H. J. Mankin, and C. W. Mays. Thorotrast associated sarcoma of bone. Cancer 44:2049-2058, 1979.

8. International Commission on Radiological Protection. Recommendations, pp. 5 and 11. Publication 26. Oxford: Pergamon Press, 1977.

9. Keane, A. T. Skeletal location of primary bone tumors in the radium cases, pp. 59-66. Argonne National Laboratory Report ANL-7860. Part II. 1971.

10. Kim, J. H., F. C. Chu, H. Q. Woodard, H. R. Melamed, A. Huvos, and J. Contlin. Radiation-induced soft-tissue and bone sarcoma. Radiology 129:501-508, 1978.

11. Lloyd, E. L., and D. Hodges. Quantitative characterization of bone. A computer analysis of microradiographs. Clin. Orthop. Relat. Res. 78:230-250, 1971.

12. Marshall, J. H., P. G. Groer, and R. A. Schlenker. Dose to endosteal cells and relative distribution factors for radium-224 and plutonium-239, compared to radium-226. Health Phys. 35:91-101, 1978.

13. Marshall, J. H., J. Liniecki, E. L. Lloyd, G. Marotti, C. W. Mays, J. Rundo, H. A. Sissons, and W. S. Snyder. Alkaline Earth Metabolism in Adult Man, p. 86. International Commission on Radiological Protection Publication 20. New York: Pergamon Press, 1973.

14. Martland, H. S. The occurrence of malignancy in radioactive persons. Amer. J. Cancer 15:2435-2516, 1931.

15. Mays, C. W. Discussion of plutonium toxicity, pp. 115-143. In Proceedings of the Symposium, National Energy Issues—How Do We Decide? Plutonium As a Test Case, American Academy of Arts and Sciences, Argonne National Laboratory, April 1979.

16. Mays, C. W. Response on comments by Alvin M. Weinberg on shape of the dose response curve, pp. 332-339. In Proceedings of the Symposium, National Energy Issues—How Do We Decide? Plutonium As a Test Case, American Academy of Arts and Sciences, Argonne National Laboratory, April 1979.

17. Mays, C. W. Risk to bone from present ^{224}Ra therapy, pp. 37-43. In W. A. Müller and H. G. Ebert, Eds. Biological Effects of ^{224}Ra. Benefit and Risk of Therapeutic Application. The Hague: Nijhoff Medical Division, 1978.

18. Mays, C. W., and R. D. Lloyd. Bone sarcoma risk from ^{90}Sr, pp. 352-375. In M. Goldman and L. K. Bustad, Eds. Biomedical Implications of Radiostrontium Exposure. USAEC CONF-710201. 1972.

19. Mays, C. W., and H. Spiess. Bone sarcoma risks to man from ^{224}Ra, ^{226}Ra, and ^{239}Pu, pp. 168-181. In W. A. Müller and H. G. Ebert, Eds. Biological Effects of ^{224}Ra. Benefit and Risk of Therapeutic Application. The Hague: Nijhoff Medical Division, 1978.

20. Mays, C. W., and H. Spiess. Bone tumors in Thorotrast patients. Environ. Res. 18:88-93, 1979.

21. Mays, C. W., H. Spiess, and A. Gerspach. Skeletal effects following ^{224}Ra injections into humans. Health Phys. 35:83-90, 1978.

22. Mays, C. W., H. Spiess, G. N. Taylor, R. D. Lloyd, W. S. S. Jee, S. S. McFarland, D. H. Taysum, T. W. Brammer, D. Brammer, and T. A. Pollard. Estimated risk to human bone from ^{239}Pu, pp. 343-362. In W. S. S. Jee, Ed. Health Effects of Plutonium and Radium. Salt Lake City: J. W. Press, University of Utah, 1976.

23. Müller, W. A., W. Gössner, O. Hug, and A. Luz. Late effects after incorporation of the short-lived α-emitters ^{224}Ra and ^{227}Th in mice. Health Phys. 35:33–35, 1978.

24. Rowland, R. E. The risk of malignancy from internally-deposited radioisotopes, pp. 146–155. In O. F. Nygaard, H. I. Adler, and W. K. Sinclair, Eds. Proceedings of the Fifth International Congress of Radiation Research. New York: Academic Press, 1975.

25. Rowland, R. E., and A. F. Stehney. Exposure data for radium patients, pp. 177–231. Argonne National Laboratory Report ANL-75-3. Part II. 1974.

26. Rowland, R. E., and A. F. Stehney. Radium-induced malignancies, pp. 259–264. Argonne National Laboratory Report ANL-78-65. Part II. 1978.

27. Rowland, R. E., A. F. Stehney, and H. F. Lucas, Jr. Dose-response relationships for female radium dial workers. Radiat. Res. 76:368–383, 1978.

28. Schales, F. Problems and results of a new follow-up study. ^{224}Ra in adult ankylosing spondylitis patients, pp. 30–36. In W. A. Müller, and H. G. Ebert, Eds. Biological Effects of ^{224}Ra. Benefit and Risk of Therapeutic Application. The Hague: Nijhoff Medical Division, 1978.

29. Spiess, H., and C. W. Mays. Bone cancers induced by ^{224}Ra (ThX) in children and adults. Health Phys. 19:713–729, 1970 (plus addendum, Health Phys. 20:543–545, 1971).

30. Spiess, H., and C. W. Mays. Protraction effect on bone sarcoma induction of ^{224}Ra in children and adults, pp. 437–450. In C. L. Sanders, R. H. Busch, J. E. Ballou, and D. D. Mahlum, Eds. Radionuclide Carcinogenesis. USAEC Symposium 29, CONF-720505. Springfield, Va.: National Technical Information Service, 1973.

31. Taylor, G. N., W. S. S. Jee, and C. W. Mays. Some similarities of radium and plutonium toxicity in the beagle and man, pp. 523–536. In W. S. S. Jee, Ed. Health Effects of Plutonium and Radium. Salt Lake City: J. W. Press, University of Utah, 1976.

32. Thurman, G. B., C. W. Mays, G. N. Taylor, A. T. Keane, and H. A. Sissons. Skeletal location of radiation-induced and naturally-occurring osteosarcomas in man and dog. Cancer Res. 33:1604–1607, 1973.

33. Yamamoto, T., and T. Wakabayashi. Bone Tumors among Atomic Bomb Survivors, 1950–1965; Hiroshima-Nagasaki. Atomic Bomb Casualty Commission Technical Report TR 26-68. Hiroshima: Atomic Bomb Casualty Commission, 1968. 14 pp.

34. Yoshizawa, Y., T. Kusama, and K. Morimoto. Search for the lowest irradiation dose from literatures on radiation-induced bone tumours. Nippon Acta Radiol. 37:377–386, 1977.

PARANASAL SINUSES AND MASTOID AIR CELLS

Among U.S. dial-painters and others internally contaminated with radium-226, carcinomas of the paranasal sinuses and mastoid air cells have been recorded in 29 persons of known body burden and in four persons of unknown dose.[7] The latent periods are long; the time from first exposure to tumor diagnosis ranged from 19 to 52 yr. These carcinomas are still appearing in the U.S. radium subjects at a significant rate. Between the end of 1969 and the end of 1977, seven of the 29 carcinomas in persons of known dose appeared. Only four of those 29 persons have survived more than 5 yr after diagnosis.

Evans[3] concluded that the accumulation of radon-222 gas (half-life, 3.8 d) in the paranasal and mastoid cavities was the important inducing agent; the appearance of these carcinomas correlated well with radium-226 (which produces radon-222), but not with radium-228 (which does not produce radon-222). Rowland *et al.*[8] and Littman *et al.*[5] have confirmed Evans's conclusion, and further support comes from the absence of these carcinomas in the German radium-224 patients followed for up to 33 yr.[6]

Rowland *et al.*[8] have used a minimal latent period of 10 yr for sinus and mastoid carcinomas in the patients exposed to radium-226 and disregarded the last 10 yr of dose. Their analyses showed that a linear dose response provides an acceptable fit for this type of cancer. Their lowest average skeletal dose in a case of head sinus carcinoma was 605 rads at diagnosis. Using only the dosage from radium-226 and its daughters, Rowland *et al.* evaluated the risk rate per microcurie of radium-226 intake into the blood and per rad of average skeletal dose. The cumulative risk was calculated to the projected end of average life span. These risk estimates are shown in Table A-28.

TABLE A-28   Risk of Induction of Paranasal Sinus and Mastoid Carcinomas by Radium-226[a]

Evaluation	Risk[b]
Risk rate per unit intake of radium-226 into blood	$\dfrac{16 \text{ carcinomas/yr}}{(10^6 \text{ persons}) (\mu\text{Ci of radium-226})}$
Risk rate per unit dose to marrow-free skeleton[c]	$\dfrac{1.6 \text{ carcinomas/yr}}{(10^6 \text{ persons}) (\text{rad})}$
Cumulative risk to end of life span[d]	$\dfrac{64 \text{ carcinomas}}{(10^6 \text{ persons}) (\text{rad})}$

[a] Data from Rowland *et al.*[8]

[b] These risk estimates pertain *only* to radium-226. They must *not* be used for other internally deposited emitters or for external radiation that produces a ratio of sinus dose to average skeletal dose different from that resulting from bone-deposited radium-226 and the accumulation of its decay products.

[c] The skeletal dose in rads from radium-226 and its decay products is averaged over the marrow-free skeleton (7 kg in a 70-kg man).

[d] On the basis of the risk rate per unit dose to skeleton and an assumed average life expectancy of 50 yr after radium-226 intake, this corresponds to a minimal appearance time of 10 yr followed by 40 yr at risk.

In a review of the literature, Fabrikant et al.[4] described 10 cases of neoplasia of the maxillary sinuses after antral injection of Thorotrast (thorium dioxide) for radiodiagnostic purposes, in which Thorotrast was left in the sinus cavity. The description of the cases indicates that prolonged contact with an alpha-emitting substance can induce carcinomas of the head sinuses. Interestingly, no cancers of the head sinuses have been observed in 192 traced Portuguese patients from whose paranasal sinuses the Thorotrast was removed within 6 d after instillation.[1,2]

REFERENCES

1. da Motta, L. C. Follow-up study of Portuguese patients given thorium dioxide in the peranasal sinuses. In R. L. Swarm, Ed. Distribution Retention and Late Effects of Thorium Dioxide. Ann. N.Y. Acad. Sci. 145:811–816, 1967.
2. da Silva Horta, J., M. E. da Silva Horta, L. C. da Motta, and M. H. Tavares. Malignancies in Portuguese Thorotrast patients. Health Phys. 35:137–151, 1978.
3. Evans, R. D. The effect of skeletally deposited alpha-ray emitters in man. Brit. J. Radiol. 39:881–895, 1966.
4. Fabrikant, J. I., R. J. Dickson, and B. F. Fetter. Mechanisms of radiation carcinogenesis at the clinical level. Brit. J. Cancer 18:459–477, 1964.
5. Littman, M. S., I. E. Kirsh, and A. T. Keane. Radium-induced malignant tumors of the mastoid and paranasal sinuses. Amer. J. Roentgenol. 131:773–785, 1978.
6. Mays, C. W., H. Spiess, and A. Gerspach. Skeletal effects following ^{224}Ra injections into humans. Health Phys. 35:83–90, 1978.
7. Rowland, R. E., and A. F. Stehney. Radium-induced malignancies, pp. 259–264. In Argonne National Laboratory Report ANL-78-65. Part II. 1978.
8. Rowland, R. E., A. F. Stehney, and H. F. Lucas, Jr. Dose-response relationships for female dial workers. Radiat. Res. 76:368–383, 1978.

BRAIN

Neural tissue has been traditionally regarded as radioresistant, and there is little information about radiation tumorigenesis of the central nervous system (CNS) in man. The 1972 BEIR report[11] did not consider neural tumors, but several epidemiologic studies of radiation and brain tumors have since appeared.

EXPERIMENTAL DATA

Radiation-induced brain tumors have been reported mainly in primates. Kent and Pickering[7] reported a glioblastoma in a monkey after about 2,500 rads of thermal-neutron irradiation. Haymaker et al.[5] reported three glioblastomas among 21 monkeys that had survived 2 yr or more after doses of 400–1,000 rads of 55-MeV protons. Another study[15] reported three malignant ependymal brain tumors in 14 animals followed for 1–5 yr after exposures of 600–1,000 rads of 55-MeV protons. In the latter two studies, whole-body irradiation was given, but the brain was the most common site of malignancies.

HUMAN STUDIES

Antenatal Irradiation

Although this general topic has been considered elsewhere in this report, the data on CNS tumors are briefly reviewed here.

Stewart Case-Control Study   Bithell and Stewart[2] have compared the reported *in utero* radiation history of 1,332 deaths from malignant CNS tumors occurring at ages 0–14 yr with that of 8,513 children without tumors. Only singleton births were included. They found that 13.4% of the children with CNS tumors versus 9.9% of controls had a history of *in utero* irradiation, for a relative risk (RR) of 1.42 (80% confidence limits, 1.3 and 1.6*). If only the 1,332 cases and their matched controls are considered in a matched-pair analysis, the RR was 1.30.

From their calculation of cumulative mortality caused by malignant CNS tumors in Great Britain for ages 0–14 (viz., 17.5×10^{-5}), one can derive an average annual death rate over that age range (1.17×10^{-5}). Their estimates of RR can be applied to this rate to derive excess risk. Using an estimate of 0.8 rad as the average dose,[11(p. 163)] the absolute excess risk is 6.1 deaths per 10^6 PY per rad (80% CL, 3.9 and 8.6). With the RR of 1.30 from their matched analysis, the estimate is 4.4 per 10^6 PY per rad.

MacMahon Study   MacMahon[8] identified over 734,000 children born in selected hospitals during the period 1947–1954 and searched death-certificate lists for the years 1947–1959 to identify cancer deaths. Birth records for all cancer deaths and for a 1% sample of the total study population were searched to ascertain the frequency of *in utero* x-ray exposure. A total of 120 CNS cancer deaths were found; 19 subjects (15.8%) had received prenatal x rays. In the 1% sample of controls, 770 (10.6%) of 7,424 had similar x-ray exposures. This yielded a crude RR of 1.57 (80% CL, 1.1 and 2.2). However, once the data were adjusted for birth order, year of birth, religion, maternal age, sex, and pay status (private or clinic), the adjusted RR was 1.33 (approximate 80% CL, 0.94 and 1.9).

---

* The 80% confidence limits (CL) reported here are equivalent to what the 1972 BEIR report called the "90% lower CL" and "90% upper CL."

The estimates of absolute excess risk for the crude and adjusted estimates are 11.2 and 6.3 per 10^6 PY per rad, respectively.

Diamond Study The cohort study of antenatal radiation by Diamond, Schmerler, and Lilienfeld[4] proved to be too small to detect an excess of brain tumors. Among about 20,000 irradiated children, three deaths from nervous-system malignancies were found at age 0–9, and eight occurred among 35,000 controls.

Atomic-Bomb Study Jablon and Kato[6] reported that, among 740 children who were *in utero* at the time of the bombing and whose mothers received a dose of 1 rad or more, there were no brain-tumor deaths in the first 10 yr of life. They estimated that these children represented at least 17,500 person-rads to the fetuses. On the basis of an average excess risk of about 6 per 10^6 PY per rad in the MacMahon and Bithell-Stewart studies, one would expect an excess of about 0.11 brain-cancer death (and the expectation of spontaneous brain-tumor mortality is only about 0.02). The difference between zero observed and 0.12 expected is not significant; therefore, although the data lend no support to the hypothesis of radiation induction of brain tumors, neither are they incompatible with it.

Postnatal CNS Irradiation

Because the malignancy of some brain tumors (e.g., astrocytomas) is often not well established, and benign brain tumors can often be serious for the patient, "brain tumors" refers to both malignant and benign intracranial tumors. There are several case reports of brain tumors after head and neck radiotherapy,[1,10,12] but these are not considered here, because they provide no quantitative information.

New York Tinea Capitis Study Shore *et al.*[14] have reported on brain tumors among about 2,200 children given x-ray therapy for ringworm of the scalp and 1,400 ringworm patients without x-ray therapy. The dose to the brain averaged about 140 rads. The children were treated at an average age of 8 yr and have been followed for 15–34 yr (mean, 25 yr). Eight brain tumors have been found in the irradiated group versus none among controls (R. E. Shore and R. E. Albert, personal communication). The tumors were of a variety of types: one malignant glioma, two astrocytomas, one hemangioblastoma, two meningiomas, and two acoustic neuromas. The minimal latency was about 5 yr, and four of the eight tumors occurred more than 25 yr after radiation. On the basis of Connecticut brain-tumor rates,[13] about 1.1 would have been expected in the irradiated group and 0.7 among controls. The observed-to-expected ratio (8:1.1) was 7.2 (80% CL, 4.1 and 12.4). The estimate of absolute excess risk was 1.3 per 10^6 PY per rad (80% CL, 0.7 and 2.4), on the basis of 5 yr or more of followup after irradiation.

Israeli Tinea Capitis Study Modan et al.[9] reported on brain tumors among about 10,900 children given x-ray therapy for ringworm of the scalp and 16,400 sibling and population controls, on the basis of death certificates and the national Tumor Registry. The mean brain dose was about 140 rads per treatment, and, inasmuch as about 10% had two treatments, the mean per person was about 150 rads. Modan et al. found 16 brain tumors in the irradiated group and three among controls, for an RR of 8.0. With approximate person-years calculated from the publication,[9] the estimate of absolute excess risk is about 0.7 per 10^6 PY per rad (80% CL, 0.2 and 2.2). This is probably an underestimate of the true incidence, in that only death certificates were available for ascertainment before 1960. Eight of the 16 brain tumors in the irradiated group were reported to be malignant.

Michael Reese X-Ray Therapy Study Of 5,166 persons given x-ray therapy to the head and neck (80% treated to the tonsils and nasopharynx), 3,108 have been followed for an average of about 22 yr since irradiation. Colman et al.[3] reported 14 intracranial tumors in this group, of which six were malignant. On the basis of age-specific brain-tumor incidence in Connecticut,[9] about 1.6 would have been expected, for an observed-to-expected ratio (14:1.6) of 8.8. A "dose-response" relationship was reported; but, because the doses used were those to the midplane of the neck, rather than the brain, this finding cannot be readily evaluated. Likewise, the reported doses cannot be used to calculate risk estimates.

Host Factors

Nothing is known about sensitivity to radiation with respect to age at irradiation, sex, or other characteristics.

REFERENCES

1. Beller, A. J., M. Feinsod, and A. Sahar. The possible relationship between small dose irradiation to the scalp and intracranial meningiomas. Neurochirurgia (Stuttg.) 15:135–143, 1972.
2. Bithell, J. F., and A. M. Stewart. Pre-natal irradiation and childhood malignancy. A review of British data from the Oxford survey. Brit. J. Cancer 31:271–287, 1975.

3. Colman, M., M. Kirsch, and M. Creditor. Radiation induced tumors. IAEA Symposium, Vienna, March 1978. IAEA-SM 224/706. Vienna: International Atomic Energy Agency, 1978.
4. Diamond, E. L., H. Schmerler, and A. M. Lilienfeld. The relationship of intrauterine radiation to subsequent mortality and development of leukemia in children. Amer. J. Epidemiol. 97:283-313, 1973.
5. Haymaker, W., L. Rubinstein, and J. Miquel. Brain tumors in irradiated monkeys. Acta Neuropathol. 20:267-277, 1972.
6. Jablon, S., and H. Kato. Childhood cancer in relation to prenatal exposure to atomic-bomb radiation. Lancet 2:1000-1003, 1970.
7. Kent, S., and J. Pickering. Neoplasms in monkey (Macaca mulatta). Spontaneous and irradiation induced. Cancer 11:138-147, 1958.
8. MacMahon, B. Prenatal x-ray exposure and childhood cancer. J. Natl. Cancer Inst. 28:1173-1191, 1962.
9. Modan, B., D. Baidatz, H. Mart, R. Steinitz, and S. Levin. Radiation-induced head and neck tumours. Lancet 1:277-279, 1974.
10. Munk, J., E. Peyser, and J. Gruszkiewicz. Radiation-induced intracranial meningiomas. Clin. Radiol. 20:90-94, 1969.
11. National Research Council, Advisory Committee on the Biological Effects of Ionizing Radiations. The Effects on Populations of Exposure to Low Levels of Ionizing Radiation. Washington, D.C.: National Academy of Sciences, 1972.
12. Raskind, R. Central nervous system damage after radiation therapy. Int. Surg. 48:430-441, 1967.
13. Schoenberg, B., B. Christine, and J. Whisnant. The descriptive epidemiology of primary intracranial neoplasms. The Connecticut experience. Amer. J. Epidemiol. 104:499-510, 1976.
14. Shore, R. E., R. E. Albert, and B. S. Pasternack. Followup study of patients treated by x-ray epilation for tinea capitis. Resurvey of post-treatment illness and mortality experience. Arch. Environ. Health 31:17-28, 1976.
15. Traynor, J., and H. Casey. Five-year follow-up of primates exposed to 55 MeV protons. Radiat. Res. 47:143-148, 1971.

SKIN

Skin cancer as a late radiation sequela was first reported in 1902, 7 yr after the discovery of x rays, and many hundreds of cases have since been reported. However, the literature on radiation-induced human skin cancer is sufficiently unsystematic that many questions remain essentially unanswered—e.g., as to the shape of the dose-response curve, the effects of dose fractionation, and ethnic and age differences in susceptibility.

EXPERIMENTAL EVIDENCE

Experimental studies of radiation-induced skin cancer have shown that, after a latent period, skin tumors are produced for the rest of the lifetime.[21] In rats, very high radiation doses decrease the latent period, but there is little indication of such a decrease at doses of 200-2,000 rads.[1]

Tumor incidence appears to depend roughly on the square of dose in both mice and rats;[4,19] therefore, the incidence at doses below 100 rads is too low to be readily studied (F. Burns, personal communication). Various studies have found an RBE for tumor formation of about 3 for high-LET, compared with low-LET, radiation, when the doses were 200 to several thousand rads.[12,47] The RBE at doses under 100 rads is unknown.

Dose fractionation reduces the tumorigenicity of electron, beta, and x radiation at total doses of over 1,000 rads,[1,13] but lower doses have not been adequately studied.

Investigations of the critical radiation penetration depth for tumorigenesis in squamous-cell (as well as hair-follicle) tumors have suggested that the dermis is prominently involved in epidermal tumors.[1,10]

Studies of low-LET radiation effects on the skin that compared a sieve pattern with a uniform pattern of irradiation (equivalent for skin dose and adjusted for skin area irradiated) showed that the sieve pattern afforded a marked protective effect against tumor formation;[2,3] this suggests the presence of macrotissue effects. No such protective effect was observed, however, when a similar experiment was performed with proton radiation.[11]

Attempts to use the rat as a model system to study the joint effects of ultraviolet and ionizing radiation have not been entirely successful. The primary type of tumors caused by ultraviolet radiation (keratoacanthomas) is but rarely encountered after ionizing radiation, so the degree of nonadditivity of the two effects cannot readily be evaluated (Burns, personal communication).

HUMAN EVIDENCE

Most of the reports in the literature of radiation-induced skin cancer have described a series of cases from some undefined population that were collected from one or more medical facilities (e.g., see Meara,[25] Pack and Davis,[31] Petersen,[33] Ridley and Spittle,[35] Sarkany et al.,[36] and Totten et al.[44]). That there are few adequate followup studies of defined exposed populations (and controls) is partly attributable to two factors: nonmelanotic skin cancers have a low case-fatality rate, and they are grossly underreported to Tumor Registries, so studies relying on mortality or Tumor-Registry data will fail to reflect most occurrences.

Case-Control Studies

Case-control studies provide supporting evidence about radiation-induced skin cancer, but the available studies have not permitted detailed quantitative statements. Martin et al.[23] compared 156 patients with skin cancer of the head or neck with 434 other patients drawn from their private practice, as to history of radiotherapy of the head and neck, and found a relative risk of 4.4 (19% versus 5%; $p < 0.0001$).* The doses were unknown, so no further estimates of risk could be derived. From an enlarged series of 368 skin-cancer cases after radiotherapy, they found that about 25% had been treated for hirsutism and 35% for acne. Information on the degree of radiation-associated skin changes was available on 314 of these irradiated skin-cancer patients, of whom 19% showed slight or no skin changes.

Takahashi[43] has reported case-control data indicating that medical radiation is associated with skin cancer among Japanese; 4.5% (14 of 308) of skin-cancer patients and 0.6% (23 of 4,067) of controls had a history of previous irradiation. The excess of cancer was significant at the lowest dose range, 500–2,000 R, and there was a highly significant dose-response relationship (χ^2 for trend = 79.3; 1 df; $p < 0.0001$). However, the number of irradiated subjects was too small and the dose estimates too uncertain to test the shape of a dose-response curve.

Epidemiologic Studies

Four epidemiologic followup studies showed excessive skin cancer after radiation exposures. These are reported next, and then several studies that did not show radiation-induced skin cancer.

R. E. Shore (personal communication) and Shore et al.[40] have followed 2,215 children given x-ray therapy of the scalp for tinea capitis at ages 3–14 and a comparison group of 1,395 children with tinea capitis who did not receive radiation treatment. Both groups have been followed for an average of 25–26 yr. The x-ray treatments consisted of exposures of five fields of the scalp at one sitting, yielding doses to various portions of the scalp of 450–850 rads (the dose varied because of overlapping fields).[37]

In the irradiated group, there were 24 patients with skin cancers (23 basal-cell and one basal-cell plus squamous-cell) on the scalp or in areas bordering the scalp after a dose of 350 rads or more. There were also five patients with basal-cell cancers on other parts of the face or neck, where the radiation doses were about 20–60 rads.[17] In the control group, there were two patients with skin cancers in locations corresponding to the heavily irradiated area and two on other areas of the face and neck.

---

* This study was reported incorrectly in the 1972 BEIR report.[30] The latter stated that there was "a relative risk of 3.74 in 649 irradiated U.S. patients." It was actually a case-control study, and only 59 of the 649 had received radiation therapy. The odds-ratio estimate of relative risk, which is appropriate for a case-control study, is 4.4, and not 3.7.

In the irradiated group, no skin cancers occurred among the 25% who were black. An analysis (controlling for interval since irradiation) showed that this difference was significant ($p = 0.02$) and indicated ethnic variation in susceptibility to radiation-induced skin cancer. The analyses were thereafter restricted to Caucasians.

Age- and sex-specific skin-cancer rates among whites from the Third National Cancer Survey[38] were used to compute expected values. In the control group, the ratio of observed to expected skin cancers was 4:4.7; in the irradiated group, it was 29:7.8. Preliminary data from the study suggest that fairness of the skin is probably a more important mediating factor in skin-cancer development than is the amount of sunlight exposure (Shore, personal communication).

The temporal pattern of skin cancers was of interest. Table A-29 shows the distribution of observed and expected skin cancers by age. An analysis of an age trend in the absolute excess risk (for ages 20–49) was significant ($p < 0.0001$). The trend in the absolute excess risk by interval since irradiation (15–34 yr), as shown in Table A-30, was also highly significant ($p < 0.0001$); this indicates that absolute risk rises with time. Whether it will eventually plateau, continue to rise, or taper off in a wave-like fashion is unknown, although Ridley and Spittle's report[35] of a mean interval of 45 yr between x-ray therapy for tinea capitis and skin-cancer diagnosis in a series of 26 cases suggests that excess risk will continue for a considerable time.

Although in this study there was a range of doses (20–850 rads) cor-

TABLE A-29   Observed and Expected Skin Cancers by Age among Caucasians Irradiated for Ringworm of the Scalp[a]

Age, yr	No. Skin Cancers		Person-years (PY)	O/E	Absolute Risk per 10^6 PY
	Observed (O)	Expected (E)[b]			
1–19	0	0.01	18,160	0	0
20–24	0	0.54	8,075	0	0
25–29	4	1.39	7,332	2.9	356
30–34	8	1.76	5,714	4.5	1,092
35–39	12	2.32	3,407	5.2	2,841
40–44	5	1.63	1,106	3.1	3,047
>44	0	0.20	100	0	0

[a] Data from R. E. Shore (personal communication).
[b] Expected values based on sex-specific and age-specific rates from the Third National Cancer Survey.[38]

TABLE A-30   Observed and Expected Skin Cancers among Caucasians by Interval since Irradiation for Ringworm of the Scalp[a]

| Interval, yr | No. Skin Cancers | | Person-years (PY) | O/E | Absolute Risk per 10^6 PY |
	Observed (O)	Expected (E)[b]			
1–9	0	0.04	15,074	0	0
10–14	0	0.40	8,239	0	0
15–19	1	1.13	7,846	0.9	0
20–24	7	1.85	6,555	3.8	786
25–29	14	2.57	4,470	4.7	2,110
30–34	9	1.86	1,710	4.8	4,175

[a] Data from R. E. Shore (personal communication).
[b] Expected values based on sex-specific and age-specific rates from the Third National Cancer Survey.[38]

responding to different locations on the head, the shape of the dose-response curve has not been assessed. Because of the small numbers, the nonsignificant excess (five observed and about 3.4 expected) at 20–40 rads is interpretively compatible with either linearity, curvilinearity, or no effect.*

In a group with x-ray therapy for an enlarged thymus, Hempelmann et al. found an excess of skin cancers.[18] In the irradiated group, nine of 2,872 persons had a skin cancer in the irradiated area (six basal-cell cancers and three malignant melanomas) versus three of 5,055 in the corresponding skin area among controls (E. Woodard and L. Hempelmann, personal communication). There was no difference between the groups in the incidence of skin cancers outside the irradiated skin area. Considering years 10–49 after irradiation, the rates in the irradiated area were nine in 50,226 PY, or 179 per 10^6 PY, in the irradiated group and three in 89,625 PY, or 33 per 10^6 PY, in the equivalent skin region in the control group (RR = 5.4; $p = 0.01$). Assuming that the skin doses were approximately 20% greater than the air doses, the skin doses would range from 40 to about 1,500 rads, with a mean of 330 rads. The radiation was given mainly in fractions of approximately 40–400 rads.

Because this is the only available study that can provide any information on the dose-response relationship at lower doses or on dose-

* In another study of tinea capitis irradiation, Modan et al.[26] did not find an excess of skin cancer. However, Modan has indicated (personal communication) that he was essentially not able to study skin cancer, because his sources of information were a Tumor Registry and a Death Registry, both of which yield gross underreporting of skin-cancer incidence.

fractionation effects, the data were (crudely) analyzed for these effects (Shore and Hempelmann, personal communication). The results should be reviewed as only suggestive, however, because of the small number of skin cancers. There was no evidence of a lower risk at lower doses, when doses under 400 rads were compared with doses over 400 rads. The absolute-risk estimate was numerically (but not significantly) higher in the lower-dose group (0.66 versus 0.32 per 10^6 PY per rad). Dose fractionation was examined according to number of fractions and to average dose per fraction. A tabulation by number of fractions yielded estimates of 0.6, 0.3, and 0.5 per 10^6 PY per rad for subjects with 1, 2, and 3+ fractions, respectively. A crude breakdown of the dose-per-fraction data yielded estimates of 0.4 per 10^6 PY per rad for 1–199 rads per fraction and 0.5 per 10^6 PY per rad for 200–400 rads per fraction. A finer analysis of these data with the Cox regression model (controlling total skin dose) tentatively suggested that both increased fractionation and smaller dose per fraction decreased the cancer yield ($p = 0.05$ for each effect).

G. M. Matanoski and R. Seltser (personal communication) and Matanoski *et al.*[24] recently reported an excess of skin-cancer mortality among radiologists, compared with three other groups of medical specialists. Table A-31 shows that the excess was especially pronounced in the earliest cohort (those who became members of the radiological society during the 1920s), who presumably had the highest radiation exposures. The average skin doses received by the various cohorts of radiologists are essentially unknown. Estimates for the earlier cohorts range from a low of about 200 rads (calculated from Lewis[22]) to a high of 2,000 rads.[9] The 1972 BEIR report[30(p.112)] used 800 rads as a tentative estimate. With that estimate, the absolute risk of skin-cancer deaths in the 1920–1929 cohort would be 0.5 per 10^6 PY per rad.

M. Sevcova, J. Ševc, and J. Thomas (unpublished data) found a significantly increased incidence of skin cancers, primarily basal-cell carcinomas of the face, in a large series of uranium miners, who had estimated cumulative alpha-radiation doses to the basal-cell layer of the skin of around 100 rads. The observed incidence (per 10^6 PY) over that expected from age-specific population rates was 374/82, or a relative risk of 4.6. Using an estimate of 100 rads (which may well be a low estimate, but is the one used in the 1977 UNSCEAR report[46]), this yields an absolute-risk value of 2.9 per 10^6 PY per rad. The rate among nonmining uranium workers (with small exposures) was not increased; this suggests that the increased rate among miners was not a methodologic artifact of better case detection in the workers than in the general population.

Of the incidence studies of skin cancer in defined irradiated populations, the Japanese atomic-bomb study is the largest. As part of its Adult

TABLE A-31 Skin-Cancer Deaths among Cohorts of Radiologists (Radiological Society of North America) and Other Medical Specialists (American College of Physicians and American Academy of Ophthalmology and Otolaryngology)[a]

Cohort (Year of Entry into Society)	Radiologists		Other Specialists	Relative Risk (90% Lower and Upper Confidence Limits)
1920–1929	Deaths	15	3	10.0
(all ages)	PY	33,367	66,679	(3.9, 28.9)
	Rate per 10^6 PY[b]	450	45	
1930–1939	Deaths	3	5	3.6
(all ages)	PY	16,575	100,309	(1.1, 11.6)
	Rate per 10^6 PY[b]	181	50	
1940–1959	Deaths	4	5	3.1
(age-limited)[c]	PY	54,808	215,634	(1.1, 9.1)
	Rate per 10^6 PY[b]	73	23	

[a] Data through 1974. From G. M. Matanoski and R. Seltser, personal communication.
[b] Use of crude rather than age- and time-adjusted rates (Matanoski et al.[24]) makes the comparison slightly conservative.
[c] Includes data up to age 74 for the 1940–1949 cohort and up to age 64 for the 1950–1959 cohort. Combining these cohorts introduces a small additional conservatism into the comparison.

Health Study evaluation program, an extensive dermatologic evaluation was performed on 9,646 subjects from Hiroshima or Nagasaki, including 1,830 with doses of 90 rads kerma or more and an additional 2,081 with doses of 10–89 rads.[20] These subjects were compared with 2,696 others in the city at the time of the bombing who had doses of 1 rad or less. The sample was carefully drawn, and the examinations were conducted 19–21 yr after irradiation, which apparently allowed ample time for latent malignant changes to begin to appear.

Little evidence was found of skin changes other than burn scars. The irradiated group (≥ 90 rads) had more patterned pigmentation and hyperpigmentation than the controls (< 1 rad), but the dose-response relationship was inconsistent with a radiation interpretation; among those aged 0–19 at the time of bombing, facial elastosis was more prevalent in the irradiated group, but the dose-response relationship was not significant, nor was an association with radiation found across the whole age range. No evidence was found of radiation-induced skin cancer; the only observed person with skin cancer received less than 1 rad.

On the surface, this study appears at variance with other studies of radiogenic skin cancer and with the Takahashi study[43] in particular. However, it has been shown that the Japanese have skin-cancer rates only one-fortieth to one-thirtieth those of Caucasians for similar environments, which suggests that they are highly resistant to the carcinogenic effects of ultraviolet radiation. It can be speculated that their skin is similarly resistant to carcinogenesis by ionizing radiation. That Takahashi[43] found a radiation effect is not wholly incompatible. His lowest dose grouping was 500–2,000 R, higher than that for nearly all the atomic-bomb survivors, and only three skin cancers were found in this range. The other doses ranged up to 10,000 R, and higher. Nor is the study statistically incompatible with other studies that have shown a skin-cancer effect at doses under 1,000 rads. Using an estimate of 0.7 skin cancer per 10^6 PY per rad as an average of the estimates found in other studies (see Table A-32), one would have expected only about 3.5 skin cancers in the irradiated group; finding none is incompatible with chance.*

Another skin-examination program was conducted by Sulzberger *et al.*,[42] who evaluated 1,000 patients treated 5–23 yr previously with superficial x rays, 90% of whom had been treated for benign dermatoses and the remainder for skin cancer. For comparison, 1,000 former patients treated for similar conditions without irradiation were evaluated. The bulk (72%) of the irradiated group received between 150 and 1,000 rads to one or more localized areas of the skin, and the remainder received larger doses. For the benign conditions, the doses were typically given in fractions of 35–85 rads. The patients were reasonably distributed over the full age range at irradiation, with the exception of a deficit under age 20. Six skin cancers were found in the irradiated group, of which only one occurred in an irradiated area; and nine were found in the control group.

Three problems arise in interpreting the study: (1) It is not known whether the age distribution of the control group was comparable with that of the exposed group. (2) Only about 10% of the population that the researchers set out to evaluate were actually examined; thus, the potential for sample-selection bias was great, and its nature and degree are largely unknown. (3) The average interval from irradiation to evaluation was only

---

* In this and later calculations of expected values based on the absolute-risk estimate from the positive studies, no account is taken of the amount of skin area irradiated. This is difficult to assess for many studies, and perhaps meaningless, in that anatomic location of the irradiated skin may be more important than the total skin area irradiated.

TABLE A-32   Risk Estimates for Skin Cancer

	Shore et al. [a]	Hempelmann et al. [b]
Study population	Irradiated for tinea capitis 1940–1959 (Caucasian)	Thymus irradiation 1926–1957
Type of radiation	X	X
Duration of radiation exposure	Minutes	Minutes to weeks
Duration of followup, yr	15–34	17–49
Mean followup, yr	26	28
Period after irradiation on which risk estimates are based, yr	10–34	10–49
No. subjects	1,685	2,878
No. PY	28,820	50,226
External dose range, rads	350–850	40–1,500
Mean dose to tissue, rads	700	330
Age at irradiation, yr	3–14	0–1.5
Mean age at irradiation, yr	8	0
Sex	Both	Both
Nature of control	1,046 nonirradiated tinea patients	5,055 siblings
Cancer/PY		
Irradiated group	24/28,820	9/50,226
Control group	2/17,124	3/89,625
Relative risk	7.1	5.4
Increase, %/rad	0.88	1.32
Absolute risk per 10^6 PY per rad	1.02	0.44

[a] Data from Shore et al. [40] and R. E. Shore, personal communication.
[b] Data from Hempelmann et al. [18] and E. Woodard and L. Hempelmann, personal communication.

9 yr; only 26 irradiated subjects were evaluated more than 15 yr after irradiation. The methodologic weakness of a short interval since radiotherapy makes it impossible to interpret the negative results as having a bearing on, for example, the effects of fractionated exposures.

No skin-cancer deaths have been found among the approximately 15,300 patients followed after radiotherapy for ankylosing spondylitis (Court Brown and Doll[14] and R. Doll and P. G. Smith, personal communication), for which the cumulative skin dose in the primary beam was in the region of 1,000–1,500 rads. On the basis of the average risk from the studies finding an effect (Table A-32), one would expect over 40 excess skin cancers to have occurred (58,014 PY 9 yr or more after irradiation × 1,200 rads × 0.7 cancer per 10^6 PY per rad = 49 excess skin cancers expected). If one assumes a 5% mortality from skin cancer, then about two or three deaths would be expected, whereas none were observed. It should be noted that others have reported basal-cell carcinomas after such therapy.[25,36]

Boice and Monson[7] followed a series of 543 tuberculous women by questionnaire for an average interval of 26 yr after irradiation. These women had a mean cumulative skin dose of about 1,300 rads from an average of 102 fluoroscopic examinations.[7,8] Three skin cancers were reported in the irradiated group and one among controls; but it was not known whether any of the skin cancers were in the area affected by the primary beam (J. Boice, personal communication). From the average risk estimate of 0.7 per 10^6 PY per rad (for 10 yr or more after irradiation) given above, one would expect about 7.7 excess skin cancers in the irradiated group.

In another multiple-fluoroscopy study, Myrden and associates[28,29] followed 300 irradiated women with questionnaires for up to 32 yr. The average cumulative skin dose was about 1,900 rads. An excess of about five skin cancers would have been expected; in fact, none were observed (J. A. Myrden, personal communication).

Delarue et al.[15] followed 269 tuberculous women for over 20 yr after multiple fluoroscopies. An average of 142 fluoroscopic examinations were given posteriorly with a skin dose of about 850 rads. Only one skin cancer was observed in the irradiated area of the back (none among controls), but this was not surprising, in that an excess of only about 1.9–2.4 would be expected on the basis of the average risk estimate (0.7 per 10^6 PY per rad) and the estimated person-years of followup.*

Shore et al.[41] found no excess skin cancer among women given radiotherapy for postpartum mastitis, in which the average skin dose in the primary beam was about 280 rads and the average followup was 25 yr. They found six skin cancers among 571 irradiated women and 13 among 993 control women. Considering the period 10–34 yr after irradiation, 7.4 skin cancers would be expected in the irradiated group, on the basis of the control-group rate; from the excess-risk calculations, one would expect an additional 1.8. Thus, the total expected was about 9.2, versus the six observed.

Host Factors and Pathogenesis

The principal classifications of radiation-induced skin cancers are squamous-cell and basal-cell carcinomas. Anatomic location is one factor determining the relative incidence of the two types: squamous-cell

---

* Another multiple-fluoroscopy study (Shore and Hempelmann, personal communication) has over 4,000 irradiated subjects, but is not reported here, because fewer than 10% have been followed by questionnaires so far.

("prickle-cell") carcinomas have been found most often on the hands, and basal-cell carcinomas clearly predominate on the head and neck.[45] An additional factor of dose has been proposed: that squamous-cell cancers occur primarily after large radiation exposures associated with severe radiodermatitis and ulceration, whereas basal-cell carcinomas predominate at lower exposures.[27] Fibrosarcomas, of dermal origin, have also occurred after radiation exposure, but the incidence is an order of magnitude less than that of squamous- or basal-cell cancers.[27] (In addition, some of the reported fibrosarcomas would be regarded as spindle-cell carcinomas today.[23,32]) Melanomas and sweat-gland tumors have also been reported as occasional radiation sequelae.[6,40,45]

The relative sensitivities of various anatomic portions of the skin to radiation-induced cancer are not established, although it is thought that the face and scalp are especially sensitive.[45] The role of superimposed ultraviolet radiation by exposure to sun (insolation) of skin that has received ionizing irradiation is not known. The carcinogenic effects of ultraviolet and ionizing irradiation may be purely additive, but the apparent sensitivity of the face to radiation-induced skin cancer suggests that ultraviolet radiation in some way potentiates ionizing radiation.

Although chronic radiodermatitis was long thought to be a prerequisite for the induction of skin cancer, it is now amply documented that basal-cell cancers especially can occur in skin with little or no evidence of radiation skin damage.[5,35,36,40] It has been suggested that the carcinogenic effect on the skin is much greater in the presence of radiodermatitis,[45] with estimates of cumulative incidence ranging from 10 to 35% (versus ≤1% in the absence of radiodermatitis). Calculations on a per-rad basis per unit of irradiated skin surface area have not been performed to test this difference. Assuming that there is a difference, it is unclear whether it results from biologic factors having to do with damage in radiodermatitis or is primarily a reflection of a quadratic dose-response relationship.

Basal-cell cancers that follow radiation exposure are often multiple, either simultaneously or over a period of some years. For instance, Martin *et al.*[23] reported an average of five lesions per patient when the skin cancer followed irradiation, but 2.5 per patient when there had been no prior irradiation. (However, the comparability of these two groups in length of followup after the first lesion was not reported, so the comparison may be biased.) In the tinea radiation study, 70 lesions have been observed in the 29 persons with skin cancer of the head or neck (Shore, personal communication). Taking into account the time since diagnosis of the initial basal-cell lesion(s), this works out to about 0.3 new lesion per case per year. Among the four control cases, there are no multiple lesions, after a comparable postdiagnosis followup time.

No information is available on whether age at irradiation modifies the risk of radiogenic skin cancer. The latent interval for skin cancer after irradiation is extremely variable. Martin *et al.* [23] reported a range of latencies from 1 to 64 yr among 357 cases. In only about 6% of the cases was the interval less than 10 yr, whereas in about 20% it was greater than 30 yr (20% is probably an underestimate, in view of the skewed distribution of followup times since irradiation in the study population).

It is well known that blacks have a far lower incidence of skin cancer than Caucasians. [16,38,39] That they are also less susceptible to radiation-induced skin cancer has recently been shown by Shore (personal communication). There are also marked differences in skin-cancer incidence between Caucasians and Japanese. A special incidence study in Hawaii [34] reported that the relative risk of skin cancer was 39 for Caucasians, compared with Japanese (90% lower confidence limit on RR was 24). The incidence rates among Japanese in Japan and in Hawaii are quite comparable. [39] It is possible that the Japanese have low susceptibility to radiation-induced skin cancer as well.

Radiation Factors

No human studies have been performed that provide a direct examination of the dose-response curve, RBE values, or dose-rate effects. The comparison of selected studies permits at best tenuous suggestions as to these effects.

The study by Takahashi, [43] although showing a trend with dose, is unsuitable for defining the shape of the dose-response curve, because it is a case-control study not based on a defined population and only three skin cancers occurred between 500 and 2,000 rads—and none was reported between 1 and 500 rads.

The negative skin-cancer results of the Japanese atomic-bomb study and fluoroscopy studies, [7,15,20] the restricted dose range of the tinea capitis study, [40] and the limited number of skin cancers in the thymus-irradiation study [18] made it impossible to assess the dose-response relationship in any followup study. The high range of skin-cancer incidence (10–35%) among patients with radiodermatitis and the much lower rates among populations with lower doses (negative to 1%) suggests a dose-squared component of the dose-response relationship. However, at least at doses under 1,000 rads, the data of Woodard and Hempelmann (personal communication) do not suggest a diminution of effect with lower doses, inasmuch as the risk at doses less than 400 rads is numerically greater than that at doses over 400 rads.

Because no skin cancers were found in the exposed group in the Japanese atomic-bomb study, RBE factors could not be estimated. No group with low-LET radiation was sufficiently comparable (i.e., protracted whole-body radiation, with cancer-incidence data) with the uranium miners with high-LET radiation to permit even a rough comparison.

It cannot be concluded from the negative findings in the two multiple-fluoroscopy series[7,15] that dose fractionation reduces skin carcinogenesis. The postpartum-mastitis study,[41] with little fractionation of the radiation doses to the skin of the chest, was also negative. The radiologist study (Matanoski et al.[24] and Matanoski and Seltser, personal communication) indicated that highly fractionated exposures at both high and lower total doses (i.e., early and more recent cohorts) had a carcinogenic effect on the skin. However, the thymus-irradiation study suggested that dose fractionation may reduce the magnitude of the carcinogenic effect (Shore and Hempelmann, personal communication).

Risk Estimates

None of the three studies of whole-body irradiation is suitable for making risk estimates. The Japanese atomic-bomb study[20] was essentially negative. The radiologist study[24] provided only mortality data, and the doses can be estimated but crudely. Insufficient details are available on the uranium-miner study (Sevcova et al., unpublished data) to use it as a primary source of risk estimates. In addition, the doses of alpha radiation to much of the body are essentially unknown, because of shielding by clothing.

Two studies[18,40] of x-ray therapy have yielded positive results and can be used for estimating partial-body radiation risks, but several qualifications should be noted: several similar studies were negative,[7,14,15,41] so these two may represent an upper bound of risk; sensitivity to radiation induction of skin cancer probably varies by body site,[45] so extrapolations to whole-body irradiation on the basis of proportion of skin surface irradiated may be improper; and these two studies had relatively high doses and high dose rates and little or no dose fractionation, so the appropriate generalizations to protracted or low-dose radiation are unknown. A summary of the two studies is shown in Table A-32. The skin-cancer rate in the heavily irradiated area for 10–34 yr of followup among Caucasians in the tinea capitis study (Shore et al.[40] and Shore, personal communication) was 833 per 10^6 PY; in the corresponding area among controls, it was 117 per 10^6 PY. With the assumption of 700 rads as the approximate average dose to the scalp,[37] this yields an absolute-risk estimate of 1.0 per 10^6 PY per rad (with 90% confidence limits of 0.2 and 4.4). On the basis of all 29 skin-cancer cases in the irradiated group and the population-based expected value of 7.8 skin cancers, the absolute-risk estimate was 1.0 per 10^6 PY per rad (90% confidence limits, 0.7 and 1.5). In the thymus-irradiation study (Hempelmann *et al.*[18] and Woodard and Hempelmann, personal communication), the irradiated group had a rate of skin cancer in the heavily irradiated area of 179 per 10^6 PY for 10–49 yr of followup, compared with the control-group rate of 33 per 10^6 PY. With an approximate average skin dose of 330 rads, this yields an absolute risk of 0.4 per 10^6 PY per rad (90% confidence limits, 0.1 and 1.5).

REFERENCES

1. Albert, R. E. Skin carcinogenesis, pp. 111–136. In R. Andrade, S. Gumport, G. Popkin, and T. Rees, Eds. Cancer of the Skin. Biology-Diagnosis-Management. Vol. 1. Philadelphia: W. B. Saunders, 1976.

2. Albert, R. E., and F. J. Burns. The sieve effect for tumor induction with low voltage x-rays in rat skin. Proc. Radiat. Res. Soc. Abstr. Cd-2, p. 24, 1973.

3. Albert, R. E., F. J. Burns, and R. D. Heinbach. Skin damage and tumor formation from grid and sieve patterns of electron and beta radiation in the rat. Radiat. Res. 30:525, 1967.

4. Albert, R. E., W. Newman, and B. Altshuler. The dose-response relationships of beta-ray-induced skin tumors in the rat. Radiat. Res. 15:410–430, 1961.

5. Albert, R. E., A. R. Omran, E. E. Brauer, N. C. Cohen, H. Schmidt, D. C. Dove, M. Becker, R. Baumring, and R. L. Baer. Follow-up study of patients treated by x-ray epilation for tinea capitis. II. Results of clinical and laboratory examinations. Arch. Environ. Health 17:919–934, 1968.

6. Black, M. M., and E. Jones. Dermal cylindroma following x-ray epilation of the scalp. Brit. J. Dermatol. 85:70–72, 1971.

7. Boice, J. D., Jr., and R. R. Monson. Breast cancer in women after repeated fluoroscopic examinations of the chest. J. Natl. Cancer Inst. 59:823–832, 1977.

8. Boice, J. D., Jr., M. Rosenstein, and E. D. Trout. Estimation of breast doses and breast cancer risk associated with repeated fluoroscopic chest examinations of women with tuberculosis. Radiat. Res. 73:373–390, 1978.

9. Braestrup, C. B. Past and present radiation exposure to radiologists from the point of view of life expectancy. Amer. J. Roentgenol. Radium Ther. Nucl. Med. 78:988–992, 1957.

10. Burch, P. R. New approach to cancer. Nature 225:512–516, 1970.

11. Burns, F. J., R. E. Albert, P. Bennett, and I. P. Sinclair. Tumor incidence in rat skin after proton irradiation in a sieve pattern. Radiat. Res. 50:181, 1972.

12. Burns, F. J., R. E. Albert, and R. D. Heinbach. The RBE for skin tumors and hair follicle damage in the rat following irradiation with alpha particles and electrons. Radiat. Res. 36:225–241, 1968.

13. Burns, F. J., R. E. Albert, I. P. Sinclair, and P. Bennett. The effect of fractionation on tumor induction and hair follicle damage in rat skin. Radiat. Res. 53:235–240, 1973.

14. Court Brown, W., and R. Doll. Mortality from cancer and other causes after radiotherapy for ankylosing spondylitis. Brit. Med. J. 2:1327–1332, 1965.

15. Delarue, N. C., G. Gale, and A. Ronald. Multiple fluoroscopy of the chest. Carcinogenicity for the female breast and implications for breast cancer screening programs. Can. Med. Assoc. J. 112:1405–1410, 1413, 1975.

16. Emmett, E. A. Ultraviolet radiation as a cause of skin tumors. CRC Crit. Rev. Toxicol. 2:211-255, 1973.

17. Harley, N. H., R. E. Albert, R. E. Shore, and B. S. Pasternack. Follow-up study of patients treated by x-ray epilation for tinea capitis. Estimation of the dose to the thyroid and pituitary glands and other structures of the head and neck. Phys. Med. Biol. 21:631-642, 1976.

18. Hempelmann, L. H., W. J. Hall, M. Phillips, R. A. Cooper, and W. R. Ames. Neoplasms in persons treated with x-rays in infancy. Fourth survey in 20 years. J. Natl. Cancer Inst. 55:519-530, 1975.

19. Hulse, E., R. Mole, and D. Papworth. Radiosensitivities of cells from which radiation-induced skin tumors are derived. Intl. J. Radiat. Biol. 14:437-444, 1968.

20. Johnson, M. L. T., C. E. Land, P. B. Gregory, T. Taura, and R. C. Milton. Effects of Ionizing Radiation on the Skin, Hiroshima and Nagasaki. Atomic Bomb Casualty Commission Technical Report TR 20-69. Hiroshima: Atomic Bomb Casualty Commission, 1969.

21. Jones, D. C., T. J. Castanera, D. J. Kimelford, and V. J. Rosen. Radiation induction of skin neoplasms in the male rat. J. Invest. Dermatol. 50:27-35, 1968.

22. Lewis, E. B. Leukemia and ionizing radiation. Science 125:965-972, 1957.

23. Martin, H., E. Strong, and R. H. Spiro. Radiation-induced skin cancer of the head and neck. Cancer 25:61-71, 1970.

24. Matanoski, G. M., R. Seltser, P. E. Sartwell, E. L. Diamond, and E. A. Elliot. The current mortality rates of radiologists and other physician specialists. Specific causes of death. Amer. J. Epidemiol. 101:199-210, 1975.

25. Meara, R. H. Epitheliomata after radiotherapy of the spine. Brit. J. Dermatol. 80:620, 1968.

26. Modan, B., D. Baidatz, H. Mart, R. Steinitz, and S. G. Levin. Radiation-induced head and neck tumours. Lancet 1:277-279, 1974.

27. Mole, R. H. Radiation induced tumors. Human experience. Brit. J. Radiol. 45:613, 1972.

28. Myrden, J. A., and J. E. Hiltz. Breast cancer following multiple fluoroscopies during artificial pneumothorax treatment of pulmonary tuberculosis. Can. Med. Assoc. J. 100:1032-1034, 1969.

29. Myrden, J. A., and J. J. Quilan. Breast carcinoma following multiple fluoroscopies with pneumothorax treatment of pulmonary tuberculosis. Ann. R. Coll. Phys. Surg. Can. 7:45, 1974.

30. National Research Council, Advisory Committee on the Biological Effects of Ionizing Radiations. The Effects on Populations of Exposure to Low Levels of Ionizing Radiation. Washington, D.C.: National Academy of Sciences, 1972.

31. Pack, G. T., and J. Davis. Radiation cancer of the skin. Radiology 84:436-442, 1965.

32. Pegum, J. S. Radiation induced skin cancer. Brit. J. Radiol. 45:613-614, 1972.

33. Petersen, O. Radiation cancer. Report of 21 cases. Acta Radiol. 42:221-236, 1954.

34. Quisenberry, W. B. Ethnic differences in skin cancer in Hawaii. U.S. Natl. Cancer Inst. Monogr. 10:181-189, 1963.

35. Ridley, C. M., and M. F. Spittle. Epitheliomas of the scalp after irradiation. Lancet 1:509, 1974.

36. Sarkany, I., R. B. Fountain, C. D. Evans, R. Morrison, and L. Szur. Multiple basal-cell epitheliomata following radiotherapy of the spine. Brit. J. Dermatol. 80:90-96, 1968.

37. Schulz, R. J., and R. E. Albert. III. Dose to organs of the head from the x-ray treatment of tinea capitis. Arch. Environ. Health 17:935-950, 1968.

38. Scotto, J., A. W. Kopf, and F. Urbach. Non-melanoma skin cancer among Caucasians in four areas of the United States. Cancer 34:1333–1338, 1974.

39. Segi, M. World incidence and distribution of skin cancer. U.S. Natl. Cancer Inst. Monogr. 10:245–255, 1963.

40. Shore, R. E., R. E. Albert, and B. S. Pasternack. Follow-up study of patients treated by x-ray epilation for tinea capitis. Arch. Environ. Health 31:17–28, 1976.

41. Shore, R. E., L. H. Hempelmann, E. Kowaluk, P. S. Mansur, B. S. Pasternack, R. E. Albert, and G. E. Haughie. Breast neoplasms in women treated with x-rays for acute postpartum mastitis. J. Natl. Cancer Inst. 59:813–822, 1977.

42. Sulzberger, M. B., R. L. Baer, and A. Borota. Do roentgen-ray treatments as given by skin specialists produce cancers or other sequelae? Arch. Dermatol. Syphilol. 65:639–655, 1952.

43. Takahashi, S. A statistical study on human cancer induced by medical irradiation. Acta Radiol. (Nippon) 23:1510–1530, 1964.

44. Totten, R. S., P. G. Antypas, S. M. Dupertuis, J. C. Gaisford, and W. L. White. Pre-existing roentgen-ray dermatitis in patients with skin cancer. Cancer 10:1024–1030, 1957.

45. Traenkle, H. L. X-ray induced skin cancer in man. In Biology of Cutaneous Cancer. U.S. Natl. Cancer Inst. Monogr. 10:423–432, 1963.

46. United Nations Scientific Committee on the Effects of Atomic Radiation. Sources and Effects of Ionizing Radiation. Report to the General Assembly, with Annexes. New York: United Nations, 1977.

47. Zackheim, H. S., E. Krobock, and L. Langs. Cutaneous neoplasms in the rat produced by grenz ray and 80 kV x-ray. J. Invest. Dermatol. 43:519–534, 1964.

CANCER INDUCED BY IRRADIATION BEFORE CONCEPTION OR DURING INTRAUTERINE LIFE

MAIN FINDINGS OF 1972 BEIR REPORT

In a large retrospective study, the Oxford survey, Stewart and associates[40,42] found an excess of leukemia and other cancers among children exposed *in utero* to diagnostic x rays. The increase in relative risk was initially reported as about 100%,[42] but was later estimated to be about 60% in a larger sample.[40] These results were supported by MacMahon,[20] who identified cancer cases retrospectively, but traced controls prospectively, and found about a 50% increase in relative risk of cancer after *in utero* x-ray exposure. According to MacMahon, the oncogenic effect of such exposure was exhausted by the age of 8 yr; according to Stewart *et al.,* the increased risk persisted to the age of 10 yr, the maximal length of followup at that time. Some additional studies have not confirmed the increased risk, but their samples are too small to contradict these results.[21]

Other factors, however, raise doubts about the oncogenicity of intra-uterine exposure suggested by these studies. There was very little specificity of effect with regard to tumor type in the Oxford survey.[40] In the tristate study,[9] preconception exposure of the mother to diagnostic x rays was associated with as great an increase in leukemia risk as was intrauterine exposure. This latter finding was contradicted by a study of leukemia in children of persons exposed to radiation from atomic bombs before the children were conceived;[12] it found no excess risk of leukemia in these children. Finally, the Oxford group[41] found a linear relationship between the number of x-ray films and the increase in cancer risk and on this basis estimated the risk at 300–800 extra cancer deaths by age 10 per million children exposed shortly before birth to 1 rad of ionizing radiation. Jablon and Kato[14] estimated that, according to this degree of risk, at least 5.2 excess cancer deaths should have been found among 1,250 Japanese children exposed *in utero* to less than 500 rads from atomic bombs, whereas in fact no excess cancer deaths were observed.

A report from the tristate study group[8] emphasized the relationship between the joint occurrence of multiple factors (preconception x-ray exposure, intrauterine exposure, childhood viral infections, and previous maternal miscarriages and stillbirths) and increased risk of leukemia. These results were extended by Bross and Natarajan,[4] who concluded that allergy-prone children, and to a lesser extent children with some infections, were more susceptible to the leukemogenic effect of *in utero* exposure to diagnostic x rays than other children.

The 1972 BEIR report[23] discounted the effects of preconception irradiation in the production of cancer and did not assess the varied sensitivity in the intrauterine population. It based its estimates of the cancer risk after *in utero* radiation on a 50% increase in deaths per rem, or an increase of 50 deaths per 10^6 PY per rem, during the first 10 yr of life, but not later.

EXPERIMENTAL STUDIES

From the studies of children exposed prenatally to diagnostic x rays, it would appear that radiation has a much greater tumorigenic effect if exposure occurs *in utero* than if it occurs during childhood or adult life (BEIR I,[23] Table 3-2). This apparent special sensitivity of the fetus has generally not been borne out by animal studies. RF mice did not develop myeloid leukemia, thymic lymphoma, or ovarian tumors when exposed to x rays *in utero,* but the incidence of these neoplasms increased significantly after postnatal exposure.[43,44] There was no overall increase in mammary neoplasms among CFI mice exposed to 100 R of x rays on different days of gestation, compared with nonirradiated controls, but there was a tendency for tumor incidence to increase when radiation occurred toward the end of gestation.[29] The incidence of mammary and uterine tumors, as well as overall tumor incidence, among Wistar rats exposed to 270 R of x rays was smaller for fetal exposure than for exposure of suckling, young, or adult rats.[28] Benign mammary tumors were significantly less common among female Sprague-Dawley rats exposed to tritiated water throughout the intrauterine period than among their simultaneously exposed mothers, and malignant mammary tumors occurred only in the exposed mothers.[6] Rat mammary tumors are hormonally dependent,[13] so these latter results might be attributable to hormonal changes during pregnancy. However, adult female Wistar rats that were irradiated while pregnant had a lower incidence of mammary tumors than nonpregnant rats irradiated at the same age.[28]

For one type of neoplasm, prenatal radiation appeared to have a greater tumorigenic effect than postnatal radiation. CBA mice exposed prenatally to iodine-131 for a dose to the thyroid of about 8,000 rads developed a significant number of thyroid neoplasms, whereas 96-d-old mice exposed to iodine-131 for a dose to the thyroid of about 9,000 rads did not develop any.[45] In another study, prenatal administration of iodine-131 to CD rats produced significantly more thyroid neoplasms than neonatal administration, but interpretation of these results is complicated by the occurrence of a substantial number of thyroid neoplasms among the nonirradiated controls in both groups.[32]

CAUSAL RELATIONSHIP

An association between prenatal exposure to diagnostic x rays and an increased risk of developing malignancy during childhood has been definitely established, but there remains the question of whether and to what extent this represents a causal relationship. It is possible that prenatal exposure entails a process of selection, resulting in an exposed population that differs from the unexposed population with regard to factors that bear on the incidence of malignancy, and it is possible that the selection process accounts for part or all of the increased incidence of malignancy in the exposed population. A number of recent studies have produced findings pertinent to this question.

Mole[22] has reanalyzed the data from the Oxford survey regarding twins.[35,36] According to his analysis, the estimated relative risk of leukemia for prenatally irradiated twins is 2.2, compared with 1.5 for prenatally irradiated singletons. For solid tumors, the estimated relative risks are 1.6 for twins and 1.5 for singletons. It may be argued that twins should be an especially apt subgroup for testing the tumorigenic effect of prenatal irradiation, inasmuch as a great many are irradiated specifically for indications directly related to twinning, and the effect of selection for adverse medical conditions might therefore be expected to be substantially diluted. Indeed, 55% of British twin births in the population studied were x-rayed, compared with 10% of singletons. It is difficult to explain why twin pregnancies in which the diagnosis of twinning was established with x-rays should have a much higher cancer risk than twin pregnancies in which the diagnosis was not established in this manner, unless one postulates a causal relationship between the radiation and the cancer.

Burch[5] pointed out that the absolute frequency of leukemia in the whole twin population in this study was only 1.98 per 10,000, in contrast with 2.35 per 10,000 for singletons, and that there was a similar though smaller deficit of solid tumors among twins. He criticized Mole for not taking this into account. Stewart[39] has offered a possible explanation for this deficit, which is noted below. In any case, we feel that this finding does not weaken Mole's argument, because the argument depends not on the underlying incidence of cancer among twins, but only on the increase in this incidence (i.e., relative risk) after irradiation. The only criticism that we can raise concerning Mole's findings is that they are based on a reinterpretation of published data with which the author did not claim direct familiarity and therefore must be regarded with some caution. Nevertheless, we consider the twin data to provide some of the strongest support for a causal relationship between *in utero* exposure to diagnostic x rays and the later increase in cancer risk, and we feel that further studies of irradiated twins are strongly warranted.

Several other recent studies have revealed potential sources of bias in the selection of irradiated subjects that may bear on the question of causality. In a large prospective study of the mortality of white and black children exposed to diagnostic x rays *in utero,* Diamond et al.[7] found the death rate from all causes (beyond the neonatal period) among exposed white children to be nearly twice that among matched controls. In contrast, the death rate among black children was not increased. Extensive analysis revealed large discrepancies between the exposed and control groups, in both white and black subjects, with respect to pregnancy complications, abnormal labors, operative deliveries, and history of previous infant loss. Discrepancies in socioeconomic status were also found, but only among the white subjects. The investigators adjusted their data for this last factor, and that reduced the relative risk of mortality after irradiation from 2.0 to between 1.4 and 1.8. They chose not to make adjustments for the other factors, on the grounds that discrepancies were present among both black and white subjects and therefore should not account for the increased mortality found only among white subjects. On the basis of earlier data from this study,[19] Oppenheim et al.[27] showed that white subjects exposed to complications of pregnancy or operative procedures had a greatly increased mortality. Adjustment for these factors should therefore have significantly reduced the excess mortality of the irradiated white subjects.

The bulk of examinations that result in fetal exposure are pelvimetry examinations,[7,42] and one must ask whether and to what extent children exposed to pelvimetry examinations differ from unexposed children. Borell and Fernstrom[2] and Russell and Richards[30] found that these examinations were strongly correlated with conditions that lead to difficult labor. One would therefore anticipate that exposed children would have an increased neonatal death rate. However, Oppenheim[26] has shown that in the study of Diamond et al.[7] the neonatal death rate of the children exposed prenatally to pelvimetry examinations was somewhat lower than that of their matched controls, although the children exposed to other types of examinations had a much higher neonatal death rate than their matched controls (Table A-33). The reasons for these findings are not apparent.

Oppenheim et al.[27] reported results of a prospective study in which medical indication played no role in the selection of the exposed subjects. About 900 children exposed *in utero* to pelvimetry examination at a time when this examination was being performed routinely were compared with about 1,300 children born before and after this time. The death rate from all causes (beyond the neonatal period) was found to be significantly lower in the exposed group than in the control groups. This finding was attributed to a significant deficit of prematures in the exposed group; the reason for that was that many of the pelvimetries were performed in the ninth month, thus decreasing the opportunity for prematures to have the examination. Such a deficit of prematures in the exposed group is probably present in other studies, inasmuch as most obstetric examinations are performed very late in pregnancy.[20,42]

TABLE A-33 Neonatal Death Rate per 1,000[a]

Group	Pelvimetry Exposed	Pelvimetry Control	Nonpelvimetry Exposed	Nonpelvimetry Control
White males	10.8	13.6	50.7	15.6
White females	9.3	9.4	45.6	10.7
Black males	17.2	26.8	40.0	22.0
Black females	10.2	23.7	43.8	18.6

[a] Data from B. E. Oppenheim, personal communication; based on Tables 20, 21, and 22 in Diamond et al.[7]

One must note that, among the various studies of prenatal exposure to radiation, the study of Oppenheim et al. [27] and the studies of atomic-bomb survivors and their offspring are the only ones in which exposure occurred in a manner clearly unrelated to medical indication.

In a series of recent reports,[1,17] the Oxford group has investigated the extent to which various factors might have introduced bias into their data. Social class, maternal age, sibship position, and fetal irradiation appeared to have exerted separate effects on the incidence of cancer;[16] however, the relative risk of cancer after fetal irradiation was unaffected or only minimally affected by each of the other factors.[1]

Kneale and Stewart[17] applied Mantel-Haenszel analysis to the Oxford data, to examine the relationship between various irradiation factors and cancer incidence (Table A-34).* Two of the factors were the reason for the examination and the x-ray finding. Although the results are difficult to analyze, there does appear to be a clear trend: x-ray reasons and findings that imply an increased likelihood of an abnormal pregnancy or delivery had negative t values, suggesting that such conditions are associated with a relatively decreased cancer risk, whereas those which do not have this implication had positive t values, suggesting that normality at birth is associated with a relatively increased cancer risk.† Stewart[39] has postulated a mechanism to account for this: that the precancerous child is in a weakened state[37] and that the added burden of an abnormal pregnancy or delivery is likely to lead to death from some competing hazard, such as pneumonia,[15] before the cancer becomes clinically apparent. Stewart invoked this mechanism[39] to explain the deficit of cancers in twins[35] and in male infants with a history of threatened abortion[10] and as a partial explanation for the absence of an increased cancer rate among

---

* The analysis was restricted to case and control children with proven prenatal x-ray exposure. Each factor in Table A-34 was tested separately by controlling the remaining factors through stratification of the sample and discarding noninformative substrata. For each level of the test factor, the number of cancer cases observed across all substrata was compared with the number that would be expected if all levels of the test factor had the same risk of cancer. The t value apparently represents the difference between observed and expected cases divided by the standard error of the difference, although the method of computation was not indicated by Kneale and Stewart.[16,17] Then the t value has a standard normal distribution, and a large positive value indicates that the cancer risk for that level is probably increased relative to the average cancer risk of irradiated subjects, whereas a large negative value indicates that the relative cancer risk is probably decreased.

† The exception is the category "hydramnios, etc.," which had a significantly increased t value. In a number of cancer cases in this category, the abnormality was produced by a congenital neoplasm. The reasons for the increased t values of the "no record" categories are unclear.

TABLE A-34 Mantel-Haenszel Analysis of Oxford Survey Data[a]

Test Factors	No. Cancer Cases		t Value
	Observed	Expected	
Films			
1	350	354.2	−0.54
2	210	216.1	−0.81
3	95	93.1	+0.22
4	51	48.0	+2.33
5 or more	45	44.8	+0.06
	Progressive component		+1.44
Exposure age			
1st trimester	22	17.2	+2.40
2nd trimester	32	40.8	−2.31
3rd trimester	267	262.8	+1.24
No record	3	3.8	−0.72
	Progressive component		−0.34
X-ray reasons			
? Twins	181	198.5	−2.40
? Breech, etc.	195	198.1	−0.46
? Disproportion	90	91.1	−0.20
Routine	71	65.1	+1.36
Maternal illness or injury	17	18.8	−0.69
Hydramnios, etc.	49	39.8	+2.30
? Fetal age	44	39.1	+1.24
? Placenta previa	30	31.8	−0.54
No record	48	42.8	+1.31
X-ray findings			
Normal pregnancy	424	421.3	+0.35
Breech, etc.	112	117.3	−0.92
Disproportion	23	50.0	−6.48
Fetal or placental abnormality	14	14.5	+0.20
Maternal illness or injury	10	9.7	−0.22
No record	116	86.2	+5.91

[a] Reprinted from Kneale and Stewart.[17]

atomic-bomb survivors who were irradiated *in utero*.[14] It must be noted that this mechanism implies that a large fraction of children in the precancerous state die from other causes before cancer is diagnosed. The evidence presented to support this[15,38] is indirect and not very compelling.

The studies of Diamond *et al.*[7] and Oppenheim *et al.*[27] indicate that prenatally exposed and unexposed populations are not strictly comparable, but differ with regard to factors (e.g., maternal illnesses, operative delivery, prematurity) that are related to neonatal health. Reports from the Oxford group[10,15,17,35,37-39] suggested a relationship between health during the neonatal period and early childhood and the later development of clinically detectable childhood cancer. Taken together, these reports suggest that the selection process underlying prenatal irradiation may, to some extent, affect the incidence of childhood cancer. The existence and nature of such an effect of selection and its significance remain to be determined.

TRIMESTER OF EXPOSURE

In an early report,[42] Stewart *et al.* noted a case-to-control ratio of 9 for first-trimester exposure in the Oxford survey. Later reports,[1,41] based on larger samples, bore out this high figure. In a more recent report,[17] a more extensive determination of the trimester of exposure was carried out, and the case-to-control ratio was 4.8 (for unadjusted data). In that report, Mantel-Haenszel analysis was applied to test the age at exposure, with adjustment for the number of films and x-ray reasons and findings (Table A-34). Although there was a significant excess of first-trimester cases (compared with the number expected if there had been no exposure-age effect), there was an equally significant deficit of second-trimester cases, and the progressive component did not indicate a trend over the three trimesters; so the interpretation of these results is unclear. It thus appears advisable to base relative-risk estimates for different trimesters of exposure on the unadjusted data given by Kneale and Stewart[17] and to group the second- and third-trimester exposures, because these had similar case-to-control ratios (1.30 and 1.41). The relative risk of cancer after irradiation *in utero*, compared with nonirradiated subjects, is estimated to be 5.0 for first-trimester exposure and 1.47 for second- or third-trimester exposure (Table A-35).

DOSE-RESPONSE RELATIONSHIP

Stewart and Kneale[41] estimated 572 extra cancer deaths before the age of 10 per 10^6 person-rads of *in utero* exposure, on the basis of linear regression of excess cancer risk on the number of films per examination and the estimated fetal dose per exposure. This corresponds to an annual absolute risk of 57 cancer deaths per 10^6 person-rems of *in utero* exposure. Further analyses of these data by Newcombe and McGregor[24] and Holford[11] supported the hypothesis of a linear relationship between fetal dose and the relative incidence of cancer. Shore *et al.*[31] questioned the magnitude of the effect, showing that the effect per rad varied markedly over different intervals and that the effect per film depended on the reason for the examination.

Kneale and Stewart[17] investigated the relationship between the number of films and the cancer incidence, using Mantel-Haenszel analysis to adjust for exposure age, x-ray reasons, and x-ray findings (Table A-34). This analysis revealed a weak and inconsistent relationship between the number of films and the risk of cancer (as measured by the discrepancy between the observed number of cancer cases and the number expected if there had been no difference in risk for differing numbers of films). The progressive component suggested a trend toward increasing risk with increasing number of films, but the value was not significant at the 5% level. The unadjusted data, however, showed a strong trend toward increasing risk with increasing number of films, with case-to-control ratios increasing monotonically from 1.31 for one film to 2.30 for five or more films. The weakening of the trend after adjustment for other factors is to be expected, in that there was a strong correlation between the number of films and the trimester of exposure[1] and the adjustment removed this influence. The annual absolute risk per 10^6 person-rems of *in utero* exposure is therefore somewhat less than 57 (the value based on Stewart and Kneale's original assessment) and is probably less than the estimate of 50 used in the 1972 BEIR report.[23]

DURATION OF RISK

In an early report of the Oxford survey,[34] the increased risk of cancer after *in utero* irradiation appeared to extend throughout the period of followup, which was up to the age of 10 yr. On this basis, the 1972 BEIR report assumed a 10-yr duration of risk, beginning at birth. In a recent report from the Oxford group,[1] the duration of risk has been reevaluated for a much larger sample, with the followup extended through the age of 15 and with the data adjusted for the year of birth. From this analysis, the increased risk for hematopoietic tumors appears to extend throughout the first 12 yr of life (ages 0-11), but not beyond (see Table IV in Bithell and Stewart[1]). For solid tumors, the risk appears to extend throughout the period of followup (ages 0-14), but it is reduced during the later years (ages 9-14). It must be noted that for ages 10-14 the yearly incidence of hematopoietic tumors was about half and the yearly incidence of solid tumors was about one-fourth the incidences of these tumors for ages 0-9.

TABLE A-35 Basis of Risk Estimates for Cancer[a] after Fetal X-Irradiation[b]

Study Population	Duration of Radiation Exposure	Period of Increased Risk after Irradiation, yr	Dose, rads		Age at Irradiation		Relative Risk (O/E)[b]	RBE	Increase in Relative Risk per rem, %
			Range, External	Mean, to Tissues[c]	Range	Mean			
Fetuses England, 1943–1967, 1st-trimester exposure	Minutes	0–12	0.5–2.0	1.62	3 mo	Fetal, 1st trimester	5.0	1	250
Fetuses England, 1943–1967, 2nd- or 3rd-trimester exposure	Minutes	0–12	0.5–2.0	0.73	6 mo	Fetal, 2nd or 3rd trimester	1.47	1	64

[a] Insufficient data are available to separate leukemia and other cancers by trimester of exposure.

[b] Data from Kneale and Stewart.[16,17] This was a retrospective study involving the entire British child population. A control was matched to each childhood cancer death below age 15. The relative risk of cancer is r_1/r_2, where r_1 and r_2 are the relative proportions (among cases versus among controls) of exposed and unexposed subjects, respectively. There were 43 case and 9 control first-trimester exposures, and 1,090 case and 779 control second- or third-trimester exposures, among 10,528 case-control pairs, for r_1 values of 4.8 and 1.40. According to the earlier report,[16] there were 8,934 case and 9,403 control unexposed subjects among 10,519 case-control pairs, giving r_2 the value 0.95.

[c] Based on estimated dose of 0.34 rad/film[41] and estimated 4.78 films for first-trimester exposure and 2.16 films for second- or third-trimester exposure. 1

MULTIPLE RISK FACTORS AND SUSCEPTIBLE CHILDREN

Gibson *et al.*[8] showed that the leukemia risk for the children in the tristate survey was significantly greater among children exposed to combinations of four factors (preconception x-ray exposure, intrauterine exposure, childhood viral infections, and previous maternal miscarriages and still-births) than among children exposed to only one factor. They postulated that the history of viral infections or maternal reproductive wastage might be an indicator of a group of children who were especially susceptible to the leukemogenic effect of irradiation. Bross and Natarajan[4] identified the children in the tristate survey with a history of allergy, and to a lesser extent the children with a history of infection, as susceptible children who had a much greater risk of developing leukemia than nonsusceptible children if exposed to radiation *in utero.* They indicated a relative risk of 8.4 for allergic children exposed *in utero,* compared with unexposed non-susceptible children. Smith and Pike,[33] using information on group size supplied to them by Bross and Natarajan, showed that the difference between the relative risk of leukemia after exposure among the susceptible children and the relative risk among the nonsusceptible children in no instance even approached statistical significance. Mole[22] further pointed out that this study was based on the entire group of children in the tristate survey whose mothers were exposed to x rays during the pregnancy, rather than the 30% of that group whose mothers had abdominal exposure.[9] It is probable that for most of the remaining 70% the fetal dose was vanishingly small.

Bross and Natarajan[3] recently estimated that for about 1% of persons exposed to radiation there is a 50-fold increase in the risk of leukemia. These values were chosen in order to make estimated frequencies under a hypothesized model fit observed frequencies of leukemia and "indicator" diseases for children in the tristate survey. Land[18] and Oppenheim[25] criticized this paper on the grounds that the precision of these values is low, inasmuch as a wide range of values will fit the data equally well under this model. Oppenheim[25] also noted that some of the parameters on which the model is based have a low precision and that the validity of the model is questionable, in that it requires the assumption that, in the absence of radiation exposure, the probability of an indicator disease and the probability of leukemia are statistically independent, and this is not supported by the evidence presented. We observe further that, despite the authors' assertion that radiation during pregnancy included radiation only to the trunk or to the fetus, the number of observed subjects and the value 0.289 for the proportion of x-rayed mothers correspond with the total number of mothers exposed during pregnancy in the tristate survey, rather than the much smaller number of mothers who had abdominal exposures.[9]

On the basis of these criticisms, the Committee does not support the view of Bross and Natarajan.[3]

SUMMARY

Although an association between intrauterine irradiation and childhood cancer has been established, the results might be to some extent attributable to factors involved in the selection process. Some of the strongest support for a causal relationship is provided by twin data from the Oxford survey, in which the increased risk of cancer persisted despite the high exposure rate. Weak support for a selection effect is provided by studies showing a relationship between prenatal irradiation and neonatal health, on the one hand, and studies suggesting a relationship between neonatal health and later development of clinically detectable cancer, on the other hand. Further investigation is required here.

The number of extra cancer deaths per person-rad of intrauterine exposure is probably somewhat smaller than originally estimated by Stewart and Kneale, inasmuch as Mantel-Haenszel analysis reveals a weakening of the relationship between the number of films and the incidence of cancer, presumably resulting from the adjustment for trimester of exposure.

On the basis of recent (unadjusted) data from the Oxford survey, the best estimates of the relative risk of cancer after *in utero* irradiation are 5.0 for first-trimester exposure and 1.47 for second- or third-trimester exposure. The period of increased risk appears to begin at birth and last for 12 yr for hematopoietic tumors and about 10 yr for solid tumors, with parallel risk coefficients of 25 excess fatal leukemias per million children per year per rad and 28 excess fatal cancers of other types.

REFERENCES

1. Bithell, J. F., and A. M. Stewart. Prenatal irradiation and childhood malignancy. A review of British data from the Oxford survey. Brit. J. Cancer 31:271–287, 1975.
2. Borell, U., and I. Fernstrom. Radiologic pelvimetry. Acta Radiol. Suppl. 191:10, 1960.
3. Bross, I. D. J., and N. Natarajan. Genetic damage from diagnostic radiation. JAMA 237:2399–2401, 1977.
4. Bross, I. D. J., and N. Natarajan. Leukemia from low-level radiation. Identification of susceptible children. N. Engl. J. Med. 287:107–110, 1972.
5. Burch, P. R. J. The Biology of Cancer. A New Approach, pp. 294–295. Lancaster, England: MTP Press Ltd., 1976.
6. Cahill, D. F., J. F. Wright, J. H. Godbold, J. M. Ward, J. W. Laskey, and E. A. Tompkins. Neoplastic and life-span effects of chronic exposure to tritium. II. Rats exposed in utero. J. Natl. Cancer Inst. 55:1165, 1975.
7. Diamond, E. L., H. Schmerler, and A. M. Lilienfeld. The relationship of intrauterine

radiation to subsequent mortality and development of leukemia in children. Amer. J. Epidemiol. 97:283–313, 1973.

8. Gibson, R. W., I. D. J. Bross, S. Graham, A. M. Lilienfeld, L. M. Schuman, M. L. Levin, and J. E. Dowd. Leukemia in children exposed to multiple risk factors. N. Engl. J. Med. 279:906–909, 1968.

9. Graham, S., M. L. Levin, A. M. Lilienfeld, L. M. Schuman, R. Gibson, J. E. Dowd, and L. Hempelmann. Preconception, intrauterine and postnatal irradiation as related to leukemia. Natl. Cancer Inst. Monogr. 19:347–371, 1966.

10. Hewitt, D., J. C. Lashof, and A. M. Stewart. Childhood cancer in twins. Cancer 19:157–161, 1966.

11. Holford, R. M. The relation between juvenile cancer and obstetric radiography. Health Phys. 28:153–156, 1975.

12. Hoshino, T., H. Kato, S. C. Finch, and Z. Hrubec. Leukemia in offspring of atomic bomb survivors. Blood 30:719–730, 1967.

13. Huggins, C., and R. Fukunishi. Cancer in the rat after single exposures to irradiation or hydrocarbons. Radiat. Res. 20:493, 1963.

14. Jablon, S., and H. Kato. Childhood cancer in relation to prenatal exposure to atomic-bomb radiation. Lancet 2:1000–1003, 1970.

15. Kneale, G. W. Excess sensitivity of pre-leukemics to pneumonia. A model situation for studying the interaction of infectious disease with cancer. Brit. J. Prev. Soc. Med. 25:152, 1971.

16. Kneale, G. W., and A. M. Stewart. Mantel-Haenszel analysis of Oxford data. I. Independent effects of several birth factors including fetal irradiation. J. Natl. Cancer Inst. 56:879–883, 1976.

17. Kneale, G. W., and A. M. Stewart. Mantel-Haenszel analysis of Oxford data. II. Independent effects of fetal irradiation subfactors. J. Natl. Cancer Inst. 57:1009–1014, 1976.

18. Land, C. E. Genetic damage from diagnostic radiation. JAMA 238:1023–1024, 1977.

19. Lilienfeld, A. M. Epidemiological studies of the leukemogenic effects of radiation. Yale J. Biol. Med. 39:143–164, 1966.

20. MacMahon, B. Prenatal x-ray exposure and childhood cancer. J. Natl. Cancer Inst. 28:1178–1191, 1962.

21. MacMahon, B., and G. B. Hutchison. Prenatal x-ray and childhood cancer. A review. Acta Unio Contra Cancrum 20:1172–1174, 1964.

22. Mole, R. H. Antenatal irradiation and childhood cancer. Causation or coincidence? Brit. J. Cancer 30:199–208, 1974.

23. National Research Council, Advisory Committee on the Biological Effects of Ionizing Radiations. The Effects on Populations of Exposure to Low Levels of Ionizing Radiation. Washington, D.C.: National Academy of Sciences, 1972.

24. Newcombe, H. B., and J. F. McGregor. Childhood cancer following obstetric radiography. Lancet 2:1151–1152, 1971.

25. Oppenheim, B. E. Genetic damage from diagnostic radiation. JAMA 238:1024, 1977.

26. Oppenheim, B. E. Memorandum to BEIR Subcommittee on Somatic Effects, January 25, 1978.

27. Oppenheim, B. E., M. L. Griem, and P. Meier. Effects of low-dose prenatal irradiation in humans. Analysis of Chicago lying-in data and comparison with other studies. Radiat. Res. 57:508–544, 1974.

28. Reincke, U., E. Statz, and G. Wegner. Tumoren nach einmaliger Rontgenbestrahlung weisser Ratten in verschiedenem Lebensalter. Z. Krebsforsch. 66:165, 1964.

29. Rugh, R., L. Duhamel, and L. Skaredoff. Relation of embryonic and fetal x-irradiation

to life time average weights and tumor incidence in mice. Proc. Soc. Exp. Biol. Med. 121:714, 1966.

30. Russell, J. G. B., and B. Richards. A review of pelvimetry data. Brit. J. Radiol. 44:780-784, 1971.

31. Shore, F. J., J. S. Robertson, and J. L. Bateman. Childhood cancer following obstetric radiography. Health Phys. 24:258, 1973.

32. Sikov, M. R., D. D. Mahlum, and W. J. Clarke. Effect of age on carcinogenicity of ^{131}I in the rat. Interim report, pp. 25-32. In C. L. Sanders, R. H. Busch, J. E. Ballou, and D. D. Mahlum, Eds. Radionuclide Carcinogenesis. Proceedings of the Twelfth Annual Hanford Biology Symposium. USAEC Division of Tech. Inform. Serv. CONF-720505. 1973.

33. Smith, P. G., M. C. Pike, and L. D. Hamilton. Multiple factors in leukaemogenesis. Brit. Med. J. 2:482-483, 1973.

34. Stewart, A. Aetiology of childhood malignancies. Brit. Med. J. 1:452, 1961.

35. Stewart, A. M. Cancer as a cause of abortions and stillbirths. The effect of these early deaths on the recognition of radiogenic leukemias. Brit. J. Cancer 27:465-472, 1973.

36. Stewart, A. M. Factors controlling the recognition of leukemia and childhood cancers, pp. 16-24. In Health Physics in the Healing Arts. Seventh Midyear Topical Symposium. U.S. Department of Health, Education, and Welfare Publication (FDA) 73-8029. 1973.

37. Stewart, A. Gene-selection theory of cancer causation. Lancet 1:923, 1970.

38. Stewart, A. Infant leukemia and cot deaths. Brit. Med. J. 2:605, 1975.

39. Stewart, A. The carcinogenic effects of low level radiation. A re-appraisal of epidemiologists methods and observations. Health Phys. 24:223-240, 1973.

40. Stewart, A., and G. W. Kneale. Changes in the cancer risk associated with obstetric radiography. Lancet 1:104-107, 1968.

41. Stewart, A., and G. W. Kneale. Radiation dose effects in relation to obstetric x-rays and childhood cancers. Lancet 1:1185-1188, 1970.

42. Stewart, A., J. Webb, and D. Hewitt. A survey of childhood malignancies. Brit. Med. J. 1:1495-1508, 1958.

43. Upton, A. C., J. W. Conklin, and R. A. Popp. Influence of age at irradiation on susceptibility to radiation-induced life-shortening in RF mice, pp. 337-344. In P. J. Lindop and G. A. Sacher, Eds. Radiation and Aging. Proceedings of a Colloquium Held in Semmering, Austria. London: Taylor & Francis Ltd., 1966.

44. Upton, A. C., T. T. Odell, Jr., and E. P. Sniffen. Influence of age at time of irradiation on induction of leukemia and ovarian tumors in RF mice. Proc. Soc. Exp. Biol. Med. 104:769-772, 1960.

45. Walinder, G., and A. Sjoden. Late effects of irradiation on the thyroid gland in mice. Acta Radiol. Ther. Phys. Biol. 12:201, 1973.

REVIEW AND ANALYSIS OF
SELECTED STUDIES ON RECORD

MANCUSO, STEWART, AND KNEALE

Mancuso, Stewart, and Kneale[40,47] have reported finding dose-related excess cancer mortality among occupationally exposed workers, monitored with radiation badges, at the Hanford works in Richland, Wash. Their risk estimates are much higher than estimates derived from studies of the Japanese atomic-bomb survivors and the populations exposed to ionizing radiation for medical reasons. A proportional-mortality analysis of death certificates for 1,336 "nonexposed" and 2,184 "exposed" male workers who died between 1944 and 1972 and who had been employed at Hanford some time after 1943 found statistically significant associations between cumulative radiation-badge dose and cancer mortality, particularly mortality from cancers of the lung, pancreas, and bone marrow. Estimated doubling doses corresponded to increases in cancer risk, per rad, of 8% for cancer in general, 14% for pancreatic cancer, 16% for lung cancer, 40% for lymphatic and hematopoietic cancers as a group, and 125% for "bone-marrow" cancers (myeloid leukemia and multiple myeloma, considered as a group).[47] Another analysis of 4,033 deaths among badge-monitored male and female workers, some as late as 1977, gave doubling-dose estimates corresponding to cancer risk increases of 3% per rad for male workers and 12% per rad for female workers.[40] Estimates for particular sites were 6% per rad for pancreas and stomach considered as a group, 7% per rad for lung, and 28% per rad for bone-marrow cancers.

The estimates from these two studies are markedly higher than those obtained from studies of the Japanese atomic-bomb survivors and medically exposed populations. The 1972 BEIR report based its estimates of risk to the U.S. population exposed at the age of 10 or older on relative-risk increases of 2% and 0.2% per rad for mortality from leukemia and from all other cancers, respectively.[64] The position taken by Mancuso *et al.* is that their risk estimates are based directly on data from a population occupationally exposed to highly fractionated, low-dose radiation and should therefore take precedence over extrapolated estimates obtained by studying populations given acute, high-dose exposures.[40]

Other analyses of mortality data from the Hanford worker population have been made. Sanders[76] found the longevity of exposed workers to be higher than that of their identified siblings, whose longevity was higher than that of nonexposed workers. Among exposed workers, cumulative badge doses at the time of death for cancer victims during the period 1944–1972 tended to be no higher than those corresponding to other causes of death or those of survivors in the same year as the death of a cancer victim or in one or two years before that death. Sanders's analysis suggests that, if there is a cancer effect, it is seen only after adjustment for the so-called "healthy-worker" effect.

Milham, in a general study of occupational mortality in Washington State, found increased proportional mortality from cancer in deaths occurring locally among Hanford workers;[52] multiple myeloma and cancers of the pancreas and colon were singled out for mention, but significant differences were not found in the small set of observations.

Marks, Gilbert, and Breitenstein[50] and Gilbert and Marks[29] analyzed mortality data on 20,842 white males hired at Hanford before 1966. Before the cutoff date of April 1, 1974, there were 2,089 deaths among 13,075 workers employed for 2 yr or more and 1,905 deaths among 7,767 workers employed for shorter periods. Their most sensitive analysis compared mortality rates from various causes, with respect to cumulative badge dose adjusted for age, occupation, and calendar time, among 7,729 white males who were employed at Hanford for at least 2 yr and whose employment extended beyond January 1, 1960. This group included 837 deaths and all but 77 of the 2,778 men with recorded total doses of over 5 rems.* The distributions, with respect to radiation dose, of death rates from cancer (171 deaths) and lung cancer (58 deaths) were virtually flat and failed to suggest any dose relationship. Rates for cancer of the pancreas (14 deaths) and multiple myeloma (four deaths) increased with increasing dose; these were the only statistically significant associations found. Linear-regression estimates of excess mortality per 10^6 PY per rem

---

* Mancuso *et al.* used "rad" and Marks *et al.* used "rem" as units for identical dose information, but without discussion of radiation quality.

computed from Gilbert and Marks's Tables 7 and 8[29] are 10 ± 8 deaths for cancer of the pancreas and 7 ± 3 for multiple myeloma, but 2 ± 43 and 2 ± 42 for all cancers and cancer of the lung, respectively.

In contrast with the cohort approach of Marks et al.,[29,50] Hutchison et al.[35] attempted to duplicate the proportional-mortality analysis of Mancuso, Stewart, and Kneale[47] while adjusting for the more obvious sources of potential bias, namely, the demographic differences between exposed and nonexposed workers and the associations between cumulative radiation dose and calendar time and between cancer risk, calendar time, and age at death. Their analysis of essentially the same deaths among exposed workers found statistically significant associations between cumulative dose and proportional mortality due to multiple myeloma (eight deaths) and cancer of the pancreas (32 deaths), but not for all cancer (449 deaths) or for other cancer sites. In particular, associations were *not* found for lung cancer (214 deaths), myeloid leukemia (six deaths), lymphatic leukemia (two deaths), or lymphoma (28 deaths). In the case of pancreatic cancer, the association hinged on five of the 32 exposed cases with cumulative doses of over 10 rads, compared with the 1.4 expected by internal consistency, assuming no dose effect. For multiple myeloma, three of eight cases had doses of over 10 rads, compared with the 0.4 expected. A statistically significant association of dose with proportional mortality for all cancers as a group was found only by investigating dose at a series of intervals before death. At death and at 5, 10, and 15 yr before death, the p values for an association with increasing values of dose were 0.13, 0.22, 0.04, and 0.07, respectively. Estimated doubling doses computed according to the method used by Mancuso, Stewart, and Kneale, but adjusted for age and year of death, corresponded to a 0.9% increase in cancer per rad of cumulative dose at death, but 67% per rad for multiple myeloma and 20% per rad for pancreatic cancer.

The adjusted analyses of Gilbert and Marks[29] and Hutchison et al.[35] failed to find statistically significant associations between dose and mortality from all cancers as a group and from lung cancer, but the original findings of Mancuso, Stewart, and Kneale with respect to multiple myelomas and pancreatic cancer were confirmed. That is, the dose relationships for these two cancers could not be attributed to the confounding of dose with age at death, with date of death, or with broad occupational classification. As in the original analysis,[47] the risk estimates for multiple myeloma and pancreatic cancer were extremely high—so high that they can be discounted on logical grounds; such high estimates imply an improbably large causal role for background radiation in the etiology of these diseases among the general population.

It is highly relevant to note that, if the risk estimates for multiple myeloma and pancreatic cancer were not extremely high, they would not satisfy conventional requirements for statistical significance. This necessary numerical relationship is a consequence of the limited sample size and low individual radiation doses of the Hanford workers. Compared with the great majority of studies of irradiated populations, the Hanford study is distinctly lacking in statistical power. That is, assuming the conventional estimates to be representative of the true risks of radiation-induced cancer, the Hanford study could be expected to yield risk estimates that are negative with probability around 40%, positive but statistically non-significant estimates with probability around 50%, and statistically significant but highly exaggerated estimates with probability around 10%. Thus, the low statistical power of the Hanford study, according to conventional risk estimates, detracts considerably from the challenge posed by the study's results and from the validity of the estimates obtained by Mancuso, Stewart, and Kneale.

Other observations support the interpretation of the Hanford study results as small-sample phenomena. The emergence of multiple myeloma and pancreatic cancer (but not myeloid or lymphocytic leukemia) as the cancers most closely related to radiation, the observed (nonsignificant) negative associations of dose with the lymphomas, lymphocytic leukemia, and stomach cancer in the first analysis[47] and with myeloid leukemia in the second,[40] and the fact that the risk estimates obtained in the second, expanded analysis were lower than those obtained in the first are all consistent with great statistical instability.

Published criticisms of the Hanford study findings have suggested alternative explanations for the observed dose associations, including confounding of radiation exposure with exposures to other carcinogens and inadequate dosimetry.[29,35,50] Only further study can determine the validity of these suggestions. Further followup of the Hanford workers and of other groups occupationally exposed to similar quantities of highly fractionated radiation may eventually tell us whether the risks and the spectrum of affected cancer sites differ markedly from what would be expected from studies of more heavily exposed populations. At present, however, there seems to be no reason to abandon the body of epidemiologic evidence on radiation-induced cancer that, although based on greater exposures, yields consistent and statistically stable estimates.

BROSS

Irwin D. J. Bross has challenged the adequacy of low-dose risk estimates extrapolated from observed excess risks in populations exposed to radiation doses above 100 rads, claiming that new analyses of data from the Hanford study[47] and the tristate leukemia survey[8,9] have shown that the risks of radiation-induced cancer from doses of around 1 rad are an order of magnitude greater than previously predicted.[5,6]

The Mancuso-Stewart-Kneale analyses of mortality data from employees of the Hanford works[40,47] do not appear to support Bross's claim. In particular, the Mancuso-Stewart-Kneale analyses did not find a dose-related excess of leukemia, whereas the analyses by Bross *et al.* deal only with leukemia case-control data.[8,9]

Bross's claim rests on analyses according to an unconventional model, inspired by a series of analytic studies of the case-control data gathered by the 1960–1962 tristate leukemia survey.[10,11,27,63] In the model, the leukemia dose response is determined by the unknown composition of the irradiated population with respect to subgroups of varied susceptibility to radiation,[8,9] and not, as in other models, by theoretical mechanisms of radiation damage to cellular material.[12,38] No relationship is assumed among risks at different doses, in marked contrast with the regression approach to risk estimation.

The basic response variable according to the Bross model is the proportion of the irradiated population "affected by radiation." Those "affected" have an increased relative risk of leukemia that must be estimated from the data and that, possibly for simplicity, is assumed to be independent of dose. Particular applications of the model have assumed increased relative risks for some infectious diseases and allergies among "affected" children irradiated *in utero*[9] and for heart disease among affected adults with histories of diagnostic x-ray exposure.[8] The additional information on the "proportions affected" yielded by data on diseases other than leukemia, as obtained from interviews of leukemia patients and population controls, is considered by Bross and co-workers to be a particular advantage of their method.[6,8,9]

The tristate leukemia survey was a case-control study, in which inferences about the relation of x-ray exposure to leukemia risk rested on a comparison of the distribution of past exposures among the leukemia patients identified during the 3-yr period 1960–1962 with that among a population random sample of similar ages. Such data sets are usually analyzed by contingency-table methods, in which a test of nonhomogeneity among exposure classes is combined with estimates of risk relative to one of the classes, usually that with the lowest exposure.[24] Enhanced power against a specific kind of nonhomogeneity, such as a linearly increasing trend in risk with increasing dose, can be obtained by regression methods.[15,16,49]

However, the "proportion affected" of the Bross model is an absolute, rather than relative, measure of exposure effect and requires additional parameters representing the population distribution with respect to exposure, which must be estimated from the random-sample data. The underlying population risks of leukemia and the other response variables (e.g., heart disease and childhood infectious diseases) must be obtained from age-specific population rates, if available, or estimated on the basis of interview data.

The number of free parameters that must be estimated from the data is considerably greater for the Bross method than for more conventional approaches. Even with the additional information provided by data on diseases other than leukemia, the number of parameters is so large that useful estimates of the "proportions affected by radiation" are difficult to obtain. For example, Bross et al. [7,8] have reanalyzed the tristate leukemia survey case-control data for men aged 65 or older, using five exposure classes based on weighted numbers of reported diagnostic x-ray exposures, with estimated average radiation doses, and using presence or absence of a history of heart disease as another response variable. In addition to five parameters representing the "proportions affected" in each dose class, other parameters estimated from the data included the population numbers in each of the five classes, the relative risks of leukemia and heart disease in those "affected," the population rates for nonlymphatic leukemia and heart disease, and an "age-adjustment factor" to allow for the fact that the random sample of population controls was not stratified by age.

The analysis by Bross et al. obtained estimates, with confidence limits, for each of the five "proportions affected," but by a two-stage procedure in which the other parameters were first estimated from the data and the "proportions affected" were then fitted by minimal chi-square, with the values of the other parameters assumed to be known constants. Because these other parameters are, of course, not constants, the net effect of this procedure is a serious underestimation of the error variance. Including even a few of the other parameters in the fitting process with the "proportions affected," so that their statistical variation would be reflected in the confidence limits obtained, increased the length of the confidence intervals to such an extent that the estimates of effect (the "proportions affected") no longer appeared to be useful. [4] By contrast, the original analysis of these data by conventional methods showed a statistically significant association of leukemia with number of reported x-ray examinations. [28]

In another analysis of childhood leukemia and fetal x-ray exposure, Bross and Natarajan estimated that 1% of irradiated fetuses were "affected" by diagnostic exposures while *in utero* and that, among those affected, the relative risk of leukemia was 50 and the relative risk of a group of other diseases was 5.[9] A reanalysis by Land showed that the uncertainty of these estimates was very great and that, in fact, an improved fit to the data was obtained when the "proportion affected" among the exposed children was estimated to be so small as to include only children who later contracted childhood leukemia.[42]

It is doubtful that the designation "proportion affected by radiation" is appropriate in the Bross model. In their analysis of adult leukemia data, Bross and co-workers[7,8] claimed to have demonstrated an unexpectedly large radiation effect at doses in the 1-rad range, on the grounds that the estimated "proportion affected" did not approach zero as the estimated dose decreased to near zero (the smallest average dose was 0.4 rad; the analysis did not include a group with negative histories of diagnostic x-ray exposure). A more reasonable interpretation of their results, if accepted as presented, is that only the *increase* in the "proportion affected" with increasing dose should be attributed to diagnostic x-ray exposure. According to this interpretation, the effect per rad of x rays in the 1-rad range was shown to be *less* than that in the range of 10–100 rads.

In contrast with the conclusions of the original tristate leukemia survey investigators,[28,32,33] the analyses by Bross *et al.* ignore serious potential biases, such as the possibility, recognized by Stewart,[82] that early leukemia and preleukemic states might lead to increased exposure to diagnostic x rays in the years immediately before the clinical appearance of leukemia or the possibility that the clinical workup of patients with diagnosed leukemia might lead to greater ascertainment of existing heart disease.[4]

It is of some interest that the new statistical method of Bross *et al.* apparently has never been published in a journal devoted to statistical methods. The "susceptible subgroup" model, although it may contain some grain of truth, nevertheless imposes so little structure on the inferences possible from analyses of dose-response data that it is unlikely that usable estimates can be obtained with it from available data. The applications by Bross *et al.* have been clearly incorrect, and they provide no evidence that the risk of cancer from low-dose radiation is greater than indicated by conventional estimates.

NAJARIAN AND COLTON

The report by Najarian and Colton[62] was based on interviews with next of kin for 525 (of 1,722) certified deaths at ages under 80 among former workers at the Portsmouth Naval Shipyard in New Hampshire. Next of kin were asked whether the deceased had worked with radiation or had worn radiation badges; for 146, the answer was "yes" or "probably yes," and for the remaining 379, the answer was "no" or "don't know." For leukemia deaths, positive answers were received for six of eight; for other lymphatic and hematopoietic neoplasms, four of 10; and for other cancers, 46 of 126. Relative risks adjusted for numbers expected according to population rates were 7.6 for leukemia ($p < 0.01$), 2.4 for other lymphatic and hematopoietic neoplasms, and 1.1 for other cancers.

After the publication of their report, the authors were provided by the National Institute of Occupational Safety and Health with employment and radiation-exposure records from the Portsmouth Naval Shipyard for the 1,722 names in the original collection of death certificates. A preliminary analysis of this new information was presented by Colton to the 1979 Annual Meeting of the Society for Epidemiologic Research (New Haven, Connecticut, June 13, 1979). This was a proportional-mortality analysis limited to 354 deaths among badge-monitored nuclear workers who died in 1961 or later, between the ages of 35 and 80. Causes of death considered were leukemia, all hematologic cancers, and total cancers.

The analysis revealed that the decedents whose next of kin were contacted in the original study did not constitute a representative sample of those actually exposed. In particular, it was more likely that the next of kin would be contacted, and that the decedent would be correctly identified as a nuclear worker, for exposed workers who died of cancer, compared with those who died of other causes. The extent of this bias is revealed by a comparison between the numbers of deaths observed and expected (according to population rates) among decedents identified as nuclear workers by next of kin and by Portsmouth Naval Shipyard records, by cause of death:

	Identification as Nuclear Worker by					
Cause of Death	Next of Kin (Najarian and Colton[62])			Portsmouth Naval Shipyard Records		
	Observed	Expected*	Ratio	Observed	Expected*	Ratio
Leukemia	6	1.1	5.5	4	2.7	1.5
All hematologic cancers	10	2.9	3.4	9	7.1	1.3
All cancers	56	31.5	1.8	99	74.7	1.3

* Computed from population rates based on the distribution by age at death of the decedents identified as nuclear workers.

The observed: expected ratios according to identification from shipyard records are not too different from what might be expected in an employed population (the so-called "healthy-worker effect"[19,51]).

Proportional mortality from all malignancies was not related to cumulative badge dose. A dose response was claimed for hematologic cancers, however. From the distribution by dose of observed and expected hematologic-cancer deaths and that of all deaths among badge-monitored workers, it appears that this conclusion could only be based on the contrast between doses above and below 1 rem:

Deaths from Hematologic Cancer among Nuclear Workers by Cumulative Badge Dose, rems

	0	0.001–0.099	0.100–0.999	1.000–4.999	≥ 5.000	Total
Observed	1	0	2	5	1	9
Expected*	1.2	1.4	2.2	1.6	0.7	7.1
Total	58	73	110	79	34	354

A more objective contrast, based on the average dose for each of the intervals (using approximate coefficients of 0, 0.03, 0.30, 2.0, and 7.0, respectively) yields a test statistic for trend having a p value of 0.14—a value too large even to be considered "suggestive" of a dose relationship by usual statistical criteria.

These successive analyses of proportional mortality among Portsmouth Naval Shipyard workers contribute little to our understanding of health risks from low-level radiation. However, they do provide a remarkable illustration of the dangers of response bias in epidemiologic studies.

STERNGLASS

Ernest J. Sternglass appeared before the Committee to present a number of comments about the effects of low-level radiation on man. Part of Dr. Sternglass's presentation alleged that fallout from Chinese bomb-testing in 1976 led to an increased amount of radioactivity in milk in some areas of the United States. He concluded that there was an increase in infant mortality in the eastern-seaboard states from Delaware to New England shortly after these events—an increase that he ascribed to the radioactivity. Although Dr. Sternglass stated that his analysis was incomplete, the Committee received no further data on this subject. We have concluded

*Computed from population rates based on the distribution by age at death of the decedents identified as nuclear workers.

that the alleged association did not fit the time course for radioisotope movement into the cow-milk food chain; nor was there clear evidence of a universally applicable change in infant-mortality rates. Thus, the Committee did not believe that the allegation was substantiated.

Most of Dr. Sternglass's material was directed at evidence, chiefly from Dr. A. Petkau of Canada, indicating effects of various kinds of radiation at low doses and low dose rates on membranes similar to cell membranes. The Committee contacted Dr. Petkau, who kindly provided reprints of his work, as well as personal comments concerning it. The following material has been developed as a result of consideration of evidence provided by Dr. Sternglass, Dr. Petkau, and others.

The experimentally demonstrated effects of ionizing radiation on cell membranes provide an alternative or conjunctive damage mechanism in addition to effects on DNA, which are generally accepted as the primary modes of damage in biologic systems. Radiation damage to cellular and intracellular membranes is manifested by alterations in permeability, which lead to altered distribution of various intracellular molecules and ions and disruption of membrane-associated biochemical processes.[83] Although it is well recognized that membrane integrity is essential for normal cell function, there is inadequate basic understanding of membrane structure and function on which to base a detailed theory of radiation-induced damage mechanisms.

Attention has recently been drawn to the potential significance of membrane-mediated damage in biologic systems as a result of experimental studies primarily with bilayer lipid membrane models. These studies have revealed that polyunsaturated membrane lipids are subject to oxidative long-chain reactions initiated by radicals and ions that are produced by ionizing radiation; these reactions ultimately lead to alterations in membrane structure. In such systems, it has been found that, as the dose rate is reduced, the dose required to elicit a given degree of alteration is reduced; this suggests a mechanism for damage to biologic systems at low dose rates approaching natural background.[68] Low-dose-rate irradiation is believed to be more effective than higher-dose-rate exposure in causing structural damage in lipid moieties, because slowly progressing long-chain reactions are initiated by ionizing radiation and, once initiated, are sustained in the absence of the radiation. Therefore, for a given dose, the radiation-induced chemical effect should increase with decreasing dose rate, owing to the increased probability of recombination of charged species at higher dose rates. Studies of the effects of x radiation on unsaturated lipid micelles have indicated that oxidative damage is characterized by high yields ($G = 10$–40) and depends heavily on dose rate in such a way that a pronounced increase in oxidative damage is encountered at dose rates below 100 rads/min.[75] Oxidative damage in sodium linoleate preparations is reported to be initiated by radiation-induced hydroxyl radicals,[75] in contrast with oxidation of phospholipid bilayers in which the superoxide anion (O_2^-) is the primary initiator;[68] the difference is attributed to the effect of the electrostatic field of the charged lipid-water membrane interface.[75]

Radical-scavenging agents have been shown to inhibit oxidative lipid damage.[75] Enzymatic dismutation of O_2 with superoxide dismutase has been shown to afford a radioprotective effect on membranes in both *in vitro* and *in vivo* systems.[66,68,70-73] Cysteine has also been reported to protect mycoplasma *Acholeplasma laidlawii* B cells from radiation-induced membrane damage.[67] A quantitative relationship was established between the rate of cell inactivation and the sulfhydryl content of the membrane.[67] Such studies provide indirect evidence of the role of lipid oxidation in membrane damage by ionizing radiation.

Petkau and Chelack[68] measured hydroperoxide formation in model membranes after exposure to x radiation, cesium-137 gamma rays, and natural background radiation at dose rates of 2.6×10^6 and 0.75×10^{-6} rad/min in the presence and absence of bovine superoxide dismutase and other radical-scavengers. Radiation-induced hydroperoxide formation, as measured by changes in membrane absorbance at 232 nm, resulted from all sources of radiation. At background radiation intensities, the membrane alterations exhibited a dose-rate dependence that was similar to that found with higher-intensity x rays or cesium-137 exposures. The radioprotective effect of superoxide dismutase was shown by a delay in the onset of hydroperoxide formation and by the limiting of its extent to a point that was independent of dose rate, but increased with time; this was suggested as being due to autooxidation of unsaturated fatty acids by ground-state oxygen.[68] In contrast with the elimination of the dose-rate effect by superoxide dismutase in model membrane systems exposed to external radiation from cesium-137 gamma photons, the enzyme did not remove the dose-rate dependence in the case of internal radiation from tritium beta particles (A. Petkau, personal communication). The difference in the effect of superoxide dismutase on model membranes between exposure to internal fields and exposure to external fields is unexplained; if this effect occurs in living systems, the effects of low-dose-rate exposures from internal emitters may involve cell-membrane alterations as a more significant damage mechanism than formerly recognized. There is an obvious need for data on the dose-rate dependence of biomembrane alterations induced by internal emitters.

The radioprotective effect of superoxide dismutase on bilayer lipid membrane models suggests that the radiosensitivity of cell membranes may, among other things, depend on the concentration of the enzyme in the cell. The D_o value of white blood cells (granulocytes, lymphocytes, and platelets) in x-irradiated mice has been found to increase with the concentration of endogenous cellular superoxide dismutase. Exogenous intravenous bovine superoxide dismutase had no effect on the D_o of granulocytes or platelets, the cells with the greatest endogenous enzyme concentrations, whereas the D_o of lymphocytes was increased.[69] Superoxide dismutase has been found to afford a protective effect on cells if administered after irradiation at a time when DNA repair is either complete or nearing completion; this suggests radiation-induced membrane damage as an alternative cellular mechanism, in addition to direct effects on DNA.

The role of radiation damage of membranes in the induction of pathologic states in living systems has not been established, although possible connections to carcinogenesis, autoimmune diseases, and aging have been proposed on the basis of the involvement of membrane lipid peroxidation in these disease entities. Malonaldehyde, which is produced during oxidative decomposition of polyunsaturated fatty acids, has been shown to induce skin cancer in mice[79] and to result in microbial mutagenesis.[57] The relationship of these findings to carcinogenesis in humans has not been determined, nor has malonaldehyde been detected as a result of irradiation of cell membranes. An analogue of malonaldehyde has, however, been shown to result from the *in vitro* exposure of DNA to x radiation.[65] The possible involvement of membranes in viral carcinogenesis is suggested by the fact that cell transformation by viruses is accompanied by membrane changes,[55] including increased ionic permeabilities.[74] Radiation-induced lipid hydroperoxidation results in increased ion permeability,[21] which in turn can lead to inhibition of host-cell protein synthesis, thus shifting control of translation to the infecting virus.[13] Although the involvement of radiation membrane damage in the induction of diseases in mammalian systems is still largely speculative, the possible significance of such effects warrants further investigation.

The available data are not adequate to assess the role of radiation damage of membranes in the induction of pathologic states in living systems. There is, however, an extensive literature on the *in vitro* and *in vivo* effects of ionizing radiation on artificial membranes and biomembranes.[83] Of particular significance from the point of view of radiation protection are low-dose or low-dose-rate effects on biomembranes. The results of studies of the effects of ionizing radiation on cell membranes are notably highly variable. Studies of the effects of x radiation on the permeability of erythrocyte membranes to potassium and sodium ions have revealed alterations in active and passive transport mechanisms. However, such effects in general have been found to require doses of about 1 krad or greater.[3,59] Low-dose-rate effects on erythrocytes have not been reported. In contrast with the relatively high radiation doses associated with alterations in the ion permeability of erythrocytes, membrane-permeability changes and metabolic disruptions in lymphocytes exhibit significantly greater radiosensitivity.

In vivo x irradiation of experimental animals has been found to alter mitochondrial structure and function in various tissues and organs, and this leads to suppression of oxidative phosphorylation. Mitochondria irradiated *in vitro* were significantly more resistant to damage—a suggestion of the modifying effects of the cellular environment.[2,86] Cells of radioresistant tissues have, in general, been found to contain relatively larger numbers of mitochondria per cell than cells from more sensitive tissues.[30] The number of mitochondria and their structural and functional integrity are important with regard to initial radiation effects, as well as cellular repair processes.[30] Mitochondria of lymphatic node lymphocytes of rats exposed to x radiation exhibited swelling, fusion of organelles, destruction of cristae, and clarification of the matrix 30 min after irradiation at a dose of 500 R, or within 4–5 min after higher doses.[48]

The synthesis of nuclear adenosine triphosphate (ATP) was inhibited after *in vivo* exposure to x-ray doses of 50–700 R, whereas no inhibition was noted for rat thymus nuclear suspensions exposed *in vitro*. There was no detectable effect on ATP synthesis in *in vivo* exposures at 25 R. The fact that both nuclear ATP and DNA metabolism in the thymus were found to be affected by low-dose x irradiation suggests the connection of these effects with mitotic inhibition.[39] The sensitivity of mitochondria to x-radiation damage was also demonstrated by effects on the intracellular distribution of catalase, a mitochondrion-associated enzyme. Catalase release from mitochondria to cell sap in both epithelial and spindle-cell tumors was detected after *in vivo* exposures to 25 R of x radiation.[31] The effect was attributed to radiation-induced mitochondrial membrane damage, which led to enzyme leakage. No alteration in liver mitochondria enzyme permeability was detected either *in vivo* or *in vitro* after x-ray exposures at doses of up to 9,000 R;[77] this suggests marked variation in the sensitivity of mitochondrial membranes of different tissues to radiation damage.

Alterations in the morphology and motility of human lymphocytes from peripheral blood have been detected after *in vitro* exposure to x rays at doses of 2, 5, and 10 R.[78,81] Statistically significant reductions in the lymphocyte counts in mice have been induced by total-body exposures of 5 and 10 R.[20,36] Significant decreases in the absolute lymphocyte counts in humans have also been reported after a 77-wk exposure at 0.2 R/wk for an average total dose of 16 R.[41]

Exposure of rat thymocytes to x radiation at 50 R leads to interphase cell death within several hours.[60,61] At doses in excess of 100 R, the earliest evidence of radiation damage of thymocytes is the loss of bound sodium and potassium from cell nuclei[17] and the suppression of nuclear oxidative phosphorylation.[18] Later changes include labilization of protein-DNA bonding[34] and separation of histones from DNA.[23] During the first 45–60 min after exposure, intracellular ATP content decreases markedly, with no apparent change in the highly condensed nuclear chromatin structure. Later, alterations in nuclear structure are seen that are related to the release of histones and the disassociation of deoxyribonucleoprotein, accompanied by the release of potassium.[58] Exposure of rat thymocyte suspensions to x radiation revealed that potassium efflux in the dose range of 0.2–4 krads is due to interference with active transport mechanisms, whereas at higher doses (12–20 krads) passive membrane permeability is increased.[14] These results suggest that the radiosensitivity of lymphocytes may be related to intracellular- or plasma-membrane damage, the nature of which depends on the absorbed dose.

The results of studies of the effects of ionizing radiation on plasma and intracellular membranes suggest that mitochondrial membranes of some tissues are sensitive to relatively low radiation doses. The absence of data on the dose-rate dependence of such effects precludes an assessment of the involvement of long-chain lipid peroxidation reactions, which have been shown to result in an increase in radiation damage from low-dose-rate exposures in model membrane systems.

The available data relative to the effects of low-dose or low-dose-rate exposures on carcinogenesis in humans and experimental animals do not, in general, support the hypothesis of an increased probability of induction at low dose rates. Increasing the duration over which a given dose of low-LET radiation is administered, either by decreasing the dose rate or by fractionating the dose, has been generally found to decrease oncogenic effects of ionizing radiation. There are, however, exceptions to this rule, when the incidence of some effects is found to be inversely proportional to dose and/or dose rate or to increased dose fractionation. Although the interaction of the temporal dose distribution with cell repair and cell-killing has been suggested as the basis for responses of this type, the basic mechanism and kinetics of such effects are uncertain; it is therefore not possible to exclude the involvement of radiation-induced membrane damage.

Protracted occupational exposure of radiation workers to maximal yearly doses of 5 rems has been reported to result in a dose-dependent increase in bone marrow cancer and cancer of the pancreas and lung.[47] An increased incidence of malignant thyroid tumors has been associated with x-ray doses of less than 10 rads.[53] In studies of the atomic-bomb survivors of Hiroshima and Nagasaki, excess morbidity[85] and mortality[56] from breast cancer were detected in dose groups as low as 10–39 and 10–49 rads. The mortality rate per rad for cancers of the trachea, bronchus, and lung decreased from 105 per million persons per rad for Hiroshima survivors exposed to 10–49 rads to 10 per million persons per rad after exposure to over 200 rads.[56] The risk of osteosarcoma resulting from the intravenous therapeutic administration of preparations containing radium-224 appears to be higher for a given total radiation dose if the dose is administered over a period of a year, rather than over a period of months.[80]

In an investigation of the effects of gamma-ray dose rate on the induction of neoplasia in mice, it was found that low dose rates were less effective than high dose rates, except for pulmonary adenomas and nonthymic lymphomas. Exposure to 313 rads at 0.0037 rad/min resulted in an increase in the age-specific incidence of lung tumors, compared with a 300-rad exposure at 6.7 rads/min, which decreased the incidence of such tumors.[84] The results of this study are consistent with those of other studies of lymphoid and pulmonary tumorigenesis in mice by protracted gamma irradiation at low dose rates.[44–46,54]

Lesher *et al.*[43] investigated tumor induction in mice as a function of age, sex, and daily dose at dose rates of less than 56 rads/d. The cumulative incidence of thymic lymphoma increased with daily doses up to 32 rads/d and remained constant at higher dose rates. In contrast with these findings, the incidence of pulmonary tumors exceeded control values in mice irradiated at 5 rads/d, but declined with increasing dose in all other dose groups. The increased effectiveness of low-dose-rate exposures, relative to higher exposure rates, in the production of leukemia in beagles continuously exposed to gamma rays has also been reported.[26]

Although a number of factors have been suggested to account for the apparent increase in incidence of specific types of radiation-induced pathologic states in humans and experimental animals with decreasing dose rates, the basic mechanisms of such effects are incompletely understood. The inverse relationship between dose rate and the induction of damage in model membrane systems and the possible relationship of such alterations in biomembranes to carcinogenesis suggest that this phenomenon may be involved in low-dose or low-dose-rate effects in living systems. Thus, there is a need for additional studies in this field.

FRIGERIO, ARCHER, AND OTHERS

A number of recent papers have dealt with the question of whether variations in background radiation can be correlated with differences in cancer rates among populations exposed to them, as a test of whether effects of radiation at low dose rates can be detected. The chief factor modifying background exposure is altitude (see Chapter III), but geomagnetic variation (which would also affect cosmic radiation) has been analyzed.

Frigerio and Stowe[25] found an inverse correlation between background radiation (including added man-made radiation) and cancer-mortality rates for all 50 states. The inverse relationship was observed not only for all cancer but for some cancers of individual sites, including cancers of the gastrointestinal tract, lung, and female breast—but not of the stomach or thyroid and not leukemia. Jacobson et al.[37] analyzed leukemia mortality in more detail and found no significant correlation by state with average background exposure determined by aerial surveys of selected areas. Eckhoff et al.[22] studied leukemia-mortality rates for 5,000 geographic areas in the United States in relation to altitude and reported a substantial increase in mortality up to 2,000 ft (610 m) and a decrease at higher altitudes.

These studies indicate that effects of differences in background radiation on cancer induction must be so small that other factors related to cancer are overwhelming. Let us assume, on the linear hypothesis, that increased risk of total cancer induction by radiation is about 0.5% per rem. The difference between high- and low-background areas is about 70 mrems/yr for regions with sufficient population to provide an adequate test. For a stationary population of all ages exposed for life, the mean cumulative dose difference would be about 4 rems (i.e., an average accumulation at age 60); thus, the difference expected in total cancer rates for all ages would be about 2% between the high- and low-background regions. This is, however, an overestimate because of migration into and out of the area. A difference of 2% or less clearly cannot be detected in the face of numerous other environmental factors that are known to affect cancer rates, with relative-risk values of several hundred percent in exposed subgroups of the population (for example, cigarette-smokers and workers occupationally exposed to carcinogens). Radiation-induced leukemia has a somewhat higher risk, e.g., about a 1.5% increase in risk per rad on the basis of current data; thus, from the same type of analysis we might expect a difference of 6% between high- and low-background exposures. For this particular cancer type, it is likely that the validity of its certification as a cause of death has at least that much variation from area to area, and of course variations from other leukemogenic factors also have effects.

Archer[1] has published an analysis in which geomagnetic variation, as well as altitude, has been taken into account as a factor modifying background from cosmic radiation in the United States. This study indicates a *positive* correlation between the cosmic radiation flux and the neonatal death rates and mortality for some cancers, notably cancer of the kidney and breast cancer in women. In effect, both measures (cosmic radiation flux and death rates) show a progressive decrease from North to South, which accounts for the positive correlation between the two variables. The author estimates that as much as about 40–50% of all cancer may be accounted for by background radiation, in contrast with the estimate of no more than a few percent based on the present Committee's analysis.

In addition to the problems of confounding factors that might account for spurious correlations, as well as problems of geographic variations in accuracy of certification of causes of death, it is noteworthy that the cancer type with the strongest correlation in Archer's analysis, kidney cancer, is not found to be markedly increased in the irradiated populations that have been studied. Moreover, cancer of the female sex organs, exclusive of the cervix, and cancer of the prostate had positive correlations, but in both cases these cancers have not been observed to be significantly increased in irradiated human groups. In short, the pattern of cancer types observed to be related to cosmic radiation in Archer's study is difficult to reconcile with the data at hand on groups exposed to higher radiation doses and currently under study.

We conclude that these types of studies, depending as they do on death-record data aggregated crudely by geographic region, do not constitute a sufficient basis for deciding whether one or another type of environmental factor, such as background radioactivity, is related to cancer rates. Thus, as a test of the effect on cancer risks of low-dose-rate lifetime exposure to radiation, this approach does not appear to be fruitful in the United States within the framework of variations in background-radiation exposure of populations large enough to provide data that would be statistically useful.

REFERENCES

1. Archer, V. E. Geomagnetism, cancer, weather and cosmic radiation. Health Phys. 34:237–247, 1978.

2. Benjamin, T. L., and H. T. Yost. The mechanism of uncoupling of oxidative phosphorylation in rat spleen and liver mitochondria after whole-body irradiation. Radiat. Res. 12:613–625, 1960.

3. Bresciani, F., F. Auricchio, and C. Fiore. A biochemical study of the x-radiation-induced inhibition of sodium transport (Na pump) in human erythrocytes. Radiat. Res. 22:463–477, 1964.

4. Boice, J. D., and C. E. Land. Adult leukemia following diagnostic x-rays? (Review of a report by Bross, Ball, and Falen on a tristate leukemia survey.) Amer. J. Pub. Health 69:137–145, 1979.

5. Bross, I. D. J. Letter to A. W. Hilberg, National Research Council, June 7, 1977 (received July 11, 1977).

6. Bross, I. D. J. Testimony to the U.S. Senate Committee on Commerce Oversight Committee for Radiation Safety and Health, Washington, D.C., June 17, 1977.

7. Bross, I. D. J., M. Ball, and S. Falen. A dosage response curve for the one rad range: Adult risks from diagnostic radiation. Amer. J. Pub. Health 69:130–136, 1979.

8. Bross, I. D. J., M. Ball, T. Rzepko, and R. E. Laws. Preliminary report on radiation and heart disease. J. Med. 9:3–15, 1978.

9. Bross, I. D. J., and N. Natarajan. Genetic damage from diagnostic radiation. J. Amer. Med. Assoc. 237:2399–2401, 1977.

10. Bross, I. D. J., and N. Natarajan. Leukemia from low-level radiation. Identification of susceptible children. N. Engl. J. Med. 287:107–110, 1972.

11. Bross, I. D. J., and N. Natarajan. Risk of leukemia in susceptible children exposed to preconception, in utero and postnatal irradiation. Prev. Med. 3:361–369, 1974.

12. Brown, J. M. The shape of the dose-response curve for radiation carcinogenesis. Extrapolation to low doses. Radiat. Res. 71:34–50, 1977.

13. Carrasco, L., and A. E. Smith. Sodium ions and the shut-off of host cell protein synthesis by picornaviruses. Nature 264:807–809, 1976.

14. Chapman, I. V., and M. C. Sturrock. The effects of x-irradiation *in vitro* on the transport of potassium ions in thymocytes. Int. J. Radiat. Biol. 22:1–9, 1972.

15. Cochran, W. G. Some methods for strengthening the common X^2 tests. Biometrics 10:417–451, 1954.

16. Cox, D. R. Analysis of Binary Data. London: Methuen, 1970.

17. Creasey, W. A. Changes in the sodium and potassium contents of cell nuclei after irradiation. Biochim. Biophys. Acta 38:181–182, 1960.

18. Creasey, W. A., and L. A. Stocken. The effect of ionizing radiation on nuclear phosphorylation in the radio-sensitive tissues of the rat. Biochem. J. 72:519–523, 1959.

19. Decoufle, P., and T. L. Thomas. A methodological investigation of fatal disease risks in a large industrial cohort. J. Occup. Med. 21:107–110, 1979.

20. Dougherty, T. F., and A. White. Pituitary-adrenal cortical control of lymphocyte structure and function as revealed by experimental x-radiation. Endocrinology 39:370–385, 1946.

21. Dravtsov, G. M., L. I. Deev, V. P. Karagedin, and I. B. Kudriashov. Izmemenie Transporta Ovmpv L^+ Cherez Fosfolipidnye Membrany Pri Peroksidatsii, Indutsirovannoi Oblucheniem. Radiobiologiia 16:762–764, 1976.

22. Eckhoff, N. D., J. K. Shultis, R. W. Clack, and E. R. Ramer. Correlation of leukemia mortality rates with altitude in the United States. Health Phys. 27:377–380, 1974.

23. Ernst, H. Radiation-induced early changes in cell nucleus proteins. II. Comparative studies on the concentration of nuclear globulins and histone in lymphatic and parenchymatous tissues. Z. Naturforsch. B 17:300–305, 1962.

24. Fleiss, J. L. Statistical Methods for Rates and Proportions. New York: John Wiley & Sons, 1973.

25. Frigerio, N. A., and R. S. Stowe. Carcinogenic and genetic hazard from background radiation, pp. 385–393. In Biological and Environmental Effects of Low-Level Radiation. Vienna: International Atomic Energy Agency, 1976.

26. Fritz, T. E., W. P. Norris, and D. V. Tolle. Myelogenous leukemia and related proliferative disorders in beagles continuously exposed to ^{60}Co gamma-radiation, pp. 170–188. In R. M. Dutcher and L. Chieco-Bianchi, Eds. Unifying Concepts of Leukemia. Bibliotheca Haematologica No. 39. Basel: Karger, 1973.

27. Gibson, R. W., I. D. J. Bross, S. Graham, A. M. Lilienfeld, L. M. Schuman, M. L. Levin, and J. E. Dowd. Leukemia in children exposed to multiple risk factors. N. Engl. J. Med. 279:906–909, 1968.

28. Gibson, R., S. Graham, A. Lilienfeld, L. Schuman, J. E. Dowd, and M. L. Levin. Irradiation in the epidemiology of leukemia among adults. J. Natl. Cancer Inst. 48:301–311, 1972.

29. Gilbert, E. S., and S. Marks. Cancer mortality in Hanford workers. Radiat. Res. 79:122–148, 1979.

30. Goldfeder, A. Cell structure and radiosensitivity. Trans. N.Y. Acad. Sci. 26:215–241, 1963.

31. Goldfeder, A., J. N. Selig, and L. A. Miller. Radiosensitivity and biological properties of tumours. XI. The effects of x-irradiation on intracellular distribution of catalase. Int. J. Radiat. Biol. 12:13–25, 1967.

32. Graham, S., M. L. Levin, A. M. Lilienfeld, J. E. Dowd, L. M. Schuman, R. Gibson, L. H. Hempelmann, and P. Gerhardt. Methodological problems and design of the tristate leukemia survey. Ann. N.Y. Acad. Sci. 107:557–569, 1963.

33. Graham, S., M. L. Levin, A. Lilienfeld, L. M. Schuman, R. Gibson, J. E. Dowd, and L. H. Hempelmann. Preconception, intrauterine, and postnatal irradiation as related to leukemia. Natl. Cancer Inst. Monogr. 19:347–371, 1966.

34. Hagen, V. Labilization of deoxyribonucleic acid in thymus nucleoprotein after whole-body irradiation. Nature 187:1123–1124, 1960.

35. Hutchison, G. B., B. MacMahon, S. Jablon, and C. E. Land. Review of report by Mancuso, Stewart, and Kneale of radiation exposure of Hanford workers. Health Phys. 37:207–220, 1979.

36. Ingram, M., and W. B. Mason. Effects of acute exposure to x-radiation on the peripheral blood of experimental animals, pp. 58–83. In R. A. Blair, Ed. Biological Effects of External Radiation. New York: McGraw-Hill Co., 1954.

37. Jacobson, A. P., P. A. Plato, and N. A. Frigerio. The role of natural radiations in human leukemogenesis. Amer. J. Pub. Health 66:31–37, 1976.

38. Kellerer, A., and H. H. Rossi. The theory of dual radiation action. Curr. Top. Radiat. Res. Qt. 8:85–158, 1972.

39. Klouwen, H. M., and I. Betel. Radiosensitivity of nuclear ATP synthesis. Int. J. Radiat. Biol. 6:441–461, 1963.

40. Kneale, G. W., A. M. Stewart, and T. F. Mancuso. Re-analysis of data relating to the Hanford study of the cancer risks of radiation workers, pp. 386–412. In Late Biological Effects of Ionizing Radiation. Vol. I. Vienna: International Atomic Energy Agency, 1978.

41. Knowlton, N. P. The Changes in the Blood of Humans Chronically Exposed to Low Level Gamma Radiation, pp. 1–15. Work performed under U.S. Atomic Energy Commission Contract #7405-eng-36. 1948.

42. Land, C. E. Genetic damage from diagnostic radiation. JAMA 238:1023–1024, 1977. (letter)

43. Lesher, S., G. A. Sacher, D. Grahn, K. Hamilton, and A. Sallese. Survival of mice under duration-of-life exposure to gamma rays. II. Pathologic effects. Radiat. Res. 24:239–277, 1965.

44. Lorenz, E. Some biologic effects of long continued irradiation. Amer. J. Roentgenol. Radium Ther. Nucl. Med. 63:176–185, 1950.

45. Lorenz, E., J. W. Hollcroft, E. Miller, C. C. Congdon, and R. Schweisthal. Long-term effects of acute and chronic irradiation in mice. I. Survival and tumor incidence following chronic irradiation of 0.11R per day. J. Natl. Cancer Inst. 15:1049–1058, 1955.

46. Lorenz, E., L. O. Jacobson, W. E. Heston, M. Shimkin, A. B. Eschenbrenner, M. K. Deringer, J. Doniger, and R. Schweisthal. Effects of long continued total body gamma irradiation on mice, guinea pigs, and rabbits. III. Effects on life span, weight, blood pic-

ture, and carcinogenesis and the role of the intensity of radiation, pp. 24–148. In R. E. Zirkle, Ed: Biological Effects of External X and Gamma Radiation. New York: McGraw-Hill, 1954.

47. Mancuso, T. F., A. Stewart, and G. Kneale. Radiation exposures of Hanford workers dying from cancer and other causes. Health Phys. 33:369–385, 1977.

48. Manteifel, V. M., and M. N. Meisel. Early ultrastructural changes in the mitochondria of the lymphocytes of irradiated animals. Radiobiologiia 2:101–104, 1962.

49. Mantel, N. Chi-square tests with one degree of freedom. Extensions of the Mantel-Haenszel procedure. J. Amer. Stat. Assoc. 58:690–700, 1963.

50. Marks, S., E. S. Gilbert, and B. D. Breitenstein. Cancer mortality in Hanford workers, pp. 369–386. In Late Biological Effects of Ionizing Radiation. Vol. 1. Vienna: International Atomic Energy Agency, 1978.

51. McMichael, A. J., S. G. Haynes, and H. A. Tyroler. Observations on the evaluation of occupational mortality data. J. Occup. Med. 17:128–131, 1975.

52. Milham, S., Jr. Occupational Mortality in Washington State, 1950–1971. U.S. Department of Health, Education, and Welfare Publication No. (NIOSH) 76-175. Vols. A, B, and C. Washington, D.C.: U.S. Government Printing Office, 1976.

53. Modan, B. D., D. Baidatz, H. Mart, R. Steinitz, and S. G. Levin. Radiation-induced head and neck tumours. Lancet 1:277–279, 1974.

54. Mole, R. H. Cancer production by chronic exposure to penetrating gamma irradiation. Natl. Cancer Inst. Monogr. 14:271–290, 1964.

55. Montagnier, L., and G. Torpier. Membrane changes in virus transformed cells. Bull. Cancer 63:123–134, 1976.

56. Moriyama, I. M., and H. Kato. Life Span Study Report 7. Mortality Experience of A-Bomb Survivors, 1970–72, 1950–72. Atomic Bomb Casualty Commission Technical Report TR 15-73. Hiroshima: Atomic Bomb Casualty Commission, 1973.

57. Mukai, F. H., and B. D. Goldstein. Mutagenicity of malonaldehyde, a decomposition product of peroxidized polyunsaturated fatty acids. Science 191:868–869, 1976.

58. Myers, D. K. Prevention of pyenosis in rat thymocytes. Exp. Cell Res. 38:354–365, 1965.

59. Myers, D. K., and R. W. Bide. Biochemical effect of x-irradiation on erythrocytes. Radiat. Res. 27:250–263, 1966.

60. Myers, D. K., D. E. DeWolfe, and K. Araki. Effects of x-radiation and of metabolic inhibitors on rat thymocytes in vitro. Can. J. Biochem. Physiol. 40:1535–1552, 1962.

61. Myers, D. K., D. E. DeWolfe, K. Araki, and W. W. Arkinstall. Loss of soluble materials from irradiated thymocytes in vitro. Can. J. Biochem. Physiol. 41:1181–1199, 1963.

62. Najarian, T., and T. Colton. Mortality from leukaemia and cancer in shipyard nuclear workers. Lancet 1:1018–1020, 1978.

63. Natarajan, N., and I. D. J. Bross. Preconception radiation and leukemia. J. Med. (Basel) 4:276–281, 1973.

64. National Research Council, Advisory Committee on the Biological Effects of Ionizing Radiations. The Effects on Populations of Exposure to Low Levels of Ionizing Radiation. Washington, D. C.: National Academy of Sciences, 1972.

65. Payes, B. Enzymatic repair of x-ray-damaged DNA. I. Labeling of the precursor of a malonaldehyde-like material in x-irradiated DNA and the enzymatic excision of labeled lesions. Biochim. Biophys. Acta 366:251–260, 1974.

66. Petkau, A., and W. S. Chelack. Protection of Acholeplasma Laidlawii B by superoxide dismutase. Int. J. Radiat. Biol. 26:421–426, 1974.

67. Petkau, A., and W. S. Chelack. Radioprotection of Acholeplasma Laidlawii B by cysteine. Int. J. Radiat. Biol. 25:321–328, 1974.

68. Petkau, A., and W. S. Chelack. Radioprotective effect of superoxide dismutase on model phospholipid membranes. Biochim. Biophys. Acta 433:445–456, 1976.
69. Petkau, A., W. S. Chelack, and S. D. Pleskach. Protection by superoxide dismutase of white blood cells in x-irradiated mice. Life Sciences. (in press)
70. Petkau, A., W. S. Chelack, and S. D. Pleskach. Protection of post-irradiated mice by superoxide dismutase. Int. J. Radiat. Biol. 29:297–299, 1977.
71. Petkau, A., W. S. Chelack, S. D. Pleskach, B. E. Meeker, and C. M. Brady. Radioprotection of mice by superoxide dismutase. Biochem. Biophys. Res. Commun. 65:886–892, 1975.
72. Petkau, A., K. Kelly, W. S. Chelack, and C. Barefoot. Protective effect of superoxide dismutase on erythrocytes of x-irradiated mice. Biochem. Biophys. Res. Commun. 70:452–458, 1976.
73. Petkau, A., K. Kelley, W. S. Chelack, S. D. Pleskach, C. Barefoot, and B. E. Meeker. Radioprotection of bone marrow stem cells by superoxide dismutase. Biochem. Biophys. Res. Commun. 67:1167–1174, 1975.
74. Ponta, H., K. Altendorf, M. Schweiger, M. Hirsch-Kaufmann, M. Pfennig-Yeh, and P. Herrlich. *E. coli* membranes become permeable to ions following T7-virus-infection. Molec. Gen. Genet. 149:145–150, 1976.
75. Raleigh, J. A., W. Kremers, and B. Gaboury. Dose-rate and oxygen effects in models of lipid membranes. Linoleic acid. Int. J. Radiat. Biol. 31:203–213, 1977.
76. Sanders, B. S. Low-level radiation and cancer deaths. Health Phys. 34:521–538, 1978.
77. Scaife, J. F., and P. Alexander. Inability of x-rays to alter the permeability of mitochondria. Radiat. Res. 15:658–674, 1961.
78. Schrek, R. Qualitative and quantitative reactions of lymphocytes to x-rays. Ann. N.Y. Acad. Sci. 95:839–848, 1961.
79. Shamberger, R. J., T. L. Andreone, and C. E. Willis. Antioxidants and cancer. IV. Initiating activity of malonaldehyde as a carcinogen. J. Natl. Cancer Inst. 53:1771–1773, 1973.
80. Spiess, H., and C. W. Mays. Protraction effect on bone-sarcoma induction of [224]-Ra in children and adults, pp. 437–450. In Radionuclide Carcinogenesis. Proceedings of the Twelfth Annual Hanford Biology Symposium at Richland, Washington, May 10–12, 1972. Washington, D.C.: U.S. Atomic Energy Commission, 1973.
81. Stefani, S., and R. Schrek. Cytotoxic effect of 2 and 5 roentgens on human lymphocytes irradiated *in vitro*. Radiat. Res. 22:126–129, 1964.
82. Stewart, A. An epidemiologist takes a look at radiation risks, p. 37. U.S. Department of Health, Education, and Welfare Publication No. (FDA/BRH) 73-8024. Washington, D.C.: U.S. Department of Health, Education, and Welfare, 1973.
83. Strazhevskaya, N. B., Ed. Radiation damage to cell membranes. In Molecular Radiobiology. Ch. 6. New York: John Wiley & Sons, 1975.
84. Upton, A. C., M. L. Randolph, and J. W. Conklin. Late effects of fast neutrons and gamma-rays in mice as influenced by the dose rate of irradiation. Induction of neoplasia. Radiat. Res. 41:467–491, 1970.
85. Wanebo, C. K., K. G. Johnson, K. Sato, and T. W. Thorslund. Breast cancer after exposure to the atomic bombings of Hiroshima and Nagasaki. N. Engl. J. Med. 279:667–671, 1968.
86. Zicha, B., K. Lejsek, J. Benes, and Z. Dienstbier. Enhancement of oxidation products of lipids in liver mitochondria of whole-body irradiated rats. Experientia 22:712–713, 1966.

VI

Somatic Effects:
Effects Other Than Cancer

SUMMARY

Among the somatic effects of radiation other than cancer, developmental effects on the unborn child are of greatest concern. Exposure of an embryo or fetus to relatively high doses of radiation can cause death, malformation, growth retardation, and functional impairment. Recent information from Hiroshima, most of it published since the 1972 BEIR report, [106] indicates that measurable damage can be produced by doses of 10–19 rads (kerma). The effects of radiation are related to the developmental stage at which exposure occurs, and correspondence has been demonstrated in this respect between man and other mammals. The laboratory data can therefore be used with some confidence to fill in gaps in human experience.

Where developmental effects of radiation can be measured at the cellular level, as in the case of oocyte-killing during fetal or early postnatal stages, thresholds may not be demonstrable. However, most of the perceived abnormalities produced by radiation probably result from damage to more than a single cell. It is therefore unlikely that such effects bear a linear relationship to dose. Threshold doses for some effects have, in fact, already been demonstrated, but these thresholds vary for different abnormalities. For a given total dose, decreases in dose rate generally lead to decreases in developmental effects. Because sensitive stages for many specific abnormalities are relatively short, dose protraction may result in lowering to below the threshold the portion of the total dose that is received during a particular critical period.

477

Acute exposure of the testis to radiation at relatively high doses—much greater than 400 rads—could result in permanent sterility. Impairment of fertility can result from acute exposure of the ovaries to about 400 rads, but this depends, in part, on age. Little is known about the effects of protracted low-dose exposure of the gonads.

For induction of cataract of the lens, there is radiobiologic and clinical evidence of a nonlinear relationship between effect and dose, at least for low-LET radiation. This effect is related to the number of cells killed in the lens. There is little or no risk of inducing such an effect at doses and dose rates approaching those from natural background radiation.

There appear to be no nonspecific effects of radiation at low doses that lead to a shortening of life span, although the existence of specific effects in addition to cancer cannot now be excluded.

EARLY DEVELOPMENT

In comparison with the adult state, the period of early development is characterized by rapid cell proliferation, cell migration, transitions from totipotency to fixed differentiation, and (in part) association with the maternal organism. Some of these attributes are also found in some localities in the adult (e.g., in stem-cell tissues), and there is no sharp demarcation between the developing and fully formed mammal. In examining the effects of ionizing radiation on development, however, this section restricts itself to intrauterine stages (from the time of conception to the time of birth) and to the early postnatal period. Both immediate and long-term effects are reviewed.

The developing organism *in utero* is potentially vulnerable to external radiation that penetrates the maternal tissues; to radionuclides that reach the conceptus after maternal ingestion, inhalation, or injection; and to indirect effects stemming from damage to the mother even when the conceptus is not itself exposed. Alterations that may be produced are morphologic abnormalities, general and local growth retardation, and functional impairments. Although work with experimental mammals has produced evidence of all these effects, it is probably incapable of revealing some of the more subtle functional changes that could be of importance in humans. However, because of the natural variability of human populations and the many other environmental influences that can act during development, it is very difficult to derive information on the effects of low-dose radiation directly from human studies. Risk estimates must therefore be derived largely from experimental data on gross effects.

EVIDENCE FROM EXPERIMENTAL MAMMALS
CONCERNING SENSITIVITY PATTERNS

The developing organism is a dynamic system in which overall, as well as localized, conditions are ever changing with respect to cell size and type, division rate, cycle times, degree of differentiation, and association with other cell types. Nevertheless, a relatively consistent relationship has been found in different mammalian species between the developmental stage exposed to radiation and the general type of effect observed. That is, there are vastly greater similarities between the results of irradiation of different species at equivalent stages than between the results of irradiation of the same species at different stages.

Several major periods can be delineated on the basis of radiation response (Table VI-1). The first is the preimplantation period (cleavage, morula, and blastocyst), when radiation can lead to death of the conceptus shortly after exposure, but concepti that survive appear unimpaired with respect to morphology, size, short- and long-term survival, and reproductive fitness.[21,135,138,141] The quantitative relationship between dose and mortality was shown to be probably related to cell-cycle stage in cleavage;[140] and recent *in vitro* experiments[65] have discovered subtle stage-sensitivity differences within the total preimplantation period with respect to probability and time of induced death. But in no case did embryos survive to a stage that corresponds to more than 2 wk in a human pregnancy; and, in a human situation, their loss would thus probably not be noted, except as an apparent failure to conceive when conception was desired. In the mouse, another effect of preimplantation irradiation has been observed: exposure soon after sperm entry causes sex-chromosome loss, which can result in XO females (Turner's syndrome, in humans). The frequency of this effect is about 4% after an acute dose of 100 R of x rays.[140] Loss of any chromosome other than X or Y probably contributes to early death.

Shortly after implantation, the mammalian embryo begins major organogenesis, when body divisions and basic organ structures are laid down. This period merges without major demarcation into the period of the fetus, during which organogenesis becomes ever more localized and detailed, and the major feature is growth. The event of birth is not a sharp dividing point in this process.

As the embryo implants in the uterus and enters the period of major organogenesis, it becomes abruptly sensitive to the radiation induction of major malformations. Mortality induced by exposure during that period is no longer only of the very early prenatal type, but occurs mainly at birth or

TABLE VI-1   Effects of 100 R of Acute X-Irradiation on Early
Development of Mouse and Rat[a]

	Effects		
	Preimplan-tation Stage	Major Organo-genesis Stage	Fetal (and Early Postnatal) Stage[b]
Days after conception			
Mouse	0–4.5	7.5–12.5	13–20
Rat	0–5.5	8.5–13.5	14–32
Corresponding human stage (approx.)	0–9	14–50	51–280
Class of effect			
Early embryonic death	+ +	+	
Neonatal or early postnatal death	0	+ +	0
Sex-chromosome loss	+	−	−
Gross morphologic malformations	0	+ +	0
Localized morphologic defects or local size reduction	0	+ +	+
CNS defects; behavioral changes	0	+ +	+
Oocyte-killing	0	+ +	+ +
Induction of male sterility	0	+ +	+
Generalized growth retardation	0	+ +	+

[a] Symbols as follows:

+ + 100 R at almost any stage during this period produces effects. At least one stage yields incidence > 25%.

+ Effect observed from treatment of only limited number of stages during this period, and/or incidence < 25%.

0 No effects observed.

− No evidence available.

[b] Early postnatal stages of mouse and rat correspond to human fetal stages.

during infancy. General growth retardation can result and may be temporary or permanent. Irradiation during the fetal period can also produce localized growth retardation, as well as effects on germ-cell populations and on the central nervous system.[16-18,23,45,132,135,136,138,141] It is clear that, although some of the radiation effects will be apparent by the time of birth, other (fertility depression, life-span shortening, neuronal depletion, etc.) find expression later.[16,20,23,41,71,104,133] Among potential delayed effects of embryonic or fetal irradiation that could be of special significance to man are neurophysiologic and behavioral changes.[27,126] However, behavior tests in experimental mammals may have little direct application to the human situation and furthermore are subject to a number of environmental influences whose effects are difficult to distinguish from those of the radiation history (see Brent[17] and Furchtgott[61]).

Results of experiments to study mechanisms of radiation effect on the embryo and fetus have indicated that the maternal organism probably does not play a major intermediary role in the production of most radiation-induced abnormalities.[20-22, 141] The complex chains of processes leading to the finally observed abnormal characteristic may be related by direct cellular descent to the initial developmental effect of radiation, or they may be secondarily caused. In turn, an initial developmental effect results from the initial cellular effect only if the regulatory power of a process is inadequate to take care of a given amount of damage.[141] The initial cellular effect may be cell death (from aneuploidy or other causes), delay in cell division or cell migration, or interference with cell interactions. Although not all the basic mechanisms that can lead to such cellular effects have yet been identified, it is clear from the regular pattern of response observed that somatic mutation (a random process) is relatively unimportant.

The all-or-none effect of radiation during preimplantation stages was explained early by the postulated totipotency of blastomeres.[141] Recent manipulative interferences with early mammalian embryos,[70] such as cell aggregation and blastocyst injection, have amply demonstrated the great developmental plasticity of blastomeres and even of early inner-cell-mass cells and have thus confirmed the original suggestion.

Most animal experiments designed to discover critical periods in development have used relatively high, single, acute doses (100 R or greater). However, once a critical period is established, effects can be demonstrated with considerably smaller exposures. Thus, a specific skeletal change readily showed the effects of 25 R, the lowest dose tried,[137] and mitotic delay in the telencephalon could be demonstrated to have a threshold dose of less than 10 R.[83] In the case of protracted exposures, low daily doses also have produced readily measurable effects, such as reduction in female reproductive capacity after continuous irradiation at a dose rate of 0.0086 R/min (12.4 R/d),[138] various organ-weight reductions after 3 rads/d from tritiated drinking water,[26,29,91] and oocyte depletion with an LD_{50} of only 5 rads during the sensitive period.[45] Different gross abnormalities have been found to follow different dose curves, some with high thresholds;[139,141] but where cellular effects can be directly scored, clear thresholds are sometimes absent.[46]

Protraction of the dose generally diminishes the overall incidence of gross abnormalities,[39,88,89,129,153,171] presumably because less than the threshold dose is received within the duration of many sensitive periods.[138]

Fractionated acute doses are about 1.5 times more effective than continuous irradiation administered during the same intervals.[88]

A question that has been only barely touched on in experimental teratology concerns possible synergistic effects of radiation exposure and other environmental influences. A recent study on mouse embryos has shown that caffeine, at nonteratogenic concentrations, significantly increases the effect of 200 R in producing morphologic abnormalities.[180] Synergisms like this are of obvious importance in deriving risk predictions, but very few experimental results are available on which to base any quantitative estimates.

EXTERNALLY ADMINISTERED INTRAUTERINE IRRADIATION IN HUMANS

Animal experiments have clearly demonstrated the extreme importance of developmental stage, dose, and dose rate in determining the response to *in utero* radiation exposure. Unfortunately, one or more of these factors are usually not accurately known in cases where human concepti have been irradiated. Such cases come from two major sources: medical exposures, particularly during the early part of the century, when hazards were not yet fully appreciated, and particularly therapeutic irradiations; and studies of atomic-bomb survivors in Japan.

The list of human abnormalities reported after *in utero* irradiation is long.[142] It includes microcephaly, mental retardation, growth retardation, hydrocephaly, microphthalmia, coloboma, chorioretinitis, blindness, strabismus, nystagmus, coordination defects, mongolism, spina bifida, skull malformations, cleft palate, ear abnormalities, deformed hands, clubfeet, hypophalangism, and genital deformities. Many of these abnormalities are similar to those observed after treatment of experimental animals; and in a few human cases where stage of irradiation was accurately recorded,[57] the correspondence is remarkable.

Most commonly reported among human abnormalities are microcephaly (often combined with mental retardation), some other central nervous system defects, and growth retardation.[44,63,64,181] The Japanese bomb studies also reported these abnormalities more frequently than any others.[29,100,174,175,179] Microcephaly is particularly associated with exposure during early stages of pregnancy. At Hiroshima, for example, it resulted almost 6 times more frequently when irradiation occurred before the sixteenth week of pregnancy than when it occurred in the second half of pregnancy.[100,174,175] A recent more detailed followup[102] showed a 28% incidence of microcephaly after exposure (all doses combined) at some time during weeks 4–13 of the gestation period, but only a 7% incidence after exposure during the remainder of gestation. For the most sensitive interval, weeks 6-11, the incidence was 11% (2/19) for air doses of 1-9 rads, 17% (4/24) for 10-19 rads, 30% (3/10) for 20-29 rads, 40% (4/10) for 30-49 rads, 70% (7/10) for 50-99 rads, and 100% (7/7) for doses over 100 rads. In the comparable zero-dose group, the frequency was 4% (31/764).[13] Although the 11% incidence for weeks 6-11 in the lowest-dose group is not significantly higher than the 6% incidence for all other stages exposed at that dose, or than the 4% control frequency, it clearly fits in as part of a dose-effect progression for the sensitive stages. In the range of 10-19 rads kerma, the average tissue dose to the fetus is estimated as 5.3 rads gamma plus 0.35 rad neutrons; and in the range of 1-9 rads kerma, as 1.3 rads gamma plus 0.1 rad neutrons.[8,86]

Because some of the affected children observed in the earlier Japanese studies did not appear at the clinics for the followup,[102] it is possible that the actual effects were greater. However, it should be noted that the Nagasaki results showed no significant increase in microcephaly at kerma below 150 rads.[102] Although the total number of intrauterine exposures at Nagasaki was substantially lower than that at Hiroshima (namely, fewer than 20 during sensitive stages at kerma below 150 rads), it is clear that the effect was less in Nagasaki than in Hiroshima (only one case observed versus seven expected if sensitivity was equal to that in Hiroshima). The differences between the cities are probably attributable to the difference in radiation quality; in the range of interest, about 20% of the kerma at Hiroshima was due to neutrons, compared with less than 1% at Nagasaki.

Deleterious effects of *in utero* radiation on body growth are clearly indicated by the Japanese data. About 80% of 1,613 children exposed *in utero* could be followed through the age of 17 (mature growth) by annual examinations.[162,176] Those who were exposed within 1,500 m of the hypocenter of the Hiroshima bomb (average kerma, 25 rads)[7] were, on the average, 2.25 cm shorter, 3 kg lighter, and 1.1 cm smaller in head circumference than those who were at least 3,000 m from the bomb.

Mental retardation was another effect found in the Japanese bomb studies. Owing to the lack of appropriate and sensitive tests for proper overall mental functioning, mental retardation must be relatively severe to be recognized in a clinical situation. In the Japanese children, the diagnosis was applied only if a person was unable to perform simple calculations, to make simple conversation, or to care for himself ("profound" mental retardation), or if he was completely unmanageable or had been institutionalized. The "profound" retardation was often associated with the more severe grades of microcephaly and was not observed below 25 rads kerma of maternal exposure.[14] Other behavioral effects of *in utero* exposure have also been reported—e.g., disturbances of coordination after irradiation during the ninth to twelfth week[154] and retarded motor development after radiation therapy of the mother during the first two trimesters.[147]

It may be questioned why microcephaly and mental retardation figure so prominently among the array of abnormalities attributed to intrauterine irradiation. Does this represent a departure from the animal results? The answer is probably no. Head circumference has not been measured in the rodent experiments; and it would, in fact, be difficult to develop an equivalent measure. Similarly, as noted earlier, no good test to detect "mental retardation" has been developed for mice and rats. Central nervous system damage has been amply demonstrated in experimental mammals[71] and is still easily measurable at 10 R.[83] During human prenatal life, central nervous system (CNS) development occurs over a considerably longer period than does major organogenesis. In rodents, however, which have a relatively much shorter fetal period than man, the two processes are much more nearly equal with respect to time occupied. Therefore, human exposure, which has been random with respect to developmental stage, is more likely to occur during some period critical for the CNS than is exposure in experimental mammals, in which work has been concentrated primarily on specific stages during the period of major organogenesis. The facts that many abnormalities in systems other than the CNS have been reported in man and that stage correspondence can be good further indicate that human results are not out of line with animal data.

Histologic correspondence was noted in a report of human fetuses studied within days after exposure to radium gamma rays from maternal therapy for cervical cancer.[49] Among effects observed were destruction of proliferative and migratory brain cells and of some hematopoietic cells, necrosis of lymphoid and mesenchymal cells, and degeneration of oocytes. These observations provide a link with animal data on the CNS and, importantly, with recent observations on the extreme sensitivity of early developing oocytes.[45] The stage most exquisitely vulnerable to the latter effect in rodents is the early postnatal period, when ovarian development corresponds closely to that of a human fetus.

Because of large genetic and environmental variables encountered in human populations, it is very difficult to measure any effects that might be produced by low doses of radiation, such as those used in diagnostic radiology. It is therefore not surprising to find conflicting reports on whether the "spontaneous" incidence of malformations or growth retardation is increased as a consequence of such exposure (some authors[87,109,158,159] report negative findings; others[68,81] positive). At present, it is impossible on the basis of human studies alone to determine with certainty a dose below which teratologic effects in man are not induced by exposure at sensitive stages in development. As discussed above, such thresholds do, however, probably exist, and they may be higher for protracted or fractionated radiation than for acute single exposures.[19,24,138]

Radionuclides Administered During Pregnancy

The effects of various radioisotopes administered to pregnant mammals have been less extensively studied than the effects of externally administered radiation. Furthermore, one cannot generalize on the effects of administered radionuclides, because, depending on the chemical form and the type and energy of the emitted radiation, they may or may not cross the placenta, they may have specific target organs, the distribution of radiation may be nonrandom, the metabolism of radioactive elements or compounds may vary greatly from person to person because of individual biologic variations or because of the disease state of a given subject, and the change in dose rate with time may be difficult to evaluate.[35] Radioisotopes administered to the mother may also affect the newborn if they are administered shortly before birth, because many are excreted in the breast milk.[10,178]

In any event, before one can estimate the potential hazard of administering a radioactive nuclide or compound to a pregnant woman, one must determine with some accuracy the total dose to the fetus or a particular fetal tissue, the dose rate and how it varies with time, and the stages of gestation during which the radiation is received.

Until recently,[151] the radioactive isotopes of iodine were the radionuclides most commonly used in nuclear medicine. The two most important ones are iodine-131 and iodine-125. Although inorganic iodide readily crosses the placenta, iodine attached to proteins, hormones, and even radioactive rose bengal is less likely to cross. However, a significant amount of iodine usually is released from the labeled compounds and becomes available to the fetus. There is probably no radioactively labeled iodine compound that does not release some iodine to the circulation after administration.

The human fetal thyroid does not take up iodine before the twelfth week;[32,51] thereafter, however, its uptake increases, and it comes to a peak in the sixth month.[51] In the mouse, there is some evidence that the fetal thyroid has a greater avidity for iodine than does the maternal thyroid.[82] Because the human fetal thyroid accumulates considerably more iodine-131 per gram than do other fetal tissues, an inadvertent therapeutic dose to the mother of 5 mCi would deliver 6,500 rads to this organ and thereby ablate it.[50] If the dose of radioactive iodine is high enough, it can even cause inhibition of growth of the underlying trachea.[151].

Pathologic effects, including thyroid destruction, have been reported in the fetus after therapeutic (ablative) doses of iodine-131 were administered to pregnant women.[67] Tracer doses of radioactive iodine have not been reported to produce a deleterious effect on the fetus. There remains, nevertheless, a concern over the possibility of inducing thyroid cancer in susceptible people by prenatal exposure to even small amounts of radioactive iodine. If administration of radioiodine is unavoidable, it is best done before the third month of human pregnancy, when the fetal thyroid has not yet developed. Even in this circumstance, the total-body dose to the embryo should be estimated and considered.

Technetium-99m is a radioactive isomer that has become, in recent years, an important radionuclide for diagnostic imaging procedures. Its usefulness depends on its almost optimal gamma-ray energy (140 keV), its short half-life (6 h), its rapid excretion, and the fact that it emits no beta rays. Although radiation doses to the embryo or fetus would thus presumably be lower from technetium-99m than from some other diagnostically used radioisotopes, there have been no direct studies on the effects of technetium-99m on intrauterine development.

Inorganic radioactive potassium, sodium, phosphorus, cesium, and strontium cross the placenta readily. Experiments with radioactive phosphorus and strontium have indicated that, if the dose is large enough, embryonic pathology and death can be induced.[60,152]

Because tritium (hydrogen-3) is a potential pollutant from nuclear-energy production, its effect on development has been the subject of a number of studies. Tritiated water (HTO) is a common chemical state of tritium, and it has easy and rapid access to living cells, including those of the embryo or fetus. HTO administered in the drinking water to rats throughout pregnancy produced significant decreases in relative weights of brain, testes, and probably ovaries[29] and increases in norepinephrine concentration[26] at doses of 10 μCi/ml (estimated at 3 rads/d) and produced weight decreases in a number of organs at higher doses.[29] Because the length of the critical period for various organs is not known, the total damaging dose cannot yet be estimated. Relative brain weight was found to be reduced at only 0.3 rad/d (1 μCi/ml of drinking water) when exposure began at the time of the mother's conception.[27,91] Even lower exposures (0.003 rad/d and 0.03 rad/d) have been implicated in the induction of behavioral damage, such as delayed development of the righting reflex and depressed spontaneous activity.[27] However, because the data fail to show a clear dose dependence, there is some doubt about the validity of this suggestion.

Tritiated drinking water has been used to study the effects of radiation on development of a sensitive cell type, the oocyte. Oocyte counts were made in serial sections of exposed and control animals. In squirrel monkeys continuously exposed from conception to birth, the LD$_{50}$ was 0.5 μCi/ml of body water, giving a fetal dose rate estimated at 0.11 rad/d. Because the sensitive period for oocyte development is probably the last trimester, the LD$_{50}$ was calculated to be 5 rads.[45] In the mouse, the sensitive period occurs during the first 2 wk after birth, and, by a similar calculation, the LD$_{50}$ from tritiated drinking water at that time is slightly below 5 rads.[46]

Background and Fallout Radiation

Information on natural background radiation is presented in Chapter III. The average abdominal exposure for the U.S. population is probably around 80 mrems/yr. It is assumed that the embryo and fetus also are exposed to natural background radiation at about the same dose rate. Radiation from remaining fallout (at present) adds less than 4% of the background dose rate; and contributions from other man-made sources (excluding medical irradiation), such as nuclear and coal-fired power plants and consumer products, add less than 1%.

It appears, therefore, that the average American receives a dose of about 60 mrads during intrauterine life, or about 0.2 mrad/d. It has been suggested that the frequency of neonatal deaths from congenital malformation is highly correlated with the background radiation resulting from geomagnetic conditions and altitude.[173] However, this claim is not supported by the experimental data on low-dose-rate irradiation of developing mammals. Where a clear correlation with altitude does exist in the human data, it has been attributed instead to effects of hypoxia on intrauterine development.[66] In general, the natural and man-made background radiation during gestation is so low in total magnitude and dose rate that it is not thought to be a factor in the normal incidence of congenital malformations, intrauterine or extrauterine growth retardation, or embryonic death.

POSTNATAL IRRADIATION

Numerous reports have indicated that radiation exposure of the neonate, infant, or child can result in growth retardation.[37,42,101,105,128,150,157,170] Followup studies on children exposed in Hiroshima, Nagasaki, or the Marshall Islands to atomic-bomb or fallout radiation indicated that the younger children were more susceptible to these growth-retarding effects than the older ones. The most conclusive evidence on postnatal radiation effects comes from a multivariate analysis of anthropometric data on children exposed to the Hiroshima bomb and examined periodically up to 8 yr later.[108] As radiation exposure increased, there were small but statistically significant decreases in body measurements and growth rates in those who had received kerma of 100 rads or more.

Among the Rongelap children exposed to radioactive fallout, two boys who were infants at the time of exposure developed atrophy of the thyroid before puberty. The resulting hypothyroidism led to retarded body growth and sluggishness of behavior. It was estimated that the whole-body dose from externally deposited fallout was 175 rads, and the thyroid dose resulting from concentration of radioiodines between 700 and 1,400 rads.[36,156,157]

Individual case reports of children who received radiation therapy have also indicated that localized irradiation can result in local retardation, especially if growth centers (such as open epiphyses) or tissues with some growth potential are exposed. These effects are more obvious when irradiation is unilateral.

It is difficult to determine whether exposures to diagnostic radiation can produce growth retardation in growing children, inasmuch as any infant or child who receives significant exposures to diagnostic radiation is likely to have an illness that in itself could be responsible for growth retardation. Animal data support the belief that whole-body or partial-body irradiation in the diagnostic dose range probably does not affect the growth of infants or children.[12,39,153]

Early postnatal exposures of rodents can have devastating effects on female fertility. The great bulk of oocyte destruction caused by continuous exposure from conception to 14 d of age is the result of irradiation received after birth. Continuous gamma radiation at the rate of 8.4 R/d from birth to weaning totally sterilized female mice, but had no effect on males.[129] The LD_{50} for oocyte-killing in the mouse during the first 2 postnatal weeks is about 5 rads.[45] For acute irradiation, Oakberg[112] found the LD_{50} of stage I oocytes in 10-d-old mice to be 8.4 R. It is likely that in primates, including humans, the corresponding stage in ovarian development occurs during the third trimester of intrauterine environment.[45] For some other organ systems, as well, it is probable that the first 2 postnatal weeks of rodent development correspond to the latter part of human pregnancy.

ESTIMATE OF RISK FROM INTRAUTERINE AND
EARLY POSTNATAL IRRADIATION

At relatively high doses and dose rates, it can be established that there is, in general, good correspondence between results obtained from work with experimental mammals and those available for man. This correspondence obtains for developmentally (but not chronologically) equivalent stages of irradiation; because of it, one may gain confidence in the extrapolation of animal data to the human situation. This is fortunate, because available results in man fail, for a number of reasons, to provide direct information on the magnitude of risk at low exposures. The genetic and environmental variability in human populations makes the measurement of small increments in a miscellany of structural or functional impairments next to impossible to measure. There are, furthermore, no good tests for some of the subtle depressions in physical or mental performance or general fitness that could conceivably result from low-level irradiation during development, especially in view of the fact that the CNS in man is vulnerable for an extended period. Finally, the random exposures that are encountered in most epidemiologic studies fail to provide sufficiently large samples for any specific sensitive period during development.

The animal data leave no doubt that readily measurable damage can be caused by doses well below 10 R applied at stages that are sensitive to the specific effect being studied (Table VI-2). Examples are oocyte-killing in primates, with an LD_{50} of only 5 rads;[45] CNS damage in the mouse, with a threshold dose below 10 R;[83] and brain damage and behavioral damage in the rat from doses that are less than 6 rads over the whole intrauterine period, and presumably only a fraction of this for the sensitive period.[27]

The Japanese atomic-bomb data for small head circumference indicate that the human embryo is sensitive down to a few rads of mixed gamma and neutron radiation, in that air kerma of 10–19 rads (i.e., fetal doses averaging 5.3 rads gamma plus 0.35 rad neutrons) produced a clearly significant increase in incidence at Hiroshima,[102] and there are indications that air kerma of 1–9 rads was also damaging to embryos that were in sensitive stages of development at the time of the bombing. Part of the effect is presumably attributable to the fast-neutron dose, inasmuch as no significant increase in microcephaly was detectable below 150 rads kerma in the much smaller Nagasaki sample. It may be noted that microcephaly is a gross abnormality and that it is possible that more subtle changes could have gone undetected.

Where cell-killing effects can be directly measured, as in oocyte-killing, there do not appear to be any clear threshold doses under some conditions.[46] For morphologic malformations, however, a generalized straight-line extrapolation from the results of acute irradiation at high or moderate doses is probably not valid. Because it is unlikely that any perceived developmental abnormality results from damage to a single target, there are probably threshold doses for all such abnormalities. Furthermore, for a given total exposure, lowering of dose rate has been shown to diminish

TABLE VI-2 Reports of Studies Using Total Dose of Less Than 10 Rads, or Less Than 10 Rads/Day, during Early Development

Organism	Source of Radiation	Stage[a]	Effect	Dose Yielding Effect	Reference
		Single Exposure			
Mouse	X ray	13[b]	Mitotic delay in telencephalon	Threshold <10 R	83
Human	Hiroshima bomb (80% gamma, 20% neutron)	Single exposure, wk 6–11[b]	11% microcephaly	<5 rads[c]	102
			17% microcephaly	5–10 rads[c]	
Mouse	X ray	29[b]	Oocyte-killing	LD_{50} = 8 R	112

Protracted Exposure

Mouse	HTO[d]	19-33[b]	Oocyte-killing	LD50 < 5 rads	46
Monkey	HTO	Last trimester	Oocyte-killing	LD50 = 5 rads	45
Rat	HTO	0-term	Reduced brain, testis, ovary wts	3 rads/d	29
Rat	HTO	0-146	Ditto; also spleen and overall 30% reduction in testis wt	6 rads/d 3 rads/d	91
Rat	HTO	0-term	Decreased brain wt in "F_2"	0.3 rad/d	27
Rat	HTO	0-term	Decreased brain wt in "F_2"	0.3 rad/d	26
Rat	HTO	0-term	Decreased brain wt, increased norepinephrine	3.3 rads/d	28
Rat	Cobalt-60	0-term	No effect on life span	3.3 rads/d	171
Mouse	Cesium-137	20-40	Prenatal and postnatal mortality	2.5 R/d	129
Mouse	X ray	0-18	Complete sterility of females No effect	8.4 R/d 2.5, 5, or 10 R/d	89

[a] Days after conception, except where otherwise indicated. Some postconception intervals listed occur after birth. 0 indicates exposures started within hours after conception.

[b] Critical period. Where this notation appears, effects apply to this stage only.

[c] Estimated dose of gamma rays plus neutrons received by embryo.

[d] HTO = tritiated drinking water.

the effect, because, with protraction, only a portion of the dose is received during a given critical period. It is therefore likely that low-dose-rate exposures (0.01 R/min or less) at total doses of less than 1 R would not have widespread effects, even though specific damage, such as oocyte-killing, could presumably still occur. Radiation at such doses in medical practice can have clear benefits to the health of individual mothers, so one must balance these benefits against the small risk to the conceptus. However, even at such low doses, indiscriminate exposures of larger populations of embryos or fetuses should be avoided. The possibility that a pregnancy exists should always be considered before women of child-bearing age are exposed to radiation appreciably above background.

Until more is known about synergisms between radiation and other environmental agents, the possibility of such interactions (as shown in the case of caffeine[180]) should add a cautionary element to risk estimates.

SUMMARY

Developing mammals, including man, are particularly sensitive to radiation during their intrauterine and early postnatal life. The effects produced are strongly related to the developmental stage at which radiation is received, and, at moderate to high doses, close correspondence has been demonstrated in this respect between man and various experimental species. The experimental data can therefore be used with some confidence to fill in gaps in the human experience, particularly with respect to extrapolations to low exposure levels, where it is very difficult to obtain direct evidence in genetically and environmentally heterogeneous human populations.

Radiation during preimplantation stages probably produces no abnormalities in survivors, owing to the great developmental plasticity of very early mammalian embryos. Radiation at later stages may, however, produce morphologic abnormalities, general or local growth retardation, or functional impairments, if doses are sufficient. Obvious malformations are particularly associated with irradiation during the period of major organogenesis, which in man extends approximately from week 2 through week 9 after conception. More restricted morphologic and functional abnormalities and growth retardations dominate the spectrum of radiation effects produced during the fetal and early postnatal periods. Some of these effects can be apparent at birth, and others may show up later; and subtle functional damage cannot be adequately measured with available techniques. Because the central nervous system is formed during a relatively long period in human development, such abnormalities as microcephaly and mental retardation figure prominently among the list of radiation effects reported in man.

Animal data indicate that readily measurable damage can be caused by doses well below 10 R of acute irradiation applied at stages that are sensitive to specific effects being studied (cns injury and oocyte-killing). Atomic-bomb data for Hiroshima show that microcephaly was induced by acute air doses in the 10–19 kerma range (average fetal dose, 5.3 rads gamma plus 0.4 rad neutrons) received during the sensitive period and suggest that it was also increased in the 1–9 kerma range (average fetal dose, 1.3 rads gamma plus 0.11 rad neutrons). However, it is likely that there are threshold doses for most maldevelopments and that these are of a variety of magnitudes. Lowering of the dose rate diminishes the damage. Until an exposure has been clearly established below which even subtle damage does not occur, it seems prudent not to subject the abdominal area of women of child-bearing age to quantities of radiation appreciably above background, unless a clear health benefit to the mother or child from such an exposure can be demonstrated. Considerably more research is also needed to explore possible synergistic interactions between radiation and other environmental agents.

FERTILITY

The literature on radiation effects on fertility and fecundity in experimental animals is extensive (see Table VI-2), but little information on the radiation response of the testis and the ovary has become available since the publication of the 1972 BEIR report.[106] Information is now being produced on the response of human spermatogenic cells to graded doses of x irradiation.[69,131] The application of cell population-kinetics studies to spermatogenesis and oogenesis in relation to germinal-cell proliferation and differentiation has also provided a better understanding of the radiation response and tissue repair in mammalian reproductive cells.[123,131] All this has led to further refinements of our understanding of mechanisms of impairment of fertility and our understanding of other matters relevant to genetic-mutation frequency in experimental animals, and possibly in the human.[69,131]

ANIMAL EXPERIMENTS

Testis

The most recent investigations on spermatogonial stem-cell renewal in the rat and mouse have provided a model (the Oakberg-Huckins model) in which the types A_s (stem cells), A_{pr} (paired cells), and A_{al} (aligned cells) are undifferentiated cells representing the sequence of development in the undifferentiated spermatogonial stem-cell compartment; differentiation probably occurs at the stem-cell level.[74-77,110,117,122] The types A_1, A_2, A_3, and A_4, the intermediate, and the type B spermatogonia are the differentiated cells that give rise to the resting primary spermatocytes and the production of mature sperm cells.

Among the undifferentiated and differentiating spermatogonia, the proliferating type A_{al}, types A_{1-4}, intermediate, and type B cells in the mouse testis appear to be most radiosensitive; the type A_s appears to be relatively resistant to x rays.[52,53,55,73,110,122] Radiation doses (acute with high dose rates) of less than 15 rads of x radiation can lead to interphase cell death and prompt depletion of the differentiating-proliferating spermatogonial-cell population in the mouse; but sterility does not result, because there is immediate tissue repair and regeneration of the seminiferous epithelium, apparently from the surviving type A_s stem-cell population. Larger acute doses, 25–50 rads, can deplete the proliferating spermatogonial-cell population drastically and effectively, with a decreased production of sperm cells. However, impairment of fertility still is not immediate; existing spermatocytes and spermatids are resistant and may not be eliminated from the system for several weeks. Only temporary sterility would result with even higher doses; sufficient numbers of spermatogonial type A_s stem cells survive doses as high as 300 R, or even more, proliferate, and differentiate sequentially, regenerating and reconstituting the seminiferous epithelium with restoration of spermatogenesis. Acute whole-body exposures of young male mice up to 8 mo old to doses as high as 1,000 R have failed to impair reproductive potential and fertility.[53,54,73,110]

Fractionated or continuous whole-body x or gamma irradiation does not necessarily impair fertility in mammals,[53,56,113,116,121] provided that the dose rate is sufficiently low (less than 2 rads/d). Permanent sterility may ensue after higher dose rates and total doses. Male dogs exposed daily to x radiation for the duration of their life maintained sperm counts at normal values at a dose rate of 0.6 rad/wk, and no evidence of deleterious changes occurred in sperm production or fertility at 0.3 and 0.6 R/wk. Progressive cellular failure and sterility ultimately developed within months with brief daily exposures at 3 rads/wk.[25,30,43,107,134]

The proliferating and differentiating spermatogonia are extremely radiosensitive under continuous exposure; there is evidence that the testis is the mouse tissue most sensitive to continuous irradiation at very low dose rates.[25,43,52-55,107] At 16.5 R/d, testis weight decreased progressively with duration of exposure;[107] after radiation-free intervals of up to 4 wk, the testis weight recovered to over 90% of control weight and was restored more slowly than fertility. Dose-dependent damage to the testis has been observed at 2–20 R/d; at 10–40 R/d, permanent sterility has occurred in mice.[25] There was progressive decrease of the germinal epithelium; after 20 R/d or more for relatively long periods, complete absence of seminiferous epithelium occurred. At 2 R/d, rats and mice maintained reproduction for 10 generations or more, although the progeny showed some evidence of life-shortening. However, at slightly more than 2 R/d, there was a continuous and serious depletion of cell population of the testis, with later sterilization.[43]

It has been demonstrated that 0.009 R/min or less is near the threshold for recovery processes, permitting maintenance of the mouse spermatogonial population.[114,115,118-120] However, with total doses greater than 300 R, a dose rate of 0.001 R/min resulted in the spermatogonial-cell population's reaching an equilibrium at 80% control.[114,119,120] Studies in the mouse testis exposed to continuous gamma irradiation at 1.8 rads/d (0.00125 rad/min) and at 45 rads/d (0.03125 rad/min) to accumulated doses of up to 630 rads demonstrated that, at extremely low dose rates, the spermatogonia are sensitive to radiation death and cellular depletion. However, even after 15 wk of continuous exposure at 1.8 rads/d, the type A_s stem-cell population could be maintained at control values, and the temporal sequence of cellular recovery to regenerate the seminiferous epithelium begins with the type A_s stem cells.[52-56,73]

Ovary

In the mammal, susceptibility to radiation-induced cell death in the ovary depends on a number of factors, including the developmental stage of the germ cell, the age of the animal, and the mitotic activity of the oogonia. In the rat, the oogonia appear most sensitive to radiation in the fetal ovary at about 15.5 d of gestation; this would correspond in the human to approximately the fifth month of gestation.[4,5,96,112] Thereafter, radiosensitivity, in the rat, appears to be relatively low during the leptotene, zygotene, and pachytene stages of meiotic prophase; it increases with the diplotene stage of prophase when the oocyte becomes surrounded by a single layer of granulosa cells to form the primary follicle. In the rat, mouse, and rabbit, the primary follicles are quite sensitive to acute exposure, but sensitivity appears to decrease as development of the follicle proceeds.[4,5,95,96,103] This response appears species-specific; in the guinea pig and monkey, the earlier stages of prophase in the primary follicle are relatively radioresistant, and sensitivity increases with follicular development. In addition, the radiation doses required to kill a given fraction of primary follicles are also species-dependent:[112,120,125] in the mouse, a single acute dose of 10 R of x rays reduced the number of primary oocytes to half; in the rat, the comparable dose was 100 R; and in the monkey, perhaps as high as 900 R.[3]

Oocytes in the mouse change in sensitivity to radiation between the period of birth and sexual maturity;[95,111,112,124,125,146] sensitivity appears to be low at birth and increases until 7 wk of age. Differences in radiosensitivity of oocytes to cell-killing form the basis for the apparent age variation in sensitivity to radiation-induced sterility. Relatively low radiation doses (such as an acute dose of 25 R of x rays) given during the second and third weeks after birth impair fertility, owing to marked depletion of the oocyte population resulting from radiation-induced cell-killing.[95,111,112,124,125,146] In mammals, there is no repopulation of cells after loss from the existing oocyte pool, because the maximal numbers are established in the fetus. Thus, infertility and sterility result when the supply of functioning oocytes, which survive radiation injury, is exhausted. Furthermore, the radiation-induced reduction in fertility is much less than the reduction in oocytes; the younger the female, the more efficiently she may use the limited oocyte supply.

Irradiation of the mouse and rat ovary results in early and progressive decline in the numbers of oocytes and ovarian follicles.[3,111,125,146] In the female mouse fetus, doses of 60–80 R/d for 5 d (to total doses of 300–400 R) given during the late development of the ovarian tissue result in permanent sterility. Continuous irradiation of female mice with gamma rays (12.4 R/d) or with fission neutrons has shown that the interval between irradiation and conception has a striking effect on the mutation frequency in the offspring.[138] Continuous gamma irradiation in mice (12.4 R/d, up to approximately 175 R) from conception to day 14 caused a significant shortening of the reproductive period. When female mice were irradiated with fission neutrons (approximately 63 rads), the mutation frequency was high in the first 7 wk after exposure; after that, no mutations were found.[138] This appears to be due, in part, to the low mutational sensitivity of oocytes in immature follicle stages.[3,45,125,138,144-146] Exposure early in the postnatal period has marked effects on fertility in females. The LD$_{50}$ of stage I oocytes in 10-d-old female mice is approximately 8.4 R; it is about 5 R in slightly younger mice. Continuous gamma-ray exposure at 8.4 R/d from birth to weaning sterilized female mice. It may be that in the monkey, and possibly the human, the stage of development of the ovary equivalent to the early postnatal stage in the mouse occurs late during intrauterine fetal development. At the lower doses, impaired fertility and fecundity were manifested as high litter mortality, decreased litter size, and diminished litter frequency. Impairment of the ovulation rate in rats appears to depend on radiation dose. Female fetuses exposed *in utero* to doses as high as 220 rads and then mated to unirradiated males showed no significant effect on fertility or fecundity.[45,138,143-146]

HUMAN STUDIES

The reproductive cells of the human testis constitute the seminiferous epithelium and are subject to a proliferating-cell renewal system consisting of four compartments: a self-maintaining stem-cell compartment, a proliferating progenitor compartment [types A_d (A-dark), A_p (A-pale), and B spermatogonia], a differentiating-maturing compartment [types R (resting, preleptotene), L (leptotene), Z (zygotene), and P (pachytene) spermatocytes], and a functional end-cell compartment (types Sa, Sb, Sc, and Sd spermatids).[34,69,131] The seminiferous epithelium is in a steady state of cell renewal; new cells are formed throughout reproductive life, replacing functional end cells that leave the system. In man, the type B spermatogonia are the most radiosensitive, and doses of only a few rads will deplete this proliferating population.[69,90,164] The spermatogonia preceding type B (types A_d and A_p) are also radiosensitive; spermatocytes are less radiosensitive, and spermatids are the most radioresistant of all.

The human ovary contains the full complement of approximately 7 million oocytes at a fetal age of approximately 5 mo; later, the oocyte population undergoes physiologic attrition until menopause in the adult.[2,53] The female is born with only 2 million, and ovulation provides only some 360–400 mature oocytes throughout her reproductive life.[2,53] There is no oocyte renewal after the degenerative sequence progresses, and the ovary therefore lacks the capacity to replace damaged or lost reproductive cells after this time. The oocytes arrest in a preovulatory meiotic diplotene prophase stage, which is relatively radiosensitive in the mouse, but radioresistant in the human. Selection processes for cells damaged by radiation or other mutagens may not be operative until ovulation occurs and the cell is later fertilized.[2,53]

Testis

Rowley and colleagues[69,131] have reported the results of their 10-yr study on the effects of acute doses of x rays on the normal human testis. Sixty-seven men, aged 25–52 yr, received acute testicular x irradiation in doses of 8–600 rads. Most received single exposures; one subject was given weekly irradiations of 5 rads for 11 wk. The conclusions on the endocrinologic and cellular response include the following: There was an initial rise in urinary gonadotropins. There was a decrease in urinary testosterone coupled with a rise in plasma luteinizing hormone; this suggested radiation interference with Leydig-cell function. Spermatogonia are the most radiosensitive and spermatids the most radioresistant cells of the germinal-cell line. Type B spermatogonia are the most radiosensitive, followed by types A_d and A_p, whereas the differentiated preleptotene spermatocyte is relatively radioresistant, in comparison with its progenitor cells. Single acute doses of 600 rads or less cause significant cellular damage in the testis; these changes are dose-dependent, with complete recovery after doses of 600 rads or less, and with the time until recovery also dose-dependent, extending up to 5 yr.

Atomic-Bomb Survivors

Information on impairment of fertility in man is available from the study of atomic-bomb survivors and from Marshallese and Japanese who were inadvertently exposed to fallout during atomic-bomb testing in the Pacific.[15,90,106,148,165] The data lack precision, but demonstrate the following: Relatively low doses can decrease production of sperm cells, but effects on spermatogenesis are transient; the sterilizing dose in the male is probably much greater than about 400–500 rads, i.e., it probably exceeds the mean lethal dose to the whole body. Fertility is impaired in the oocyte population only after moderately high doses—200–400 rads. Little is known regarding the delayed effects of radiation on fertility in these exposed populations, nor is there information on the extent of impairment, if any, in the male and female populations exposed *in utero* and in the F_1 populations of exposed parents.[15,78,148] Followup studies of the Japanese atomic-bomb survivors and the Marshallese women exposed to fallout have failed to demonstrate any long-term effect on fecundity.[6,15,148,165]

Radiotherapy Patients and Victims of Nuclear-Reactor Accidents

Clinical data are available on male radiotherapy patients and men exposed during criticality accidents at nuclear-reactor installations.[90,165] Careful sperm-count studies after limited partial-body radiation exposure have indicated that, if sterility occurs, normal sperm counts can return in about 1 yr after doses of 100 rads and even in 3 yr after exposures in the near-lethal range.[90,165] Acute whole-body exposure has not been shown to cause permanent sterility in males.[165] The sterilizing dose therefore exceeds the lethal whole-body dose for acute radiation. Similarly, sterilization of the human testis has never been shown to result from continuous or fractionated (protracted) low-dose exposure.[30,90,144,148]

In women, radiotherapy experience has suggested that acute doses of 300–400 rads or slightly higher doses given in two or three fractions result in permanent sterility.[2,15,45,53,165] If fractionation is protracted over a 2-wk period, much larger doses (possibly 1,000–2,000 rads) are required for sterilization, depending on the age of the woman.[2,15,45,165] The ovaries of younger women are much less radiosensitive; permanent sterility is more likely as the menopause is approached.

Conclusions

Populations of mature spermatozoa in the human testis are maintained by proliferating spermatogonial stem cells. Provided that the dose remains below 400 rads (low-LET radiation, acute exposure), radiation depletion of the spermatogonial-cell population is only temporary, and the seminiferous epithelium is repopulated and regenerates from surviving and proliferating spermatogonial cells in the damaged tissue. Exposure much greater than this (perhaps by an order of magnitude) directed only at the testis could probably result in permanent sterility.

Impairment of fertility can result from absorbed doses to the human ovary in the range of 300–400 rads (low-LET radiation, acute exposure), but this depends, in part, on age. Radiotherapeutic experience has shown that women approaching the menopause may have long-term impairment of fertility or permanent sterility, whereas in younger women only transient infertility associated with amenorrhea may result. This may be associated, in part, with oocyte populations, which decrease primarily by physiologic atresia (and to a much lesser extent by ovulation) with age.

CATARACTS

A causal involvement of radiation-induced damage of epithelial cells in the germinative zone of the lens in radiation cataractogenesis has not yet been proved. However, the available evidence from animal studies strongly suggests this mechanism, on the basis of the differentiation of the affected cells into abnormal lens fibers and the time coincidence between the appearance of lens opacification and the rate of migration of lens epithelial cells into the posterior lens cortex. Accumulation of aberrant cells in the posterior cortex causes alteration in the lens cytoarchitecture, resulting in a loss of transparency.[177] There is no direct evidence that lens opacification depends on the killing of epithelial cells in the germinative zone. The sigmoid cataract dose-response curves and the protective effect of partial lens shielding provide evidence that other factors are involved in radiation cataractogenesis in addition to cell-killing.

The available data suggest a sigmoid dose-response relationship with an apparent threshold for lens opacification. Threshold doses in man for x rays and gamma rays delivered in a single exposure vary from 200 to 500 rads, whereas the threshold for doses fractionated over periods of months is around 1,000 rads.[78] Continuing observations of lens changes in survivors of Hiroshima and Nagasaki have been reported.[47,48,58,72,84,160,161] The subjective nature of the lens assay techniques used by the several investigators involved in these studies, as well as the limited dose information, precludes a quantitative assessment of dose response or of the relative effects of fission neutrons and gamma radiation on cataract induction in humans. These data are, however, consistent with a sigmoid dose-response relationship in the dose range from 20 to 450 rads, with a dose threshold of about 200 rads or greater for the induction of vision-impairing lens opacification.[48] The latent period for cataract induction has been estimated to be some 10 mo after exposure. A comparison of cataract incidence for all periods of observation with the incidence in the sample group followed for 25 yr after exposure[160] suggests the possibility of an interaction of radiation cataractogenesis with age, although the statistical significance of the difference in incidence cannot be established, owing to the above-mentioned limitations on the sample data.

Data derived from an investigation of the age-related sensitivity to the development of radiation cataracts in the rat do not support the hypothesis of hypersensitivity of young lenses.[99] The minimal dose for the induction of cataracts and the rate of opacification were greater in adult than in young rat lenses in the x-ray dose range of 200–300 rads. In the dose range of 300–900 rads, opacities occurred earlier in young lenses, but the rate of progression was greater and severe opacities developed sooner in adult animals. At doses greater than 900 rads, cataracts occurred sooner and progressed more rapidly in young lenses. Exposure of 2- to 4-wk-old lenses to doses of 400, 800, and 1,200 rads revealed a greater radiosensitivity than older or younger lenses with respect to the development of incipient cataracts, but the lenses of animals irradiated at these ages were the last to develop complete or severe opacification at these doses.[99] It is thus suggested that age-dependent factors are involved in the rate of progression and the extent of radiation-induced opacification, but young lenses do not appear to differ significantly from adult lenses in sensitivity to radiation cataract induction. A comparison of the incidence of cataracts in atomic-bomb survivors exposed at all ages with the incidence in persons exposed during infancy provides no evidence of greater susceptibility of young lenses.[160]

Data on the effects of chronic exposure of the human lens to low dose rates of ionizing radiation are lacking. Detectable but minor degrees of lens opacification have been reported in radiosensitive species, such as the mouse,[166] but the relationship of such changes to cataract formation in this or other species, such as man, has not been established. The incidence of minor non-vision-impairing lens changes has been reported to increase linearly with age in man,[33] but such alterations have not been shown to be related to cataract formation.

In general, the RBE for high-LET radiation for single cataractogenic exposures has been found to be in the range of 2–9, and the RBE for protracted exposures is assumed to be somewhat higher. A quality factor of 10 has thus been proposed for lifetime exposure of the lens.[79]

AGING

On the basis of animal experimentation, the hypothesis has been advanced that radiation exposure induces premature aging, one consequence of which is dose-dependent life-shortening. In considering this hypothesis, it is necessary to define aging, a phenomenon that involves a complex set of biologic alterations. Walburg[172] has defined aging as a progressive loss of functional capacity in all members of a population after they have reached reproductive maturity, which leads to an increased probability of disease and death.

The conditions under which irradiation may be regarded as being responsible for premature aging are as follows:[31]

• Radiation causes the force of mortality to increase more rapidly in exposed than in unexposed subjects without altering the shape of the cumulative mortality curve.

• Exposure results in a proportionate decrease in the age of onset and in the time of onset of all diseases or causes of death that affect the control group, without altering the degree, sequence, or absolute incidence of the diseases and causes of death.

• Radiation causes all the morphologic and physiologic manifestations of the aging process to appear and develop at proportionately earlier chronologic ages, to degrees and rates in the various organs proportional to the degrees and rates in organs of unexposed subjects.

• A difficulty encountered in determining whether radiation exposure decreases the life span by the same mechanisms that are involved in natural senescence is that neither the mechanisms of natural senescence nor the mechanisms of radiation-induced life-shortening have yet been established.[172] Effects of radiation on aging have involved studies of mortality, pathology and disease incidence, subclinical histopathology, and physiologic and biochemical changes in both humans and experimental animals.

One test of the hypothesis that radiation exposure induces premature aging would be the demonstration that the onset of all diseases is advanced to the same extent and by a factor related to the degree of life-span shortening. Neither the time of appearance nor the incidence of benign hepatomas in CBA mice was found to be influenced by exposure to 1,100

R, in spite of the fact that the occurrence of benign hepatomas is correlated with natural aging.[38] In other investigations in which the causes of death were analyzed in strains of mice and the carcinogenic action of radiation was not the predominant cause of death, it was determined that the general pattern was not altered by radiation, but that it was advanced in time; this suggested that radiation accelerated natural aging processes.[62,94,168] The results of such studies may, however, be misleading, in that they are based on postmortem examinations of animals that died of old age; those examinations do not generally yield reliable information on the cause of death, unless death was due to an easily detected cause, such as a tumor or leukemia.[1] Serial killing and determination of the incidence of a number of diseases in irradiated and nonirradiated control CBA male mice have revealed that radiation exposure produces complex patterns of late somatic changes in which the variations in latency, time course, and incidence are not consistent with the hypothesis that radiation advances all diseases in time and hence leads to accelerated aging.[1]

The additivity of radiation and natural senescence has not been demonstrated in animal experiments with mathematical modeling of mortality-rate data.[172] Lethal diseases have not been shown to be equally advanced by radiation; this suggests that the effects of such exposure are not directly equivalent to natural senescence. Although it is apparent that radiation advances the time of onset of some neoplastic diseases, the only nonneoplastic diseases that have been shown to be accelerated by radiation are nephrosclerosis, which occurs only at high doses,[167] and amyloid deposits in LAF$_1$ mice.[92] Mortality data statistically adjusted for competing risks by the method of Kaplan and Meir[85] strongly suggest that nonneoplastic diseases are not advanced in time in animals exposed to radiation at doses that result in life-span shortening of less than 15%.[172] On the basis of an empirical estimate of a 3–5% reduction in life span per 100 rads of whole-body exposure,[11,40,93,169] no significant increases in the rate of induction of nonneoplastic diseases would be anticipated at doses of less than 300 rads. Biochemical and physiologic studies of radiation effects on senescence phenomena—such as changes in collagen, pigment accumulation, and neuromuscular function—have not provided evidence of radiation-accelerated aging in experimental animals.[172] Radiation has, however, been shown to accelerate the development of increased interstitial fibril density and arteriocapillary fibrosis, phenomena that form the basis of a histopathologic theory of aging.[31] It is thus suggested that there are common factors in senescence and radiation-induced changes, but it is not known whether radiation causes fibrotic alterations via the same mechanisms that are involved in normal senescence.[172]

Mortality studies have indicated that radiologists experience increased mortality rates from cardiovascular-renal diseases, as well as from cancer, relative to other medical specialists[149]—a finding that supports the hypothesis of radiation-accelerated aging. Cohort mortality studies of these medical specialists over a 50-yr period have revealed a persistent excess mortality in radiologists from diseases other than neoplasia.[97,98] The 1920-1929 cohort of radiologists had the highest mortality for several chronic diseases; subsequently, radiologists ranked highest only for cancer mortality, but the initially observed excess risk of leukemia has been found to be decreased in younger cohorts. However, mortality from lymphoma, and especially from multiple myeloma, has increased, with a significant excess of deaths from this cause among radiologists who entered the specialty between 1930 and 1949.[98] Radford *et al.*,[127] in contrast, have reported that ankylosing-spondylitis patients who were not treated with radiation have shown no significant excess cancer mortality, but have experienced excess mortality from the same nonneoplastic diseases observed in ankylosing-spondylitis patients treated with x radiation—findings that do not support the hypothesis of radiation-accelerated aging.

The results of a continuing mortality study of atomic-bomb survivors of Hiroshima and Nagasaki for the period 1950-1970 have not revealed any consistent evidence of excess mortality other than that due to neoplasia as a result of radiation.[80] The analysis of postmortem data and clinical testing and observations have provided no indications of accelerated aging among atomic-bomb survivors, with the exception of chromosomal aberrations and capillary abnormalities.[59] In response to the question of the possible biasing effect of acute mortality on the experience of atomic-bomb survivors,[130,155] with respect to late effects, Beebe *et al.*[9] have reexamined the mortality experience with tabulated information on deaths through September 1974. They concluded that there is some evidence that the selection of atomic-bomb survivors for fitness by October 1, 1950, favorably influenced later mortality from nonneoplastic diseases in Hiroshima, but not in Nagasaki. Table VI-3 summarizes the age-specific regression estimates of absolute risk in terms of excess deaths for both cities per 10^6 PY per rad for all diseases except neoplastic diseases. Because none of the dose regression estimates was statistically significantly greater than zero, there is no evidence of acceleration in disease among survivors in any part of the age range.[9] Table VI-4 indicates the excess deaths from all diseases except neoplasia per 10^6 PY per rad by calendar period and by city for the period 1950-1974. These data suggest that, in Hiroshima, deaths from nonneoplastic diseases during 1971-1974 may be greater among the high-dose groups than among the low-dose groups, although this is not the case for the entire period of 1950-1974. The

TABLE VI-3   Excess Deaths among Atomic-Bomb
Survivors from All Diseases Except Neoplastic per 10^6 PY
per Rad by Age in 1945, Both Cities, 1950–1974[a]

Age in 1945, yr	Estimated Excess Deaths[b]	Significance[c]
0–9	−0.02 (−1.06, 1.02)	$p > 0.10$
10–19	0.52 (−0.67, 1.72)	$p > 0.10$
20–34	−1.32 (−3.20, 0.56)	$p > 0.10$
35–49	3.00 (−0.84, 6.86)	$p > 0.05$
>50	−10.82 (−24.64, 1.19)	$p > 0.10$
All ages	−0.24 (−1.68, 1.19)	$p > 0.10$

[a] Data from Beebe et al. [9]
[b] Numbers in parentheses are 90% confidence limits.
[c] In test for linear trend.

regression estimate for the 1971–1974 period was significant at $p = 0.05$.
In the highest-dose group (>100 rads) for this period, 99 deaths were
recorded, compared with an expected 90.6, but three of the deaths were
certified as being caused by diseases of the blood and blood-forming
organs.[9] The effect in Hiroshima, however, is indicated to be quite small,
compared with other situations where the effects of selection have been
well-documented, and no significant effect was detected in Nagasaki sur-

TABLE VI-4   Excess Deaths among Atomic-Bomb Survivors from All
Diseases Except Neoplastic per 10^6 PY per Rad, by Calendar Period and
by City, 1950–1974[a]

Calendar Period	Hiroshima Estimated Excess Deaths[b]	Significance[c]	Nagasaki Estimated Excess Deaths[b]	Significance[c]
1950–1954	−2.36 (−6.37, 1.65)	$p > 0.10$	−0.90 (−5.64, 3.84)	$p > 0.10$
1955–1958	−2.30 (−6.83, 2.22)	$p > 0.10$	0.16 (−5.01, 5.33)	$p > 0.10$
1959–1962	−0.98 (−5.59, 3.61)	$p > 0.10$	−2.11 (−7.29, 3.05)	$p > 0.10$
1963–1966	1.88 (−3.07, 6.83)	$p > 0.10$	1.63 (−3.91, 7.17)	$p > 0.10$
1967–1970	1.91 (−3.38, 7.21)	$p > 0.10$	−3.37 (−8.94, 2.19)	$p > 0.10$
1971–1974	5.66 (−0.13, 11.45)	$p > 0.05$	1.14 (−4.87, 7.16)	$p > 0.10$
TOTAL	0.14 (−1.82, 2.12)	$p > 0.10$	−0.57 (−2.76, 1.61)	$p > 0.10$

[a] Data from Beebe et al. [9]
[b] Numbers in parentheses are 90% confidence limits.
[c] In test for linear trend.

vivors. It is concluded that the data from the atomic-bomb survivors strongly suggest that the effects of ionizing radiation on mortality are specific, focal, and principally carcinogenic. There is no firm evidence that exposure to ionizing radiation causes premature aging in man or that the associated increased incidence of carcinogenesis is due to a general acceleration of aging.[9] It may be concluded from the available data that ionizing radiation induces or accelerates some but not all diseases, depending on the genetic susceptibility of the subject and the exposure conditions. For doses of less than approximately 300 rads of low-LET radiation, the principal mechanism of life-shortening is the induction or acceleration of neoplastic diseases.[172] This conclusion is essentially in accord with that of the International Commission on Radiological Protection that the evidence of life-shortening from effects other than tumor induction is inconclusive and therefore cannot be used for quantitative risk estimates.[79] The United Nations Scientific Committee on the Effects of Atomic Radiation has taken a similar position that, with the possible exception of high-dose exposures, life-shortening depends almost entirely on the induction of neoplasia.[163]

REFERENCES

1. Alexander, P., and D. I. Connell. Differences between radiation-induced life-span shortening in mice and normal aging as revealed by serial killing, pp. 277–283. In R. J. C. Harris, Ed. Cellular Basis and Aetiology of Late Somatic Effects of Ionizing Radiation. New York: Academic Press, 1963.
2. Baker, T. G. Radiosensitivity of mammalian oocytes with particular reference to the human female. Amer. J. Obstet. Gynecol. 110:746–761, 1971.
3. Baker, T. G. The sensitivity of oocytes in post-natal rhesus monkeys to x-irradiation. J. Reprod. Fertil. 12:183–192, 1966.
4. Beaumont, H. M. Radiosensitivity of oogonia and oocytes in the fetal rat. Int. J. Radiat. Biol. 3:59–72, 1961.
5. Beaumont, H. M. The radiosensitivity of germ cells at various stages of ovarian development. Int. J. Radiat. Biol. 4:581–590, 1962.
6. Beebe, G. W., and H. B. Hamilton. Future research (on atomic-bomb survivors). J. Radiat. Res. (Tokyo) 16(Suppl.):149–164, 1975.
7. Beebe, G. W., H. Kato, and C. E. Land. Life Span Study Report 5. Mortality and Radiation Dose. October 1950–September 1966, p. 90. Atomic Bomb Casualty Commission Technical Report TR 11-70. Hiroshima: Atomic Bomb Casualty Commission, 1970.
8. Beebe, G. W., H. Kato, and C. E. Land. Life Span Study Report 8. Mortality Experience of Atomic Bomb Survivors 1950–74. Radiation Effects Research Foundation Technical Report TR 1-77. Hiroshima: Radiation Effects Research Foundation, 1978.
9. Beebe, G. W., C. E. Land, and H. Kato. The hypothesis of radiation-accelerated aging and the mortality of Japanese A-bomb victims, pp. 3–27. In Late Biological Effects of Ionizing Radiation. Vol. 1. Vienna: International Atomic Energy Agency, 1978.

10. Berke, R. A., E. C. Hoops, J. C. Kereiakes, and E. L. Saenger. Radiation dose to breast-feeding child after mother has 99m-MAA lung scan. J. Nucl. Med. 14:51-52, 1973.

11. Berlin, N. I. An analysis of some radiation effects on mortality, pp. 121-127. In B. L. Strehler, Ed. The Biology of Aging: A Symposium. Washington, D. C.: American Institute of Biological Sciences, 1960.

12. Billings, M., J. Yamazaki, L. Bennett, and B. Lamson. Late effects of low dose whole body x-irradiation of four-day-old rats. Pediatrics 38:1047-1056, 1966.

13. Blot, W. J. Growth and development following prenatal and childhood exposure (to atomic radiation). J. Radiat. Res. (Tokyo) 16(Suppl.):82-88, 1975.

14. Blot, W. J., and R. W. Miller. Mental retardation following in utero exposure to the atomic bombs of Hiroshima and Nagasaki. Radiology 106:617-619, 1973.

15. Blot, W. J., and H. Sawada. Fertility among female survivors of the atomic bombs of Hiroshima and Nagasaki. Amer. J. Hum. Genet. 24:613-622, 1977.

16. Brent, R. L. Effects of radiation on the fetus, newborn and child, pp. 23-60. In R. M. Fry, D. Grahn, M. L. Griem, and J. H. Rust, Eds. Late Effects of Radiation. London: Taylor & Francis, Ltd., 1970.

17. Brent, R. L. Environmental factors. Radiation, pp. 179-197. In R. L. Brent and M. Harris, Eds. Prevention of Embryonic, Fetal and Perinatal Disease. Vol. 3. Department of Health, Education, and Welfare Publication No. (NIH) 76-853, 1976.

18. Brent, R. L. Irradiation in pregnancy, pp. 1-32. In J. J. Sciarra, Ed. Davis's Gynecology and Obstetrics. Vol. 2. New York: Harper & Row, 1972.

19. Brent, R. L. The response of the 9 1/2-day-old-rat embryo to variations in dose rate of 150 R x-irradiation. Radiat. Res. 45:127-136, 1971.

20. Brent, R. L., and B. T. Bolden. Indirect effect of x-irradiation on embryonic development. V. Utilization of high doses of maternal irradiation on the first day of gestation. Radiat. Res. 36:563-570, 1968.

21. Brent, R. L., and B. T. Bolden. The indirect effect of irradiation on embryonic development. III. The contribution of ovarian irradiation, uterine irradiation, oviduct irradiation, and zygote irradiation to fetal mortality and growth retardation in the rat. Radiat. Res. 30:759-773, 1967.

22. Brent, R. L., and B. T. Bolden. The indirect effect of irradiation on embryonic development. IV. The lethal effects of maternal irradiation on the first day of gestation in the rat. Proc. Soc. Exp. Biol. Med. 125:709-712, 1967.

23. Brent, R. L., and R. O. Gorson. Radiation exposure in pregnancy, pp. 1-48. In R. D. Moseley, Jr., D. H. Baker, R. O. Gorson, A. Lalli, H. B. Latourette, and J. L. Quinn, III, Eds. Current Problems in Radiology. Vol. 2. Chicago: Year Book Medical Publishers, Inc., 1972.

24. Brizzee, K. R., and R. B. Brannon. Cell recovery in fetal brain after ionizing radiations. Int. J. Radiat. Biol. 21:375, 1972.

25. Brown, S. O. Effects of continuous low intensity radiation on successive generations of the albino rat. Genetics 50:1101-1113, 1964.

26. Bursian, S. J., D. F. Cahill, and J. W. Laskey. Some aspects of brain neurochemistry after intrauterine exposure to tritium. Int. J. Radiat. Biol. 27:455-461, 1975.

27. Cahill, D. F., L. W. Reiter, J. A. Santolucito, G. I. Rehnberg, M. E. Ash, M. J. Favor, S. J. Bursian, J. F. Wright, and J. W. Laskey. Biological assessment of continuous exposure to tritium and lead in the rat. In Biological and Environmental Effects of Low-Level Radiation. Vol. 2. Vienna: International Atomic Energy Agency, 1976.

28. Cahill, D. F., J. F. Wright, J. H. Godbold, J. M. Ward, J. W. Laskey, and E. A.

Tompkins. Neoplastic and life-span effects of chronic exposure to tritium. II. Rats exposed *in utero*. J. Natl. Cancer Inst. 55:1165–1169, 1975.

29. Cahill, D. F., and C. L. Yuile. Tritium. Some effects of continuous exposure *in utero* on mammalian development. Radiat. Res. 44:727–737, 1970.

30. Casarett, G. W. Long-term effects of irradiation on sperm production of dogs, pp. 127–146. In W. D. Carlson and F. X. Gassner, Eds. Effects of Ionizing Radiation on the Reproductive System. New York: Macmillan, 1964.

31. Casarett, G. W. Similarities and contrasts between radiation and time pathology. Adv. Gerontol. Res. 1:109–163, 1964.

32. Chapman, E. R., G. W. Corner, Jr., D. Robinson, and R. D. Evans. The collection of radioactive iodine by the human fetal thyroid. J. Clin. Endocrinol. 8:717–720, 1948.

33. Cleary, S. F., and B. Pasternack. Lenticular changes in microwave workers. A statistical study. Arch. Environ. Health 12:23–29, 1966.

34. Clermont, Y. Kinetics of spermatogenesis in mammals. Seminiferous epithelium cycle and spermatogonial renewal. Physiol. Rev. 52:198–236, 1972.

35. Cloutier, R. J., S. A. Smith, and E. E. Watson. Dose to the fetus from radionuclides in the bladder. Health Phys. 25:147–161, 1973.

36. Conard, R. A. Medical survey of the people of Rongelap and Utirik Islands, thirteen, fourteen, and fifteen years after exposure to fallout radiation (March 1968, March 1969). BNL 50220 (T-562). New York: Brookhaven National Laboratory, 1970.

37. Conard, R., and A. Hicking. Medical findings in Marshallese people exposed to fallout radiation. JAMA 192:457, 1965.

38. Connell, D. L., and P. Alexander. The incidence of hepatomas in irradiated and non-irradiated CBA male mice as a criterion of aging. Gerontologia 3:153–158, 1959.

39. Coppenger, C. J., and S. O. Brown. Postnatal manifestations in albino rats continuously irradiated during prenatal development. Tex. Rep. Biol. Med. 23:45–55, 1965.

40. Covelli, V., P. Metalle, B. Biganti, B. Bassani, and G. Silini. Late somatic effects in syngenic radiation chimaera. II. Mortality and rate of specific diseases. Int. J. Radiat. Biol. 26:561–582, 1974.

41. Cowen, D., and L. M. Geller. Long term pathological effects of prenatal x-irradiation on the central nervous system of the rat. J. Neuropathol. Exp. Neurol. 19:488–527, 1960.

42. Dawson, W. D. Growth impairment following radiotherapy in childhood. Clin. Radiol. 19:241–256, 1968.

43. de Boer, J. The effects of chronic whole-body irradiation on the reproduction of C57 Black mice, pp. 59–72. In W. D. Carlson and F. X. Gassner, Eds. Effects of Ionizing Radiation on the Reproductive System. New York: Macmillan, 1964.

44. Dekaban, A. S. Abnormalities in children exposed to x-irradiation during various stages of gestation. Tentative timetable of radiation injury to human fetus. J. Nucl. Med. 9:471–477, 1968.

45. Dobson, R. L., C. G. Koehler, J. S. Felton, T. C. Kwan, B. J. Wuebbles, and D. C. L. Jones. Vulnerability of female germ cells in developing mice and monkeys to tritium, gamma rays, and polycyclic hydrocarbons. In Proceedings of Symposium on Developmental Toxicology of Energy-Related Pollutants. U.S. Department of Energy Symposium Series. (in press)

46. Dobson, R. L., and C. Kwan. The tritium RBE at low-level exposure—variation with dose, dose rate, and exposure duration. Curr. Top. Rad. Res. Qt. 12:44–62, 1977.

47. Dodo, T. Observations and treatment performed at Dept. of Ophthal. Hiroshima

Univ. during the 4 year period of 1957-1961. J. Hiroshima Med. Assoc. 15:878-890, 1962.

48. Dodo, T. Cataracts. J. Radiat. Res. (Tokyo) 16(Suppl.):132-137, 1975.

49. Driscoll, S. E., S. P. Hicks, E. H. Copenhaver, and C. L. Easterday. Acute radiation injury in two human fetuses. Arch. Pathol. 76:113-119, 1963.

50. Dyer, N. C., A. B. Brill, S. R. Glasser, and D. A. Goss. Maternal-fetal transport and distribution of 59Fe and 131I in humans. Amer. J. Obstet. Gynecol. 103:290-296, 1969.

51. Evans, T. C., R. M. Kretzschmar, R. E. Hodges, and C. W. Song. Radioiodine uptake studies of the human fetal thyroid. J. Nucl. Med. 8:157-165, 1965.

52. Fabrikant, J. I. Radiation effects on rapidly and slowly proliferating cell renewal system, pp. 57-78. In J. M. Vaeth, Ed. Frontiers of Radiation Therapy and Oncology. Vol. 6. Basel: Karger, 1972.

53. Fabrikant, J. I. Radiobiology. Chicago: Year Book Medical Publishers, 1972.

54. Fabrikant, J. I. Spermatogonial cell renewal. In Symposium on Cell Renewal Systems. Radiat. Res. 19:46-47, 1971.

55. Fabrikant, J. I. Spermatogonial cell renewal under continuous low dose irradiation. Amer. J. Roentgenol. Rad. Ther. Nucl. Med. 114:792-802, 1972.

56. Fabrikant, J. I. The effects of irradiation on the kinetics of proliferation and differentiation in cell renewal systems, pp. 525-576. In Y. Wang, Ed. CRC Critical Reviews in Radiological Sciences. Vol. 2, No. 4. Cleveland: CRC Press, 1971.

57. Feldweg, P. Ein ungewohnlicher Fall von Fruchtschadigung durch Rontgenstrahlen. Strahlentherapie 26:799-801, 1972.

58. Fillmore, P. G. Medical examination of Hiroshima patients with radiation cataracts. Science 116:322-323, 1952.

59. Finch, S. C., and G. W. Beebe. Aging (in atomic bomb survivors). J. Radiat. Res. (Tokyo) 16(Suppl.):108-121, 1975.

60. Frolen, H. Genetic effects of ^{90}Sr on various stages of spermatogenesis in mice. Acta Radiol. 9:596-608, 1970.

61. Furchtgott, E. Behavioral effects of ionizing radiations. Psychol. Bull. 60:157-200, 1963.

62. Furth, J., A. C. Upton, and A. W. Kimball. Late pathologic effects of atomic detonation and their pathogenesis. Radiat. Res. 1(Suppl.):243-264, 1959.

63. Goldstein, L., and D. P. Murphy. Etiology of ill-health in children born after maternal pelvic irradiation. II. Defective children born after postconceptional maternal irradiation. Amer. J. Roentgenol. 22:322-331, 1929.

64. Goldstein, L., and D. P. Murphy. Microcephalic idiocy following radium therapy for uterine cancer during pregnancy. Amer. J. Obst. Gynecol. 18:189-195, 281-283, 1929.

65. Goldstein, L. S., A. I. Spindle, and R. A. Pedersen. X-ray sensitivity of the preimplantation mouse embryo in vitro. Radiat. Res. 62:267-287, 1975.

66. Grahn, D., and J. Kratchman. Variation in neonatal death rate and birth weight in the United States and possible relations to environmental radiation, geology and altitude. Amer. J. Hum. Genet. 15:329-352, 1963.

67. Green, H. G., F. J. Gareis, T. H. Shepard, and V. C. Kelley. Cretinism associated with maternal sodium iodine I131 therapy during pregnancy. Amer. J. Dis. Child. 122:247-249, 1971.

68. Heinonen, O. P., D. Slone, and S. Shapiro. Birth Defects and Drugs in Pregnancy. Littleton, Mass.: Publishing Sciences Group, Inc., 1977.

69. Heller, C. G., P. Wooten, M. J. Rowley, M. F. Lalli, and D. R. Brusca. Action of radiation upon human spermatogenesis. Excerpta Med. Found. Int. Congr. Ser. 112:408-410, 1966.

70. Herbert, M. C., and C. F. Graham. Cell determination and biochemical differentiation of the early mammalian embryo. Curr. Top. Dev. Biol. 8:151-178, 1974.

71. Hicks, S. P., and C. J. D'Amato. Effects of ionizing radiations on mammalian development, pp. 196-250. In D. H. Woollam, Ed. Advances in Teratology. Vol. I. London: Logos Press, Ltd., 1966.

72. Hirose, I., and A. Okamoto. Lenticular opacities of posterior pole and estimated Nagasaki A-bomb exposure dose. Nagasaki Igakkai Zasshi 35:1715, 1960.

73. Hsu, T. S. H., and J. I. Fabrikant. Spermatogonial cell renewal under continuous irradiation at 1.8 and 45 rads per day, pp. 157-168. In Biological and Environmental Effects of Low-Level Radiation. Vol. 1. Vienna: International Atomic Energy Agency, 1976.

74. Huckins, C. Spermatogonial stem cells in rodents, pp. 395-421. In J. T. Velardo and B. A. Kasprow, Eds. Biology of Reproduction—Basic and Clinical Studies. III Pan American Congress of Anatomy, New Orleans, 1972.

75. Huckins, C. The spermatogonial stem cell population in adult rats. I. Their morphology, proliferation, and maturation. Anat. Rec. 169:533-558, 1971.

76. Huckins, C. The spermatogonial stem cell population in adult rats. II. A radioautographic analysis of their cell cycle properties. Cell Tissue Kinet. 4:313-334, 1971.

77. Huckins, C. The spermatogonial stem cell population in adult rats. III. Evidence for a long-cycling population. Cell Tissue Kinet. 4:335-349, 1971.

78. International Commission on Radiological Protection. Radiosensitivity and Spatial Distribution of Dose. Publication 14. Oxford: Pergamon Press, 1969.

79. International Commission on Radiological Protection. Recommendations. Publication 26. Vol. 1, No. 3. 1977.

80. Jablon, S., and H. Kato. Studies of the mortality of A-bomb survivors. 5. Radiation dose and mortality, 1950-1970. Radiat. Res. 50:649-698, 1972.

81. Jacobsen, L., and L. Mellemgaard. Anomalies of the eyes in descendants of women irradiated with small x-ray doses during age of fertility. Acta Ophthalmol. 46:352-354, 1968.

82. Jacobsen, A. G., and R. L. Brent. Radioiodine concentration by the fetal mouse thyroid. Endocrinology 65:408-416, 1959.

83. Kameyama, Y., K. Hoshino, and Y. Hayashi. Effects of low-dose x-radiation on the matrix cells in the telencephalon of mouse embryos. In Proceedings of Symposium on Developmental Toxicology of Energy-Related Pollutants. U.S. Department of Energy Symposium Series. (in press)

84. Kandori, F., and Y. Masuda. Statistical observations of atom-bomb cataracts. Amer. J. Ophthalmol. 42:212-214, 1956.

85. Kaplan, E. L., and P. Meir. Nonparametric estimation from incomplete observations. J. Amer. Stat. Assoc. 53:457-481, 1958.

86. Kerr, G. D. Organ Dose Estimates for the Japanese Atomic-Bomb Survivors. ORNL-5436. Springfield, Va.: National Technical Information Service, 1978. 46 pp.

87. Kinlen, L. J., and E. D. Acheson. Diagnostic irradiation, congenital malformations and spontaneous abortion. Brit. J. Radiol. 41:648-654, 1968.

88. Konermann, G. Die Keimesentwicklung der Maus nach Einwirkung kontinuierlicher Co60-Gammabestrahlung während der Blastogenese, der Organogenese und der Fetalperiode. Strahlentherapie 137:451-466, 1969.

89. Kriegel, H., and H. Langendorf. Wirkung einer fraktionierten Rontgenbestrahlung auf die Embryonalentwicklung der Maus. Strahlentherapie 123:429-437, 1964.

90. Langham, W. H., Ed. Radiobiological Factors in Manned Space Flight. Washington, D.C.: National Academy of Sciences, 1967.

91. Laskey, J. W., J. L. Parrish, and D. F. Cahill. Some effects of lifetime parental exposure to low levels of tritium on the F_2 generation. Radiat. Res. 56:171-179, 1973.
92. Lesher, S., D. Grahn, and A. Sallese. Amyloidosis in mice exposed to daily gamma irradiation. J. Natl. Cancer Inst. 19:1119-1127, 1957.
93. Lindop, P. J. Radiation and life-span. Sci. Basis Med. Annu. Rev. (Lond.) 1965:91-109.
94. Lindop, P., and J. Rotblat. Long-term effects of a single whole-body exposure of mice to ionizing radiation. II. Causes of death. Proc. R. Soc. Ser. B 154:350-368, 1961.
95. Mandl, A. M. The radiosensitivity of germ cells. Biol. Rev. 39:288-371, 1964.
96. Mandl, A. M. The radio-sensitivity of oocytes at different stages of maturation. Proc. R. Soc. Ser. B 158:119-141, 1963.
97. Matanoski, G. M., R. Seltser, P. E. Sartwell, E. L. Diamond, and E. A. Elliott. The current mortality rates of radiologists and other physician specialists. Deaths from all causes and from cancer. Amer. J. Epidemiol. 101:188-198, 1975.
98. Matanoski, G. M., R. Seltser, P. E. Sartwell, E. L. Diamond, and E. A. Elliott. The current mortality rates of radiologists and other physician specialists. Specific causes of death. Amer. J. Epidemiol. 101:199-210, 1975.
99. Merriam, G. R., and A. Szechter. The relative radiosensitivity of rat lenses as a function of age. Radiat. Res. 62:488-497, 1975.
100. Miller, R. W. Delayed radiation effects in atomic-bomb survivors. Science 166:569-574, 1969.
101. Miller, R. W. Effects of ionizing radiation from the atomic-bomb on Japanese children. Pediatrics 41:257-263, 1968.
102. Miller, R. W., and J. J. Mulvihill. Small head size after atomic irradiation. Teratology 14:355-357, 1976.
103. Mole, R. H., and D. G. Papworth. The sensitivity of rat oocytes to x-rays. J. Int. Rad. Biol. 10:609-615, 1966.
104. Murphree, R., and H. Pace. The effects of prenatal radiation on postnatal development in rats. Radiat. Res. 12:495-504, 1960.
105. Murphy, W. T., and D. L. Berens. Late sequelae following cancericidal radiation in children. A report of 3 cases. Radiology 58:35-42, 1952.
106. National Research Council, Advisory Committee on the Biological Effects of Ionizing Radiations. The Effects on Populations of Exposure to Low Levels of Ionizing Radiation. Washington, D.C.: National Academy of Sciences, 1972.
107. Neary, G. J., R. J. Munson, and R. H. Mole. Chronic Radiation Hazards. New York: Pergamon Press, 1957.
108. Nehemias, J. V. Multivariate analysis and the IBM 704 computer applied to ABCC data on growth of surviving Hiroshima children. Health Phys. 8:165-183, 1962.
109. Nokkentved, K. Effect of Diagnostic Radiation upon the Human Foetus. Copenhagen: Munksgaard, 1968.
110. Oakberg, E. F. A new concept of spermatogonial stem-cell renewal in the mouse and its relationship to genetic effects. Mutat. Res. 11:1-7, 1971.
111. Oakberg, E. F. Effect of 25 R of x-rays at 10 days of age on oocyte number and fertility in female mice, pp. 293-306. In P. J. Lindop and G. A. Sacher, Eds. Radiation and Ageing. London: Taylor and Francis, 1964.
112. Oakberg, E. F. Gamma-ray sensitivity of oocytes of immature mice. Proc. Soc. Exp. Biol. Med. 109:763-767, 1962.
113. Oakberg, E. F. Gamma ray sensitivity of spermatogonia of the mouse. J. Exp. Zool. 134:343-356, 1957.
114. Oakberg, E. F. Initial depletion and subsequent recovery of spermatogonia of the

mouse after 20 R of gamma rays and 100, 300, and 600 R of x-rays. Radiat. Res. 11:700-719, 1959.

115. Oakberg, E. F. Radiation response of the testes. Excerpta Med. Found. Int. Congr. Ser. 184:1070-1076, 1968.

116. Oakberg, E. F. Sensitivity and time of degeneration of spermatogonic cells irradiated in various stages of maturation in the mouse. Radiat. Res. 2:369-391, 1955.

117. Oakberg, E. F. Spermatogonial stem-cell renewal in the mouse. Anat. Rec. 169:515-532, 1971.

118. Oakberg, E. F. The effects of dose, dose rate, and quality of radiation on the dynamics and survival of the spermatogonial population of the mouse. Jap. J. Genet. 40:119-127, 1964.

119. Oakberg, E. F., and E. Clark. Effects of dose and dose rate on radiation damage to mouse spermatogonia and oocytes as measured by cell survival. J. Cell. Comp. Physiol. 58:173-182, 1961.

120. Oakberg, E. F., and E. Clark. Species comparisons of radiation response of the gonads, pp. 11-24. In W. D. Carlson, and F. X. Gassner, Eds. Effects of Ionizing Radiation on the Reproductive System. New York: Macmillan, 1964.

121. Oakberg, E. F., and R. L. DiMinno. X-ray sensitivity of different stages of primary spermatocytes of the mouse. Genetics 42:386-387, 1957.

122. Oakberg, E. F., and C. Huckins. Spermatogonial stem cell renewal in the mouse as revealed by ^{3}H-thymidine labeling and irradiation, pp. 287-302. In A. B. Cairnie, P. K. Lala, and D. G. Osmond, Eds. Stem Cells of Renewing Cell Populations. New York: Academic Press, 1976.

123. Pederson, T., and H. Peters. Proposal for a classification of oocytes and follicles in the mouse ovary. J. Reprod. Fertil. 17:555-557, 1968.

124. Peters, H., and E. Levy. Effect of irradiation in infancy on the fertility of female mice. Radiat. Res. 18:421-428, 1963.

125. Peters, H., and E. Levy. Effect of irradiation in infancy on the mouse ovary. A quantitative study of oocyte sensitivity. J. Reprod. Fertil. 7:37-45, 1964.

126. Piontkovskii, I. A. Certain properties of the higher nervous activity in adult animals irradiated prenatally by ionizing radiations. The problem of the effect of ionizing irradiation in offspring. Byull. Eksp. Biol. Med. 46:77-80, 1958. (in Russian)

127. Radford, E. P., R. Doll, and P. G. Smith. Mortality among patients with ankylosing spondylitis not given x-ray therapy. N. Engl. J. Med. 297:572-576, 1977.

128. Reynolds, E. L. Growth and Development of Hiroshima Children Exposed to the Atomic Bomb. Three-Year Study (1951-1953). Atomic Bomb Casualty Commission Technical Report TR 20-59. Hiroshima: Atomic Bomb Casualty Commission, 1959.

129. Rönnbäck, C. Effects of continuous irradiation during gestation and suckling periods in mice. Acta Radiol. 3:169-176, 1965.

130. Rotblat, J. The puzzle of absent effects. New Sci. (UK) 75:475-476, 1977.

131. Rowley, J. J., D. R. Leach, G. A. Warner, and C. G. Heller. Effect of graded doses of ionizing radiation on the human testis. Radiat. Res. 59:665-678, 1974.

132. Rugh, R. Major radiobiological concepts and effects of ionizing radiation on the embryo and fetus, pp. 3-26. In T. J. Haley and R. S. Snider, Eds. Response of the Nervous System to Ionizing Radiation. New York: Academic Press, 1962.

133. Rugh, R., and M. Wohlfromm. X-irradiation sterilization of the premature female mouse. Atompraxis 10:511-518, 1964.

134. Rubin, P., and G. W. Casarett. Clinical Radiation Pathology. Vol. 1, pp. 374-422. Philadelphia: W. B. Saunders, 1968.

135. Russell, L. B. X-ray induced developmental abnormalities in the mouse and their use

in the analysis of embryological patterns. I. External and gross visceral changes. J. Exp. Zool. 114:545–602, 1950.

136. Russell, L. B. X-ray induced developmental abnormalities in the mouse and their use in the analysis of embryological patterns. II. Abnormalities of the vertebral column and thorax. J. Exp. Zool. 131:329–395, 1956.

137. Russell, L. B. Effects of low doses of x-rays on embryonic development in the mouse. Proc. Soc. Exp. Biol. Med. 95:174–178, 1957.

138. Russell, L. B., S. K. Badgett, and C. L. Saylors. Comparison of the effects of acute, continuous and fractionated irradiation during embryonic development, pp. 343–359. In A. A. Buzzati-Traverso, Ed. Special Supplement to International Journal of Radiation Biology. Immediate and Low-Level Effects of Ionizing Radiation Conference held in Venice. London: Taylor & Francis, 1960.

139. Russell, L. B., and M. H. Major. Radiation-induced presumed somatic mutations in the house mouse. Genetics 42:161–175, 1957.

140. Russell, L. B., and C. S. Montgomery. Radiation-sensitivity differences within cell-division cycles during cleavage. Int. J. Radiat. Biol. 10:151–164, 1966.

141. Russell, L. B., and W. L. Russell. An analysis of the changing radiation response of the developing mouse embryo. J. Cell. Comp. Physiol. 43:103–149, 1954.

142. Russell, L. B., and W. L. Russell. Radiation hazards to the embryo and fetus. Radiology 58:369–376, 1952.

143. Russell, W. L. Effect of the interval between irradiation and conception on mutational frequency in female mice. Proc. Natl. Acad. Sci. USA 54:1552–1557, 1965.

144. Russell, W. L., and E. F. Oakberg. The cellular basis and aetiology of the late effects of irradiation on fertility in female mice, pp. 229–232. In R. J. C. Harris, Ed. Cellular Basis and Aetiology of Late Somatic Effects of Ionizing Radiation. London: Academic Press, 1963.

145. Russell, W. L., L. B. Russell, and E. F. Oakberg. Radiation genetics of mammals, pp. 189–205. In W. D. Claus, Ed. Radiation Biology and Medicine. Chapter 8, Part 3. Reading, Mass.: Addison-Wesley Publishing Co., 1959.

146. Russell, W. L., L. B. Russell, M. J. Steele, and E. Phipps. Extreme sensitivity of an immature stage of the mouse ovary to sterilization by irradiation. Science 129:1288, 1959.

147. Scharer, K., J. Muhlethaler, M. Stettler, and H. Bosch. Chronic radiation nephritis after exposure in utero. Helv. Paediatr. Acta 23:489–508, 1968.

148. Seigel, D. G. Frequency of live births among survivors of Hiroshima and Nagasaki atomic bombs. Radiat. Res. 28:278–288, 1966.

149. Seltser, R., and P. E. Sartwell. The influence of occupational exposure to radiation on the mortality of American radiologists and other medical specialists. Amer. J. Epidemiol. 81:2–22, 1965.

150. Shurygin, V. P. Changes in the long bones of children after radiotherapy. Med. Radiol. (Moskva) 12:37–42, 1967.

151. Sikov, M. R., D. D. Mahlum, and E. B. Howard. Effect of age on the morphologic response of the rat thyroid to irradiation by iodine 131I. Radiat. Res. 49:233–244, 1972.

152. Sikov, M. R., and T. R. Noonan. Anomalous development induced in the embryonic rat by the maternal administration of radiophosphorus. Amer. J. Anat. 103:138–162, 1958.

153. Stadler, J., and J. W. Gowen. Observations on the effects of continuous irradiation over 10 generations on reproductivities of different strains of mice. In W. D. Carlson

and F. X. Gassner, Eds. Proceedings of an International Symposium on the Effects of Ionizing Radiation on Reproductive Systems. New York: Pergamon Press, 1964.

154. Stettner, E. Ein weiterer Fall einer Schadingung einer menschlichen Frucht durch Rontgen Bestrahlung. Jahrb. Kinderheilk. Phys. Erzieh. 95:43–51, 1921.

155. Stewart, A. The carcinogenic effects of low level radiation. A re-appraisal of epidemiologists methods and observations. Health Phys. 24:223–240, 1973.

156. Sutow, W. W., and R. A. Conard. The effects of fallout radiation on Marshallese children, pp. 661–672. In M. R. Sikov and D. D. Mahlum, Eds. Radiation Biology of the Fetal and Juvenile Mammal. Ninth Annual Hanford Biology Symposium. U.S. Atomic Energy Commission Symposium Series. No. 17. CONF-690501. Richland, Wash., 1969.

157. Sutow, W. W., R. A. Conard, and K. M. Griffith. Growth status of children exposed to fallout irradiation on Marshall Islands. Pediatrics 36:721–731, 1965.

158. Tabuchi, A. Fetal disorders due to ionizing radiation. Hiroshima J. Med. Sci. 13:125–176, 1964.

159. Tabuchi, A., S. Nakagawa, T. Hirai, H. Sato, I. Hori, M. Matsuda, K. Yano, K. Shimada, and Y. Nakao. Fetal hazards due to x-ray diagnosis during pregnancy. Hiroshima J. Med. Sci. 16:49–66, 1967.

160. Toda, S., Y. Hosokawa, K. Choshi, A. Nakano, and M. Takahashi. Ocular changes in A-bomb survivors exposed during infancy. Folia Ophthalmol. Jap. 15:96–103, 1964.

161. Tokunaga, T. Atomic bomb radiation cataracts in Nagasaki. Acta Med. Nagasaki. 5:24–42, 1960.

162. United Nations Scientific Committee on the Effects of Atomic Radiation. General Assembly Document, 24th Session, Supplement No. 13 (A/7613). New York: United Nations, 1969.

163. United Nations Scientific Committee on the Effects of Atomic Radiation. Sources and Effects of Ionizing Radiation. Report to the General Assembly. New York: United Nations, 1977.

164. Upton, A. C. Effects of radiation on man. Annu. Rev. Nucl. Sci. 18:495–528, 1968.

165. Upton, A. C. Somatic and genetic effects of low-level radiation, pp. 1–40. In J. H. Lawrence, Ed. Recent Advances in Nuclear Medicine. Vol. 4. New York: Grune & Stratton, 1974.

166. Upton, A. C., K. W. Christenberry, G. S. Melville, J. Furth, and G. S. Hurst. The relative biological effectiveness of neutrons, x-rays, and gamma rays for the production of lens opacities. Observations on mice, rats, guinea-pigs, and rabbits. Radiology 67:686–696, 1956.

167. Upton, A. C., M. A. Kastenbaum, and J. W. Conklin. Age-specific death rates of mice exposed to ionizing radiation and radiomimetic agents, pp. 285–294. In R. J. C. Harris, Ed. Cellular Basis and Aetiology of Late Somatic Effects of Ionizing Radiation. New York: Academic Press, 1963.

168. Upton, A. C., A. W. Kimball, J. Furth, K. W. Christenberry, and W. H. Benedict. Some delayed effects of atom-bomb radiations in mice. Cancer Res. 20(No. 8, part II):1–93, 1960.

169. Van Cleave, C. D. Late Somatic Effects of Ionizing Radiation, p. 25. TID-24310. Washington, D. C.: U.S. Atomic Energy Commission, Division of Technical Information, 1968.

170. Vaughan, J. The effects of skeletal irradiation. Clin. Orthoped. 56:288–303, 1968.

171. Vorisek, P. Einfluss der kontinuierlichen intrauterinen Bestrahlung auf die perinatale Mortalitat der Frucht. Strahlentherapie 127:112–120, 1965.

172. Walburg, H. E., Jr. Radiation-induced life shortening and premature aging, pp. 145–179. In J. T. Lett and H. Adler, Eds. Advances in Radiation Biology. Vol. 5. New York: Academic Press, 1975.

173. Wesley, J. P. Background radiation as the cause of fetal congenital malformation. Int. J. Radiat. Biol. 2:97–112, 1960.

174. Wood, J. W., K. G. Johnson, and Y. Omori. *In utero* exposure to the Hiroshima atomic bomb. An evaluation of head size and mental retardation: 20 years later. Pediatrics 39:385–392, 1967.

175. Wood, J. W., K. G. Johnson, Y. Omori, S. Kawamoto, and R. J. Keehn. Mental retardation in children exposed *in utero* to the atomic bomb in Hiroshima and Nagasaki. Amer. J. Public Health 57:1381–1390, 1967.

176. Wood, J. W., R. J. Keehn, S. Kawamoto, and K. G. Johnson. The growth and development of children exposed *in utero* to the atomic bombs in Hiroshima and Nagasaki. Amer. J. Public Health 57:1374–1380, 1967.

177. Worgul, B. V., G. R. Merriam, A. Szechter, and B. D. Srinivasan. Lens epithelium and radiation cataract. Arch. Ophthalmol. 94:996–999, 1976.

178. Wyburn, J. R. Human breast milk excretion of radionuclides following administration of radiopharmaceuticals. J. Nucl. Med. 14:115–117, 1972.

179. Yamazaki, J., S. Wright, and P. Wright. Outcome of pregnancy in women exposed to the atomic bomb in Nagasaki. Amer. J. Dis. Child. 87:448–463, 1954.

180. Yielding, L. W., T. L. Riley, and K. L. Yielding. Preliminary study of caffeine and chloroquine enhancement of x-ray induced birth defects. Biochem. Biophys. Res. Commun. 68:1356–1361, 1976.

181. Zappert, J. Uber roentgenogene fetale Mikrocephalie. Monatsschr. Kinderheilk. 34:490–493, 1926.

Glossary

ABCC   Atomic Bomb Casualty Commission (see *RERF*).

Absolute risk   Expression of excess risk due to exposure as the arithmetic difference between the risk among those exposed and that obtaining in the absence of exposure.

Absorption coefficient   Fractional decrease in intensity of a beam of x or gamma radiation per unit thickness (*linear absorption coefficient*), per unit mass (*mass absorption coefficient*), or per atom (*atomic absorption coefficient*) of absorber due to deposition of energy in the absorber; *total absorption coefficient* is sum of individual energy absorption processes (Compton effect, photoelectric effect, and pair production).

Accelerator, particle   A device for imparting large kinetic energy to electrically charged particles, such as electrons, protons, deuterons, and helium ions; common types of particle accelerators are direct-voltage accelerators, cyclotrons, betatrons, and linear accelerators.

Alpha particle   A charged particle emitted from atomic nucleus, with mass and charge equal to those of helium nucleus: two protons and two neutrons.

Angstrom (symbol, Å)   Unit of length = 10^{-8} cm.

Anion   Negatively charged ion.

ANL   Argonne National Laboratory.

Atomic mass (symbol, μ)   The mass of a neutral atom of a nuclide, usually expressed in atomic mass units; atomic mass unit is one-twelfth the mass of one neutral atom of carbon-12, equal to $16,604 \times 10^{-24}$g.

Attenuation   Process by which a beam of radiation is reduced in intensity when passing through material—combination of absorption and scattering processes, leading to a decrease in flux density of beam when projected through matter.

Average life (mean life) Average of lives of individual atoms of a radioactive substance; 1.443 times radioactive half-life.

BEAR Committee Advisory Committee on the Biological Effects of Atomic Radiation (precursor of BEIR Committee).

BEIR Committee Advisory Committee on the Biological Effects of Ionizing Radiations.

Beta particle Charged particle emitted from the nucleus of an atom, with mass and charge equal to those of an electron.

Bone-seeker Any compound or ion that migrates in the body preferentially into bone.

Bremsstrahlung Secondary photon radiation produced by deceleration of charged particles passing through matter.

Carrier Nonradioactive or nonlabeled material of the same chemical composition as its corresponding radioactive or labeled counterpart; when mixed with the corresponding radioactive or labeled material, so as to form a chemically inseparable mixture, the carrier permits chemical (and some physical) manipulation of the mixture with less loss of label or radioactivity than would be possible in the use of undiluted label or radioactive material.

Cation Positively charged ion.

Chamber, ionization An instrument designed to measure quantity of ionizing radiation in terms of electric charge associated with ions produced within a defined volume.

Cost-effectiveness The economy with which a given task, program, or policy is carried out.

Curie (abbr., Ci) Unit of activity $= 3.7 \times 10^{10}$ nuclear transformations per second. Common fractions are:

Megacurie One million curies (abbr., MCi).

Microcurie One-millionth of a curie (abbr., μCi).

Millicurie One-thousandth of a curie (abbr., mCi).

Nanocurie One-billionth of a curie (abbr., nCi).

Picocurie One-millionth of a microcurie (abbr., pCi).

Decay, radioactive Disintegration of the nucleus of an unstable nuclide by spontaneous emission of charged particles, photons, or both.

Decay product (synonym, daughter) A nuclide resulting from radioactive disintegration of a radionuclide, formed either directly or as a result of successive transformations in a radioactive series; may be either radioactive or stable.

Dose A general term denoting the quantity of radiation or energy absorbed; for special purposes, must be qualified; if unqualified, refers to absorbed dose.

Absorbed dose   The energy imparted to matter by ionizing radiation per unit mass of irradiated material at the point of interest; unit of absorbed dose is the rad.

Cumulative dose   Total dose resulting from repeated exposure to radiation.

Dose equivalent (abbr., DE)   Quantity that expresses all kinds of radiation on a common scale for calculating the effective absorbed dose; defined as the product of the absorbed dose in rads and modifying factors; unit of DE is the rem.

Genetically significant dose (abbr., GSD)   The gonad dose from all sources of exposure that, if received by every member of the population, would be expected to produce the same total genetic effect on the population as the sum of the individual doses actually received; can be expressed algebraically as $GSD = \Sigma D_i \hat{N}_i P_i / \Sigma N_i P_i$, where D_i = average gonad dose to persons age i who receive x-ray examinations, $\hat{N}_i$ = number of persons in population of age i who receive x-ray examinations, P_i = expected future number of children of persons of age i, and N_i = number of persons in population of age i; in 1964, the GSD was computed to be 55 mrads per person per year for the United States; an estimated 55% of the population were receiving x rays at that time; thus, the average dose to those receiving medical radiation could be computed to be approximately 80 mrads.

Maximal permissible dose equivalent (abbr., MPD)   The greatest dose equivalent that a person or specified part shall be allowed to receive in a given period.

Median lethal dose (abbr., MLD)   Dose of radiation required to kill, within a specified period, 50% of the individuals in a large group of animals or organisms; also called LD_{50}.

Permissible dose   The dose of radiation that may be received by an individual within a specified period with expectation of no substantially harmful result.

Threshold dose   The minimal absorbed dose that will produce a detectable degree of any given effect.

Doubling dose   The amount of radiation needed to double the natural incidence of a genetic or somatic anomaly.

Dose fractionation   A method of administering radiation in which relatively small doses are given daily or at longer intervals.

Dose protraction   A method of administering radiation in which it is delivered continuously over a relatively long period at low dose rate.

Dose rate   Absorbed dose delivered per unit time.

Electron volt (abbr., eV)   A unit of energy $= 1.6 \times 10^{-12}$ ergs $= 1.6 \times 10^{-19}$ J; 1 eV is equivalent to the energy gained by an electron in passing through a potential difference of 1 V. 1 keV $= 1,000$ eV; 1 MeV $= 1,000,000$ eV.

Exposure   A measure of the ionization produced in air by x or gamma radiation; the sum of electric charges on all ions of one sign produced in air when all electrons liberated by photons in a volume of air are completely stopped in air, divided by the mass of the air in the volume; a unit of exposure in air is the roentgen (abbr., R).

Acute exposure   Radiation exposure of short duration.

Chronic exposure   Radiation exposure of long duration, because of fractionation or protraction.

FDA   Food and Drug Administration.

Fission, nuclear   A nuclear transformation characterized by the splitting of a nucleus into at least two other nuclei and the release of a relatively large amount of energy.

Fission products   Elements or compounds resulting from fission.

Fission yield   The percentage of fissions leading to a particular nuclide.

FRC   Federal Radiation Council.

Fusion, nuclear   Act of coalescing of two or more nuclei.

Gamma ray   Short-wavelength electromagnetic radiation of nuclear origin (range of energy, 10 keV to 9 MeV).

Gram atomic weight   Mass, in grams, numerically equal to atomic weight of an element.

Gram molecular weight (synonym, mole)   Mass, in grams, numerically equal to molecular weight of a substance.

Gram-rad   Unit of integral dose $= 100$ ergs.

Gray (abbr., Gy)   Proposed unit of absorbed dose of radiation $= 1$ J/kg $= 100$ rads.

Half-life, biologic   Time required for the body to eliminate half an administered dose of any substance by regular processes of elimination; approximately the same for both stable and radioactive isotopes of a particular element.

Half-life, effective   Time required for a radioactive element in an animal body to be diminished by 50% as a result of the combined action of radioactive decay and biologic elimination $=$ [(biologic half-life) (radioactive half-life)]/[(biologic half-life) $+$ (radioactive half-life)].

Half-life, radioactive   Time required for a radioactive substance to lose 50% of its activity by decay.

Incidence   The rate of occurrence of a disease within a specified period; usually expressed in number of cases per million per year.

Ion   Atomic particle, atom, or chemical radical bearing an electric charge, either negative or positive.

Ion exchange   A chemical process involving reversible interchange of ions between a solution and a particular solid material, such as an ion-exchange resin consisting of a matrix of insoluble material interspersed with fixed ions of charge opposite to that in solution.

Ionization   The process by which a neutral atom or molecule acquires a positive or negative charge.

Ionization density   Number of ion pairs per unit volume.

Ionization path (track)   The trail of ion pairs produced by ionizing radiation in its passage through matter.

Primary ionization   In collision theory, the ionization produced by primary particles, as contrasted with "total ionization," which includes the "secondary ionization" produced by delta rays.

Secondary ionization   Ionization produced by delta rays.

Isotopes   Nuclides having the same number of protons in their nuclei, and hence the same atomic number, but differing in the number of neutrons, and therefore in the mass number; chemical properties of isotopes of a particular element are almost identical; term should not be used as a synonym for "nuclide."

Kerma (*K*inetic *E*nergy *R*eleased in *Ma*terial)   A unit of quantity that represents the kinetic energy transferred to charged particles by the uncharged particles per unit mass of the irradiated medium.

Labeled compound   A compound consisting, in part, of labeled molecules; by observation of radioactivity or isotopic composition, this compound or its fragments may be followed through physical, chemical, or biologic processes.

Latent period   Period of seeming inactivity between time of exposure of tissue to an injurious agent and response.

LD$_{50}$ (radiation dose)   See *Dose, median lethal.*

Linear energy transfer (abbr., LET)   Average amount of energy lost per unit of particle spur-track length.

Low LET   Radiation characteristic of electrons, x rays, and gamma rays.

High LET   Radiation characteristic of protons and fast neutrons. Average LET is specified to even out the effect of a particle that is slowing down near the end of its path and to allow for the fact that secondary particles from photon or fast-neutron beams are not all of the same energy.

Particle	Mass, amu	Charge	Energy, keV	Average LET, keV/μm	Tissue Penetration, μm
Electron	0.00055	−1	1	12.3	0.01
			10	2.3	1
			100	0.42	180
			1,000	0.25	5,000
Proton	1	+1	100	90	3
			2,000	16	80
			5,000	8	350
			10,000	4	1,400
Deuteron	2	+1	10,000	6	700
			200,000	1.0	190,000
Alpha	4	+2	100	260	1
			5,000	95	35
			200,000	5	200,000

Linear hypothesis   The hypothesis that excess risk is proportional to dose.

LSS   Life Span Study of the Japanese atomic-bomb survivors; sample consists of 109,000 persons, of whom 82,000 were exposed to the bombs, mostly at low doses.

Man-rem   See *Person-rem.*

Maximal credible accident   The worst accident in a reactor or nuclear-energy installation that, by agreement, need be taken into account in deriving protective measures.

Medical exposure   Exposure to ionizing radiation in the course of diagnostic or therapeutic procedures; as used in this report, includes:

1. Diagnostic radiology (e.g., x rays).
2. Exposure to radioisotopes in nuclear medicine (e.g., iodine-131 in thyroid treatment)
3. Therapeutic radiation (e.g., cobalt treatment for cancer)
4. Dental exposure.

Micrometer (symbol, μm)   Unit of length $= 10^{-6}$ m.

Morbidity   1. The condition of being diseased.
                   2. The incidence, or prevalence, of illness in any sample.

Neoplasm   Any new and abnormal growth, such as a tumor; "neoplastic disease" refers to any disease that forms tumors, whether malignant or benign.

Nonstochastic   Describes effects whose severity is a function of dose; for these, a threshold may occur; some nonstochastic somatic effects are cataract induction, nonmalignant damage to skin, hematologic deficiencies, and impairment of fertility.

Nuclide A species of atom characterized by the constitution of its nucleus, which is specified by the number of protons (Z), number of neutrons (N), and energy content or, alternatively, by the atomic number (Z), mass number $(A = N + Z)$, and atomic mass; to be regarded as a distinct nuclide, an atom must be capable of existing for a measurable time; thus, nuclear isomers are separate nuclides, whereas promptly decaying excited nuclear states and unstable intermediates in nuclear reactions are not.

Person-rem (synonym, man-rem) Unit of population exposure obtained by summing individual dose-equivalent values for all people in the population. Thus, the number of person-rems contributed by 1 person exposed to 100 rems is equal to that contributed by 100,000 people each exposed to 1 mrem.

Plateau A period of above-normal, relatively uniform incidence of morbidity or mortality in response to a given biologic insult.

Prevalence The number of cases of a disease in existence at a given time per unit population.

Quality factor (abbr., QF) The LET-dependent factor by which absorbed doses are multiplied to obtain (for radiation-protection purposes) a quantity that expresses the effectiveness of an absorbed dose on a common scale for all kinds of ionizing radiation.

Rad Unit of absorbed dose of radiation = 0.01 J/kg = 100 ergs/g.

Radiation 1. The emission and propagation of energy through space or through matter in the form of waves, such as electromagnetic waves, sound waves, or elastic waves.

2. The energy propagated through space or through matter as waves; "radiation" or "radiant energy," when unqualified, usually refers to electromagnetic radiation; commonly classified by frequency—Hertzian, infrared, visible, ultraviolet, x, and gamma ray.

3. Corpuscular emission, such as alpha and beta radiation, or rays of mixed or unknown type, such as cosmic radiation.

Background radiation Radiation arising from radioactive material other than that under consideration; background radiation due to cosmic rays and natural radioactivity is always present; there may also be background radiation due to the presence of radioactive substances in building material, etc.

External radiation Radiation from a source outside the body.

Internal radiation Radiation from a source within the body (as a result of deposition of radionuclides in tissue).

Ionizing radiation Any electromagnetic or particulate radiation capable of producing ions, directly or indirectly, in its passage through matter.

Secondary radiation Radiation resulting from absorption or other radiation in matter; may be either electromagnetic or particulate.

Radiation sickness A self-limited syndrome characterized by nausea, vomiting, diarrhea, and psychic depression; follows exposure to appreciable doses of ionizing radiation, particularly to the abdominal region; its mechanism is unknown, and there is no satisfactory remedy; usually appears a few hours after irradiation and may subside within a day; may be sufficiently severe to necessitate interrupting a treatment series or to incapacitate the patient.

Radioactivity The property of some nuclides of spontaneously emitting particles or gamma radiation or of emitting x radiation after orbital electron capture or of undergoing spontaneous fission.

Artificial radioactivity Man-made radioactivity produced by particle bombardment or electromagnetic irradiation.

Natural radioactivity The property of radioactivity exhibited by more than 50 naturally occurring radionuclides.

Radioisotopes A radioactive atomic species of an element with which it shares almost identical chemical properties.

Radionuclide A radioactive species of an atom characterized by the constitution of its nucleus; in nuclear medicine, an atomic species emitting ionizing radiation and capable of existing for a measurable time, so that it may be used to image organs and tissues.

Radiosensitivity Relative susceptibility of cells, tissues, organs, and organisms to the injurious action of radiation; "radiosensitivity" and its antonym, "radioresistance," are used in a comparative sense, rather than in an absolute one.

Ray Alpha: Beam of helium nuclei (two protons and two neutrons).

Beta: Beam of electrons or positrons.

Delta: Beam of electrons ejected by ionizing particles in passage through matter.

Gamma: Beam of high-energy photons from radioactively decaying elements.

X: Beam of mixed lower-energy photons.

Neutron: Beam of neutrons.

Proton: Beam of protons.

Reactor, breeder A reactor that produces more fissile material than it consumes, i.e., has a conversion ratio greater than unity.

Reactor, converter A reactor that produces fissile atoms from fertile atoms, but has a conversion ratio less than unity.

Reactor, nuclear An apparatus in which nuclear fission may be sustained in a self-supporting chain reaction.

Recovery rate The rate at which recovery takes place after radiation injury; recovery may proceed at different rates for different tissues; among tissues recovering at different rates, those having lower rates will ultimately suffer greater damage from a series of successive irradiations, and this differential effect is considered in fractionated radiation therapy if neoplastic tissues have a lower recovery rate than surrounding normal structures.

Relative biologic effectiveness (abbr., RBE)   A factor used to compare the biologic effectiveness of absorbed radiation doses (i.e., rads) due to different types of ionizing radiation; more specifically, the experimentally determined ratio of an absorbed dose of a radiation in question to the absorbed dose of a reference radiation required to produce an identical biologic effect in a particular experimental organism or tissue; the ratio of rems to rads; if 1 rad of fast neutrons equaled in lethality 3.2 rads of kilovolt-peak (kVp) x rays, the RBE of the fast neutrons would be 3.2.

Relative risk   Expression of risk due to exposure as the ratio of the risk among the exposed to that obtaining in the absence of exposure.

Rem   A unit of dose equivalent = absorbed dose (in rads) times quality factor times distribution factor times any other necessary modifying factors; represents quantity of radiation that is equivalent—in biologic damage of a specified sort—to 1 rad of 250-kVp x rays.

RERF   Radiation Effects Research Foundation. Japanese foundation chartered by the Japanese Welfare Ministry under an agreement between the United States and Japan. It is the successor to the ABCC.

Roentgen (abbr., R)   A unit of exposure = 2.58×10^{-4} coulomb/kg of air.

Sievert (abbr., Sv)   Proposed unit of radiation dose equivalent = 100 rems.

Sigmoid curve   S-shaped curve, often characteristic of a dose-effect curve in radiobiologic studies.

Softness   A relative specification of the quality or penetrating power of x rays; in general, the longer the wavelength, the softer the radiation.

Specific activity   Total activity of a given nuclide per gram of a compound, element, or radioactive nuclide.

Stochastic   Describes effects whose probability of occurrence in an exposed population (rather than severity in an affected individual) is a direct function of dose; these effects are commonly regarded as having no threshold; hereditary effects are regarded as being stochastic; some somatic effects, especially carcinogenesis, are regarded as being stochastic.

Target theory (synonym, hit theory)   A theory explaining some biologic effects of radiation on the basis that ionization, occurring in a discrete volume (the target) within the cell, directly causes a lesion that later results in a physiologic response to the damage at that location; one, two, or more "hits" (ionizing events within the target) may be necessary to elicit the response.

Thermography A noninvasive diagnostic radiologic imaging technique that uses infrared radiation to picture the heat emitted by the surface, which characterizes the temperature distribution in the various underlying organs and tissues of the body.

Threshold hypothesis The assumption that no radiation injury occurs below a specified dose.

Ultrasonography A noninvasive diagnostic radiologic imaging technique that uses acoustic radiation and the acoustic properties of biologic structure to picture the structure and function of various organs and tissues of the body.

UNSCEAR United Nations Scientific Committee on the Effects of Atomic Radiation.

Working level (abbr., WL) Any combination of radon daughters in 1 liter of air that will result in the ultimate emission of 1.3×10^5 MeV of potential alpha energy.

Working-level month (abbr., WLM) Exposure resulting from inhalation of air with a concentration of 1 WL of radon daughters for 170 working hours.

X ray Penetrating electromagnetic radiation whose wavelength is shorter than that of visible light; usually produced by bombarding a metallic target with fast electrons in a high vacuum; in nuclear reactions, it is customary to refer to photons originating in the nucleus as gamma rays, and those originating in the extranuclear part of the atom as x rays; sometimes called roentgen rays, after their discoverer, W. C. Roentgen.